# The Course
# *of* Empire

## BOOKS BY
## BERNARD DeVOTO

THE CROOKED MILE

THE CHARIOT OF FIRE

THE HOUSE OF SUN-GOES-DOWN

MARK TWAIN'S AMERICA

WE ACCEPT WITH PLEASURE

FORAYS AND REBUTTALS

MINORITY REPORT

MARK TWAIN AT WORK

THE YEAR OF DECISION: 1846

THE LITERARY FALLACY

MOUNTAIN TIME

ACROSS THE WIDE MISSOURI

THE WORLD OF FICTION

THE HOUR

THE COURSE OF EMPIRE

THE JOURNALS OF LEWIS AND CLARK

THE EASY CHAIR

# The Course *of* Empire

BY BERNARD DeVOTO

*with an introductory essay on the author*
BY WALLACE STEGNER

*maps by Erwin Raisz*

A MARINER BOOK
HOUGHTON MIFFLIN COMPANY
BOSTON · NEW YORK

*Library of Congress Cataloging-in-Publication Data*
DeVoto, Bernard Augustine, 1897–1955.
The course of empire / by Bernard DeVoto ; with an introductory
essay on the author by Wallace Stegner ; maps by Erwin Raisz.
Bibliography : p.
Includes index.
ISBN 0-395-92498-7 (pbk.)
1. United States — Territorial expansion. 2. United States —
Discovery and exploration. I. Title. II. Series.
E179.5.D4 1989 88-34456
923.1 — dc19 CIP

Printed in the United States of America
QUF 10 9 8 7 6 5 4 3 2

*for*
*Helen*

# Introduction

## by Wallace Stegner

ONE MORNING in 1925, when I was a freshman at the University of Utah, I came into the English building just as a professor yanked open his office door and hurled a magazine into the hall. The offensive journal was *The American Mercury*, about which even I, and even in Salt Lake City, had heard, and the obviously offending part of it was an article entitled "Utah," by Bernard De-Voto. It was a calculated piece of mayhem that out-Menckened Mencken. Though I have not looked at it for thirty-five years, I have a vivid recollection that it said dreadful things about the state where I lived. Apparently we were the ultimate, final, and definitive home of the Boob. We were owned by Boston banks and the Copper Trust, and bamboozled by the Church of Jesus Christ of Latter-day Saints, an institution whose beliefs would bring a baboon to incredulous laughter but whose business acuteness had made it a director of a hundred corporations. This Utah occupied the fairest mountains and valleys on the footstool and grew the best peaches ever grown and had a history of dedication and heroism, in however dubious a cause. But it lived now, as in its founding years, by fantasy, myth, and wishful thinking, and it demonstrated all too faithfully the frontier curve from piety through property to vulgarity. It was conventional, stuffy, provincial, hypocritical, deluded; it had never produced a writer, sculptor, painter, statesman, soldier, scholar, or distinguished man of any sort, it was terrified of any expression of mind, there were not fifteen people in the whole state for whom signing their own names was not an effort.

And so on, a considerable mouthful, spit out with a vigor and venom that only the twenties or Bernard DeVoto, or the two in combination, could have generated. I did not then know anything about the Village Virus or any of the robust literary antibiotics that were being used against it; nor did I know any of the personal reasons that made Mr. DeVoto so angry at Utah and especially his home city of Ogden. I simply let myself be swept up in the happy vehemence of his rhetoric, though even then I suppose I must have understood that a lot of what he said was exaggerated and unjust. If he got a few innocent bystanders, I was willing to sacrifice them for the pleasure of looking upon the more deserving corpses.

As an introduction to DeVoto, that article could have been improved upon, and DeVoto himself never thought enough of it to include it in any of his collections of essays. Nevertheless it expressed him. It had his chosen and almost compulsive subject, the West. It had his habit of challenge and overstatement set off from a launching pad of fact. It had his frequent mixture of the lyrical (when dealing with scenery, say, or peaches, or the beauty of Utah girls) and the vituperative (when dealing with most aspects of society, education, institutions, interests, delusions, or public characters). It had also his incomparable knack of infuriating people. But learn one trick, which most of his readers and all of his friends learned quickly enough: learn to discount him ten to twenty per cent for showmanship, indignation, and the inevitable warping power of his gift for language, and there remained one of the sanest, most acute, most rooted-in-the-ground observers of American life that we have had. He wrote dozens of essays out of indignation or with short-term objectives; these he seldom collected, but let them die. The ones that were close to his considered convictions he kept, and often revised. When he caught himself in exaggeration or error, as he did when he came to collect some of his early essays on education, he cheerfully ate crow. For despite a reputation as a wild man or an ogre, he was open always to the persuasion of facts. He seldom dealt in the outrageous merely for the sake of outraging, though he knew the dramatizing value of shock, and used it. His exaggerations were likely to be extensions of observed truth; and when he was wrong, as he surely sometimes was, he was wrong in the right directions. If he said, in effect, that American civilization was sleeping with every bum in town, there was almost certainly *someone* in her bed.

He had a gift for indignation — which means only that he believed some things passionately and could not contain himself when he saw them endangered by knaves or fools; and however ironic and detached he tried to be, he could become a Galahad in a cause that enlisted as much of him as did the conservation and public lands fights of the late 1940's and 1950's. He began and remained an unfriendly critic of Mormonism, but his half-Mormon heritage and background had bred a good deal of Mormon moralizing into him. Even his peculiar brand of eloquence, at once biblical, orotund, and salty, is related to the eloquence of some celebrated Mormon preachers such as J. Golden Kimball. Like Kimball, Benny could thunder colloquially. Once at Bread Loaf, Vermont, I heard him deliver a sort of lay sermon — let us say it was on the necessity of acknowledging what is under our noses, one of his recurrent themes — and in the course of his talk he got so worked up he brought on a terrible electrical storm that knocked the lights out. No one was in the least surprised; it seemed the most natural thing in the world, the anticipable consequence of the pulpit oratory that went on thundering out of the dark.

But that was much later, a full quarter century after that angry boy, talented and educated beyond his native environment, had fled Ogden with the manuscript of his first novel under his arm. When he wrote his diatribe against Utah, and by extension the whole Rocky Mountain West, he was a young instructor at Northwestern, the author of two novels, beginning to be known as a writer for the magazines, an angry young man full of the heady rebellions of the twenties, with talents that he knew were notable, and with fears of himself that sucked him down into spells of despair and darkness. He was as hot with aspiration as a turpentined mule, an ardent, extravagant, romantic, idealistic, indignant young man with a future. And despite his fears about himself and his fits of depression, he had some notion what that future would involve.

Among the DeVoto Papers at Stanford University is a letter to Melville Smith dated October 22, 1920. In it young DeVoto's natural ebullience is compounded by the fact that when he wrote it he was recovering from a disappointment in love and from a perhaps consequent nervous illness. Nevertheless it may be taken seriously. It was Benny's lifetime habit to bounce high when he was thrown hard, and to fall farther the higher he bounced. "I burst," he said:

> I burst with creative criticism of America — I have at last found a kind of national self-consciousness. Not the mighty anvil-on-which-is-hammered-out-the-future-of-the-world. Still less the damned-bastard-parvenu-among-the-nations. But I have begun to see American history with some unity, with some perspective, with some meaning . . . to dare to think from cause to effect, from the past to the present and future, always with this curious new sense of yea-saying youth.
>
> I do not commit the historic folly, from Washington Irving to Van Wyck Brooks, of hearing fiddles tuning up all over America . . . But I have dared at last to believe that the Nation begins to emerge from adolescence into young manhood, that hereafter the colossal strength may begin to count for

the better as well as for the worse. That indeed we have come to say yea, at last.

   And in the facts which alone can show whether we take the turn, or in the study of them, I shall, I think, spend my vigorous years . . . I believe I have found something into which I may pour that arresting, God-awful emulsion that is I.

Even the friendliest reader will find both self-dramatizing and turgidity in that manifesto, and Benny is surely a figure of the twenties as he says yea to America with one side of his mouth while saying nay to Ogden, the Village, with the other. The earnestness, however, is real, the dedication is real, the repudiation of certain effusive ideas is real (even a favorite whipping boy is anticipated), and the impulse toward history, toward the study of cause and effect in the making of American civilization, is significant. Young Milton impatiently strengthening his wings for broader flights, hypnotized by the novel that was the mirror of his emotional turmoil (its title was "Cock Crow," and the manuscript is among his papers at Stanford) he countered emotional strain by dreaming of an effort more stringently intellectual than fiction. Constitutionally a believer and a yea-sayer, but already suspicious of literary criticism and other forms of "beautiful thinking," he had even at twenty-three a faith in knowledge, in facts; but there were some facts that he did not yet know, one of them being that it was the West, which he scorned, that he most wanted to say yea to. During the next thirty-five years, the "vigorous years" of his dedication, he would be many things — novelist, professor, editor, historian, pamphleteer, critic, and under a half dozen aliases, hack writer — with such range and in such profusion that no neat classification can hold him. Visible in his God-awful emulsion along with scissors and snails and puppy dogs' tails, as real an ingredient as the irritable idealism and the scorn and the skinless self-doubt, would be a belligerent professionalism. He would pride himself on being a pro, would wear the discipline of deadlines and editorial specifications like a hair shirt, because he despised literary phonies, narcissistic artists, public confessors, gushers, long-hairs, and writers of deathless prose; and he would despise these because he feared them in himself. All through Benny's life, a submerged romantic, a literary Harvard boy from Copey's class, would send up embarrassing bubbles of gas, and one way to cover these moments would be the overt belch of professionalism. Professionalism would lead him, too, to take on many kinds of literary jobs — articles, introductions, magazine stories and serials, political speech-writing, reviews — and he would work himself punishingly. In self-defense he would affect to mistrust the imagination and value the hard head, hard work, hard facts; but certain things would not change, he would retain his contradictions.

His repudiation of his western birthright would not stick; the West would not let him go so casually. He would have to come back and worry its complacent provincials and its Two-Gun Desmonds, deflate its myths, expose its economics and its politics, and tell its story in half a hundred essays, half a dozen novels, several works of criticism, and a monumental series of histories. Debunking or correcting western myths, scorning the things the West had become, he would continue to love, to the point of passion, western openness, freedom, air, scenery, violence; and would accept some of the myths as eagerly as the most illiterate cowhand reading *Western Stories* in the shade of the cookhouse. In all his literary jobs he would be a wholly competent workman; in some, as in his hymn to alcohol called *The Hour,* he would be delightful; in many, as in his pamphleteering essays, he would be splendid. But when he wrote history, when he brought together the whole story of the West as frontier, as dream and discovery, exploration and confrontation, he would be magnificent.

The variety of DeVoto's literary work reflects intellectual and physical vitality, not a groping. Though he did many things on the side, he found his field early, worked at it steadily, and brought it to a triumphant bumper crop in his trilogy of histories. But in thus fulfilling the somewhat incoherent program he had set for himself in 1920, he did not quite comprehend all America as cause and effect, promise and payoff. Not even his appetite for work could accomplish so much — and nostalgia, moreover, is a local emotion. Nostalgia, the release of dammed images, memories, feelings, would be necessary for his imaginative re-creation of the frontier. Whenever Benny's mind and emotions could be brought into phase, when Cambridge could bridge time and space back to Weber Canyon, he wrote as he was manifestly born to write, and produced what a continued residence in the West would have made difficult, but what exile made inevitable.

Without committing the error of imagining an Ordeal of Benny DeVoto, one must insist on the importance of alienation and exile. He was born on a frontier as the frontier was passing, when there was no future for frontiersmen, and division and doubt were in him from birth, for his mother was the daughter of a pioneer Mormon farmer (see "The Life of Jonathan Dyer") and his father, the son of an Italian cavalry officer, was an intellectual, a Catholic, "a man of great brilliance and completely paralyzed will," a total outsider. To go to a Catholic school in a Mormon community, to be the only boy in a roomful of girls, to be brilliant and bookish where brilliance and bookishness had (he came to think) neither audience nor function nor reward — these were only aggravations of a dislocation already begun. Of the seven children of his grandfather, not one stayed on the farm or in the Mormon Church; of all the children and grandchildren, "only the novelist, a romantic," ever revisited the home place after the death of the man who had grubbed it out of the sagebrush and irrigated it into fertility.

A divided inheritance can give a boy parallax, he has a base for triangulation and judgment, but he will never be quite at home in his home. And brilliance, especially when associated with insecurity and assertiveness, can isolate a child as effectively as if the disapproving community had shut him in the closet of his mind. It was in his mind that he lived, there and in the canyons of the Wasatch that let him play at independence, freedom, and self-reliance. The limitations and frustrations of Ogden could be transcended on lonely expeditions, by the practice of partly imaginary survival skills, by feats that tested him where there was no chance of being seen and humiliated, by the comfort of wilderness, by the whistle of marmots in mountain meadows, by the eternal sound of mountain water always passing and always there. The most lyrical of his 1920 letters to Melville Smith describes just such a healing expedition. Many years later, in Cambridge, I used sometimes to get desperate telephone calls from Benny: "Come on and walk me around Fresh Pond!" And when we were spending summers in Vermont there would come letters: "Is there anywhere on your place where a man can walk — walk a long way, off the roads, in the woods? Is there any place where a man can shoot a .22?"

It is entirely possible that like many lonely children young Bernard DeVoto didn't recognize his trouble as loneliness, but I am sure it was. That shooting, for instance. Shooting was his favorite boyhood sport — and there is no lonelier sport unless it is rowing. But where rowing is mindless, subduing body and mind to a rhythm, lulling identity to sleep, shooting is another thing. Marksmanship can not only fill empty time, it can feed fantasy, for you can put any face you want on the black dot of the bull's-eye, and the process of shooting holes in it is not only skill, which is comforting, but revenge, which is sweet.

Because I grew up on a frontier even cruder than Weber Canyon, and was halved and quartered away from wholeness at least as early as Benny DeVoto, I think I can imagine him as a boy. Probably he was a little brattish, he showed off before girls, teachers, and God; he was contentious, captious, critical, quick to scorn; he affected superiorities and was constantly in controversies and in his bad spells was desolated that some people didn't like him. He was devoted to certain people, ideas, books, with a fanatical devotion, an ardent, unreasonable extremity. He got crushes on girls and for a week or a month was in a frenzy to immolate himself, and within another week or month had discovered that the blond goddess was dimwitted. Nevertheless his real friendships were affectionate, lasting, and utterly loyal — there have been few people to whom friendship meant more. He read furiously, always beyond his years, and if he used his reading to impress people, that would not be strange. The fact is, he couldn't *not* read. His brain was as busy as a woodpecker, it pecked at him all day and at night perched ominously just above the lintel of sleep, as disturbing as Poe's raven. Probably even at ten or twelve he had migraines — certainly he had them later. He was a problem to his parents and a terror to the conventional and a despair to his teachers, and he learned early, being an outsider, and different, to fling the acid of his scorn into the face of provincial Ogden, already too small a world for him.

He tried the University of Utah for a year, and that too, in 1914–1915, was such a small world that he stirred it like a porpoise in a Paddock Pool. After one year he was off to Harvard, and now he found himself for the first time in a world profoundly, self-consciously, intellectual. If he did not quite grapple with Greek and Hebrew verbs all day and then take a walk in Mount Auburn Cemetery for recreation, he did the equivalent. He devoured books. He was inspired by great teachers, including Copey, and he met people of his own age whom he could not cow: boys as bright as he, better read, better disciplined, boys who tamed some of his exuberances and shamed him out of some of his provincial prejudices, but who at the same time — and this would be sweet — found him something new and special, amusing or arresting, a wild man from the West, a sensitive intellectual out of the howling wilderness of Utah. He never quite got over the role — and the role, observe, was a double one.

At Harvard he crossed a sort of intellectual South Pass the way it was first crossed in fact — eastward — and saw something of what opened up beyond. On the dubious side, he was corroborated in his literary and idealistic posturing: ("Thank God for Harvard. It has given me a thorough-going contempt for these externals. Harvard took me and turned me inward, showed me the heart of things, set burning a lamp in the sanctuary of the ideal. Harvard has shown me that the flesh is nothing beside the spirit . . ."). But he learned too that his native West was interesting, even romantic, to Cambridge eyes and to the eyes of nostalgia, and that it was a splendid place for illustrating pure aspiration dragged down by philistinism. He began to play aficionado about his country, as a hundred thousand western boys have played it in eastern colleges. Being Benny DeVoto, he would have taken the trouble to back up his brags or his diatribes with reading; somewhere very early he began to read western history, geography, exploration, travel, and he read them avidly all his life. (Once he told me cynically that as a journalist he had learned to make a fact go a long way, which was true. It was also true that few people ever collect as many facts about their specialty as he collected about the West. He was loaded, a learned man, and he had the ease with his information that great familiarity brings. In Robert Frost's phrase, he could swing what he knew.)

At Harvard too (I am guessing, but I would back my guess with a bet) he was confirmed in a habit that regional folklore and personal insecurity had already formed in him: the belief or pretense that true vigor, including intellectual vigor, is always a little bit crude and aggressive. During the course of his career he offended some people with the consequences of that assumption, which remained part of his critical method. He scared a lot of others with his cartwheeling and shooting, and some of them never did comprehend that often he was only riding up in his warpaint to shake hands.

Finally Harvard must have encouraged him, in his role of western wild man, to hunt the picturesque and sulphurous phrasing of a man who split a plank every time he spit. Impolite western literature furnished him excellent models, including Mark Twain; there have been few if any moderns who could handle spreadeagle invective as Benny could.

Whatever poses he adopted or roles he assumed, the self-doubting youth was still there, and would remain there. Among the DeVoto Papers is a series of letters written to his parents when he was in OCS camp at Camp Lee, Virginia, in 1918. Even to one who knew and loved Benny well, they are a revelation. As a son, however much unrest he might have given his parents (and he was fond of remarking later that the only torture worse than being a parent was being a child) he was an agonizedly affectionate son. In those letters he throbs and yearns and aspires in a way that begins by being almost embarrassing and ends by being touching. If there was ever a boy who needed love, faith, praise, reassurance, and who hoped to deserve them, hoped humbly to earn them, it was he. The need for the safety and reassurance of friends, too, persisted throughout his life. His letters were often intensely personal, a pouring out of aspiration, confession, self-analysis, self-blame, a release of gas from the submerged affection-craving boy, freed in the privacy of love or friendship from the necessity of deceptive belches and hoots and catcalls.

One of the tenderest things I know about Benny — one of the tenderest things I ever heard about anybody — is unfortunately sequestered in the Stanford Library vaults for a good many years because of the personal nature of some of its references. This is the correspondence between Benny and Kate Sterne, a shut-in tubercular patient who had written him a fan letter to which Benny, as he always did, courteously replied. She wrote again, he replied again, and before long he was periodically relaxing after a hard day's work by writing her long eruptive midnight letters. He took down his hair and said what he thought about those who had offended or slighted or pleased him, he wept for lost battles and exulted for victories, he analyzed people and events and policies and regions, assessed reputations, let air out of balloons. The correspondence continued from 1933 until 1944, when Kate Sterne died. Some day it will make a touching and wonderful book, because the invalid became for Benny the most intimate confidante, and his letters constitute a secret, indiscreet, uninhibited diary of eleven of his most active years. What is more to the point here, they kept Kate Sterne alive and mentally engaged long after she might have been expected to be dead. He dedicated *The Year of Decision: 1846* to her, he told her things that he would never have put in print and that he perhaps never told another living soul. And he never met her, not so much as to shake hands.

To young Bernard DeVoto, wild intellectual from the Rocky Mountains, rebel, iconoclast, and idealist, the war came like a tornado that uproots trees and houses. He enlisted in a tumult of patriotism. It is hard to remember and believe the faith that young men once had that they might save the world for democracy. Benny had that faith, had it like *paralysis agitans*. Quite seriously, he was ready to die with an inspiring phrase on his lips. Accentuating the spirit, he immediately began a novel, writing it in his head because boot camp gave him no time for pen and paper, "a novel of my own country, the wide and ample theatre of the hills, the peaks and valleys, the mountain streams, the railroads, above all the people . . . If ever there was a labor of love it is the construction of this book, into which I am putting all the knowledge that has come to me . . . Already I have written much of a unique book, of a book which touches depths of feeling I had not know[n], which is the expression of my most outstretched aspiration . . .''

It does not sound like the DeVoto we know. And yet this was the tender thing that never quit living in Benny's shell. From Camp Lee he wrote asking his family to preserve his letters for their possible historical value, and he corresponded thereafter with posterity's perhaps critical eye on him: "What kind of a man do these missives show me to be? Is there any hope, do you think, for my ultimate salvation?" He took pride in bearing up under heat and fatigue, and we see suddenly that in Ogden this wild man was probably thought a sissy: "I am no weakling after all, despite the sneers I have known in the past." As the camp progressed and he found he could measure himself against the others without apology, he grew cocky, and out of one letter scrawled in a tent in the breathless Virginia heat bursts a salvo of pure DeVoto, in color and tone, exuberance and hyperbole, a foretaste of the bumptiousness of his maturity:

> What need shall I have of a wife? I have learned to cook my meals and wash my dishes, to make my bed and sweep the floor, to clean, wash, and mend my clothes, to arrange all my belongings and possessions with a neatness and accuracy unknown to women. Why should I ever call a doctor? I know how to keep myself trim and healthy, know the infallibility of iodine and C.C. pills — the Army's own medicines — and know all the approved methods of treating every injury from dysentery to disembowellment, from sunstroke to a cracked knuckle. Practical rubbing up against all kinds of men, the lectures of many officers, and the clean thoughts of a healthy body have taught me more ethics and morality than any minister possibly could. I am or am becoming astronomer and surveyor, indian scout and clerk, statistician and prospector, woodsman, artist, farmer — every trade and employment within the seven seas. I dare go alone into the wilderness confident that, within a week, I could build me a forty room villa with lawns, garden and garage: fit it out with furniture; grow, kill, and cook my own food; set up a religion and code of laws; establish industry and train an army; organize departments of public health and finance — and, in short, build me a State with no other tools than an intrenching spade and a cartridge belt.

Posterity might wink, but the cockiness, like some other DeVoto exaggerations, was built on facts. He graduated seventh — apologizing to his parents for not doing better — in a class of 150. But he did not get overseas; he was too good a rifle shot, he had improved the loneliness of Ogden to too good purpose, and he wound up in Camp Perry, Ohio, as a musketry instructor.

He did not know it, but that was as close as he would ever get to going abroad. He would make certain Prohibition-years excur-

sions into Canada in search of the rye and bourbon that he called
the two greatest American inventions ("How is it that every time
I go to Canada the word gets around the neighborhood by osmosis
and I have a hundred solicitous friends?") and in the later years of
World War II he would suffer himself to be inoculated against
typhoid, tetanus, yellow fever, and other plagues, preparatory to
going to Africa to write the history of the African campaigns for
the War Department. But something happened to that. He did not
go, and never got off the continent of North America. It was just as
well. On that continent he could be an expert and a specialist.

After the Armistice, Lieutenant DeVoto went back to Harvard
to finish, and was told by a member of the faculty not to take himself too seriously, and resolved to follow that advice. After that he
returned to Ogden to write his first novel ("a disgustingly immature
production for one who asserts so much maturity as I"), to fall in
love and be jilted ("The fiction of romantic love is not likely again
to impose on me"), to address the University Club on American
liberty and get himself expelled, to conduct with vigor his war
against all that Ogden stood for ("Do not forget that at best I am a
spore in Utah, not adapted to the environment, a maverick who may
not run with the herd, unbranded, given an ill name. These people
are not my people, their God is not mine . . ."), to suffer a nervous
collapse ("I have a peculiar capacity for suffering in those areas of
personality which neither anatomy nor psychology has yet been able
to describe — areas inextricably tangled with religion and sex and
faith and poetry"), and in general to take himself very seriously
indeed. He was emancipated by an offer to teach English at Northwestern, and so fled the Babylonian captivity of his native city.
He found places where his mind could work effectively, if not in
peace — Evanston, New York, especially Cambridge — and settled
into furious pursuit of the distinction that he yearned for. From
there on, it is the career we know.

The weight and effect of a maverick's career is not immediately
assessable, and in DeVoto's case the assessment, when it comes, will
have to be a composite one, for no single biographer or critic is
likely to be able to follow him into all the corners of American
life where he had authority or exerted influence. Nevertheless the
outlines of his lasting reputation seem already reasonably clear.
At least it may be worth while to indicate one's personal preferences,
to state what seems most important as one looks back over the work
of a man who remains so stubbornly alive. When he died, one's first
thought after the shock of loss was "Who will do his work? Who
can carry on what he did?" For it was never so apparent as when
death stopped it that for many years he had done at least three

men's work, and that it would take three very good men indeed to replace him — that in fact there was not an adequate replacement for any fraction of him.

As a novelist he does not seem truly important, though it was novelist he set out to be. He used to tell students that until a man had written five novels he had no right to call himself a novelist, by which he meant a pro. By that definition Benny qualified, for under his own name he published six and wrote a seventh and part of an eighth, and under the pseudonym of John August he wrote four others. He made a clear distinction between his serious novels and those he wrote as serial entertainments for the magazines — not that he wanted to play demi-virgin, but that his magazine serials were frankly written for money and he did not value them. His serious novels he did value; the inscriptions to his father in copies of his first three books indicate awareness that performance has not quite lived up to desire, but they insist that these are honest books, as true to fact as he can make them, as good as he can do. They also reflect his preoccupation with the West, for *The Crooked Mile* and *The House of Sun-Goes-Down* chronicle the development of a western town not unlike Ogden, *The Chariot of Fire* is the story of a frontier prophet and martyr not unlike Joseph Smith, and even the last one, *Mountain Time,* which begins in a New York hospital, brings its hero Cy Kinsman back to his western birthplace for a sort of reconciliation of the exile that Benny himself had gone through.

These are all honest books and competent ones, and except for *The Crooked Mile,* which sprawls rather badly toward the end, they are well carpentered, witty, packed with observation and ideas. But for me at least the thrill of life is not in them. The eloquence sounds a little wrong when put in the mouths of fictional characters, the dialogue often glitters but seldom falls exactly right on the ear. There is some dreams-of-glory posturing, especially in the early ones: Benny's heroes have a facility for attracting gorgeous women and for knocking down stupids, sots, and other denizens of the modern West. *Mountain Time,* which seems to me the best of the five novels, contains a detailed and persuasive look at the medical profession, besides characters who move and talk with a great deal more naturalness. But Cy Kinsman in that novel tends to repeat Gordon Abbey of *The Crooked Mile,* and the paralysis of the will that marks them both seems more a contrivance to delay a denouement than something the characters couldn't help.

In short, the fault in these novels seems to me to be that they lean too hard on contrivance, they never quite become life, they are tainted — *et ego peccavi,* Benny! — with the literary, a thing that

Benny himself despised. The romantic idealist of the youthful letters, the literary young man from Copey's class, shows through more clearly in the novels than in any of the other writings, and it is an inescapable fact that DeVoto is less sure in his handling of emotional situations in the novels than he is, say, in the reporting of Mark Twain's *Wanderjahre,* or the hardships of the Mormon migration, or the ecology of fur hunters. *Mark Twain's America,* which came between his third and fourth novels, showed him for the first time at something like his full powers. He must himself have recognized that he wrote much better, more authoritatively, more pungently, more importantly, when he could not only write out of the western experience that he knew best, but when he could speak in his own tone of voice, without the ventriloquisms of fiction. Then information, lyricism, irony, indignation, the habit of hyperbole and picturesque phrase-making, could all come together, and Benny's emotional attitudes, though they are evident, are evident at some distance; they inform the facts but are not dwelt on; they produce something very like the tone and inflection of a voice, the stop and flow and rise and fall and thunder and hush of a knowing, intelligent, committed, and unremittingly interested observer.

DeVoto's first significant notice came not because of his novels but because of his essays in *American Mercury, Harper's,* and other magazines. They are the beginning of his lifelong career in social criticism and pamphleteering; the essay on Utah that I read in 1925 was one of them, and it, like dozens of others, was never collected and is not likely to be. Typical is what happened to a whole series of angry essays on education written during the five years at Northwestern. Reviewing them in 1936 for possible inclusion in *Forays and Rebuttals,* Benny found most of them "outrageously over-simplified" and others, such as the much anthologized "The Co-Eds: God Bless Them," to be "in some part untrue, in greater part obvious and irrelevant, and in no part profound." Less than a third of his essay production up to that time seemed to him worth reprinting, and his judgment was probably right. But even with a casualty rate of 66 per cent, DeVoto's essays in social criticism, including the magnificent series on western land problems and conservation reprinted in *The Easy Chair* just before his death, retain an astonishing vitality. I doubt that any body of like essays from the twenties, thirties, and forties would prove, on examination, to have dated so little, and some of them in their time were of robust usefulness in causes that I cannot but think good.

Who spoke any more forthrightly or effectively against irresponsible Congressional red-hunting than Benny did in "Guilt by

Distinction," a mordant undressing of the Reece Committee? Who among us did not cheer when in "Due Notice to the FBI" Benny spoke our minds? ("Representatives of the FBI and of other official investigating bodies have questioned me, in the past, about a number of people and I have answered their questions. That's over . . . If it is my duty as a citizen to tell what I know about someone, I will perform that duty under subpoena, in open court, before that person and his attorney . . .") Whether he was fighting the battle of freedom or protesting the pasteurization of cheese, exposing the land-grab plans of western stock interests or bringing to bear lessons of history upon the problems of the present, DeVoto as essayist performed public services greater than those of most public servants, and during his twenty years in *The Easy Chair* he built up not only an effective information organization but an enormous and respectful countrywide audience. When students of the future come to sift the scores of essays that he threw off at white heat all his life, a good many are going to be found to be not only reprintable but as near as such things come to being permanent, a part of the tradition, a part of the literature. "I stand on the facts," Benny said about his rejected education articles. "I should not care to stand on all the conclusions." Of a couple of fat bookfuls of his total production he could stand on the conclusions as well, and on some of them he would not even have to grant the customary discount.

The essayist who began as an amusing wild man grew into one of the most respected voices of the public conscience. His parallel and interwoven career as a literary critic does not show the same upward and rising curve. He fell away from literary criticism, in fact, as he fell away from fiction, because at bottom he was suspicious of it. His first successful book, *Mark Twain's America,* he called neither history nor biography nor literary criticism, though it was in some part all three, but an "essay in the correction of ideas." In that book his mind was speaking not only against a critical and psychological theory and not only against Van Wyck Brooks, but against a whole habit of mind and — never forget it — against the irresponsible literary romantic in his own house. An anti-literary bias, a sometimes belligerent philistinism, marked much of his criticism and marred some of it. Whether he was attacking Malcolm Cowley, in a review of *Exile's Return,* for assuming that the expatriates constituted a whole American writing generation, or whether he was tempestuously rejecting Thomas Wolfe's undiscipline in "Genius Is Not Enough," he hammered at the need for knowledge, information, facts, and on top of those, professional discipline. Never having been abroad, he looked with skepticism upon the exiles; too often feeling truly lost, he did not want to be part of any lost generation. And anyway, his exile did not reach so far. He had found in New England an intellectual climate that he liked and in which he could work; and in the American scene, past and present, he had found adequate subject matter. Not to find these things seemed to him a literary affectation; and he knew something about literary poses, for he had been there.

So the best of Benny's criticism is related to the most American of all our writers, Mark Twain, of whose papers he was curator from 1938 to 1946. Already addicted to the disciplines of history, he based *Mark Twain's America,* as he said, solidly on the works themselves, as he based *Mark Twain in Eruption* and *Mark Twain at Work* solidly on the manuscript papers. Even in his examination of the despairing backgrounds of *The Mysterious Stranger,* one of the most speculative of his literary studies, he built his speculation on a foundation of manuscripts, false starts, scraps, letters; and when he started what other scholars have turned into a continuing search for the key to the composition of *Huckleberry Finn* he proceeded from evidence, not from a theory.

All his life, that is, he had a quarrel with the habit of making literary judgments about life, what he finally came to call the "literary fallacy." The little book by that title, first given as a series of lectures at the University of Indiana, was the summation of ideas implicit or explicit in all his criticism from *Mark Twain's America* on, and it more or less marked DeVoto's retirement from literary criticism. It is a book which must be taken at the customary discount, and it precipitated a painful literary quarrel. Also, in some eyes it marked Benny as a philistine. If being a philistine means valuing facts and suspecting attitudinizers, he was; a belligerent one of a kind it is healthy to have around. Presented with dream boats, he was likely to make pragmatic tests such as stepping on the starter to see if the motor ran.

Philistine or not, he was a healthy and skeptical influence in a profession likely to be full of hot air; Mark Twain criticism could use him right now. Though he was perhaps somewhat less important as a literary critic than as a gadfly of the public conscience, he was still a critic of range, depth, vigor, and a consistent point of view, and in the area which he made his specialty he was major. But in neither social criticism nor literary criticism was he so important as in history. There he brought off something monumental, massive, grandly conceived and beautifully controlled, a three-volume history of the West as imagination and reality and realization. *The Course of Empire, Across the Wide Missouri,* and *The Year of Decision: 1846* seem to me to warrant all the superlatives that they have consistently won; they belong on the shelf that contains only Prescott, Bancroft, Motley, Adams, and Parkman, and they are not unworthy of the company they find there. In every way they were the climax of Benny DeVoto's career, and with them he won the absolute distinction that he aspired to. The novels seem, in view of this achievement, like experiments in the tricks of dramatizing action and revealing character; the literary criticism like a course in the estimating of documents; the historical essays like finger exercises; the edition of Lewis and Clark's *Journals* like an encore. The real program of this career was the trilogy of histories.

They are, for one thing, incredibly learned. Their pages are a web of cross reference and allusion, packed with facts, crowded with brilliant historical portraits. A lifetime of reading and study is distilled in them, and the mistrust with which Benny regarded his romantic and literary lesser half led him to work by preference from original documents, to the sharp intensification of dramatic effect. There is, moreover, more than mere information; these are not merely history as record, they are history as literature. And here the frontier boyhood and the personal acquaintance with country and weather, landscape and coloring and quality of light, drouth and distance, paid off. These histories are related to Parkman's in their quality of personal participation, in the way history can be felt on the skin and in the muscles because the author himself has been able to imagine it that way, having taken the trouble to live as much of it as possible himself.

This way, at least, the exile who never fully admitted he was an exile came home. This is a better and fuller reconciliation than he arranged for his character Cy Kinsman, who rather unrealistically wound up teaching physiology in a cow college. Reconciliation might have proceeded even further if Benny had lived a little longer, for when he died he was working on the manuscript of a book to be called *The Western Paradox*. Except for one *Easy Chair* on literary cowboys, it had got no farther than a rough draft, and Mrs. DeVoto decided, wisely I think, not to let it be published. Reconciliation or not, the full fusion of western past and western present, of the local realities and the exiled intelligence, it would have been anticlimactic unless he had been able to take it through the second and third drafts that put the sting into his prose and the bite into his ideas. And even without it, he had done enough. More than enough.

# Contents

# List of Maps

# Acknowledgments

MR. ELERS KOCH, a retired official of the United States Forest Service who lives in Missoula, Montana, is a learned student of Montana history. In 1946 when I spent the summer following the trail of Lewis and Clark in the interest of this book, Mr. Koch traced for me on Forest Service maps (which are the most accurate of all generally available maps) various sections of that trail in Montana and Idaho as in years of careful work he had been able to establish them. In 1951 he visited with me the section of the trail that leads through Lolo Pass and down to the headwaters of the Clearwater River, considerably adding to my knowledge as we traveled. In the course of that trip we discussed a conjecture of mine: that the falls in Clark's Fork which the captains decided on hearsay must keep the salmon from reaching Bitterroot Valley were probably Metaline Falls and the rapids in Z Canyon. A couple of days later Mr. Koch brought me a profile of Clark's Fork at Metaline Falls and Z Canyon which he had painstakingly plotted on graph paper after getting the data from Geological Survey water-supply papers, together with a smaller-scale profile of Albany Falls for comparison.

Similarly, in 1949 Mr. Gerald Fitzgerald, the chief of the Topographic Division of the Geological Survey, worked out for me and entered on a map of the United States the distances that would be required by various hypothetical routes discussed in my text. I have used his figures repeatedly, checking against them many of the misconceptions I deal with. For three weeks in 1950 Brigadier General S. D. Sturgis, Jr., commanding the Missouri River Division of the Army Engineers, put at my disposal the knowledge and facilities of his Division, to call on as I might see fit in my study of the Missouri. Mr. Lewis Hanke directed me to a monograph that enabled me to complete an inquiry I had got stuck in, and Mr. John Dos Passos pointed out to me the significance I had missed in an action that turned out to be pivotally important in the book.

I specify these instances as typical of a very great many. Writing history is hardly private enterprise; it is public business of the republic of learning and letters, whose citizens desire that any job undertaken shall be done well. During the past eight years I have often felt that I was no more than a clearing house for the knowledge of others who were writing my book, so often have I appealed to friends and strangers for assistance, direction, counsel, instruction, information, clarification, and labor I could not do myself. To thank some of them by name is to slight scores of others — engineers and lock tenders on the Columbia, engineers and pilots on the Missouri, guides and wranglers and trappers and wilderness travelers, librarians and museum curators in many cities, historians and botanists and zoologists and ethnologists and geographers at many colleges — and yet I am under special obligations that must have the faint return of acknowledgment.

Other officials of the Forest Service besides Mr. Koch, in the West and in Washington, have facilitated my researches and field work and I must specially thank Walt L. Dutton and James K. Vessey. I have often called on two historians of the National Park Service, Ronald F. Lee and Herbert E. Kahler, and have used researches of their colleagues which they made available to me. At the Smithsonian Institution I am indebted to two anthropologists, Waldo R. Wedel and Frank M. Setzler.

I have done most of my work at three Harvard libraries, Widener, Peabody Museum, and Houghton, at the last of which Carolyn Jakeman guided me through the collections as she did with two earlier books. I have constantly used the collections of the Missouri Historical Society, where I have had invaluable help from Brenda Gieseker and Charles van Ravenswaay. I want also to thank officers and attendants of the American Geographical Society, the Boston Athenaeum, the Massachusetts Historical Society, Yale University Library, the Library of Congress and especially its Map Division, the New York

Public Library, the New York Historical Society, the Newberry Library, the Chicago Natural History Museum, the American Museum of Natural History, and the American Antiquarian Society and its amiably omniscient Clarence S. Brigham.

For many kinds of help I am indebted to Ansel Adams, Harold E. Anthony, John Bakeless, Theodore Blegen, Herbert E. Bolton, Catherine Drinker Bowen, John O. Brew, Lyman Butterfield, John Ciardi, Carvel Collins, Julie Jeppson, Wallace Kirkland, Clyde Kluckhohn, John Kouwenhoven, Godfrey Leland, Stewart Mitchell, Samuel E. Morison, Richard Neuberger, Stanley Pargellis, Mildred Pelmont, Fletcher Pratt, Hugh Raup, Henry Reck, Charles G. Sauers, Arthur M. Schlesinger, Jr., Wallace Stegner, and Walter Whitehill.

Two friends of mine whose knowledge I called on repeatedly for many years died before I finished writing the book. I am the poorer in that Joseph Kinsey Howard and Charles P. Everitt cannot read what I have made of what they taught me.

Everyone who works in the history of the westward movement owes more to Frederick Merk than can be repaid, and that debt will always be a fixed charge on those who may work in it hereafter. For years I have derived security from the knowledge that he was at hand, and when I have called on him he has always come to my help.

Parian Temple worked on this book for four and a half years, less as my secretary than as one lobe of my research brain. She digested scores of books and hundreds of articles for me, made thousands of pages of notes, shared most of the investigations I made during that time, conducted others without direction from me, always ingeniously and exhaustively, always with good humor that the irritations of research could not disturb. Circumstances then denied me her help for two years but good fortune restored it to me in time for her to prepare the final typescript. The checking of that typescript and the proofreading that followed, a labor so heavy and repulsive that I am sure only the marriage tie could procure it, were done by Avis DeVoto.

Dr. Erwin Raiz's maps originated in specifications which I drew up. But in giving my ideas visual expression Dr. Raiz so far improved on them that the maps not only interpret my text but enhance it. I am very greatly indebted to him.

Anne Barrett and Paul Brooks of the Houghton Mifflin Company read my manuscript at various stages, as editors but also as friends who understood what I was trying to do and willingly sacrificed their time to put their skill and ingenuity at my service. The work which their colleague Helen Phillips did as copy editor, bringing my habitual chaos into order, was unerring, devoted, and little short of miraculous.

Garrett Mattingly and Helen Everitt knew that I was going to write this book as soon as I did, perhaps before I did, and have followed it with belief through all heavy weather of preparation and writing. Both listened to me with a courtesy only seasonably begemmed with profanity, both read the manuscript chapter by chapter as I wrote it, and I guided myself by what they said about it. That long ago became the standard operating procedure for the books I write. My singular good fortune is that I have learned history from Mr. Mattingly. In the long association during which Mrs. Everitt has been successively my agent, my coach, my collaborator, and my editor and publisher I have learned the integrity of my trade. She knows what the writing of this book has been and has meant. So she knows that the "For" of the dedication is a pallid word.

# Preface

ONE OF the facts which define the United States is that its national and its imperial boundaries are the same. Another is that it is a political unit which occupies a remarkably coherent geographical unit of continental extent. The two facts are not discrete but inseparably related, and they have affected all the other facts that go to define the United States. And they cannot be separated from a feeling which, historically, the American people have always had, a feeling that properly they must become what they have become, a single society occupying the continental unit. That feeling was a powerful force in the creation of the American nation and the American empire. From the first coming of Europeans to North America the continental reality, whether understood or misunderstood, felt as the wilderness beyond the trading posts and settlements, helped to shape the experience of men and societies.

History abhors determinism but cannot tolerate chance. Why did we become what we are and not something else? Or, as a smaller question, why did the American political experiment establish the shape it has maintained? Or, as a much smaller one, why does the United States consist of just the land area the map now shows: why are there not two or more nations in that area, why does it not include parts or all of Canada or Mexico? Long ago historians decided that any effort to answer those questions must assign absolute importance to the force of continentalism. Some have said that of all the forces in American history it is the most powerful. I do not believe that history's great equations can be factored to mathematical results: I do not know how powerful a force, absolutely or relatively, our continentalism has been. Enough that, though American history could be written truthfully if only in part as the history of the continental experience, it could not be written at all if that experience were left out of account.

This book completes a design I formed many years ago, to describe some closely related parts of the continental experience. It is the last of three narrative studies, which circumstances have forced me to write in an order opposite to that of the events they deal with. In the two earlier books I was able to focus a large field by means of a small lens, to describe the historical process at decisive turning points. The time-lapse of one is less than two years, that of the other seven years What my purpose required of this one, however, prevented the use of so biddable a narrative instrument: the time-lapse here is two hundred and seventy-eight years. This difference of scale has necessitated the use of different conventions, historical as well as literary, and a different method.

All treatments of history, even the abstracted ones of narrowly limited monographs, are simplifications and yet even the most simplified treatments are exceedingly complex. On the scale imposed here complexity becomes formidable. I have sought to manage it by means of a thematic structure. Of necessity the themes are treated separately for a space but soon, I trust, they are so treated that the development of one produces the development of some, and with good fortune all, of the others. The themes are meaningful in themselves but separately they are only instruments: the meaning of the book is their meaning in combination.

There are minor themes but the principal ones are these: the geography of North America in so far as it was important in the actions dealt with; the ideas which the men involved in those actions had about this geography, their misconceptions and errors, and the growth of knowledge; the exploration of the United States and Canada, so much of it as was relevant to the discovery of a route to the Pacific Ocean; the contention of four empires for the area that is now the United States; the relationship to all these things of various Indian tribes that affected them. Let me repeat: the meaning is not the themes in themselves but their combination.

The scale of the book being determined, certain consequences followed. In the long period of the American past reviewed here only a portion of the historical facts have been established beyond question. But on this scale a writer must write, for his reader's sake no less than for his own, as if all except those that lie directly in his path may be accepted as given. A good many passages in my book, I dare say a good many single paragraphs, present as thus given summaries of events which are incompletely known or are in dispute. There is no help for that: a paragraph must say what it can about the subject, though it be one to which several or perhaps many students have devoted books. I have made the statements that seem to me most justified. Usually I have made them with reservations or have told the reader that uncertainty exists. But one who presumes to write about events is under obligation to account for them and where I thought it necessary to take one side of a dispute among historians I have done so.

A dismaying amount of our history has been written without regard to the Indians, and of what has been written with regard to them much treats their diverse and always changing societies as uniform and static. Indian history has, as it were, fallen between two specialties. Anthropologists have preferred archeological and ethnological inquiries to historical ones; in most of their treatises "the period of white contact" is likely to be the one most perfunctorily explored. Moreover, in such historical inquiries as they make, they tend to overvalue Indian traditions, which are among the least trustworthy of human records. Historians are wary because the records are scanty, very often unreliable, and sometimes nonexistent, and the axiom that without documents there can be no history yields but reluctantly to the other axiom that when documents are lacking history must find other instruments to use. But also, I believe, there is some inherent tendency to write American history as if it were a function of white culture only.

My narrative has had to deal with Indians throughout. One force which it has had especially to keep in mind is the movements of the tribes, especially the migrations that resulted from the pressure of other tribes and of white men. At every stage of the white man's experience I have tried to make clear to the reader which tribes affected it, how they did, just where they were, and what their experience was. I have had to bring together indications from a very great many sources, most of which could not easily be less satisfactory. If in these passages the probability of error increases, I would have used more help had more been available.

Throughout the action too I try to keep the reader aware of what the geographical reality was as well as what the people with whom I deal thought it was. I do not apologize for this insistence: too many treatises have erred through forgetting or ignoring our geography — and some from being ignorant of it. The forces which our continental unit has exerted on our historical experience cannot be understood unless the unit is understood. Lincoln reminded us that though a people and their laws change, a third part is bound up with them in the nation, their land, and that the earth abideth forever.

This is not to set up a determinism: no one who reads my book can suppose that I believe in historical predestination. Men are masters of their societies, society's will is free, and history is not geography, it is men and the events they produce. But the natural conditions in which men live help to shape their societies. They can and do live in the desert and on the Arctic ice, but on terms which desert and ice impose. A river or a mountain range will not stop a society that has a strong enough desire to cross it, or a sufficiently compelling dream. Yet there are places where the river can be bridged and places where it cannot be; a road can be built across the mountains by some routes only. Some soils will not grow wheat, steel cannot be made where iron and coal cannot be brought together, some areas lack resources to support large populations, currents and channels take shipping near one headland and away from another. In such elementary ways geography admittedly conditions history. There are more complex ways, larger conformations, reciprocals of greater generality, and some day we must try to state them. Meanwhile though no one can say to what degree the physical and psychological forces of our continentalism have conditioned our history, no one can avoid seeing that they have conditioned it to some degree. Nothing in history is more visible than the transformation, in response to the continent, of Europeans into Americans.

In a climactic passage herein I quote Lincoln on the national homestead and its adaptations and aptitudes. The trilogy which this book completes is a gloss on that passage in his

Second Annual Message to Congress. I have tried to express some things it has meant to the people of the New World to occupy the national homestead while developing the adaptations and aptitudes that enabled them to occupy it. The sober melancholy anyone must feel on finishing so long a job is intensified by another consequence of the scale on which I have had to write this book. An expanse of two hundred and seventy-eight years foreshortens individual figures and the experience of individual men must be fragmented and can be allotted but little space. Furthermore one who spends many years reading the records of wilderness men is in danger of taking for granted the labor, strain, hardship, weariness, hunger, thirst, passion, fear, anger, pain, desire, and wonder that were their fare. They cannot for a moment be taken for granted nor, on any scale, is one who writes about the past permitted to let the people who compose it diminish from his touch. May the omen be absent.

BERNARD DEVOTO

Cambridge, Massachusetts
December 1 1951

*I say*
*When night has fallen on your loneliness*
*And the deep wood beyond the ruined wall*
*Seems to step forward swiftly with the dusk*
*You shall remember them. You shall not see*
*Water or wheat or axe-mark on the tree*
*And not remember them.*
*You shall not win without remembering them,*
*For they won every shadow of the moon,*
*All the vast shadows, and you shall not lose*
*Without a dark remembrance of their loss*
*For they lost all and none remembered them.*

STEPHEN VINCENT BENÉT

Even more than other history the Age of Discovery is men come upon strangeness and traveling it mostly in dream. The historian works in an obscurity that ranges from twilight through dusk to dark. By sea or land he will seldom know whether the peaks and headlands on the horizon are real or creations of fear, hope, or faith; most will retreat before him with the horizon; the clouds will seldom open and disclose Polaris. It was not otherwise with the men he has chosen to follow. The archipelago was fable but they threaded it and found a channel. The ridges that hemmed them round were myth but one day they felt the west wind on their faces and had found a pass. Indeed facts are the historian's accidentals; his chronicle is the illusion, for which the Western star is a better guide than Polaris. It was a wanderer and it was desire, therefore illusory but therefore also constant, and it led them at last to the Indies, with the Americas an accident along the way.

EMERY HEYWOOD

A nation may be said to consist of its territory, its people, and its laws. The territory is the only part which is of certain durability. "One generation passeth away and another generation cometh, but the earth abideth forever." It is of the first importance to duly consider and estimate this ever-enduring part. That portion of the earth's surface which is owned and inhabited by the people of the United States is well adapted to be the home of one national family, and it is not well adapted for two or more. Its vast extent and its variety of climate and productions are of advantage in this age for one people, whatever they might have been in former ages. Steam, telegraphs, and intelligence have brought these to be an advantageous combination for one united people.

ABRAHAM LINCOLN

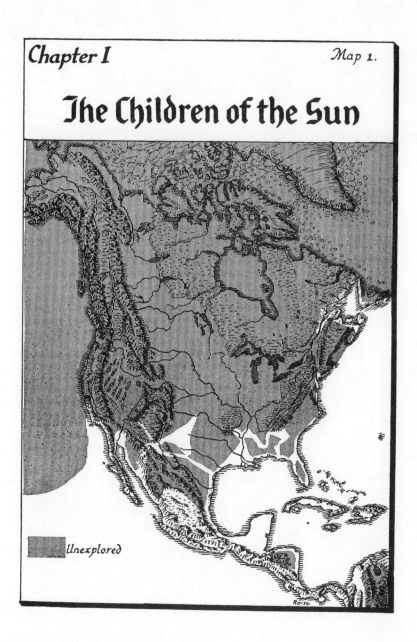

# Chapter I
Map 1.

# The Children of the Sun

Unexplored

# I

# The Children of the Sun

EARLY IN the eighth century a mixed people whom history was to call the Moors crossed the Strait of Gibraltar from Africa. They went as mercenaries to break a stalemate in one of the civil wars which the people of the Iberian Peninsula have been fighting at intervals for two thousand years. But they stayed as conquerors. Within a quarter of a century they had subjugated Spain to the Pyrenees. There were no Spanish people then; the inhabitants of the Peninsula were more heterogeneous than the Moors. The eighth century was nearly over before complexities of politics, trade, religious allegiance, and philosophical speculation permitted combinations capable of military opposition. From then on the people of the Peninsula drove the Moors southward, were driven northward before resurgences of Moorish power, and paused in their war on the infidels only when domestic developments made it necessary again for them to fight one another.

The conquest, or reconquest, lasted almost exactly seven hundred years. Seven centuries of hardly intermitted war created the Spanish people and they, the most medieval people in western Europe, created the Kingdom of Spain. The seven centuries forged the Spanish soul of valor, honor, and chivalry, and tempered it in fanaticism, cruelty, and treachery. In 1492 the army of Ferdinand of Aragon and Isabella of Castile broke the last remnant of Moorish power in Spain and the conquest was complete. But there would still be a use for conquistadores: in the same year a navigator in the employ of Isabella discovered the New World.

3

❧

*Most Christian, most exalted, most excellent and powerful
Princes, King and Queen of the Spains and of the islands of the
sea: In the present year I saw the banners of your Highnesses
raised on the towers of the Alhambra in the great city of
Granada, and I saw the Moorish king go out to the gates of
the city and kiss the hands of your Highnesses and of my lord
the Prince. In the same month your Highnesses told me about
a prince called the Grand Khan, which means in our Spanish
tongue the King of Kings, who many times had sent to Rome
for teachers of our religion but the Holy Father had never sent
any to him, wherefore many cities were lost through idolatry
and belief in hellish sects. Therefore your Highnesses, as Catholic Christian Princes and propagators of the Holy Faith and
as enemies of the sect of Mohammed and of all other idolatries
and heresies, determined to send me, Christopher Columbus, to
the countries of India, so that I might see what they were like,
the lands and the people, and might seek out and know the
nature of everything that is there. And you ordered me not to
travel to the East, not to journey to the Indies by the land route
that everyone had taken before me, but instead to take a route
to the West, which so far as anyone knows no man had ever
attempted. . . .*

❧

The Moors were Mohammedan; Spain's long night was the
more bitter in that the Cross had fallen. Yet there were miracles
and one legend told how a Christian archbishop and six bishops
led their people out of the infidel's path. They went to the seacoast, where they found boats and sailed westward into the
Atlantic, the Sea of Darkness. Somewhere beyond the horizon
they found an island, and there each prelate established a city
that flourished in Christian peace. As the Island of the Seven
Cities it had a holy radiance in the minds of men whose conquerors despised their God.

The island had acquired the name Antillia by the fourteenth
century, when some of the darkness had gone from the Ocean

Sea, though not much, and cosmographers drew maps showing what they thought lay westward from the coast of Europe. Antillia moves in an arc about these maps, appearing anywhere in the Atlantic as men's whims or belief may dictate. There are other islands too of the same substance, which is guess and desire and dream — and underneath them something more solid but not to be identified. The Island of Brazil moves similarly about the maps, and Mayda, the Green Island, the Island of Demons, St. Brendan's Island, Plato's continent of Atlantis shrunk to convenient size, many more.

We know only that someone had been somewhere. No one can say whether any of the mariners who reported that they had coasted Antillia had sailed farther than the Azores. There is no Antillia now, except as the two island crescents that frame the Caribbean Sea preserve its name in theirs, the Antilles. But in the fifteenth century it was as real as Malta to all who sailed westward and to all who studied the accounts of what they found. The king of Aragon granted it to a petitioner in 1474. Christopher Columbus (who had sailed as far as the Island of Thule) wrote to a learned doctor of Florence for confirmation of his theory that the world was so small the cheapest way to reach the Indies was to sail west. Answering him, the physician inclosed a copy of a letter he had written to an earlier inquirer, which said that one route to China — a voyage of only five thousand miles — passed by Antillia, which, the letter remarked, "is well known to you."

Whoever should reach Antillia would find the Seven Cities there. Since they were cities beyond the horizon and of miracle, the sands they stood on would be gold.

❧

In 1536 Charles I, the founder of the Hapsburg line, was king of Spain. As Charles V he also reigned over the paradox of tradition and geography called the Holy Roman Empire. As Emperor he was the champion of Christendom against the expanding Turkish power, and in 1536 Spanish armies were fighting for the Cross in North Africa. As Emperor he had also to defend his inheritance against France and in 1536 Spanish armies were fighting the French in the Italian Peninsula. Under his son Philip II, Spanish armies would fight two more wars with France, would begin the effort to keep the Netherlands subjugated that ended in defeat many years later, and would suppress one of the civil wars that recur like the equinoxes in Spain. Under Philip too Spain as a naval power would turn back the Turk at Lepanto on that day when Don John of Austria burst the battle line. It would lose a fleet in its own harbor of Cadiz when Sir Francis Drake, a pirate of great genius, caught them at anchor. Then Spain would raise a fleet from the whole Empire to crush the British power that was bringing another empire into existence — but the Armada lost. Finally under Philip III Spain would enter the Thirty Years' War.

These were the principal wars that the bullion of the New World was called on to finance but did not. (The crown's share and its levies on private loot barely covered the expense of the Spanish New World, the Department of the Indies.) It is the century of Spain as a world power, as the first nation of Europe, as the organizer of a new hemisphere. It is the century of Calderón, Lope de Vega, and Cervantes; of El Greco, Murillo, and Velásquez. The century of the expulsion of the Jews, of the Inquisition, of Ignatius Loyola and the Jesuits. A century when the local, simple economies of Europe were shaping toward larger ones, partly because of the bullion from the Americas that expanded specie elsewhere but in Spain only inflated the currency. The Spanish genius burned to incandescence and the modern world came in as the creation of a people who were still medieval when, a century later, they plunged from their highest pinnacle into the dark gulf. They created the finest armies since Caesar and spent them in a century-long effort to unify Europe, but they could not keep Spain at peace with itself. Their vision of a world order turned into a fanaticism that loosed from the unconscious mind fantasies of horror whose like the world was not to see again until our time. They discovered and in three-quarters of a century possessed a hemisphere richer than all the myths with which Europe had compensated its poverty and misery — and had to mortgage it to enemy bankers who used the wealth to finance their defeat. With the Sword of God and for His glory they drove from Spain the greater part of Spain's learning, science, and administrative talent. They gave two continents organization and, slowly, government — but they destroyed the agricultural base of Spain to fund the Spanish Empire. They laid waste their land. In the seventeenth century Great Spain broke hardly less spectacularly than the Empire of the Incas had broken and the star of the morning sank forever. Spain has been hungry ever since and has had no peace.

But there were the Americas. For two hundred years after the collapse they bore Spain on their shoulders, just high enough to nourish its dream and preserve the threat it was to Europe. The century of its greatness rested on them. Not on the culture that Spain developed in the cities it built in the Americas, though it became a brilliant culture, but on their gold and silver. First there was the loot, then there were the mines. There was never enough for the crown that was trying to dominate Europe. But especially there was never enough for the gentlemen of Spain whom, for a percentage of the take, the crown licensed to conquer new lands and exploit their treasures. They, the gentlemen of Spain, the hidalgos, explain much of the history of the world. Spain had developed only a rudimentary middle class. Instead, seven hundred years of war forced an evolution which produced as the principal caste and class of Spain a specialization, the conquerors. They had or won for themselves the status of gentry. They had each a horse, armor, arms, honor, courage, and an anarchic soul, and early in life most of them had little more. For seven centuries the way to lands, a competence, and distinction had been to go out and conquer them from the Moors. Ride off into the marches, make a conquest, assassinate your leaders, betray your companions, massacre and enslave the infidels, and push on farther. When the Moors were conquered, there were the Indies, there was the Carib coast, there were Mexico and Peru, there were the Amazons or the Cathaians or the Cipanguese or the Gilded Man. These, it was certain, had much gold, of which there could never be enough.

☙

In the summer of 1521 Hernando Cortés and the conquistadores whom Prescott calls "a mere handful of indigent adven-

turers" occupied Tenochtitlán for the second time, with their
legions of native allies whom the Spanish genius for treachery
had inspired to revolt against Montezuma. This time they
razed the city and had broken the Aztec Empire, whose gold
was already becoming an energy only less powerful than the
voyages of Columbus in the revolution of the world. Cortés,
the least bloody of all the conquistadores in the Americas and
now governor of New Spain, pushed westward and northward
out of the Valley of Mexico. He hoped to find the water
passage across the island — or the continent — of North
America that had eluded Columbus and all other mariners.
And he had heard that to the northwest lay the country of the
Amazons, whom Columbus had sought and missed. "That
they cut off the right dug of the brest [in order to draw a bowstring better] I doe not finde to be true," Sir Walter Raleigh
was to say but he did find that they had "great store of these
plates of golde." They always had, in whatever direction their
country lay, and so Cortés had long known. But his agents
found on the west coast of Mexico no passage to India, no
Amazons, and no gold.

When the jealousies, treacheries, and rebellions that were a
part of all conquests threatened Cortés at court, he had to
abandon the west coast for a time and go to Spain. In 1530
there came to that region his most formidable enemy, the president of the high court that had been named to give Mexico a
government, Niño de Guzmán. Guzmán cannot be classed with
Pizarro as a thug and a murderer, and his massacres, assassinations, pillage, rapine, and destruction were less grievous than
those which his countrymen had inflicted along the shore of the
Caribbean and those which they were about to begin in the
interior of South America. But Mexico never saw another
who waded so deep and joyfully in blood. This was a good
country to which the Spanish had now come down, in climate
and productivity much like the Mediterranean littoral of Spain.
The Indians there lagged far behind the Aztec civilization but
they were of peaceable demeanor and they lived well. In two
years Guzmán had devastated and depopulated much of the
area that is now the states of Nyarit and Sinaloa. By the end of
six years his atrocities against not only the natives but the
Spanish who had followed him had become too revolting to be
endured in New Spain.

Meanwhile, however, he had organized his conquest, founded settlements, and made exploring expeditions northward. He went north "to finde the Amazons, which some say dwell in the Sea, some in an arme of the Sea, and that they are rich, and accounted of the people for Goddesses, and whiter than other women." [1] He did not find them but he heard news of the greatest importance to a conqueror whose conquest had wrung but little gold from its victims. Even earlier and in widely separated places the Spanish had heard stories of rich tribes who lived in caves . . . in seven caves. Now Guzmán heard that beyond the northern mountains there were seven cities. Indians on trading journeys, the reports said, had brought back large quantities of gold and silver from them, and had seen streets lined with the shops of smiths who worked the precious metals. Guzmán's efforts to find them broke against the barrier of the Sierra Madre Occidental in Sonora, and late in 1536 his career in massacre and land piracy ended with his arrest.

Far up in Sinaloa he had established the frontier post of San Miguel de Culiacán, the northernmost Spanish settlement in all Mexico. His and other gangs made forays from it into the mountains of Sonora to capture Indians for slaves. Meanwhile the piety and the frenzy for gold which in equal intensity the Spanish brought to the New World had merged the native account of seven distant cities with the myth of the Seven Cities. And meanwhile the Empire of the Incas had shattered at Pizarro's touch. It yielded a fabulous treasure, the greatest Spain ever looted, and it made sure that for a century of unimaginable daring and cruelty Spain would go looking for another like it.

In April of 1536 a small gang of slave hunters were coursing the foothills of Sinaloa. Something not to be understood happened. They met three Spaniards and a Negro, all naked, and more than six hundred Indians from the far side of the mountains, who were escorting them in the belief that they were gods. They were the survivors of an expedition to explore Florida for gold that had ended in shipwreck on the coast of Texas eight years before. One of them was named Alvar Núñez Cabeza de Vaca and under his leadership they had come from the Gulf coast to Culiacán.

After six years of captivity these four men, between 1534 and 1536, made the first transcontinental crossing north of central

Mexico, the first that touched any part of the United States. They traveled across Texas, New Mexico west of the Rio Grande, and Arizona till they turned the Sierra Madre and could start southward through Sonora.

They had learned much in eight years. Some of it must be given in Cabeza de Vaca's words.[2] "We learned," he says and so he told the slave stealers and repeated across New Spain up to the viceroy, "We learned that on the coast of the South Sea are pearls and great riches, and the best and all the most opulent countries are near there." He is designating supposed countries near the Pacific Ocean in places about which no man knew anything. Where they were, how far away they were, in what latitudes and longitudes they lay and on what compass bearing from Culiacán, no one knew; no one could guess. Cabeza de Vaca's remark that they were the richest and most opulent countries is probably the minimum statement he made about them. To those who heard him it meant that they were richer than Peru. Moreover he had found evidence that proved the existence of these rich lands. Near Rincon, New Mexico, or thereabout he met a tribe of clean and genial Indians who in reverence gave him deerskins and fine cotton shawls, corals, turquoises, and five stones shaped into arrowheads which he thought were emeralds. "I asked whence they got these, and they said the stones were from some lofty mountains that stand toward the north, where were populous towns and very large houses."

Cities! And all his words were momentous.

For when Cabeza de Vaca walked naked out of a miracle (having, he tells us, shed his skin there twice a year like a snake) imperial ingredients were waiting in solution for just such a precipitant as he at once became. He had emeralds; beyond the deserts and mountains there were Seven Cities, countries richer than Peru, and much gold. So now the conquistadores would push the frontier of the Spanish Empire out of Mexico into the area of the United States. This slight increment of force, itself a minute integer of experience in a sum of fantasy, made the first whorl in what would be a vortex of forces, and that vortex would become the contention of four empires. And it produced the first entrances into the Mississippi Valley and the farther West, which would be the final stakes of that imperial struggle.

No European, no white man, knew anything about the Great Valley or the West. They were as completely beyond the reach of knowledge or conjecture as the Americas had been till Columbus touched the island screen. They existed in men's minds like a dream born of a high fever. They were a dream, but were born of mythology and desire. The Spanish entered the Valley from the East; after Cabeza de Vaca they entered the West from the south and the west. They were looking for what steel-fingered compulsion told them must be there but was not there. Romantic, histrionic, cruel, and trance-bound, they marched in rusty medieval armor toward the nonexistent.

❧

The story of Cabeza de Vaca is incredible and would have to be considered myth except that it is true.

By 1527 the entire shoreline of the Gulf of Mexico had been coasted and, after a fashion, mapped. Portions of the map on paper and in men's minds were misconceived; in particular there were serious condensations of distances. The Atlantic coastline of Florida and in fact the entire coast of North America as far as Newfoundland had been investigated more carefully than the Gulf coast. In some areas the results were already amazingly accurate but the preconceptions behind the investigations affected all results. The errors of thought that followed were inevitable but they were also enormous.

In 1527 the Spanish usually applied the name Florida to the vague, misconceived, unknown extent of North America westward from the peninsula to the unknown northern reaches of Mexico. No one knew whether the peninsula of Florida was an island or part of the continental mainland which, by then, North America was widely though by no means universally believed to be. The extent and shape of the continental mainland, if such it should turn out to be, were unknown but variously conjectured. That continental North America was a peninsula of Asia was widely, perhaps generally, believed. Its relation to Asia was of first concern to all who were concerned with the Americas, but arbitrary acts of logic or guess were the only means of determining what the relation was. The Spanish had penetrated the peninsula of Florida inland from both coasts a little way. They had done the same in the Carolinas and Virginia and perhaps in Maryland. Several perfunctory efforts to establish colonies, or rather bases for conquest, on the Atlantic seaboard had been wrecked by ignorance, disease, Indians, or the lust for gold. Many slave-hunting raids, from Cuba or Santo Domingo, had acquainted the coastal Indians with metal and the white man's artifacts, and had flashed along the Indian underground word that these pale, bearded beings might be gods and were as dangerous as the evil gods.

Of the interior: nothing. Nothing was known for there was no experience. The mind will not tolerate a vacuum. When there is no knowledge there will be data; wish, desire, fear, and deductive thinking will provide them. Variously as people thought about the interior, everyone thought it far smaller than it was, thought it but tenth or a hundredth of its real size. Whatever size and shape it might have, no one doubted that it was cut across somewhere by a strait or some other kind of water passage that led from the Atlantic or the Gulf to the Pacific Ocean. (To the South Sea; the Pacific was called that not because Balboa had first seen it from the north, as is sometimes thought, but because it was visualized as south of Asia.) No one doubted that it contained many marvels and much gold.

In 1527 Pánfilo de Narváez got himself named governor of the elastically bounded big unknown called Florida, which meant that he was licensed to conquer and exploit it. He had had some experience at conquest but more at administering conquests others had made. In the islands he had displayed the requisite treachery and barbarity and had made himself rich. He was arrogant, vain, red-bearded, and one-eyed. He had lost an eye when the governor of Cuba sent him to steal the conquest of Mexico from Cortés. Cortés easily parried the double cross, bribing and exhorting most of Narváez's force away from him, then defeating the rest of it. Now he was staking his fortune on the unknown. . . . Cabeza de Vaca was appointed treasurer of the expedition, the king's agent to make sure that the crown got its stipulated percentage of the take.

Preliminary disasters were an omen that the dark star which shone on so many ventures in the New World would govern this one. In the islands treachery, betrayal, disease, and a hurricane reduced the expedition by a third. But in 1528 Narváez set out for Florida with four hundred men and eighty horses.

(Horses were by far the best weapon for offensive warfare that the Spanish had.) He understood that he was sailing for the southern and western boundary of his domain, which, though without stated location, would be somewhat north of the northernmost outpost town on the east coast of Mexico, Pánuco. (It can be located on a map near Tampico.) Storms drove the ships off their course and nearly half the horses died. When they made a landfall the pilots knew that they were still east of Pánuco but not how far. They decided that they were only a few days' sail from it, say two or three days. They were in fact just south of Tampa Bay and so erred by the width of the Gulf of Mexico.

This was April of 1528. As soon as they landed they met Indians but they killed none yet and in fact the conquest of Florida did not last long enough to be very bloody. At once they saw, or thought they saw, minute quantities of gold — the focus of desire. They could talk to the Indians only by gesture.[3] Desire translated gestures to mean that people to the north and west — a numerous and populous people who lived in big towns — had gold in bulk. The Indian pastime and expedient of getting rid of visitors by waving them on begins here. Well, there was gold in Georgia and Carolina but these peninsula tribes knew nothing of it — some did not know what gold was and none valued it — and the white men would not exploit it for nearly three centuries yet.

They were in the New World. The European mind had as yet but few concepts for dealing with it, and such as it had were based on the conquest of the islands, Mexico, and the littoral of the Caribbean. Here even the most accurate of those ideas were unrealistic to the verge of fantasy. Thinking made incapable by assumption of dealing with the realities at hand was fully as responsible as the climates, diseases, and strange food of the New World for the disasters that overtook Narváez and many who came after him. Learning of Indian towns which were temporary aggregations of a few bark huts, they expected them to be like Spanish cities, Seville, say, or Cadiz. News of a chief who held together a small band of underfed neolithic savages meant (and could only mean) a prince or a king, which implied a court, nobles, ceremony, and an economy which could produce the wealth required to maintain them.

And the Spanish had come here to find great cities, such as they had found in Mexico. These cities must be close to the South Sea or on the water passage that led to it. They would therefore always be west or northwest, and not much farther on. They had come to find at least such quantities of gold as they had found in Mexico. They knew, they expected, they wished, they dreamed, and so they always found confirmation or would find it a few days from now. Finally, the mind was stretched taut by dealing with the strange landscape and terrain, the strange weather, the strange world. Strangeness was everywhere and it was roofed over by the unknown. Among whose components have always been mirage and fear.

Narváez was without inner substance; he lacked the steel a conquistador must have. Two weeks of swamps, rivers, forests, and strangeness crumpled him. His logistics were so fantastic that there were only a couple of pounds of food per man: Was not Florida the land of plenty? He ordered the ships, with a quarter of his command aboard, to find the big harbor which his chief pilot thought must be near at hand while he marched the expedition overland toward it. The decision insured his ruin and his nerve had already broken. They set off by land and for two months marched, waded, forded, and swam. They had no notion where they were going, except that Pánuco on the east coast of Mexico could be only a few days farther on, and no plan but to get to the salt water on their left flank and then work west. They captured the town they had heard of and got food but not gold.

This was the end of June and now they reached Indians who were better fed, more developed, and more warlike. They had to fight the rest of their way, learning that neither cuirasses nor shirts of mail would turn an arrow. They knew nothing about living in the wilderness and so never had enough to eat. By August a third of them were sick from malnutrition and swamp fevers. They had traveled a wavering course through fearful country, from a little north of St. Petersburg to the Apalachicola River (in the western extension of Florida) somewhat above its mouth. They were sick, hungry, frightened, licked, and in despair. The ships on which they had counted for food and rescue were gone forever. They could get out of this country, Cabeza de Vaca thought, "only through death, which from its coming in such a place was to us all the more terrible."

They did a tremendous thing: they made boats. Almost all of the condemned were hidalgos, gentlemen, whose pride it was to be incompetent at everything but fighting. But a carpenter and a smith taught them to make woodworking tools from their arquebuses and swords and belt buckles, sails from their shirts, and rigging from horse hair and palm fibers. They ate the forty-two horses that were left. Two hundred and forty-two conquistadores who had survived so far embarked in five indescribable craft. It was the end of September and they intended to follow the Gulf coast to Pánuco, near Tampico. They would reach it in just a few days.

The sun against which they had no shelter was as bad as the storms that blew them out to sea. The thirst was worse; once they had no water for five days. When they landed, the Indians were almost always hostile; in one attack they wounded all forty-nine of the boatload that Cabeza de Vaca captained. They were traveling toward their deaths but one day destiny burned bright. Cabeza de Vaca's boat rounded a promontory at the mouth of a wide river whose current swept them out from the shore, and two and a half miles out they "tooke fresh water within the Sea, because the River ranne into the Sea continually and with great violence." So he and his crew saw the Mississippi thirteen years before De Soto.

They may have been the first white men who ever saw it. The chart of this coast, for what it was worth, had been made by a navigator named Pineda who in 1519, having been blown into the Gulf when he had intended to go up the east coast of Florida, coasted all the way to the mouth of the Pánuco River, which he named and where the presence of Cortés frustrated his designs. (Like everyone, Pineda wanted gold and hoped to find the water passage to the South Sea.) The map showed the mouth of a big river somewhere in the vicinity, allowing for the condensed distances, of where Cabeza de Vaca unquestionably saw the Mississippi. Either he or his chief named it the Espíritu Santo, the River of the Holy Ghost. Thereafter the Spanish knew that a big river reached the Gulf somewhere in these parts, but no one can tell now whether Pineda had seen the mouth of the Mississippi or the estuary of the Mobile.

Four days after they took fresh water within the sea another storm separated the boats. Cabeza de Vaca's caught up with one, lost it, and after an undeterminable time but probably early in November, was driven ashore by a storm. Hunger, thirst, disease, apathy, and despair had so worn them down that only he and another had been able to stand to an oar; none were far from death. They thought that they were near Pánuco; it was more than five hundred miles away along the coastline but in comparison with the route by which he eventually reached his countrymen they were near it. Their landfall was Galveston Island, off the northeast coast of Texas.⁴ By an amazing chance another of the boats was wrecked near by. (Even more amazingly, they were able, eventually, to make sure that the other three had been lost.) Eighty naked, unarmed men, already dragged into death's undertow by the failure of their culture, then, were here the last muster of the expedition of Pánfilo de Narváez "to conquer and governe the Provinces which lye from the River of Palmes [the Rio Grande] unto the Cape of Florida."

The Indians of the region, good-natured but of so wretched an economy that half the year they could barely hold starvation away, took them in. Soon all but five of them were dead: some drowning in a heroic effort to go on to Pánuco, some killed by the Indians for a few remaining personal trinkets or just idly, some executed in Indian revulsion for eating corpses of their companions, some of starvation, more of an epidemic, even more of the fading-out of the soul that may occur in the wilderness when hope has failed. In five the will to live prevailed but in one of them not quite enough. When the time came this one made the start with Cabeza de Vaca, in whom the will was absolute, but fled back to the comfort of slavery in terror of the unknown.

That was in 1532; Cabeza de Vaca, the king's treasurer, had been the slave of a small band of Indians for four years. So, with other bands, had the man of broken will and the other three whom he eventually found and who dared to make the cast with him. The bands moved with the seasonal food supply — they had no agriculture and little game — and sometimes came together to conduct a miserable trade in trinkets, holy objects, and a small surplus of weapons and clothing. They met once a year for gluttony when the cactus (*Opuntia*, the flowering prickly pear) was ripe for eating and there was enough food for everyone. At one of these annual feasts Cabeza de Vaca met the other three, at the next one they made their plans. Meanwhile his owners worked him sorely and with the caprice of Indians alternately beat or tortured him and treated him as one of themselves, which was not much better than torture. But he had two periods of ease. For a time he was a healer, curing the sick and the possessed. At another time he was an itinerant peddler making long journeys, perhaps as far as Oklahoma, trading such shells, feathers, flints, ochre, and sinews as he could collect and prepare. He kept his manhood firm in this squalor of heat, cold, mud, sickness, insects, reptiles, semistarvation, and hope deferred. And three times he saw buffalo, the first white man who ever saw them in the area of the United States.

In early fall of 1534 the four fled from slavery, going south for a space in hope of finding their countrymen at Pánuco. They knew neither where they were, nor how far their island was from their starting place or their destination, nor what the lay of the land was. Almost to the end of their journey they supposed that the Gulf of Mexico was at their left.

The names of Cabeza de Vaca's two white companions do not matter here but that of a slave who belonged to one of them does. He was called Estéban; Cabeza de Vaca says he was "an Arabian black," which means that he had once been owned by Moors. For the last six years he had been the slave of a slave. Now he was to become a god.

Cabeza de Vaca understood that the four had survived by miracle, for the expiation of their sins and for some other, still unrevealed purpose of God. He believed that he had been saved so that he could lead the others; his was certainly the dominant intelligence and will and without him they would have succumbed to the wilderness. Not long after the beginning of their journey the favor of God was manifest and they began to work miracles. They were strangers to the tribes they met, as strangers they were given welcome, and their strangeness was strong magic. The demented, the sick, and the dying were brought to them and they healed them of divers diseases. They thought they healed by prayer and faith, but as word ran ahead of them the Indians knew that they were children of the sun.

Their suffering was over now — except for marching that

never ended, bloody feet, backs cut by pack-straps, long thirst, desert sun, mountain cold, tempest, and hurricane. But a pillar of cloud went ahead of them by day and the tribes came to pay homage and be healed.

They reached the Colorado River of Texas, which empties into Matagorda Bay halfway down the state's long coastline, and turned inland up its valley, traveling north of west. Sometimes they were by themselves but there was always an Indian trail and, if they had no notion where they were going, they were never lost. (Almost always when white men are crossing the American wilderness for the first time, they are following an immemorially old highway system of Indian trade, hunting, and war.) More often there were reverent guides, embassies coming out to meet them, and other embassies to take them on when they would stay no longer. When cold weather came on they were nearing the West Texas hills. They met a tribe who were better fed than any they had known before, though still half starved in the winter time, and spent eight moons with them. Here Cabeza de Vaca learned another language — he knew six by the time he reached Culiacán and Estéban knew even more. Here too they were able to check off the last two of the original company whose deaths they had not heard about.

But the radiance begins in the summer of 1535, when they set out again. They traveled on up the valley of the Colorado, then left it and turned toward the Pecos River, marching west into the fourth house of the sky, where their father slept at night. Since Cabeza de Vaca was living a myth, his account is majestically unregardful of landmarks and geography but eventually they saw wraiths of peaks on the horizon and these must have been the Davis and Guadalupe Mountains of New Mexico. He thought that they came down from the North Sea. . . . Another judgment from the geography beyond men's knowledge, and a revealing one. The North Sea was the one that hypothesis said must lie above Asia. So the continents drew together and the world was small.

Cabeza de Vaca thought that the mountains walled them off to the southward and he understood the Indians to say that the coast, the Gulf of Mexico, was only a short distance to the south. There were many tribes in that direction and the Indians, who were charging high fees in native goods for the miracles worked by their guest-divinities, wanted them to turn eastward long enough to reach these untouched fields. But the coastal tribes they had met had been cruel: best avoid them. Moreover, the apathy of lost, starved, enslaved men had ended; they were proud and expectant. Who knew what they might find beyond this new horizon? The children of the sun would keep to their westward bearing. The Indians told them, truly, that this meant harsh mountains and canyons, worse deserts, and long stretches between tribes. No matter: "We ever held it certain that going toward the sunset we would find what we desired."

In four centuries no one ever said it more fully.

Traveling toward the sunset took them a long, devious, bone-wearying journey. They found all the hazards they had been promised and, again, were often alone but never for long. And now they were masters of men, they commanded and were obeyed. Sometimes runners went ahead of them, sometimes they sent Estéban, sometimes seekers came to meet them, sometimes throngs traveled with them. It was the privilege of these companions, the escorts of gods, to loot the next village and when the deities moved on the looted accompanied them to become looters in turn. Feasts were held for them, religious dances, ceremonial purges and vomitings — from tribe to tribe, "so many sorts of people the memory fails to recall them." All bearing gifts. One gave them some of its sacred symbols, medicine rattles made of gourds, and one of these had fate as well as pebbles shut up in it. Another gift was even more fateful: "a hawk-bell of copper, thick and large, figures with a face." (A bell doubtless, but not a hawk-bell.) Copper was momentous and they heard that there was much of it to the northward. They guessed foundries. And big cities.

More gifts than they could carry, more food than they could eat, and more worshipful, singing escorts than they could count, for Cabeza de Vaca sometimes numbers them to three or four thousand. . . . Presumably the king's treasurer, who must have been an accountant, is to be trusted but throughout the history of primitive America the statistics at hand must be treated with caution. Strangeness, surprise, wonder, or fear could multiply a number lavishly, especially the number of enemies or game animals. Beyond that is the fact that in the sixteenth and even

the seventeenth century very few men knew arithmetic. Even
a learned clerk might not be able to add a column of figures, a
gentleman might scorn so menial an ability, and the unlettered,
which meant nearly everybody, had to use their fingers. Besides, are not travelers permitted to make a good story of it?
Indians could count by tens on their two hands but how far is
not clear, and they had little conception of number or exactness. A large number, whether reported by white men or
Indians, commonly means from "quite a few" on up to "a lot."
The state of mind a man is in will determine how many make
quite a few.

They crossed the Sacramento Mountains of New Mexico,
went south along the barren wastes at their western base to the
Diabolos, crossed them, and went on to the Rio Grande, which
they must have reached somewhere near El Paso. They traveled
upstream along its banks and met a new kind of Indians. Their
houses and towns were permanent, which seemed to make certain that there would be cities farther on. They were clean
and though this was a drought year they had fine gardens.
Their goods were plentiful and well made; their craftsmanship
in leather was admirable. The daily tribute to the gods began
to contain robes made of cowhide. They were an article of
trade from tribes farther north and the cows were the "bunch-
backed oxen," the buffalo.

They worked great miracles here. "All held full faith in our
coming from heaven. . . . We possessed great influence and
authority; to preserve both we seldom talked with them. The
Negro was in constant conversation; he informed himself about
the ways we wished to take, of the towns there were, and the
matters we desired to know." Now they fell into step with the
destiny of the New World. The gifts included cotton shawls
which were superior to any made in Spain. There were corals
— and they had come so far by now that they realized the corals
must be from the South Sea and that it could not be far distant
from where they were. There were turquoises, which came
from the north. And there were those five arrowheads, also
from the north, which Cabeza de Vaca knew must be emeralds.
(Malachite or smithsonite.) To the north, to the west, to the
northwest was the South Sea, and pearls and great riches. The
best and most opulent countries must be near there.

They did not quite find them. As the year wore on toward the winter of the high plateaus, going west from Rincon they had reached the Gila River and quartered south across the Peloncillo and Chiricahua Mountains of Arizona to the San Bernardino valley of northern Sonora. They were south of it at Christmas time; wearily they followed worn Indian trails, among tribes who seemed more worshipful than any before them. Now it was April 1536. One day they met an Indian from whose neck hung the buckle of a sword belt and a horseshoe nail. A great gush of hope rose through eight years of desperation, but they were afraid lest the white men had been adventurers on the South Sea and had by now sailed away. The fear ended a few days later when they met hundreds of Indians fleeing from enslavement and death. Christians could not be far away.

Cabeza de Vaca was older than the others but had more fire in him; with Estéban he went ahead. So they met a slave hunter and his gang, and eight years ended in tears, ecstasy, and the glory of God. They who had been dead lived again and Spain, which had entered the wilderness at Tampa Bay, came out of it in Sinaloa.

The slavers were exhausted, famished, and without captives. Cabeza de Vaca's Indians brought them food, though skeptical that murderers traveling out of the sunset could be of the same religion as the miracle-workers they had guided toward the sunset. The chief land-pirate confirmed their judgment. He sent his countrymen onward by one route and, taking another one himself, fell on some of their worshipers and captured them.

But Cabeza de Vaca and Estéban and the two others came down to Culiacán. And began to tell New Spain about the dangers they had passed . . . and the cities they had almost seen.

℮∾ჿ

Events had marched fast in New Spain. Returning there from court, Cortés sent a series of mariners to explore the mist and fable of the seacoast. In 1533 one of them discovered what he took to be an island and part of an archipelago and after he had been murdered, in a routine way, his successor reported that there were pearls there, if not Amazons. Presently this supposed island would get a name from knightly romance, California, and the discoverer had seen the tip of its lower peninsula. In 1535 Cortés led an expedition there and established a small base where the wealth in pearls was to be garnered. He went back to New Spain to prepare supplies and reinforcements but was forced to stay there and, toward the end of 1536, to send for the survivors. The first viceroy of New Spain had ordered him to stop exploring.

The viceroy was Antonio de Mendoza; he had been sent to give New Spain more orderly government, but principally to make sure that the great Cortés did not become too great. He organized the northwestern marches, where Culiacán was and where already veins of gold and silver were being mined, as the province of New Galicia. He had held his office less than a year when Cabeza de Vaca came out of the wilderness in 1536. The travelers reached him in Mexico City in July. What they meant to him, as to the whole Spanish world, was that another Peru, a new greatest of all treasures, lay somewhere north of New Galicia. Mendoza prepared to reconnoiter and conquer it. One of Cabeza de Vaca's white companions refused at once to return to the northern wilderness, the other temporized and then refused. Mendoza bought the slave Estéban to serve as a guide but Cabeza de Vaca would form no alliance with him. In 1537 he went back to Spain with the tidings of a new Peru near the South Sea. He wanted the king's license to conquer it himself.

He did not get it. New Peru was in Florida and another man had already been licensed to make the conquest that Narváez had not made. Cabeza de Vaca drew instead a commission to strengthen the slight hold Spain had on the River Plate and to go on up it to the darkness and glory beyond. So he joined the conquest of interior South America, the longest, most anarchic, and most incredible of the wars of the Spanish Empire in the New World — and the search for the one-breasted women who were fairer than other women and had much gold, and for that other mirage El Dorado, the Gilded Man. Already he had lived a great story to triumph. Now he would live another one almost as fantastic, and very honorable in that his years among Indians had given him a charity toward them that shines bright against the bloody background of the South American wars. But it is not within the compass of this narrative.

☙

"Hernando De Soto was the son of an esquire of Xeréz de Badajóz and went to the Indias of the Ocean Sea. . . . He had nothing more than his sword and buckler. For his courage and good qualities Pedrárias appointed him to be captain of a troop of horse, and he went by his order with Hernando Pizarro [Francisco's brother] to conquer Peru." So says the Portuguese known only as The Gentleman (or Knight) of Elvas, who wrote the principal account of De Soto's North American failure. De Soto was a brother-in-law of Balboa, who first heard of Peru but got no chance to conquer it. He became one of Pizarro's ablest commanders, led the van of the army through much of the conquest, and was one of those who in the name of honor protested the treacherous murder of the Inca. The sword that was his only possession won him a great fortune and when the usual betrayals turned the generals to fighting one another he had the wisdom to return to Spain. There was no more representative conquistador: he was rich, famous, young, gallant, handsome, ambitious, a courtier, a spendthrift. Add the comment of the contemporary historian Oviedo, that he "was very fond of this sport of killing Indians."

De Soto had been made governor of Cuba and *adelantado* of Florida and was preparing to conquer the unknown interior when Cabeza de Vaca brought to Spain the nacreous, the opalescent news of its great cities, identified by now as the Seven Cities. He tried to ally their celebrant with his expedition but Cabeza de Vaca would not accept a subordinate command. He made what he could of the story before he sailed, though there would have been no trouble getting conquistadores without this bait. He took his expedition to Cuba in 1538 and in April 1539 sailed thence for Florida, with the most lavishly equipped force so far dispatched for conquest toward the New World. He had about 720 men and 237 horses.[5] He was traveling toward the gold of the Seven Cities, he knew not by what route.

The expedition landed at Tampa Bay on May 30 1539. De Soto is usually said to have led it brilliantly. Certainly he did not make the lethal blunders that Narváez did, there was no flaw in his courage, he could hold men to his will, and his will did not soften for a long time. But he was aimless, drunk on fantasy, and as compared with Coronado, toward whom he marched, futile and blind. And he was very fond of killing Indians. . . . Spain's first claim to the Great Valley rested on him, and his expedition lighted new fires across Europe. It inspired, especially, the first Englishmen who dreamed of challenging the Catholic power overseas and winning "a better Indies for her Majestie than the King of Spaine hath any." The narratives are the first descriptions that have any value, so that anthropologists and naturalists have cherished them.

But though the long pursuit of a mirage ranks as an exploration to those who came afterward, it was hardly one to De Soto or his companions. They were supposed to colonize Florida: the ships that brought the colonists never landed and were not even met. They were charged with finding the water passage to the Pacific (and, if it should prove convenient, China) but its possible existence seems never to have figured in their thinking till they were broken and it might provide an escape. They made little of the country they marched through except that it had too few fresh-water pearls and no gold. Upheld in a mysterious wilderness, they had scant time even for mystery for their minds strained toward the horizon and beyond it to the nonexistent: one hardly finds a sense of reality even when they fight or are wounded. Two centuries later it would be important that, on the way to the "discovery" of the Mississippi and a tableau for American iconography, they traversed lands no white man had ever seen before. After spending the winter of 1539–40 in Florida, near Tallahassee, De Soto, following the rumor of gold that fled ahead of him, led the army across Georgia, parts of interior North and South Carolina, across the Great Smoky Mountains, into Tennessee, south across Alabama almost to the Gulf, and northwest into Mississippi where they spent the next winter. It is not very important here. He slaughtered Indians and he reached the Mississippi River.

He told the scrubby tribes of lowland Florida that he was a child of the sun and backed the claim by killing some of them. Like all pioneering conquistadores he found that his best armament was his horses, which terrified the natives, but the big dogs too were excellent for pursuit and for the sport of torture. They were the breed of modified wolfhounds that guarded the flocks of merino sheep and chewed the throats of Moors in Spain; they were called greyhounds; mastiffs of today most resemble drawings of them. Some of the troops had crossbows, a good weapon but not much better than the Indian bows; it took a long time to shoot them and their mechanism was fragile. Some had arquebuses, whose principal usefulness as in earlier conquests was that the flash, smoke, and noise terrified Indians. It probably took about two minutes to fire one, during which time an Indian might shoot up to thirty arrows that would carry twice as far.[6]

In Cuba De Soto had heard that one Juan Ortiz, who had been on a ship sent by the widow of Narváez to search for traces of the lost company, had survived in Florida. Coastal Indians, who by now had a feeling for Spaniards, had enticed him and several others ashore. They killed the others but spared him (the story is so like that of John Smith's deliverance by Pocahontas that it may be the original) and he had spent ten years in captivity. Now soon after the expedition landed Ortiz ran up to one of the scouting parties. He became invaluable for he knew several Indian languages.

"Florida was so wide, in some parts of it there could not fail to be a rich country." It was one of the richest countries in the world but it was a desert to De Soto because it had no gold and the native custom of heating the trumpery river pearls had ruined many of them. He ever held it certain that going toward the sunset he would find what he desired. So in the second summer, after the long loop to the northeast, he traveled toward the Seven Cities. (Back in Cuba word came from New Spain that Mendoza's scouting party — Estéban and a Franciscan named Marcos — had found them in a country called Cíbola.) And he traveled with a routine cruelty whose horror degenerates into sick disgust. The figures which the Gentleman of Elvas gives must be scaled down, there is no knowing how far. (Still more Garcilaso's; he has 11,000 Indians killed at Mabilla, for instance, where Elvas conservatively makes it 2500.) Even so it was a fearful passage. And once they got inland they reached tribes that could make war. He massacred all comers but on his way to the Mississippi he fought with Choctaws, Chickasaws, and Alabamas, who were as warlike as his troops, as well or better armed, healthier, far more adept at ambushes and skirmishes in forests, creeks, and swamps. They were so formidable that only the horses turned the balance in De Soto's favor. Since they killed Spaniards, there must be reprisals after a battle.

Everywhere the conquerors went they demanded carriers, though soon there was little besides food to be carried. Everywhere they made slaves, long coffles of them, hundreds ironed together, though there were no slave markets and little or no work for slaves to do. Everywhere they burned fields and villages, gathered in women, tortured, maimed, and killed. "The land over which the Governor had marched lay wasted and was without maize." . . . "The Governor ordered one of them to be burned." . . . "Many dashing into the flaming houses were smothered and, heaped upon one another, burned to death." . . . "Two the Governor commanded to be slain with arrows and the remaining one, his hands having first been cut off, was sent to the cacique [chief]." . . . "The Governor sent six to the cacique, their right hands and their noses cut off." . . . "Many were allowed to get away badly wounded, that they might strike terror into those who were absent." . . . Many killed, a hundred men killed, many men and women killed, twenty-five hundred killed. Till the Portuguese narrator at last protests: "Some were so cruel and butcher-like that they killed all before them, young and old, not one having resisted little nor much; while those who . . . were esteemed brave broke through the crowds of Indians, bearing down many with their stirrups and the breasts of their horses, giving some a thrust and letting them go, but encountering a child or a woman would take and deliver it to the footmen."

And no good thing done, no New Peru or Seven Cities found, except that they lay a few days farther west. Some, perhaps many, of his gentlemen were convinced already. But when word came that his ships had returned from Cuba — perhaps to Mobile Bay, southward from one of his bloodiest battles — De Soto forbade Ortiz, the interpreter, to let anyone know. The news might light mutiny in the army and there was no gold to send back to Cuba, where his subjects, colonists, and investors were. He was "an inflexible man and dry of word"; he marched and murdered onward. For the winter 1540–41 the army camped in northeastern Mississippi.[7] It was a harsh time for men unused to cold, on short rations of captured corn and

pork from the herd of swine which had been driven all the way
from the ships, with constant Indian raids and occasional
battles. In the spring it was war all the way to the Mississippi.
They reached it in May 1541, spent a month building boats, and
on June 18 made the historic crossing. The spectacle in our
murals has nothing to justify it and the armor is too antique.
But Hakluyt gives the scale: "If a man stood still on the other
side of it, it could not be discerned whether he were a man or
no."

It was a very great river. Some of the army guessed it was the
Espíritu Santo of Pineda's chart, but mostly they called it Rio
Grande, the big river.

Now there were other fighting tribes, Quapaws, Tulas,
Tunicas, Caddos, perhaps at the farthest west some wandering
Wichitas. They too, though at first not ill-disposed, reacted
murderously to the murdering children of the sun. Their
minds fogbound and prismatic, the army came to the valley of
the Arkansas River and marched up it. Gold, mines and
smelters of gold, tribes weighted down with gold, were to the
northwest. Objects of trade new to the Spanish were coming
from that direction, robes made of hides of the bunch-backed
oxen, pemmican, cotton cloth. They marched toward the new
country and the fighting got harder, for these Indians could
hold off the Plains tribes. Rumors of gold led De Soto to send
a scouting party northeastward, toward the Ozark Mountains,
but it did not get out of Arkansas. And, as reality began to
break through the fever dream, he was "anxious to learn if we
could take the northern route and cross to the South Sea." But
gold kept glimmering in other places and after going up the
valley of the Arkansas some distance past Little Rock, he left it
and turned southwest. His farthest west was in Montgomery
County, at the foot of the Caddo Mountains, forty-odd miles
east of the Oklahoma line. By now the fever had lifted and De
Soto faced an ineluctable fact: he had failed. He turned southeast to the Ouachita River, marched down it, and built a
stockaded camp for the winter of 1541–42.[8] Here, not far from
Louisiana, he heard that the gold was to the south. But fear
that they would not get back to Spain had begun to clutch the
army's heart.

Spain had come close to inking a line across the United

States in 1541. At De Soto's farthest western reach in the valley of the Arkansas, he was less than three hundred miles from Coronado, who had reached the province called Quivira farther up the same valley in Kansas. Their possible meeting had caused fear in Spain, for the two forces would surely fight over the treasuries of gold, and in Mexico, for if De Soto should reach them first Mendoza would lose his investment. Neither army knew of the other's presence, though later on a refugee Indian did make a kind of contact between them.

De Soto had come a long way: from Seville to Cuba, to the Gulf coast, to Nowhere. All this distance his mind had held a picture of a room in a palace in the provincial city of Caxamarca in Peru. The room was, Prescott says, seventeen feet broad and twenty-two feet long. The Inca Atahualpa stood on tiptoe and reached as high as he could, nine feet for he was a tall man. To buy his freedom from Pizarro he would fill the room nine feet deep with gold. And before they garroted him he did, nine by seventeen by twenty-two: gold panels from the temples, "ewers, salvers, vases . . . curious imitations of different plants and animals. . . . Indian corn in which the golden ear was sheathed in broad leaves of silver from which hung a rich tassel of threads of the same precious metal . . . a fountain which set up a sparkling jet of gold." All this won by the sword and to be melted down for bullion, with only ten thousand natives killed in the streets of Caxamarca, and the whole empire of Peru beyond it still untouched. God! only eight years ago.

Beyond his rabbit-hutch on the Ouachita, a river of rivulets now frozen at the edges, there was nothing for De Soto. In the northwest from which he had turned back, he made out, there was no maize for men and horses and the Indian towns petered out. Many "cattle" were there, the bunch-backed oxen which none of them had seen, but few houses and they always farther apart, which meant no cities and therefore no gold. Seemingly there was no route to the South Sea there, though there might be one to the south, where the army had come faintly to doubt that there was gold. He had done no good thing in two and a half years' passage through a country whose rivers, mountains, and trails had no logic. At least two hundred and fifty of his men had died of fever or been killed by Indians, and a hundred and fifty of the horses. The few arquebuses and crossbows were wearing out — were worn out. For clothing there were only deerskins and buffalo hides. Much of the armor had been to begin with so poor and rusty that the elegant Gentleman of Elvas laughed at it, and most of it would not stop an arrow. It ended here. He would get back to Cuba — if he could. Then, if he could, he would make another attempt on the Seven Cities.

In March 1542 they started back, defeated, and the interpreter Juan Ortiz died. They spattered the landscape with more blood. By now failure had devoured De Soto's bowels. In Louisiana, on the west bank of his river, probably about opposite Natchez, he took to his bed, appointed his successor, and died. His estate in horizon-land was "two male slaves and three females slaves, three horses, and seven hundred swine."

They must make Pánuco or all would die but, remembering Narváez, they were afraid of the sea. So Luís de Moscoso, a qualified understudy in massacre, led them in a dead-reckoning march across Louisiana and into Texas, perhaps as far as the Trinity River. They heard sunny legends of Cabeza de Vaca and his cloudland and taught the Indians that he had not been a representative child of the sun. Here too they captured an Indian woman whom one of Coronado's officers had captured before them and who had escaped from her master when they too were in Texas, far to the west, before starting north to Quivira. The endlessness of the land made them despair again and Moscoso took them back to the Mississippi, almost to where they had left it. Like Narváez they invented boats, but better ones for they had seen Indian dugouts, and after killing most of the horses that remained they took off downriver. Their firearms and crossbows were worthless by now and the Indians, their superiors in everything except courage, harried them all the way and came close to exterminating them. Reaching the mouth, they made a wild voyage on the Gulf and three hundred and eleven of the original company came into Pánuco in deerskins, their hands empty.

They had accomplished nothing. The western entrance to the interior of the continent had been opened by Coronado while they wandered and the Spanish would push through it, slowly, as the years passed. But no one came through the eastern door again for a hundred and forty-two years, when La Salle did. His enterprise too ended in tragedy but it alarmed Spain into taking up again where De Soto had left off, though from the west and south. In 1684 La Salle landed on the Texas coast, somewhere between Galveston Island where Cabeza de Vaca's boat had floundered in 1528 and the Colorado River which he had reached in 1534, and built a small stockade there. The purposes of his expedition were ambiguous but the Spanish were right in believing them hostile.

De Soto and his company bequeathed to those who came after them a legacy of knowledge. It was of two kinds.

The three hundred survivors had traveled almost as far west as Oklahoma, and had crossed Louisiana and part of Texas. Of what they had seen they understood something and misunderstood more. Mingled with their understanding were true and false data given them by Indians, false data given them honestly by Indians who were in error, true data distorted by mistranslation and by the sheer inability of the Spanish mind and language to deal with Indian concepts, and a mass of rumors, legends, myths, and mere lies. All this they themselves bent and twisted with logic, fantasy, and fear. These veterans who had been there, as logical and realistic as might be, talked to others and they to still others, and all of them corrupted knowledge with desire, fantasy, and fear. There was here a sum of empirical knowledge but the facts in it were suspended in a great bulk of error. Such as it was, the empirical knowledge was closely held. It was passed on to illiterate men, or educated ones, in Europe and the New World who wanted to go farther than men had yet gone. It was only a little published but it was circulated and it would affect public and secret ventures from now on.

The other kind was formal knowledge. It began as the same stuff with the addition of what the chroniclers wrote about the expedition. It reached the learned, the scientists and literary scholars. Cosmographers, cartographers, and geographers took the mixed stuff and fitted it as well as they could, but by violence mainly, to their knowledge and theories of what the world was like. They too had knowledge, fantasy, and preconceptions, and the *must be* of deductive reasoning was even stronger. Hence their charts of the Atlantic, the South Sea, and presently the Western Sea, and world maps, maps of the Americas, maps of North America. They were made by minds that were among the best in Europe, and they represented the highest reach of knowledge. On them human intelligence pushes the line of the known outward into blank space. On the eastern coastline they began, not long after De Soto, to be tolerably good. But for the interior, long after empirical knowledge was making its way confidently from known place to known place, they were fantasy. In fact they were almost entirely fantasy for two hundred years. Suspended in the fantasy were grains of fact but no one knew certainly what they were. Shapes could be glimpsed through a thick mist but men who had to act according to what was known had no way of telling whether they were truth or illusion.

૯✧৩

When Mendoza, the viceroy, selected the Franciscan who is always known as Fray Marcos to take Estéban and reconnoiter the cities Cabeza de Vaca had heard about, he may have been using the Church as a leverage against his rivals. The conqueror of Guatemala was currently building a big navy on the west coast, which he threatened to use in exploring the rich new lands. Cortés too, who was too powerful to be controlled, was preparing a naval expedition for that explicit purpose and got it started four months after Fray Marcos set out. This was 1539 and the three ships were commanded by Francisco de Ulloa, who made a notable voyage. He sailed up the Gulf of California and proved that it was a gulf, that Lower California which had been supposed to be an island was in fact a peninsula. Shallows and the tremendous tidal rise at its upper end stopped him and he was not sure whether the fresh water that came down was from a river or a series of lakes, so perhaps it is not entirely accurate to call him the discoverer of the Colorado River. The gulf would be called the Sea of Cortés now but Ulloa's demonstration that California was part of the continental mainland did not long impress the cartographers. Before the end of the century it was an island again and it grew to be a big one.

The conquest of the Seven Cities, however, begins with Fray Marcos and Estéban. (Originally there was another friar too but either he fell sick or Marcos quarreled with him and he dropped out.) Marcos had been one of the religious attached to the conquest of Peru and he had been in New Galicia when Cabeza de Vaca came down to it. He was skilled in celestial navigation; he was devout and very gifted at fantasy. And for the greater glory of God, the Franciscan order, and most of all the Viceroy he was a great liar.[9]

Marcos set out from Culiacán in March 1539 with Estéban, a number of tame Indians, and for a time his fellow priest. He says that he sent Estéban on ahead while he paused to determine whether an island offshore was a rich one of which Cortés had had tidings, but it seems likely that the Negro himself urged it. He was to push beyond the desert and mountains to Cabeza de Vaca's cities, and eventually he was leading Marcos by more than two weeks.

The arrangement was that if Estéban should hear of "a rich country, something really important," he was to send back to Fray Marcos a cross that was a span long. If he learned of something even more important, make the cross two spans. "And if it were something greater and better than New Spain [which shows the size of the dream] he should send me a large cross."

On the fourth day after Estéban went ahead, his first messengers reached Fray Marcos. They were bringing a cross as tall as a man. The new country was richer than the Aztecs, then. The holy man hurried on. Estéban had set up still larger crosses for him along the trail, and had left word that the name of the rich country was Cíbola. (The origin of the word is uncertain; it may be Pima or Aztec. As *cíbolo* it became the Spanish word for buffalo.) Then there were more huge crosses and news that beyond Cíbola lay two other rich domains, each of which had seven cities.

Estéban's servitude had taken him from Algiers to Spain and on to Florida and Texas. Then with Cabeza de Vaca he had been one of the children of the sun and had crossed the continent to Culiacán. Now he was alone in godhead. To assert his sanctity he wore bells and dyed feathers on his wrists and ankles; he carried painted, cabalistic things. He strode in majesty through the strong sunlight and the medicine rattle given him four years ago far to the east swung from his belt. As he crossed northern Sonora, the resident tribes brought him gifts and food, went along with him in throngs, and gave him turquoises and women. He kept demanding more turquoises and more women and they gave him more. It was passing brave to be a god and march in triumph on Cíbola. He sent Marcos word that the wonders were increasing and that he was nearing the goal. He reached it when Háwikuh, one of six pueblos of the Zuñi Indians, thrust up its adobe-plastered walls, golden with sunset light, from the plain. Regally he had messengers carry his sacred rattle to the elders and demand entrance. The elders identified it as medicine from a tribe who had warred on them. They broke it and snarled at Estéban's messengers, ordering him and his retinue away from Háwikuh. The god would not be commanded. He went in. The elders confined him while they discussed his ambiguous, probably threatening invasion of their peace. Either he tried to escape or, as one of Mendoza's commanders (Alarcón) heard it later on, he demanded a god's tribute in turquoises and women. So they killed him.

His escort fled back across the desert and reached Fray Marcos, still in Sonora Valley, south of the Arizona border. The friar did not stand on the order of his going: he was back at his starting place by the end of June. He reported that after learning of Estéban's death he had gone forward amidst horrible dangers till he too, from a hilltop, saw the jeweled walls of one of the cities of Cíbola shining in the sunset. He had seen Cíbola, he said. So he had, with the eyes of faith, though it may have been the faith of the viceroy, who was going to conquer the rich country and had rivals to forestall. And the name of one of the farther kingdoms Estéban had heard about was Tontonteac, and Marcos had heard of two others, Marata and Acus.

His report to Mendoza shows that the Indians had told the truth about the country which they had learned on trading trips. They had accurately described the Zuñi pueblos, the turquoises set in doorposts, the cotton, the buffalo robes, the trade with nomadic tribes from the east. It was not their fault that, reaching minds spellbound with mythology and the lust for gold, what they said meant much more and blew fantasy to flame. Marcos' report is circumspect, only lightly brushed with rainbows, but his conversation across New Spain was, possibly on Mendoza's instructions, much more extravagant than Cabeza de Vaca's had been. So now all that had been heard about the horizon cities had been proved true.

Their country had a name, a gift from Fray Marcos. It was called Cíbola, and it had seven rich cities ripe for conquest and Christianity. The circle of desire had been closed. The Island of the Seven Cities, Antillia, had been found at last, only two months' march away.

❧

Though the threat manifested by De Soto's expedition and Cortés' plans was urgent, Mendoza organized the conquest of Cíbola with unhurried thoroughness. Following the return of Marcos, he sent a detachment of cavalry to reconnoiter the route again and fraternize with the Indians. It was led by a first-rate frontier commander, Melchior Díaz, the alcalde of Culiacán who had welcomed Cabeza de Vaca when he arrived there; it went all the way to the Arizona desert south of the Gila River. He provided three vessels under Hernando de Alarcón, to sail up the Gulf of California, get in touch with and support the main expedition, and explore northward or wherever by sea — for Cíbola must be in or on the edge of Asia. To head the conquest, which was set for 1540, he appointed his young friend Francisco de Coronado, whom he had already made governor of New Galicia.

Not yet thirty, well born, married to a woman of even higher station, Coronado was a soldier but had not been a conquistador. He put down an Indian revolt in his province, firmly but not vindictively — a promise of better times in New Galicia. That finished, in 1539 he was off to find another El Dorado in the Sonora mountains, east of the route Fray Marcos took. Nothing; and the mountains turned him back. Then as tidings of the rich Cíbola overspread Mexico, he was commissioned captain-general — absolute and independent head — of the whole enterprise. Fray Marcos, who had been made father provincial of the Franciscans in Mexico on Mendoza's insistence, was named to accompany him. Marcos would lead the friars of the Church spiritual and would guide Coronado, the deputy of the Church militant.

There was no dearth of recruits to conquer the Seven Cities. To a humble settler of Culiacán who joined the expedition and became its annalist, it seemed "the most brilliant company ever collected in the Indies to go in search of new lands." They were splendor when Mendoza reviewed them at Campostela, south of Culiacán, in February 1540. Two hundred and thirty caballeros, some with many horses, and sixty-two footmen were

mustered in there. There were five friars with humbler assist-
ants, a military guard, and their private Indians. (Eventually
the total force reached 336.)¹⁰ Nearly a thousand Indians went
along as servants and auxiliaries, and were a great strain on the
commissary. There were at least fifteen hundred horses, mules,
and beef cattle. Most of the arms were those of medieval hand-
to-hand fighting, advanced technology being represented by only
nineteen crossbows, seventeen arquebuses, and a few bronze
popguns on wheels that were no good. Nearly everyone wore
armor, though a lot of it was only leather cuirasses, or shirts of
mail. Coronado, riding down the line of pennons at review,
with the music playing and the trumpets like Caesar's and the
gentlemen of Spain shouting to St. James, looked like El
Dorado. His armor was gilded and two plumes waved from his
helmet.

The expeditionary force which looked like a parade moved
so slowly that on April 22 Coronado set out ahead with a light
column of about one hundred. Among the optimisms of Fray
Marcos had been statements that the Indian trails through the
mountains were excellent roads and nearly level, and that there
were food and forage all the way. It proved far otherwise and
from the foothills on the troopers found the going vile. The
gallant idlers of Mexico City were seasoned very soon; they
gaunted on short rations, mountain slopes, and waterless
marches, and so did their horses. Their bellies were concave
and their eyes sunken by the time they reached the chromatic
Arizona desert. Horses broke down and died, only a little corn
could be bought from the occasional Indians, and water holes
were far apart. But the trails led to better country and the last
days were easier, though rations ran entirely out.

The Indians had them under observation and on July 6 1540
they met some. The commander of the advance point assured
the natives that the invasion was peaceful but that night they
attacked the camp, causing confusion but doing no damage. On
July 7 the almost starving column, plodding under its dust
across the plain, saw a building that no doubt was shining in
the sun. It appeared to be made of stone, it was four stories
high with setbacks and terraces, and it was large enough to
accommodate something less than two hundred families: it was
a pueblo. The army had two bushels of corn left and needed a
lot more. So Coronado parleyed with the defending chiefs,

offering them peace for submission. They were not a warlike people but they didn't care to submit. The army shouted the "Santiago!" with which Spanish troops had launched the charge for centuries and it was soon over. Coronado was twice knocked down by stones from the walls, he got an arrow in his foot, two or three others were wounded, three horses were killed. The chiefs surrendered with quarter and Spain had conquered the Seven Cities.

This was again the Zuñi pueblo called Háwikuh, a few miles from the surviving Zuñi of today, southwest of Gallup. There were five other Zuñi pueblos hereabout.[11] The six of them, with one that had vanished or was legendary or imaginary, were Cíbola, which Fray Marcos had named and with which Guzmán's informants had traded.

The Spanish had reached an Indian culture of which Cabeza de Vaca had heard, some of whose products he had seen. The pueblo-dwelling Indians of Arizona and New Mexico were of several tribes. They had a higher civilization than any the Spanish had previously met except in Peru and the Valley of Mexico. They had lived here for at least six hundred years and had developed a magnificent scientific agriculture. Their arts, expressed chiefly in turquoise jewelry, music, pottery, and weaving, have delighted white men ever since Cabeza de Vaca. Their religion was and is noble. A gentle people, they were flourishing when Coronado's soldiers came to count seventy-one pueblos and bring most of them under subjugation, but perhaps they had passed the noon mark. For they were less numerous than they had been, many pueblos had been mysteriously abandoned, and the fierce Athapascan tribes that had migrated like Tartars from the north were pushing their frontiers in on them.

They were dignified, industrious, and peaceful. Too bad that Spain, here reaching one of the farthest frontiers of the Empire, could not attach them in friendship. It could not: the nature of the Spanish soldier and of the Church spiritual forbade. Coronado was no Guzmán or Pizarro but an honorable man who wanted a peaceful conquest and tried hard to get one; he almost held his army in check but not quite. When he took them back to Mexico, they left behind them an ineradicable memory of senseless killing. It was a foundation on which, coming back years later, the Spanish built notable cruelties, and

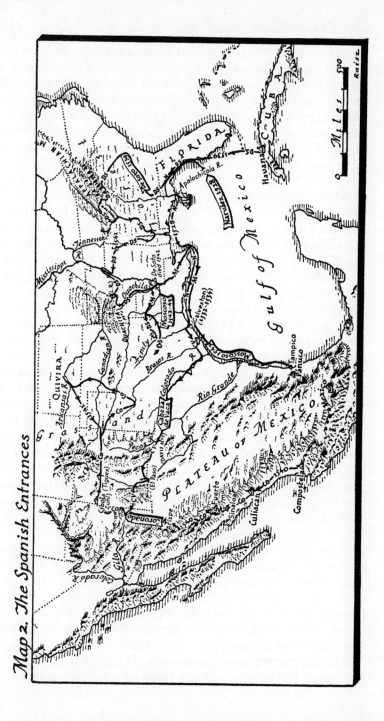

Map 2. The Spanish Entrances

for the love of Christ the priests worked cruelties of their own. The Church in Spanish America cannot be summed up in an adjective. After the blood-bath of the first half-century of conquest, heroic priests did steadily bring the conquerors into some kind of control and steadily moderated their inhumanity. The priest was the only defender the Indian had (outside the law courts of Spain itself) against massacre and torture — and yet there were always the doctrines of punishment and obedience, and when heresy or even obstinacy endangered souls which murder might deliver into Heaven, just enforcement of the laws was true mercy. . . . The pueblo dwellers too, like all Indians, had a genius for cruelty and revenge, so for generations the history of New Mexico was stained with blood. But they also had a genius for spiritual and cultural resistance. It has served them well. Today they are still a people, as few other tribes of the United States can be said to be.

But to its conquerors Cíbola was a shock so great as almost to unhinge the mind. They saw that these were superior Indians and, thank God, had plenty of corn and beans, but they had expected rooms corded nine feet deep with gold and emeralds. Fray Marcos found it expedient to leave with the first express for Mexico and Coronado wrote to Mendoza that "he has not told the truth in a single thing he said." This was a Spanish New World army: it had expected to be rich by the end of its first charge. To keep it from wreaking its anger on the Indians, which was the way of conquerors, required the finest leadership but Coronado kept his men in hand.

He set about exploring the land and hunting for the treasure that must be somewhere. One reconnaissance reached the kingdom of Tontonteac north of Zuñi, about which Estéban had heard. Its seven rich cities turned out to be a cluster of Hopi pueblos, again with full granaries but no treasure. Another one set out for a great river which the Hopis had described and, under García López de Cárdenas, a violent man but a brilliant commander, reached the rim of the Grand Canyon. A third marched eastward to another kingdom, called Tiguex, which turned out to be the center of the pueblo culture, the valley of the Rio Grande upstream from Cabeza de Vaca's passage. Thence, it pushed on to the eastern outpost of the culture, the pueblo of Pecos, where it got news of the buffalo herds and made a momentous captive, an Indian who was a prisoner of the pueblo and whom the Spanish called "the Turk." A fourth, commanded by Melchior Díaz, went west to the Colorado River, where it found messages from the ships commanded by Alarcón.

This last meeting of the lines which Mendoza had sketched on the unknown is a tremendous feat of planning and exploring, though less marvelous than it seems since Alarcón had buried his message under a monument which he set up where a customary trail from Cíbola reached the river. Alarcón's journey requires notice. He sailed up the Gulf of California to its head and, reaching the bars and shallows there, realized that a river lay beyond them, about which Ulloa had been uncertain the year before. With some expert piloting and the big tide he got the ships over the zone of shallows and well into the mouth of the river. Thus he rather than Ulloa was the true discoverer of the Colorado, a year before De Soto saw the Mississippi. He named it Buena Guía, the River of the Good Guide; Díaz, who recognized it as the one Cárdenas had seen from the mile-high rim of the Grand Canyon, called it Tízon, the Firebrand River, from the torches of a downstream tribe. ... It was usually Buena Guía on maps but likely to be Tízon in Mexico and would not be the Colorado for a long time yet. But it was a big river, so was the Mississippi, and so, farther down its course, was the one Coronado wintered on. All three were frequently referred to as just that, the Rio Grande, and this simple fact was to confuse cartographers in Europe for nearly three centuries, and governors and explorers in America for nearly two.

Alarcón could not take the ships up the fierce current and so anchored them and set out with two boats, which had to be pulled by hand. Soon he met Indians. He dealt with them so wisely that they welcomed him as what he said he was, a child of the sun bringing them a new god to worship and a commandment from his father to make peace with their enemies. They were Yuman tribes, fine people, tall, strong, and genial, and they made a community sport of pulling the boats for him. He went (on his second trial, for he had to go back to the ships for supplies) as far up it as the mouth of the Gila — two hundred miles up a river that was to resist exploration longer than any

other in the United States. No white man was to see as much
of it as he for more than a century, or travel it upstream for
more than three.

The Indians liked him so much that one tribe asked him to
stay and rule them in the name of the sun and another denounced neighbors who had word of slaughter in Cíbola, saying that he could not be of the same breed as the Christians.
And he liked them, which was a novelty among Spaniards and
almost unique among conquistadores. He heard about Cíbola
and in fact was inquiring about it when word came along the
Indian underground that the bearded white men called Christians were there. (Brisk journalism; it was about thirty days'
journey away and the news arrived in less than six weeks.) He
tried to send a report to Coronado and then to make a quick
march overland to Cíbola. But he could get only one volunteer
for either project, a Moorish slave, and so many enemy tribes
would be encountered that the Indians refused to carry a
message. He made no trouble in this far, strange land and he
met none. The Indians had an experience they were never to
repeat: they were sorry to see these white men leave. He was
a good officer and a wise conqueror. The history of the Spanish
and American empires would have been much different if there
had been more explorers of his kind.

Like most firstcomers, whatever their nationality, Alarcón
thought that the Indians he saw were instinctive Christians, in
part at least because they recognized the cross as a sacred
symbol. Much misconception and many endeavors were to turn
on this fact as white men traveled the continent. But it was a
native symbol in many tribes. It stood for the four houses of
the sky and the four winds that come out of them, or for the
morning and the evening star, or for the path of the sun, or in
the form of the swastika for the rain god who gave life, and his
bird that was the thunder, and the snake that was sometimes
his avatar.

☙

Coronado moved his advance column over to the province
they called Tiguex, the Rio Grande valley between Albuquerque and Bernalillo, which his lieutenant had recommended

for winter quarters. He dedicated it to Our Lady and he and many of his companions found in this high valley, which is also a high plateau with strongly colored mountains shining through thin air at its edge, something that spoke to their loneliness. It was like Granada, it was like Castile, it was like Estremadura — like the hard sparse land that was home. But no gold or emeralds, and it was like Spain too in that it was poor. Bitterly disenchanted, Coronado called it sterile, and when he reported to Mendoza the most he could say for it was that these Indians made the best tortillas he had ever eaten.

The main body of the army joined him here. He billeted them in several pueblos, evicting the owners. To feed them he levied on the corn stored against drought years, and thus put a strain on the economy of the whole region. The winter came on very cold. The troops took the clothes and robes they needed, from such shoulders as had them on. And the customs of Spanish soldiers among Indians could not be wholly checked, nor the frustration of finding only apartment houses of mud brick instead of kings' treasuries. So before spring many Indians rebelled — to make the first item in New Mexico's sum of hatred. There was some minor fighting and two pueblos were sacked. At one of them Coronado's adjutant, the Cárdenas who had found the Grand Canyon, taught the Indians about conquests. He offered quarter, then when the chiefs accepted it he set up stakes and burned thirty or more of those who had surrendered and slaughtered sixty others when they tried to break out of their pen.

But the renewal of hope warmed the winter. "We shall go Always a little further: it may be Beyond that last blue mountain barred with snow . . ." The army now heard that the stories were true after all, no matter the hard shock of Cíbola. It was only that they had not gone far enough. The new Peru certainly did exist; it was northeastward from here and it was called Quivira. Thus a greater, more luminous, more enduring name begins to echo in desirous minds. . . . The Turk told them all this, the captive Pawnee whom they had found at the pueblo of Pecos and had brought back to the Rio Grande.

The Turk could converse forever in the sign language of the Plains tribes but could not speak Spanish. His nightly audience could speak no Indian tongue. They understood him to say that in his country [which eventually turned out to be not Quivira but another great kingdom called Harahey] there was a river in the level country which was two leagues wide [that would be seven miles but there was some fact behind it, whether the Platte at its mouth or the Mississippi] in which there were fishes big as horses [well, catfish beyond a hundred pounds, but from here on the listeners are airborne] and large numbers of big canoes, with more than twenty rowers on a side, and that they carried sails, and that their lords sat on the poop under awnings, and on the prow they had a great golden eagle. He said also that the lord of that country took his afternoon nap under a great tree on which were hung a number of little gold bells, which put him to sleep as they swung in the air. He said also that everyone had their ordinary dishes of wrought plate, and the jugs and bowls were made of gold.

Bells of gold, such as chime faint and far away in all cathedrals beyond the horizon or under the sea. They must not take heavy packs to the Gran Quivira, the Turk said, for the horses would need all their strength to bring back the gold. . . . From the beginning two other prisoners who were from Quivira told the Spaniards that the Turk was lying. It was impossible that any should listen to them.

The conquest, then, could still be saved, the rich countries discovered, the gold gathered up. So in the spring of 1541 Coronado took his troops in a magnificent foray out of the Rio Grande valley into the country that would be the stake of empires. Far beyond the boundary that Mexico could hold, well along toward the eastern boundary of what for a time would be Spanish territory on loan from France. From New Mexico out into a new, strange province, the Great Plains, seen before them only by Cabeza de Vaca and his three, and then seen along a southern fringe. They were plainsmen now and, if in armor that their successors would learn to do without, they at once showed the talent for plainscraft that seems to have been native to the Spanish.

East across the southern spur of the mountains and across New Mexico and the Canadian River. Southeast to the Staked Plain and so into Texas, and east again. They and their Indian auxiliaries and prisoners made a big caravan — fifteen hundred men, a thousand horses, five hundred beef cattle, five thousand sheep — but immensity swallowed them. As immensity lengthened out to no imaginable end there was also no sign or tidings of Quivira. They touched the headwaters of the Colorado

River of Texas and found Indians who had heard of Cabeza de Vaca and some who had seen him. They were near the eastern edge of the Staked Plain, disturbed, bewildered, frustrated, growing angry. Vastness in itself seemed a danger and bewildered ignorance was another one — but how could a man acknowledge he had failed to find the best and richest country? Moreover, they knew now that the Turk had misled them; they handcuffed him and promoted to his place one of the two prisoners who had always denounced him and who now said that Quivira, of which he was a native, was to the north — and not rich. Coronado ordered the army back to the Rio Grande and, again, himself led a force of picked men into the farther unknown, traveling due north by the compass needle. The march took him across the Brazos, the Red River, the Canadian again, and on to a river which he called St. Peter's and St. Paul's. It was the Arkansas and when he reached it De Soto was in its lower valley.

This was Kansas and Coronado had reached the river and crossed to its north bank in Ford County, at the town of Ford.[12] He turned down the wide, shallow valley and knew that he had entered the great land of Quivira for he met a hunting party of its inhabitants, whose language the new guide spoke. (They were Wichitas.) Under the burning-glass July sun they went on across the green, rolling land-ocean under the unbounded sky, past the Great Bend of the Arkansas, left the river, traveled northeast till they reached the Smoky Hill River, and came to a Quivira village. A hunting camp, its palaces the grass huts of the eastern culture from which the Wichitas had migrated, it was singularly barren of kings and gold plate. It was in McPherson County, Kansas, near the village called Lindsborg, and here Coronado ended his penetration of Quivira.

All this crossing of the plains had meant new landscapes, new experiences, new peoples, and they were all strange. "The land is the shape of a ball," the annalist says, the first man who ever wrote the thought so many have had since, "wherever a man stands he is surrounded by the sky at the distance of a crossbow shot." Everywhere the sun mocked the eye with unearthly distortions. Seared eyes could find no trees for solace except the willows and cottonwoods that marked watercourses and sometimes a small, hidden ravine choked with smaller stuff.

Only the earth and the sun and the arch of the sky, buffalo grass everywhere and then taller grasses. Ahead of them the grass bent as the wind trod it; the line of horsemen bent it too as they crossed; it rose again from wind and hoof and closed behind them and no sign of their passing had been left. Scouts, stragglers, the column itself might get lost in the tranced emptiness except that they piled stacks of buffalo chips to mark the way. Those same chips were the only fuel; their punk-like pungency for the first time prickled the noses of white men cooking supper.

That meant the herds, the buffalo at last, in numbers so large as to forbid belief. And buffalo meant new, strange Indians, nomads, the Plains tribes: first various Apache tribes, chiefly those to be called the Vaqueros or Cowboy Apaches later on, then tribes hard to identify but grouped together as the Texas, and finally the Quivirians, the Wichitas, sometimes to be called Pawnee Picts or called, erroneously and with strange consequences, by the name properly applied only to the Comanches, the name Padoucas.

> As for those hunch-backed Kine [this is Purchas quoting the contemporary Spanish historian Gómara, who was quoting the first-comers] they are the food of the natives, which drank the bloud hot, and eate the fat and often ravine the flesh raw. They [the natives] wander in companies, as the Alarbes and Tartars following the pastures according to the seasons. That which they eate not raw, they rost, or warm rather at a fire of Oxe-dung, and holding the flesh with his teeth, cut it with Rasors of stone. These Oxen are of the bignesse of our Bulls, but their hornes lesse, with a great bunch on their foreshoulders, and more haire on their foreparts than behind, which is like wooll; a mane like a Horses on their backe bone, and long haire from the knees downward, with a store of long haire at the chinne and throat, a long flocke also at the end of the males tailes. The Horses fled from them, of which they slue some, being enraged. They are meat, drinke, shooes, houses, fire, vessels [dishes], and their masters whole substance.

This is as accurate a description of the buffalo and its place in the Plains culture as anyone could write today, even though it is at third hand. Misconception comes back two sentences later, with a statement that the Indian dogs fought the bulls. But Coronado's annalist, an eyewitness, reports the dogs truly, the travois they drew, and their miserable howling when the packs slipped. He describes the methods of the hunt, gaping at the power of a bow which could drive an arrow clear through the shoulders of a bull, as the Gentleman of Elvas and Biedma gape at the bows of the Alabama tribes. He describes the tribes, their costumes, something of their handicrafts, something of their trade. In his pages are the first prairie dogs and prairie wolves, the first jack rabbits, and a number of other firsts.

A single instance adequately illustrates the paradox of minds which were as logical as any today and, further, were here working not with myth but with the most concrete realities and yet were betrayed by sheer strangeness. The principal reason why Coronado went no farther than he did but turned back was that his supplies of corn were short and his men and horses might starve. For the horses there was only buffalo grass, the most nutritious grass in the New World, and for the army only buffalo meat, the most complete single food that mankind has ever known.

℆

But there was no gold, no silver, no emeralds, no lords of the country lolling in gondolas and soothed asleep by golden bells, no golden plates and ewers. Quivira was not a new Peru but only Kansas. The shock, more stunning than at Cíbola, was hardly to be endured. They forced a confession from the Turk. A captive in a far country, he had wanted to go back to his own — Harahey, and they were still south of it — and the Pueblos had shown him how to procure his freedom with a useful lie. The Pueblo hope was that if the conquerors could be lured into the vastness, Apaches would kill them or they would starve. Told the truth at last, the Spaniards wound a cord round the Turk's throat and strangled him. Some remained certain that the new Peru and all the storied gold were only a little farther on — admittedly not in Harahey, the next kingdom, but perhaps in one called Haya, or in many places, or in the mountains the Quivirians talked about which must be close at hand. But Coronado, facing his failure and thinking that he faced starvation, turned back. The Wichitas guided him home, up the Arkansas and then by an old trail that diverged from it to the southwest, a much shorter passage than his outward one. This was an ancient route of trade and war; it became the road between New Mexico and St. Louis and would be called the Santa Fe Trail.

The army spent another winter where it had spent the last one. A severe injury sustained in an accident warped Coronado's mind and disposition. There was some trouble with the local Indians but not much. But guard stations on the route from Culiacán had been badly commanded, the tribes had raided them, the line of communication and supply was endangered. Many of the gentlemen still believed that gold would be found in a greater Cíbola or a farther Quivira, and some thought that Coronado was concealing information for his private gain. But there was no sane thing to do except to take the army home and he did so in 1542. When it neared Culiacán it began to melt away, its discipline broken, and Coronado had to face the viceroy as a failure.

Considering that the Spanish had had only one purpose in 1540, the expedition was indeed an absolute failure. It so shattered one particular dream of gold that the still luminous myth of Quivira had to flee a thousand miles northwest to the shore of the Pacific, and no one started toward Cíbola again for forty years. The frontier of settlement in Mexico, however, continued to move northward; a less glorious frontier than that of Cíbola, though the force that moved it was the silver mines. At last the sterile land was remembered, and in 1581 a frontier captain made a swift journey to Coronado's winter quarters on the Rio Grande. The next year another soldier on a similar errand of reconnaissance, salvation, and some punishment followed his trail, went on to make a wide compass, and got as far as Zuñi. Finally, in 1598, fifty-six years after Coronado, Juan de Oñate came over the last marches into Tiguex, under orders to colonize New Mexico and hold it for Spain and New Spain.

But only in that Coronado found no treasure did his expedition fail. Few explorations in all American history were better led, few dealt more successfully with the problems of wilderness travel. And no one had preceded it, it was the first that passed this way, it could call on no distillation of other men's experiences, every emergency it met was strange and every place it traveled to was new. Thus early the exploration of interior North America sounds a high clear note. Coronado was so aware of what he was doing and so competent at meeting the challenge of the wilderness that, although mirage and myth had led him here, he seems a critical intelligence supported by

exact knowledge. Whereas at every moment that you touch
De Soto, in the same year, he seems delirious. And this holds
true for all Coronado's lieutenants and chroniclers.[13] In relation to the large geography, the map of North America behind
the mist, they could not escape the distortions implicit in guess
and in deduction from things misconceived. They were, that
is, men of their time; moreover the bestiaries and the fabulous
ethnologies of the lands beyond the horizon held good for them,
so that they thought of giants and dwarfs a little farther away,
of monsters, supernatural or magical beings, dragons and mis-
begotten serpents and fish, and creatures that had names but no
habitation outside fantasy. But what they saw, they saw with
surprising clearness and they reported it well. They handed on
knowledge to those who came later — knowledge intimately
combined with error, but a good deal of it firm. It went into
the empirical thinking of those who turned their attention
toward this land. It spread over Mexico and on to Europe and
so to enterprises of the future. More slowly and much twisted
by conjecture, it worked into the thinking of the learned. So
the mist was a little thinner here and there; it opened in some
places on factual solidity that could be put to use.

The annalist is Pedro de Castañeda de Najera, a humble man
and, when he wrote, a man growing old.[14] He was a settler, not
a conquistador, a man of the land and its people. And a man
with a proper feeling for the fine young gentlemen who, though
he saw that they had faults and could do foolish and evil things,
were nevertheless the chivalry of Spain as he was not. He took
pride in them and their exploits, and he concluded that such
noble adventurers who "went in search under the Western
star" (the famous phrase is his) could not have failed unless
God had willed them to. He was tranquilly sure that God's
purpose would be clear some day and that the undiscovered
lands would be discovered, and then men would know "for
whom He has guarded this good fortune." And Castañeda,
growing old, remembered his share in all this as a great glory.
Once life had leaped up for him like a flame and had shone
brighter than any imagining. The great country, pure and
untouched, stretched ahead and in his strength and of his sharp
wit he, a young man, strode through it proudly, knowing he
was a discoverer. As he looks back to his splendor there is in

him just such fullness as there was in that old soldier who had
fought with Cortés to Tenochtitlán. Bernal Díaz, who remembered the weariness and the fighting, the fear and the labor,
the thudding of the heart in battle, and friends bound together
against death in the sorrowful night of slaughter. He had
noticed, Castañeda says, that we do not value a thing highly
when we have it in our hands, we do not imagine how we shall
miss it when it is gone. But, so he knew now, "after we have
lost it and miss the advantages of it, we have a great pain in
the heart, and we are all the time imagining and trying to find
ways and means by which to get it back again."

So, toward evening, feel all adventurers in new lands. But
there is something more in Castañeda's reverie, a color and
emotion that fairly entitle him to be called the archetypal
Pioneer. Cabeza de Vaca traveled the land with no thought of
it except to live off it and get it behind him. Sometimes though
not often Biedma or Ranjel or the Gentleman of Elvas,
De Soto's chroniclers, will note that here is a good place to build
a fort for the protection of the slave ships or that the land is
so fertile that a garrison could support itself with little expense
to his majesty, but though they were forerunners of an intended
colony there is only the slightest regard for the landscape as a
home — all told much less than there is in the fourth
chronicler, the Inca Garcilaso, who never saw it but could understand. But Castañeda, the emigrant, the settler, the home-
steader, had his eyes always on the land, the fine things it grew,
the great wealth of crops — not gold — it might produce. It
was a good land in his sight and he remembered it as good,
even when he thought "of that better land we did not see." He
told his memories like a friar's beads, and he felt the comrade-
ship of the men who had gone with Coronado. "Granted that
they did not find the riches of which they had been told, they
found a place in which to search for them, and the beginning
of a good country to settle in, so as to go farther from there.
Since they came back from the country which they conquered
and abandoned, time has given them a chance to understand
the direction and locality in which they were, and the borders
of the good country they had in their hands, and their hearts
weep for having lost so favorable an opportunity."

# The
# Spectrum of Knowledge

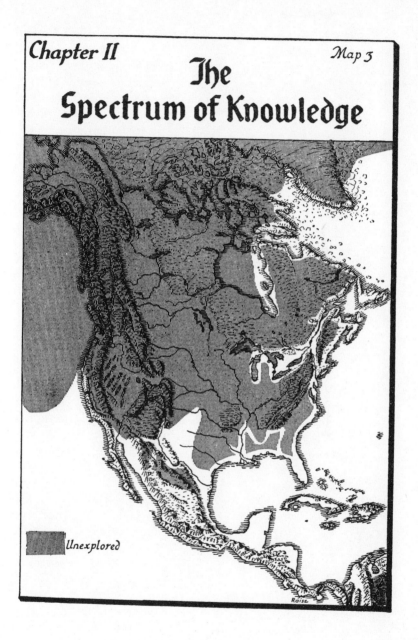

Unexplored

# II

# The Spectrum of Knowledge

THE SPANYARDS have notice of seven cities which old men of the Indians shew them should lie towards the Northwest from Mexico," Hakluyt wrote. "They have used and use dayly much diligence in seeking of them, but they cannot find anyone of them. They say that the witchcraft of the Indians is such that when they come by these townes they cast a mist upon them, so that they cannot see them." There was indeed a mist which hid from view not only Cíbola, and not only all the other horizon lands and Islands of the Sea, but the honest earth itself wherever one had not gone. Not witchcraft created the mist, however, but human thought working at the unfamiliar with the familiar for a gauge. Columbus, ending the prologue of his journal with a notation that he must forgo sleep and diligently observe his navigation in order to carry out the great undertaking, addressed his sovereigns: "I propose to make a new map on which I shall draw the Ocean Sea and all its lands in their true positions and under their winds. And I desire to compose a book in which I will make drawings to represent everything truly, by the latitude from the equator and by the longitude from the west." In respect of the continent he discovered, the task thus stated is not yet finished; even the peripheral outline of his map was not completed for more than three hundred years.

We speak of the white light of knowledge but rainbows playing along the mist that hides the Islands of the Sea are there because the components of knowledge have different wave lengths. In the infinitely difficult act of thinking nothing is

51

more difficult than to separate what is known from what is not known — unless it be to understand that the separation must be made. The pitfalls ready-made in the material with which the intelligence must work are not more formidable barriers to the achievement of knowledge than the traps intelligence sets for itself. There is an interaction and parts of it, parts of the process of human thinking about the geography of North America, are an immediate concern of this narrative.

၏၈

We do not know when Castañeda wrote his account of Coronado's expedition, though it must have been around 1555. He could therefore have read Oviedo's general history of the Indies, but not any firsthand account of De Soto's expedition. But he uses in his own book ideas which De Soto's three hundred veterans contributed to the intellectual estate of New Spain.

In the second division of his book Castañeda writes a descriptive chapter about Quivira, "the most remote land seen." It contains a passage of absolute importance:

> The Great Spiritu Santo river that had been discovered by Don Fernando De Soto in the land of Florida [the Mississippi] flows from this region [Quivira: Kansas]. It runs through a province called Arache [Harahey, north of Quivira], according to information which was considered reliable, though its sources were not seen, because it was said that they come from very far, from the land of the southern cordillera and comes out at the place where it was sailed by Don Fernando De Soto's men. This is more than three hundred leagues from where it empties into the sea. On account of this and its many tributaries, it becomes so mighty when it reaches the sea that they lost sight of the land, and the water was still fresh.[1]

The commanding fact revealed by this passage is that in 1541 Coronado learned from Indians of the existence of the Missouri River. That it was entirely blended in the Mississippi, that two rivers were understood to be one, does not matter. In the course of one year the Spanish had encountered two fundamental features of the geography of North America, one empirically and the other by hearsay, and had, though by misconception, postulated a true relation between them. The misconceptions that followed were as important as the realities in the history of the next two and a half centuries. De Soto had crossed the Mississippi some four hundred and fifty miles above its mouth, Moscoso had traveled the last hundred and seventy-five miles of its course, and by the time Castañeda wrote the river so crossed and traveled was identified as the Espíritu Santo, whose wide mouth had been variously located in charts of the Gulf coastline and ideas about it.

Coronado hears of two solid facts, the existence of the river now called the Missouri and its origin somewhere in the northwest, and its continuity with De Soto's river is postulated. Castañeda makes a single river of the Missouri and the Mississippi. He understands that it rises in a "southern" range of mountains, but he nevertheless has the orientation correct. For he knows that these mountains are, vaguely, north of Quivira: he calls them the southern cordillera only because they are somewhere in the vicinity of the Pacific, the South Sea. How much west there is in his south and north no one can tell, for the data he used were only what Cabeza de Vaca and the exploring parties Coronado sent westward had guessed. The Missouri-Mississippi, then, flows down from mountains that are northwest of Quivira and across the plains which Coronado had seen. And somewhere in a latitude north of Quivira, Castañeda does not guess how far, a range of mountains comes in from the east, from the Atlantic here called the North Sea. This big river cuts through this northern range more than seven hundred and fifty miles above its mouth, and thence flows to the Gulf of Mexico.

Castañeda believes that the mountains in "these western regions" lie near to India, of which he thinks Cíbola and the related countries are an extension. It is clear that he did not think of them as connected with the New Mexico ranges which Coronado's expedition had seen. It is clear too that he had no conception of the great system which we call the Rocky Mountains. In fact that mountain system would be the last of the basic features of our continental geography to be conceived, understood, or explored.

But though Castañeda's western mountain range cannot be assigned an approximate location, his northern range can be exactly delineated. De Soto had reached the Appalachian system and crossed its most southerly range, the Great Smokies.

It is the fundamental eastern mountain chain of the United States and it trends from southwest to northeast. What Castañeda did, and what cartographers interpreting De Soto's data and some later data did for many years, was to pivot the Appalachian system on its southern tip and turn it west. In Castañeda's mind, and on maps for more than a century after him, a mountain range crosses North America from east to west (at various distances north of the Gulf). On the maps it continues as a broken or intermittent range after it has ceased to be a continuous one.[2]

On the 35th parallel the distance between the Appalachians and the Rockies is upwards of fifteen hundred miles. Castañeda guessed that it was about a thousand miles. But he also, and very likely at the same time, thought of it as much less — just as Cabeza de Vaca made it both five thousand miles and seven hundred miles. Also, in Castañeda's mind the continent north of Coronado's farthest north spread out east and west like an opened fan. In its high latitudes it was enormously wide, reaching almost or quite to China in one direction and to Iceland or Europe in the other. (Many world maps for more than a century after him would express the same idea.) Nevertheless, it was at the same time small.

That paradox does not matter. In 1541 Coronado heard that a great river, which behind the mist was the Missouri, headed in the eastern slopes of a western range of mountains. He had heard of a basic feature of our continental geography and a fundamental key to it. The year before, 1540, De Soto had determined the existence of a big river, the Mississippi, which was another basic feature and fundamental key of the continental geography, though all but a short stretch of it was hidden in the mist. These features were to involve the destinies of empires.

Even earlier than this the white men had heard of a third fundamental feature and key of continental geography. In August 1535 (Cabeza de Vaca was in the second year of his westering) a great navigator in the service of Francis I of France, Jacques Cartier, discovered that a wide opening into North America which he thought might lead across the land mass to the Pacific Ocean was not a strait but a river. He referred to it as the Great River of Canada but eventually it was to be known as the St. Lawrence. After healing the sick as Cabeza de Vaca was doing, he took his little ship as far as he could up the narrowing corridor of the river, at its most beautiful with the summer sunlight carrying the earliest premonition of autumn, and then went on in boats. Rapids stopped him at the foot of a height which he named Mount Royal. The local Indians took him up the peak and showed him the wilderness stretched westward, a green tapestry sewn with silver thread. Far to the west, they said, were several seas — but seas of fresh water. The way to them made a long journey, many suns. The largest of them was the farthest west and it was on the rim of the world; the Indians did not know if "there were ever man heard of that found out the end thereof." Cartier thought that this distant fresh-water sea might open on the South Sea, or else would lead to it.

Europe thus heard of the Great Lakes and the water route to them.

&⭒⭗

Logic and desire imperatively insisted that there must be a water passage across North America, across the United States or Canada or both. The Americas were a dismaying interruption of Europe's course westward to the East. Though they soon began to provide gold as compensation, they remained a barrier. Logic made them out to be so narrow, however, that perhaps they were only a barrier reef. It was soon clear that South America was wide. But the continent — if it were a continent, and who knew? — thinned to a narrow isthmus at Panama and again at Tehuantepec. Thereafter it widened again but no one knew how much or for how long: there might be other isthmuses. So clearly the water passage from the Atlantic to the Pacific must be in North America. It led westward and it must lie somewhere north of Florida . . . or north of Alabama, Louisiana, Texas . . . or north of Virginia, Long Island, New England, Newfoundland. Hence the name of the *ignis fatuus* whose existence was established by the most searching analysis of experience and the most controlled extrapolation of theory. It was the Northwest Passage.

It could be a strait, a salt-water corridor across North Amer-

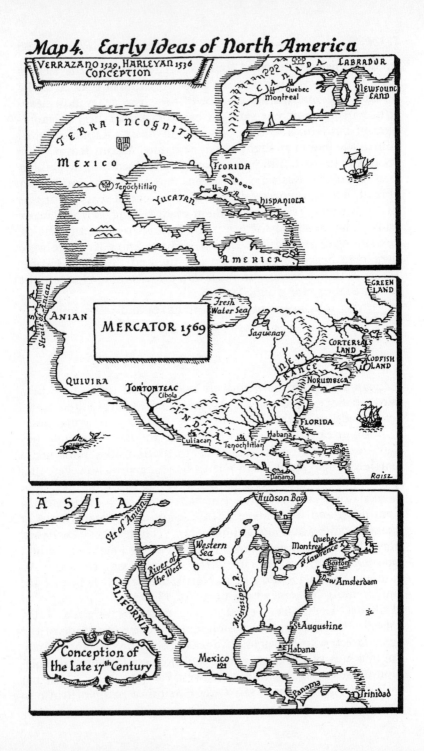

# Map 4. Early Ideas of North America

**VERRAZANO 1529, HARLEYAN 1536 CONCEPTION**

Anian???  CANADA  LABRADOR

Quebec
Montreal

NEWFOUND LAND

TERRA INCOGNITA

MEXICO

FLORIDA

Tenochtitlán

YUCATAN  CUBA  HISPANIOLA

AMERICA

---

**MERCATOR 1569**

ASIA

Strait of Anian

ANIAN

GREENLAND

Fresh Water Sea

Saguenay

CORTEREALS LAND

CODFISH LAND

NEW FRANCE

QUIVIRA

TONTONTEAC
Cibola

NORUMBEGA

FLORIDA

Culiacan  Tenochtitlan  Habana

Panama

*Raisz*

---

ASIA

Str. of Anian

Hudson Bay

Western Sea

River of the West

Quebec
Montreal
St Lawrence
Boston
New Amsterdam

CALIFORNIA

Mississippi R.

St Augustine

Habana

Conception of the Late 17th Century

Mexico

Panama  Trinidad

ica, and this was the great hope. Or less conveniently it might
be a route up some river that emptied into the Atlantic. By
sailing upstream you would come to a mingling of the waters
and so could descend to the Pacific. (The actual interoceanic
mingling of the waters, or the actual interlocking of fundamental watersheds which amounted to the same thing, proved
to be the most durable of all misconceptions.) This transfer
point might be a small or large lake, from which the river you
had ascended flowed east and the one you would descend flowed
west. Or it might be a landlocked sea of the interior, from
which the same kind of westward-flowing river led to the
Pacific. Or it might be an arm of the Pacific that came crank-
ing in, you did not know from what direction or for how many
leagues.

By the beginning of the seventeenth century this last conception had created an entity that must be differentiated from
the Northwest Passage, though very often that Passage was connected with it. The inland waters, whether landlocked or a
gulf, whether salt or fresh, were thought of as a separate, sharply
individual sea. A protean body of water, as extensible and as
migratory on the map as it was illusory, it was called the
Western Sea.[3] It was not far away. It was only a few days'
journey farther on.

The Florentine Giovanni Verrazano was not responsible for
the belief that North America was just a hoot and a holler wide
or that, at least, somewhere it narrowed to convenient thinness.
No other idea was possible to the European mind, for the sum
of human knowledge necessitated the conclusion that the North
American land mass was narrow. This conclusion was reinforced by the tremendous power of reasoning from observed
things and from hypothesis, and by the greater power of wish.
Nevertheless Verrazano gave the idea fixation. Also, and this
is more important, he gave it a shape that could be visualized
and therefore elaborated.

Like Cartier, Verrazano sailed for Francis I of France. In
1524 he made a voyage along the coast of North America that
brought considerable new knowledge out of the mist. (It was
the first year of the conquest of Peru. The first circumnavigation of the globe had ended in 1522.) In the course of it he
coasted a long stretch of the United States — he says two hun-

dred miles — so narrow that from the deck of his ship he could see water on the western shore that was unquestionably the South Sea. Historians have debated whether he was north of Cape St. Charles on the Eastern Shore of Virginia and Maryland, in which case his South Sea would be Chesapeake Bay, or north of Cape May, which would make it Delaware Bay. But the distance he says he sailed and the extreme narrowness of the isthmus he thought he was coasting make it likely that he was off North Carolina, looking across the barrier sandspits to Pamlico Sound.[4]

At this isthmus the continent was less than a mile wide. So it is to Verrazano that we owe a map-image of North America which is roughly like a pair of kidneys lined up northeast-southwest. The northeast lobe contains (in modern terms) the middle United States, New England, and Canada and on some maps it is attached to Europe. The southwest lobe is Florida and Mexico and at first it stretches all the way to India. A narrow isthmus connects the lobes and west of it is the Sea of Verrazano. Sometimes this sea runs south of Asia as the South Sea; sometimes it runs north of Asia as the Atlantic or eventually the Arctic. This conception has a number of variations but more important is what happened to the continent when the conception had been proved erroneous. As the eastern coast of North America became known in detail, as its western coast became a little understood, the width and interior configuration of the continent remained unknown. The Sea of Verrazano then began to push inland from the Pacific, eastward, in various latitudes and to various penetrations. As a visualized conception of the Western Sea, it affected other conceptions.

Every inlet, estuary, bay, and river mouth on the Atlantic coast might be a strait that would lead to the Pacific — might be the Northwest Passage. Failing that, it might be the Northwest Passage at one remove — it might lead to the Western Sea, from which one could go on to the Pacific. Every inland body of water that one heard about from the Indians might be the Western Sea. (Or the Sea of Verrazano or some way station.) The Penobscot River, for instance: the Englishman George Weymouth sailed up it bound for Asia in 1605, having failed to get there three years before in the high latitudes. Or the Hudson: Henry Hudson who discovered it in 1609 forthwith learned that it was not the Northwest Passage but sailed up it to Albany because it might lead to the Western Sea, which he had heard might be near there. He no doubt heard about it from Captain John Smith, who himself had gone up the James and Chickahominy in 1607 to reach the Pacific. The next year Smith nosed into inlets of Chesapeake Bay on the same errand, found the Potomac, and sailed up it hopefully. His explorations left him convinced that the Western Sea lay just west of the mountain range whose piedmont he had seen — in effect De Soto's mountains and those which Castañeda pivoted toward the west. Long before him French and Spanish captains had tried the rivers farther south and though they did not reach the Pacific that way, for a long time the assurance that they could lingered on, fed by persistent rumors of Indian metropolises whose houses were made of gold. After Smith others investigated Chesapeake Bay again, and Delaware Bay, the Delaware River, the Susquehanna, the Connecticut, the Merrimack, the Kennebec. And Long Island Sound, for Hell Gate might open on China.

༄

A beginning.

The Pacific was so near that the charters of Massachusetts (1628 and 1691) granted the colony the full extent of the land west to the ocean. Virginia (second charter 1609) extended from sea to sea. Connecticut too ran to the Pacific (1662) and included the islands adjacent to the shore. So did both Carolinas. None of these grants were absurd to English minds but the later ones seem fantasy in the light of what the French, who had pressed beyond Lake Superior and the Mississippi, knew when they were made.

They *were* fantasy and this geography of fantasy played a part in forging the United States. When the Articles of Confederation were ratified (1781) these western boundaries had been realistically amended; they were now the Mississippi. But under the Articles the first national possession of the United States was the lands that had been granted by English kings according to the geography of fantasy. They were the nation's most valuable asset and its strongest cohesive force. They were a prime mover in fulfilling the geographical, political, and psychological destiny that required us to be a continental nation. Even more may be said. In the geography of fantasy the English colonies, which in the first moment of settlement began to become the United States, extended to the Pacific Ocean. The usage and expectation that were born of fantasy are the first bud of the psychological component. We completed the continental nation when we fulfilled the fantasy by pushing the western boundary to where it had first been drawn.

꒰ↄ

On his first voyage, 1534, Cartier, cruising the Gulf of St. Lawrence, thought that both Chaleur Bay (north of New Brunswick) and Belle Isle Strait (between Newfoundland and Labrador) might be the Northwest Passage. The next year the St. Lawrence lighted the hope again. When it turned out not to be the Northwest Passage, he hoped that it might lead to the Western Sea. On the top of Mount Royal he learned that it probably did lead there, by way of the Ottawa River and a series of fresh-water seas. This moved the Western Sea many days' travel farther inland than the Virginians were to estimate the distance to the South Sea in 1609, though how far Cartier had no way of calculating. Sixty-eight years later, in 1603, a greater man, Samuel de Champlain, followed him to Mount Royal. Champlain got the same information from the tribes but now there was a detail of closer approximation: the water of the untraversed far-western sea, they told him, was known to be salty.

Champlain was one of the greatest of all explorers. He estimated that at Montreal he was nine hundred miles from the Pacific. And independently of the stories that had the Great Lakes opening on the South Sea, he heard of another inland lake which may be identified as Lake Erie. He sensed that it might be part of the great water route he had been told about, but on the other hand it might be the Western Sea. Captain John Smith thought it was, when he heard of this same lake a little later, and it was Lake Erie as the Western Sea that Henry Hudson thought might be near Albany. Indeed the rumored sea underneath whose fabulous shape Lake Erie lay concealed had reached theoretical geography before Champlain and Smith heard of it, probably by way of a man who claimed to be a survivor of the crews which John Hawkins had abandoned after his battle with the Spanish at Vera Cruz in 1568. This man claimed to have made his way up the continent to New Brunswick and from there to England. If his story was true he had made a journey almost as amazing as Cabeza de Vaca's, but there is no evidence except his own and all that can be said is that somehow he had picked up some rumors about the lay of the land. How much truth his story contained does not matter for in England he came to the attention of Sir Humphrey Gilbert, who believed it all.

Gilbert was one of the giants of a great age, a geographer, a colonizer, an explorer of the Northeast Passage (Atlantic to Pacific north of Asia), an explorer of the Northwest Passage and the author of the most famous work demonstrating its existence, the *Discourse of a Discoverie for a New Passage to Cataia*. Raleigh's half brother and an associate of Richard Hakluyt, he was at the center of the tremendous Elizabethan movement to challenge "Charles the 5. who had the maidenhead of Peru" and all the Spanish Empire, to find in the New World a domain for England where "the common souldier shall fight for golde and pay himself, instead of pence, with plates of halfe a foot groad." . . . Gilbert assimilated the rumored lake to the fresh-water seas that Cartier had heard about, and also to another, entirely hypothetical sea from which he thought the Saguenay River must flow down to the St. Lawrence. So he and his associates worked out the most grandiose conception of the Western Sea that is recorded — and, in doing so, for the first time postulated the Continental Divide. The evidence, he decided, indicated that in the interior of the continent there must be a height of land. Along its crest there must be a congeries of inland seas, from which the rivers of the continent flowed down to the two oceans.

๛

These ideas originated in experience, however incomplete or misunderstood the experience was, however logic distorted or fantasy enlarged it. In the mind's innocence men could pass from ideas that had a little empirical fact in them to ideas that had none, without ever knowing that they had crossed a line. (Who could draw the boundary between China, the actual land that was a little known, and Cathay, the land of fable and fantasy?) Logic, desire, or mendacity could invent geography that would influence thought and action quite as much as real geography.

The Strait of Anian was a fixture on maps from the 1560's down to the last quarter of the eighteenth century. It was in the high latitudes on the west coast of North America and gave its name to the adjacent land. The province called Anian was sometimes located on the shore of Asia but was usually on the American side. . . . Dreams have their own gravity and exert an attractive force on one another. As the mist crept over New Mexico again following Coronado's failure to find the corded gold, Cíbola stayed where Estéban first saw it, but Quivira floated off through the rainbows. It came to rest on the hypothetical upper Pacific coast and there the maps show it, hard by its kindred myth the province of Anian. And Tontonteac could not be merely the Hopi pueblos, though the Hopis were among the last tribes to be known, and so it too soared into the sunset and the cartographers showed it neighboring with Anian and Quivira. Coronado's Thayguayc is there sometimes too and so are other names he made known. "Maps are a precise index of geographical knowledge," one recent student says.[5] Far from it. Down to the eighteenth century maps of North America make excellent indexes to geographical theory but they lag so far behind knowledge that frequently they have little relation to it.

The Strait of Anian was supposed to cut through the continental land mass and connect the Pacific Ocean with the Atlantic, with the Arctic, or with some switch-route to one or the other. It was entirely imaginary. For a long time students supposed it had some connection with the voyages of the Cortereals (in 1500 and 1502) and there have been other theories about its origin, but no one knows. It may have been someone's guess at what lay, the width of the continent farther on, west of Hudson Bay, or beyond some other arm of the sea which Hudson perhaps had entered without identifying it. More likely the necessities of logic invented it as a Northwest

Passage. Logic's *must be* has unlimited creative power, and as facts piled up to demolish the strait, it was logic that moved it northward. Not even the voyages of Vitus Bering, not even Cook's last voyage, demolished it or robbed it of influence on thought, exploration, and commerce. But the strait which Bering almost discovered and which was named for him lies so close to where the Strait of Anian was supposed to be that it compels admiration for a mysterious power. Human fantasy, then, can create out of no material, on no grounds, for no reason, and knowing nothing, approximations of reality.

Or facsimiles. The Strait of Juan de Fuca originated not as a geographical theory but as a creation of the artistic imagination. In 1596 a Greek seaman who may have been a navigator convinced important thinkers that four years earlier he had made a notable voyage which was entirely imaginary. His name was rendered Juan de Fuca and he said that he had sailed from the coast of California to a sea, the Atlantic or the Arctic, by way of a strait that led through the continent. This strait too was accepted, laid down on maps, sought for, and sometimes traveled by other liars. (Liars of record traveled the Strait of Anian too and some mariners who, being on the king's business, kept their routes secret were supposed to have traveled it.) De Fuca gave an exact location for the Pacific entrance of his strait: he said it was at 47° N. That is just more than a degree below the entrance to Puget Sound, which, when at last someone actually saw it, received his name.

Items of both kinds occur and recur in the history of the wilderness as the mind tries to understand it. There was, for instance, the Great River of the West, which is central in the plot of this narrative. It must exist because it had to. The logic of deduction from known things required it to, and so did the syllogism of dream — both on no grounds whatever. So it did exist in personal narratives and speculative treatises, in treaties and on maps, under various names, flowing in various directions. Born much later and of much shorter life, a river, usually called the Buenaventura, flowed west to San Francisco Bay from the desert of Utah or Nevada. There was no warrant at all for that river; there was some for the Multnomah, which flowed north to the Columbia from California and northwest to it from Idaho.

These rivers existed by reasoning and in response to need. The Baron de Lahontan et Heslèche provided geography with another kind. He was a first-rate soldier and a courageous explorer, a trustworthy historian, a shrewd and sharp critic, a penetrating and cynical intelligence — but a literary man. With a single chapter he befogged a large area of geography for half a century. Poetry or a formula for bestsellers came upon him and he created a big lake of salt water in the interior West and discovered on its shores an Indian culture as ornate as the Byzantine. Fact graciously fulfilled this fiction with the Great Salt Lake, but no Byzantines, more or less where he had said it was. It was otherwise with the Long River, which Lahontan caused to flow among impossible tribes across a landscape from the dark side of the moon. The Long River was art but it ministered to desire, for it might be the water route to the Pacific, which was even more urgently wanted in 1703 than before. So it twisted across the maps and the printed page, sometimes paralleling the Missouri River, whose exploration it affected, sometimes creating from its own substance large provinces for the truth to get lost in.

༄

The dawn of knowledge is usually the false dawn. The Spanish set out to conquer Cíbola, a country whose cities were walled with gold, on the word of one man that five emeralds which were not emeralds had come from there. They set out to conquer the even richer Quivira on the word of one man that the common supper plates there were made of gold. . . . Mankind has always thought objectively about the exchange of goods and the skills of warfare, but not much else. Francis Bacon's *Advancement of Learning* was published in 1605. Not many men in that year, or in 1540, cared or were able to inquire into the nature of evidence, or to advance learning by perceiving that there is a difference between the real and the unreal. Not many care or can now, but there were fewer in an age when the medieval mind, with its daily bread of miracle and sorcery, had only begun to turn away from wonder. My son's wife's brother knows a man who slew a unicorn and in the country of Prester John are men with three legs. Men whose heads do grow beneath their shoulders, men with the heads of dogs, and trees whose fruit is living sheep.

The substance need not be sought as far as the Lyonesse that had sunk beneath the waters or St. Brendan's stone boat touching an island-whale where he talked with Judas out of hell; it is in the mirages at hand. Those that led the Spanish across North America were not a tenth part as powerful as those in South America. Through a third of a century and across half that huge continent they went looking for El Dorado, losing more Spanish lives, perpetrating more mutinies and assassinations, and killing more natives than in any other conquest. There was a fact here: deep in Colombia a tribe appeased the spirit of its dead chief by blowing gold dust on the anointed nude body of his successor, who then purified himself by bathing in a sacred lake. To men steeped in dream and miracle this meant a country richer than Cathay. So the conquerors, performing prodigies of courage and endurance, turned into the dogheaded men themselves, and went rabid in a pack.

After a generation at full stretch of mind and sinew they were purged at last and that one delirium ended, though it came back at intervals well into the eighteenth century. (The same Gilded Man took Sir Walter Raleigh, after the failure of Roanoke, to Guiana and on to the high politics that covered him all over with these two narrow words, *hic jacet*.) In their quest they circled to hunt with dogs the rich one-breasted female warriors — and to give their name to the continent's greatest river. They died like the flies they sometimes had to feed on, scouring the tropical rain forest to find the Land of Cinnamon, a migratory version of the Spice Islands. Southward from Chile was the secret, enchanted City of the Caesars, where even the bedsteads were carven gold; they stopped looking for it at the end of the seventeenth century but believed in it for another hundred years. Their foreign banker-conquistador went looking for the House of the Sun, all gold and jewels, in Colombia. There was something of jacinth and porphyry called El Gran Moxo in Paraguay. Balboa lost half an army in Darien looking for the temple of Dobayba, which worshipers had filled with gold offerings. The Islands of Solomon — the Sepulchers of Zenu — the Enchanted Islands . . . but make it lower case, the enchanted islands, the enchanted lands, the clouds above the sunset. Beyond the horizon, for the Spanish and all others, there were many provinces where as Raleigh said "the graves have not bene opened for golde, the mines not broken with sledges, nor their Images puld downe out of their temples."

❧

One of Bienville's company who in 1718 founded New Orleans was a Fleming named Le Page du Pratz, who lived there and elsewhere in Louisiana for sixteen years. After going back to France he published, in 1758, his *Histoire de la Louisiane*. He had a volatile mind and was furiously interested in Indians and geography. He combined these interests to write a gorgeous fairy tale. An acquaintance of his named Dumont borrowed it and, acknowledging that he got it from du Pratz, scooped him by publishing a much diluted version of it in his own book, *Mémoires Historiques sur la Louisiane*, in 1753.

Du Pratz called the hero of his fiction Moncacht-Apé, explaining that the name meant "he who overcomes difficulty or weariness." He was an old man, a kind of Ulysses among his people, who sought out strange things in order to understand them, and his supposed journeys must be dated not later than 1700. Moncacht-Apé longed to know where his forefathers had come from. In order to find out he spent seven years in travel; he was alone but all tribes welcomed him. He went from the vicinity of Natchez first to the St. Lawrence and the Atlantic, then to the Pacific. That would make him the first person known to have crossed the continent in the area of the United States.

Moncacht-Apé's eastern tour is not relevant here. When he set out for the Pacific he went up the Missouri River. (So far as is known in 1700 no white man had yet ascended it as far as the mouth of the Kansas River.) He met the tribe called the Missouris and after them a tribe which he called the Nation of the West; du Pratz identifies them as the Kansas. They sketched for him a route that would take him to the sea. (There is no evidence that the Kansas had any idea where the sea was.) Following it, he traveled up the Missouri for another month, during much of which he had mountains in sight to the west. (In a month's walk from Kansas City there are no mountains anywhere along the Missouri. They are shown, however, on maps which du Pratz had studied, although the map he published has them east of the river.) The Kansas had told him to look for a tribe whom they called the Nation of the Otters, and he supposed that they were the Indians he now fell in with. They weren't but his new friends traveled with him up the Missouri for nine days more. Then, as the Kansas had directed him to do, they left the river and headed due north. After five days' northing the great mystery of the interior was solved. For the fifth day brought them to a westward-flowing river. It was a magnificent stream, so clear and pure that the Otters, whom Moncacht-Apé now found living on its banks, called it La Belle Rivière, Beautiful River.

A deputation of Otters took a pipe to a tribe downstream and Moncacht-Apé went with them, a voyage of eighteen days. He spent the winter with this new tribe and in the spring set out alone down the Beautiful River. Again all the tribes were friendly; the last one lived either two and a half miles or five days' travel (for he says both) from the Great Water, the Pacific. While staying with them he learned that white men came regularly to the coast in ships and fought with them. Joining a war party that waited to ambush these whites, he helped his hosts win a skirmish. The white men proved to be neither French, Spanish, nor English; Moncacht-Apé had seen no one that resembled them. They were stunted of stature and had extraordinarily white skin; they wore long, untrimmed beards; their clothes were made of materials and in a design that were strange to him. Their purpose in coming to this coast was to get a yellow dyewood; by now they had pretty well used up the forests. They had firearms but neither the guns nor the powder were as good as those used in Louisiana.

Moncacht-Apé now traveled northward with another tribe. The final leg of his journey lasted just long enough to confirm the thesis of du Pratz that the continent curved out far to the northwest, and to verify the discoveries of Vitus Bering, which were made known some years after du Pratz got back to France and some years before he published his book. Then he came back to his home near Natchez, at the mouth of the Yazoo River. . . . He had made by a good deal the fastest crossing of the continent before the stagecoach. It was all the more remarkable in that according to du Pratz the distance from the Yazoo River to the Pacific was 4800 miles in a direct air line, which would work out close to 6000 miles by the rivers.

The journey of Moncacht-Apé appears to have been invented in the interest of a particular geographical theory and with maps of a particular school on the inventor's desk. It got rid of the Western Sea, which would seem to have been the end in view, offering exploration instead an equally convenient route to the Pacific, by way of the Great River of the West. This hypothetical stream had been heatedly debated by the Louisiana French during du Pratz's stay, many rumors about it circulated, and nothing could have been more important to the French Empire than its discovery. The Beautiful River which du Pratz invented was merely one of its many shapes. He located it some sixty-five miles north of the Missouri: there is no westward-flowing river anywhere north of the Missouri. The two rivers flowed parallel courses, one east, the other west. It took Moncacht-Apé to the Pacific but, after he left the vicinity of Omaha where there are no mountains though he found some there, nowhere did he see any mountains at all, not the Rockies, the Cascades, the Sierra, nor the Coast Range.[6]

There is still another wave length in this spectrum.

The challenge of Elizabeth's England to the Spanish in the New World took Drake round the world with a pause at California on the way, Hawkins to the Caribbean, Raleigh to Guiana, and John Smith to Virginia. The great imperialists accepted whatever help was offered, and in 1583 one of them published a pamphlet which was dedicated to one of the fieriest of them all, Sir Francis Walsingham, then Secretary of State. The pamphlet told a story which, although it is wholly fictitious, has never since faded from belief. The story was told again the next year in a history of Wales. It was retold by Hakluyt, Purchas, Raleigh, and other contemporary celebrants of the English destiny. It recounted the discovery of America by a Welsh prince in the year 1170.

His name was Madoc and, so the story said, he was the son of the last independent Lord of North Wales, Owen Gwynned. (Owen had no son named Madoc.) Loathing the civil wars that followed his father's death, he sailed into the ocean and "came to a land unknowen . . . where he saw many strange things." Liking what he saw, he went back to Wales, gathered a company of colonists, and returned to America. As the tale was told and retold after 1583 these colonists numbered up to three thousand and they landed, according to the storyteller's version, at Newfoundland, Virginia, Florida, the Gulf coast, Yucatan, the Isthmus of Tehuantepec, Panama, the Caribbean coast of South America, the West Indies, or the mouth of the Amazon.[7]

This is the core of the story that was to be elaborated down to the twentieth century. If it is legendary and not invented a few years before it was published, perhaps we may see in it the news of Columbus's voyages, or Cabot's, reaching the hills of Wales, whose people were Celts. It is a race for whom the insubstantial world has always been more real than the visible one, for whom the little people have always shaken their milk-white arms in a ring by moonlight and the towers of Avalon have always glimmered in the sunset — and a people who though they have always gone forth to battle have always died. Like all Celts the Welsh had been the hillmen who raided the imperial frontiers and for them the exploits of Madoc were a stay and a glory in the defeat for which their hearts still grieved. The story, however, had a value for the English expansionists that can be seen in Hakluyt's words: "This land must needs be some part of that Country of which the Spanyards affirme themselves to be the first finders since Hanno's time. Whereupon it is manifest that that countrey was by Britaines discovered long before Columbus led any Spanyards thither." It became by far the most widespread legend of pre-Columbian discovery. In the United States it became our most elaborate historical myth and exercised a direct influence on our history.[8]

In the primitive state, the myth had Madoc's colony surviving for a time, then being assimilated by the Indians but leaving traces of Welsh culture to be discovered later on. This, however, was too humble for the Cymric dream and the myth flowered in wonder. We are to see the Madocians multiplying tremendously in the New World. Soon they numbered many thousands. But they had many thousands of neighbors too, who were warlike though they lacked the culture of Wales. They were forced westward down the Ohio Valley and on their migration built the mounds that so engrossed the curiosity of our forefathers. But very soon after the founding of the colony a vigorous pioneering branch of this people reached Mexico, where they spread southward and established the Aztec Empire. (Was not Quetzalcoatl's name Welsh? Was he not white? Did he not teach his people civilization and then sail in a Welsh ship eastward into the Atlantic?) They sent out a colony which founded Mayan civilization. They sent another one down the Pacific coast to Peru and created the Empire of the Incas. All this was proved by the remains they left behind.

Meanwhile the main body of the Madocian nation slowly retreated westward before its unnamed enemies. It left behind it such objects as turn up at spring plowing to bemuse the cross-roads mind with wonder what they are and how they got there. Reaching the Mississippi, they broke up into groups and their chiefs led them in various directions — for in various directions there are mounds, earthworks, and mysterious artifacts that signalize their passing. But one splendid variant, earlier than this, brought them to a sad ending. Reduced to thirty thousand now, the story says without noting where the number is recorded, the Madocians faced their foes in a climactic battle on an island near the Falls of the Ohio, where Louisville now stands. There almost all of them were killed. But the extermination was not quite complete: a few survived. These of course had the learning and culture of their race; they would be the seed of Welsh civilization in some safer place. They went down the Ohio, started up the Mississippi, and came to the mouth of the Missouri. They turned up that great river and disappeared behind the mist.[9]

This may be called the first cycle of the myth. The evidence adduced was almost all archeological and linguistic. Indians here and there recognized the cross as a sacred symbol or made ritualistic gestures that were palpably Christian: Welshmen must have taught them to. Certain words in the languages of tribes, languages which otherwise sounded uncouth, seemed like Welsh words and meant the same thing or very near it. (The Welsh word *pengwynn* meant "white rock"; clearly it was the same as a South American Indian word, *penguin*, a bird with a black head. The word America itself was Welsh, it meant "on the high and farthest seas"; clearly Vespucci had assumed it to attest the truth of Madoc.) There were piles of what seemed to be masonry which were beyond the ability of Indians to construct and therefore must be the work of Madoc's men. Copper objects were found scattered across the wilderness and when found by a Welshman meant that a Madocian had passed this way. Copper was one of truth's most eloquent witnesses for the Indians did not know how to smelt it — and did not need to since for hundreds of years before even Madoc virgin copper from Michigan moved along the trade routes that crisscrossed the continent. Or someone found an axehead or a piece of pottery and it must be Welsh.[10] Or someone kicked up a button from the dust and it was stamped with the mermaid of Welsh heraldry.

In this first cycle an occasional tale reported that someone had seen an Indian who spoke Welsh, but almost always the Welsh Indians were thought of as an extinct tribe who had left some evidence of their culture. As the eighteenth century came on, however, and the English colonists met more tribes and more Welshmen came to America, the stories changed. The most fecund one, though not published till 1704, circulated for some time before that wherever there were Welsh colonists and was alleged to date back to 1686. It told how a Welsh parson en route from Carolina to Virginia was captured by Indians who had white skins and spoke Welsh, and how to their great joy he spent four months among them preaching the gospel in Welsh. So it was certain that a tribe of civilized, Welsh-speaking Indians lived somewhere hereabout or thereabout. This story became many stories and soon many people had visited this tribe or met some of its members. They had white skins, blue eyes, blond and even red hair — a surprising ethnological change for Madoc's colonists, being Welsh, must mostly have been swarthy and black-haired.

Like many other tribes, they steadily moved west, ahead of the advancing frontier. Though they were always farther on, a surprising number of people saw them or their trail. John Sevier, the founder of Tennessee, understood that they had been the firstcomers to Alabama but that the Chickasaws drove them out. During the last French war Francis Lewis, who was to sign the Declaration of Independence as a delegate from New

York, was captured by Montcalm's troops and turned over to the Indian mercenaries. Among them he found a chief, doubtless from the far country, who conversed with him in Welsh. James Girty, of the renegade woods-runners, knew so many Welsh Indians that he was able to help a linguist compile a vocabulary. Daniel Boone knew that their moccasins had trod the trails before him, and so did another great frontiersman, George Croghan, though he knew only that they were a very great distance to the west. George Rogers Clark recognized as theirs certain mysterious fortifications near Kaskaskia.

Always they could be identified by their white skins, their language, and their British ways. They reverenced their ancestral relics, especially a Bible which Madoc had brought with him. (They were right to reverence it for it had reached America three centuries before Gutenberg and four centuries before the Scriptures were translated into Welsh.) And always they were farther west, in some place we had not reached yet. They were the Delawares, the Tuscaroras, or the Conestogas till those tribes were better known, when they moved into the Ohio Valley as the Shawnees and on down it. At least thirteen actual tribes were at one time or another supposed to be the Welsh Indians.[11] Besides these at least five imaginary tribes with names made up to fit were designated, and at least three not named but vividly described. One of these last maintained in lower Louisiana an earthly paradise with all the gold the Spanish had looked for, the luxury Lahontan had imagined, and as much sensibility as Chateaubriand caressed in swooning prose.

As early as 1710 it was rumored that they had withdrawn beyond the Mississippi and as the years passed this became a certainty. They first appear in this region as the Padoucas. That was the name the French gave to the Comanches, whom the English did not know at all, but it was sometimes erroneously applied to the Pawnees, who were the people of Harahey. But as the Pawnees came to be known it was clear that they were not Welsh and the truth came out at last, that Madoc's people lived far up the Missouri. Their migration was not to stop there, for later they were to flee to the Rockies, to the Southwest, to Nevada, to the Canadian Rockies. But the golden age of this great people was on the Upper Missouri.

So far up the Missouri that no one could estimate the dis-

tance, or tell how near their country was to the Western Sea
or the River of the West. For a space now they had no tribal
name, but word of their white skins, splendid culture, and
majestic towns kept reaching traders and the settlements.
Clearly they had kept their Welsh blood pure and had retained
their language and civilization but it was not known if they
still practiced Christianity. If they didn't, and in any event for
the glory of Wales, it was desirable to seek them out. But there
was a much more important reason: they would know the
secrets of the wilderness and therefore the water route to the
Pacific, which in fact might cross their country.

A great French explorer reached them from Canada in 1738
but no one came up the Missouri to their country till more
than fifty years later. In the meantime, however, they had been
identified: the Welsh Indians were the Mandans.[12]

⊛

The historian Hubert Howe Bancroft used the excellent
phrase "Northern Mystery" to designate the portion of the
Pacific coast above San Francisco Bay that resisted knowledge
and bred error for so long a time. It will be useful here to
designate as the Northwestern Mystery an area that provided
one of the climaxes of the imperial struggle for the continent
and that contained the key features of continental geography
which were the last to be discovered and understood. This area
begins with the Upper Missouri River.

Historically the Upper Missouri begins at the mouth of the
Platte River, which from the time it was first seen constituted
a kind of equator in men's thinking. Geographically, however,
it begins farther north, in South Dakota, where the river cuts
through the escarpment called Pine Ridge. Here one subdivi-
sion of the High Plains ends and another begins; geologists call
the new one the Missouri Plateau. The White River flows eastward along its northern base to empty into the Missouri, which
a little farther north curves through more than 360°, creating
the Grand Detour, a stretch of more than twenty-five miles of
river with a neck of land within the loop which is less than a
mile wide at its narrowest. The Grand Detour is forty-odd
miles south of Pierre.

The White is the first of a series of rivers entering from the west which one comes to as he travels the Missouri upstream. They originate in or flow through the western Dakotas, where the soil is so unstable that wind erosion has created the bizarre shapes from which the country gets its name, the Bad Lands. They are thick with silt and till the 1870's they were what gave the Missouri the name it has in our poetry, the Big Muddy. (The Yellowstone and its affluents which are equal carriers of silt today were comparatively clear before the Cattle Kingdom.) [13] East of the Missouri, at a distance varying from twenty-five to sixty miles in the Dakotas, another though gentler escarpment comes curving down from Canada in an arc whose bearing is southeast at first but thereafter nearly due south. Geologists make this the western boundary of the Central Lowlands, which lie east of it, but the boundary of the Northwestern Mystery is still farther east.

Ascending the Missouri one enters, some sixty miles beyond Bismarck, North Dakota, on the long sweeping curve that eventually changes the direction of travel to due west. This is the Great Bend of the Missouri. At its sharpest corner the loop which the Souris River drops down from Canada is forty-odd miles north. From this farthest south point the Souris turns north again, crosses the international boundary into Manitoba, and presently empties into the Assiniboine River. Here the Assiniboine is flowing east and has been for a long stretch; about a hundred and twenty miles beyond the mouth of the Souris, at the city of Winnipeg, it empties into Red River. In the time of the fur trade this exceedingly important junction was called the Forks of the Red.

The Red River ("of the North") is the boundary between North Dakota and Minnesota and roughly the boundary between two geological regions, though it is in the transition zone. It may be taken as the eastern boundary of the Northwestern Mystery. It is the only large river in the United States east of the Rockies that flows north. Its northern flow and the lay of the adjacent land are the most significant aberration in the remarkable symmetry and integration of the continental unit which the United States occupies. It flows past Winnipeg for about forty miles and empties into the big, muddy, shallow, tempestuous Lake Winnipeg.

Manitoba west of Lake Winnipeg is a watery region. Of a

series of lakes so close together and so intricately connected by rivers and sloughs that they may almost be called continuous, the two biggest are Lake Manitoba, whose southern end is only a few miles north of the Assiniboine River, and Lake Winnipegosis whose southeastern bay almost touches it. North of Winnipegosis and separated from it by only a narrow neck of land is the smaller Cedar Lake, by means of which the great Saskatchewan River reaches Lake Winnipeg. The Saskatchewan comes all the way down from the Rocky Mountains (as the Assiniboine does not), mostly in a sunken valley much like that of the Missouri from Great Falls eastward to the Breaks. It may be taken as the northern boundary of the Northwestern Mystery. There is no need to specify a western boundary: the Northwestern Mystery was all West.

The key features of the area thus delimited were, historically, the Upper Missouri, its Great Bend, Lake Winnipeg, and the Saskatchewan River. When the area came into history the Assiniboin Indians, a Plains tribe once a part of the Sioux (Dakota) nation, roamed along the Assiniboine River, performing the economic function of middlemen to the tribes farther west and south. The populous, far-ranging Crees were in contact with the Assiniboins and on trading journeys and war parties penetrated almost as far south as they did. The Crees were an Algonquin tribe, forest dwellers whose country stretched far to the east, who were already getting European goods from Hudson Bay, and who traded and warred far to the north. A division of them had forsaken the forests and learned the Plains culture. Another great Algonquin tribe, the Chippewas (sometimes called the Ojibways), had been making a long, slow migration westward and a fringe of the divisions that were leading it had reached the Red River, beyond which they were to go no farther. The Red was a war road between the countries of the Crees and Chippewas and that of the Sioux, Minnesota and Wisconsin. Farther south, the edge of the westward migration of the Sioux was extending toward the Missouri River but was not to get there for a long time. A convenient date for the beginning of the historic period here is 1690, but it is entirely arbitrary.

When the French at St. Louis began to trade upriver, the peoples whom they called "the Upper Missouri tribes" include

## Map 5. The Northwestern Mystery

the Omahas, the Otos, and the Poncas. But the phrase designated more particularly three tribes who lived farther upriver,
not far from one another, and had interchanged much of their
cultures. All were sedentary, living in permanent villages composed of substantial earth lodges which were partly sunk in
the ground, walled with big posts, and roofed with poles
covered by a thick layer of dirt. All had a flourishing agriculture but also hunted the buffalo and, because they did, had
assimilated much of the economy and culture of the Plains
tribes west of them. One tribe, the Arikaras, were a Caddoan
people, closely related to the Pawnees; they are the "Rees" of
the literature. Another tribe, called the Hidatsa by ethnologists, were known to their neighbors as the Minnetarees but
usually appear in the literature as the Gros Ventres. They
belonged to the Siouan family but this is a linguistic classifica-

tion and does not ally them with the Sioux of this narrative,
:he Dakota. They were closely related to the Crows, a Plains
tribe.

The third of the Upper Missouri tribes also belonged to the
Siouan family. They were the Mandans.

Mandan cosmology taught that the First People lived in the
underworld. The roof had a hole opening into the sunlight; a
grapevine grew toward it. One day the Mandans started to
climb the vine but only half the tribe got through the hole,
for the vine broke under the weight of a fat woman. The Old
Ones who got through found themselves at the heart and center
of the world and so the stream they settled on is named Heart
River. Their actual arrival on the Upper Missouri cannot be
dated but they reached it from the east, probably before 1400.
A populous tribe, they soon dominated the region. When white
men first heard of them (the French at the western end of Lake
Superior) they were old settlers, firstcomers, holdovers from a
culture older than that of the parvenus who surrounded them.

The Welsh Indian myth that at last centered on them got a
happy assist from fact, for many of them had comparatively
light skins and hair. Some had cheeks fair enough to show a
blush, brown and occasionally red hair, brown and sometimes
even blue eyes. The hair of some turned gray with age and
there were cases of albinism. Indians vary as much as white
men in their coloring and these same characteristics were shown
by the Minnetarees, the Arikaras, and the Crows, though less
often. The variations were entirely Mendelian: the Mandans
had not the slightest trace of any but Indian blood or Indian
culture. The literary imagination has found that they were not
only Welsh but Irish, Scandinavian, and Jewish as well but they
were as pure Americans as their distant relatives, the Quapaws.

Theirs was a fine hunting and farming country and it lay
square across a great four-corners of the trading routes. The
rich, well-watered soil of the Missouri bottom grew corn, beans,
squash, pumpkins, sunflowers, and tobacco so abundantly that
their storage pits were always full and they had a big surplus
for the trade. They added to it by drying berries and wild fruit.
They made the best, in fact the only, pottery hereabout.
(Coarse, utilitarian stuff, not to be compared with that of the
Southwest.) They were superbly skillful at working and or-

namenting leather: dressing skins, making garments and robes and shields and medicine objects. Their food, pottery, leather work, and such things as pigments for paint, and their strategic situation between tribes that got European goods and those that did not, enabled them to dominate trade westward and for a considerable distance northwestward and downriver. But they were a homekeeping folk and never became carriers like the Hurons, the Ottawas, and the Crees. All the nonperishable items of the prehistoric trade except turquoises have been found in their sites. When the white men reached them iron and steel that came to them along the trade routes had almost completely displaced stone in their weapons, and they had kettles, axes, knives, and other manufactured goods. At just this moment too they got horses.[14]

By the time the first rumors about them had reached the French, the Mandans had passed the crest of their national life. They were richer and more powerful than their neighbors, more skillful at handicrafts, and in their trade at least more intelligent. Their villages, the equivalent of metropolises for a centralizing force had concentrated them in a few big ones, made them secure against attack. The earth lodges, in palisaded groups with moats and earthworks, were impregnable. If they were outnumbered on the plains they had only to retire into these fortresses; presently they were able to stand off the ferocious Sioux for a good many years. No wonder that the Assiniboins and Crees who told the French about them spoke of them as superior beings who lived in luxury. The Assiniboins were Plains nomads who had not yet got horses, therefore found the buffalo economy precarious, and were undistinguished for either intelligence or skill. The Crees, though they were getting English goods and guns, were at the mercy of the Canadian winter and under annual threat of famine. They thought of the well-fed Mandans as millionaires.

Even so, the Mandans had begun to decline. Two powerful disintegrating forces were at work on them. Far to the west, the Comanche migration southward had disturbed the international equilibrium all the way from Wyoming to New Mexico and from the Rockies to the Pawnee nation below the Platte. The other force was stronger: the revolution had come to the old world we call the New and there would never be an end to it. As soon as the Indians got manufactured goods their life and culture were made over. West of the Mandans, tribes had got horses and the Plains way of life had begun the flowering that was to last for two centuries and make them the most vigorous and powerful of all Indians. East of the Mandans, the Iroquois wars, the arrival of British trade at Hudson Bay, and the arrival of French trade at Lake Superior and Lake Michigan had displaced all the tribes, flung them into a social maelstrom, and created new economies and new sequences of wars. Of this chaos the Hundred Years' War between the Chippewas and the Sioux most affected the Mandans. The retreating Sioux pushed weaker tribes ahead of them westward and then a full half of the Sioux nation itself headed toward the Missouri River.

By 1690 or thereabout, then, the Mandans were becoming a social anachronism and historical processes had faced them with a challenge which they were to prove unable to meet. When at last the white men reached their country they found an admirable, genial, friendly people who liked and welcomed them. Neither the French, the English, the Spanish, nor the Americans ever had trouble with the Mandans. But if they liked the newcomers and liked their goods still more, they got the idea soon. Indian myths can seldom be dated but are likely to be less ancient than you think. It may have been the end of the eigheenth century when the priests added another chapter to the Mandan cosmology, in explanation of the beings who had come to their country, the center of the world. It was an addition to the story of creation. In the time of beginnings, this new story said, the first Mandans were preyed upon by fierce wolves, which seemed likely to destroy them. The culture hero and the friendliest god crossed the river to save the people. They killed all the old wolves. They taught the young ones not to attack men but to eat only the flesh of other animals. The rotting bodies of the old wolves remained. They threw these into the Missouri and as they floated downstream they turned into white men.

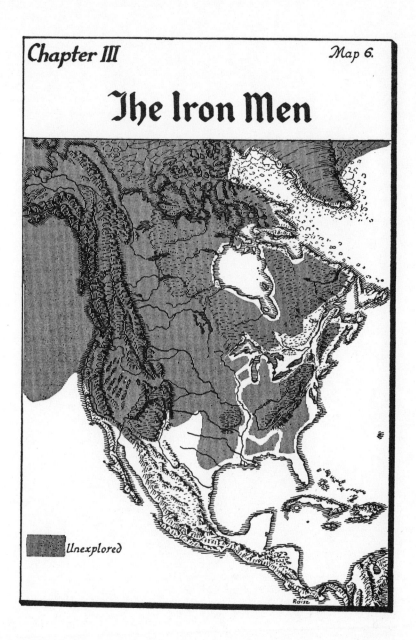

Chapter III — Map 6.

# The Iron Men

Unexplored

# III

# The Iron Men

HIS NAME was Nicolas Perrot but the Potawatomis and their neighbors in Wisconsin called him Metaminens, the Little Maize or Little Corn. In 1669, with years of wilderness life behind him, he was twenty-five years old. That year the Mascoutens, the Fire Nation, sent an embassy to the Potawatomis, who then lived on the shore of Green Bay, asking them to pay a visit and to bring some of the Spirits (so the Jesuits translated the word) who were among them, the Frenchmen.

There is a moment when the Mascoutens welcome Perrot. The elder who spoke for them carried the symbol of peace, a calumet hung with feathers. Rain had fallen recently and so when he tried to get fire to light the ceremonial tobacco, the pointed stick he twirled between his palms did not ignite the damp wood. But Perrot was bringing to this province of the neolithic world the industrial technology of the modern world. With flint and steel he struck sparks into a tinderbox. The priest understood that here was medicine, a spiritual power, and when he had lit the pipe he blessed the pocket lighter as well as its owner.

The Fire Nation sang and danced for their guest, commended him to their supernaturals, and stuffed him with the banquet meats it would have been hostile to refuse. Finally it was his turn to practice the oratory that Indians valued no less than prowess in battle. "I am the dawn," he said. He handed them his musket: the young men were to use it against their enemies. Here was his iron kettle: cook their meat in it and carry it anywhere for it would not break like their earthen pots. He tossed knives and awls to the women: throw away their bone needles and their blades of stone so laboriously ground or flaked to a dull edge. As casually as one might throw confetti, he flung them an armful of beads to decorate their apparel and that of their children.

He was a god scattering largess. He was an agent of the big combine forestalling the attempt of the Potawatomis to establish a distributor's monopoly in these parts. He was the extreme point of the economic system of Europe, whose industry had already drawn nourishment for a century and a half from the market which he was thus bringing to the Fox River. He was a representative of Louis XIV helping to incorporate the richest area of the world in the French Empire. And as all these, a thousand miles west of Quebec, he was one of the breed of men who already had made the French our greatest wilderness people.

The breed must be seen in another aspect. . . . A hundred of the most formidable Indians of this region, the Sioux, found Perrot and half a squad of his men at the farthest French trading post, a log hut on the Mississippi where it is the boundary of Wisconsin. They were painted for war against the Foxes, so their hearts were bad. They would kill these whites and pillage the trade goods at the post, which was as rich to them as a subtreasury. They didn't. Perrot filled a cup with brandy and, while haranguing them, sent one of his men to the river with it to bring him a drink of water. He paused, seemed about to drink, and then thrust into the cup an ember from the fire. The braves leaped back from the flame and Perrot had cowed them. If they should lay a finger on him or his goods, he said, he would burn up the lakes they fished in and the marshes where they got their wild rice. Similarly Pierre Radisson threw gunpowder in the fire to show what would happen if a Frenchman were molested.

The Foxes had a surly, belligerent independence that outlasted the English, once saved the Americans from defeat, and then repeatedly defied them till the war in which A. Lincoln made a field campaign. In the desperate time preceding Frontenac's second governorship, when New France was in danger of being destroyed, Iroquois diplomats worked on the Foxes to attack the Chippewas, who were among the most loyal

French Indians. Perrot appeared among them. He had heard, he said, that they wanted Frenchmen to eat. Well, get out the kettles: he had brought them himself and a few young men. Drawing his sword, he pointed it at his body. "My flesh," he said, "is white and savory but it is very salt. If you eat it, it will scarcely pass your Adam's apple before it pukes you." He stood for the power of Louis XIV, as the most Christian king and within a span of mastering Europe. But Louis XIV had no power in Wisconsin and the foremost nation of Europe meant less to the Foxes than any skulking band of Sioux. Yet they dared not puke themselves with this man, and the Iroquois plot shattered against his moral strength. The chiefs begged forgiveness and piled before him the gift of beaver pelts with which Indians announced another fidelity that would last a while. He had given them birth, the elders said: he had brought them iron.

He met a war party of Sioux on a raid toward their hereditary enemies, the Foxes. They were the terrors of the Northwest but they held him and his goods in awe and they began the ceremony of the calumet. He seized the pipe, the sacred symbol of peace, and threw it at their feet. Would he accept peace from dogs, he asked, would he so much as listen to dogs singing the peace song? From where the sun now stood he closed the road: they were to go no farther and no trader would bring them goods. They could have snuffed out him and his handful of companions like a candle but they implored him to let his face shine upon them, and when they forswore their raid he did. . . . All these tribes had the ferocity of spitting cougars, and like all Indians they were also easily stampeded into panic. He found the Ottawas worked up to an inconvenient war. French authority over the wasp-swarms was in danger and he ordered them back to their nests. Their fathers, he said, had used bows, stone axes, and earthen cooking pots till he came among them bringing the great Governor's goods. What if he should now abandon them to hunt with bows and defend themselves against tribes who obeyed the French spirits and therefore had guns?

Such scenes display a quality that all white men who would succeed with Indians had to have. One sees it in the young Radisson directed by his companion Groseilliers, who had exhausted himself with oratory, to take up the speech and carry on. These were Hurons at Lake Superior, afraid to take the canoe fleet with its beaver pelts to Montreal since the Iroquois lay in wait for them. But the survival of New France depended on the arrival of those furs. The youth who was not old enough to grow a beard stood up, seized a beaver pelt, and began to beat the warriors with it across the face. "Do you not know the French way? We are used to fight with arms, not robes. . . . Shall your children learn to be slaves among the Iroquois for their fathers' cowardice? . . . Do what you will, for mine own part I will venture choosing to die like a man than a beggar." And likewise in Duluth, though embarked at last on his repeatedly postponed project of discovering the Western Sea, abandoning it because he had learned that the Sioux had made prisoners of three Frenchmen. He took two men to paddle his canoe and after forty-eight hours of continuous travel reached them. The prisoners were two of La Salle's *voyageurs* and Father Hennepin. Duluth kicked two calumets away and faced down the sons of the white wolf till they gave up the prisoners and restored the last shred of Hennepin's vestments, which they had torn up and distributed.

In such moments the Indians are to recognize the dominance of a mind subtler than theirs. And they are to see the disregard of death and the will to use force without stint regardless of the consequences which the civilized mind could focus instantly but to which the primitive mind could be worked up only by a long series of religious exercises. The white man who would control Indians must respect and even like them. He must understand their infinite capriciousness, brag, and instability of motive and emotion. He must understand their pride. He must be adept in their logic, which was on a wide tangent from the logic of the European mind. But most of all he must have the moral strength to commit himself. For good or ill it is an attribute of civilized man that, disregarding loss, defeat, and death, he can instantly decide to shoot the works. It has always been basic in his ascendancy over primitives.

New France had fewer than three thousand inhabitants in the period of its greatest peril. The lives of all these depended on the Indian trade and therefore on a specialized type that the New World had formed from French stock to conduct that trade. A few hundred were enough to make the seventeenth century the century of the French in North America as it was in Europe. "Spanish civilization crushed the Indian," Parkman says: "English civilization scorned and neglected him; French civilization embraced and cherished him." The generalization is something less than true. Though their first generation showed little sign that they could or would, the Spanish succeeded in making societies of Indians and no one else ever did. Perhaps the French might have done so too and they might have made better societies. (For such men as La Salle regarded them as citizens of the human commonwealth, the Jesuits envisioned a theocracy that had as much Plato as Augustine in it, and nowhere in New France was there anything properly to be called race prejudice.) But they failed to because they had an incomparably greater skill than either the Spanish or the English in using the Indians as instruments of imperialism.

〜〜

The incidents just described are of 1669 and later, except for the one involving Radisson. They reveal the French at the western end of the Great Lakes and on the upper Mississippi. Their westering had reached a longitude perhaps a hundred and twenty-five miles west of New Orleans. No English colonist of record crossed the Appalachian divide to westward-flowing waters till 1671.[1] No English trade goods reached the *pays en haut* till 1685. Before 1700 both Virginia and Carolina traders had turned the southern end of the Appalachians and were trading with transmontane Indians but deep penetrations were so rare that they hardly matter. There was no English post on the Great Lakes till 1722.

These late dates, however, give no measure of the celerity with which the French penetrated the continent by the great water route. Champlain founded Quebec in 1608 — two years before Santa Fe, almost on the farthest frontier the Spanish were able to hold. In 1613 he reached Allumette Island, far up the Ottawa River. The most adventurous of the young men he had sent to live among the Indians may have reached Lake Huron the year before, 1612. (That was five years after Jamestown; Plymouth was 1620; the Dorchester adventurers reached Cape Ann in 1626; the Great Migration began to come to Massachusetts Bay in 1630.) In 1615 Champlain went up the Ottawa again and on to Lake Huron. This immemorial Indian route was to become the main highway of New France: up the Ottawa to the Mattawa River, up that to Lake Nipissing, and down French River to Georgian Bay, *la mer douce*. The journey had taken him across more than four hundred of the nine hundred miles which in 1603 he had estimated to be the distance from Montreal to the Pacific Ocean. Now he heard from Indians that the inland sea of which Georgian Bay was a part stretched on for a thousand miles farther and that a civilized people lived on its western shore. It must therefore lead to China or at least to some part of the Orient, though the way was longer than he had thought. But he was not to make the great journey for meanwhile he had created New France and he had to govern it. His young men must make the search. In 1623 one of them, Etienne Brulé, reached the Sault, at the passage between Lake Huron and Lake Superior. It was a hub of the Indian world and one of the prime centers of North American geography. He went on to precipitate a splendid reality out of a resplendent myth, for it is all but certain that he reached the farthest and largest inland sea, the upper lake, Superior.

And there was Jean Nicolet. Like Brulé he had heard of Indian tribes far to the west, Chippewas, Mascoutens, Potawatomis, and especially the Winnebagos. The Indians whom Nicolet knew called the Winnebagos the People of the Sea — that would be of the Big Water or some such phrase — and said that the sea they lived on smelled bad. The sea was Lake Michigan, whose existence was still unknown to the French because they had misunderstood Indian allusions to it. We do not know whether the arm of it called Green Bay smelled bad because of mud flats or mineral springs, or whether perhaps it was the Winnebagos who, because they were fisheaters, smelled bad. But nobody could escape the conclusion: the bad smell meant salt water and therefore the Pacific could not be far beyond the Sault. The fair-haired people whom Champlain had heard about must live there, and now there were tales that they had big ships, shaved their heads, and wore strange costumes made of strange stuffs. Unquestionably the Indians were describing Chinese. So in 1634 Champlain sent Nicolet to complete the voyage to China that he himself had begun at St.-Malo in 1603.

The Jesuits had taken over the mission two years before and Nicolet had to conduct a company of them to the Hurons. (The first western mission had been established by the Recollects but converting Indians was expensive and they, being Franciscans of the strictest observance and vowed to poverty, had to ask the rich, crusading, imperialistic Jesuits to take their place.) He went on to Georgian Bay and then, accompanied by seven Hurons, started for the unknown. He retraced Brulé's line to the Sault. He turned south there, found the Strait of Mackinac, and brought Europe into Lake Michigan. Its far shore led him to Green Bay, the stinking water of the stories, near whose lower end lived the Winnebagos, the People of the Sea. They might be subjects of the Grand Khan, to whom he was a fully accredited envoy of France: they must be Tartars, Chinese, or a nation tributary to them. So he sent them formal notification that he was coming and put on a robe of Chinese silk, "all strewn with flowers and birds of many colors." But when he stepped ashore from the canoe the Chinese proved to be only another tribe of Indians who were prepared to greet a Frenchman as a spirit. They made him welcome and told him many brave, amazing stories about the country farther west. It was full of marvels but no other was so great as the news he carried back to the dying Champlain. Here at the mouth of Fox River, with Green Bay rippling in its shallows, with the Second Lake of the Hurons just across Green Bay Peninsula to the east — here, the Winnebagos said, he was only three days' journey from the Big Water. He was in fact not much more than that from the Mississippi River, which is what the Winnebagos were truthfully telling him about. (Theirs was a Siouan language, whereas he had learned only Algonquin dialects.) But to Champlain and everyone else in New France the Big Water meant either the Pacific or what would do almost as well, the Western Sea.

Here for twenty years it rested. Lake Superior had been glimpsed. The Sault and the Strait of Mackinac had been traversed. Lake Michigan had been entered, although no one knew how far it extended and its relation to the system of inland seas was misconceived. It was not clearly differentiated from Green Bay, which hereafter would be the Baie des Puants (Stinkers) on the maps. Green Bay itself had been explored and it may be that Nicolet had gone up Fox River to Lake Winnebago. In any event, France reached the 88th meridian in 1634. just a year later the General Court of Massachusetts Bay sanctioned the establishment of the first town beyond the frontier. It was sixteen miles out from Boston at the site of an Indian village known as Musketaquid, and because the Indians were friendly it was called Concord.

ҩↄ

The revolution that destroyed the neolithic world began with the voyage of John Cabot in 1498. No evidence has ever been found that before that date fishermen — Basque, Spanish, Portuguese, Breton, or English — visited the Grand Banks and went on to the Canadian mainland. But many a historian has had to suppress the twitch of a nerve that comes from certain inexplicable data, of which the most galling one is this: that as early as there are accounts of visits to these shores there are also Indians offering to trade furs for manufactured goods.

Cartier's voyages of 1534 and 1535 found fishermen in the Gulf of St. Lawrence and up the river, and found Indians who were habituated to the fur trade which had been begun by fishermen who landed to dry their catch. During the next seventy years the intertribal trading system — which was prehistoric, as old as the oldest archeological discoveries — was converted into the trade for furs. By the time Champlain arrived in 1603 the tribes that Cartier had met had been displaced, a complex system of collecting furs and distributing goods had developed, a practically continuous intertribal war had begun, and Indian life within reach of the trade had been shifted to a new basis. By Champlain's time too the fur trade had become the beaver trade. Down to 1763 the history of Canada is primarily the history of the trade in beaver pelts. Down to about that date, moreover, the fur trade was the principal objective of imperial competition, and war, everywhere in the continent north of Mexico.

That statement, however, must be phrased in a different way for it emphasizes the currency — furs — whereas what counted was the market for European manufactures. The New World was a constantly expanding market; no limit to its development could be foreseen and indeed there was no limit. Its value in gold was enormous but it had still greater value in that it expanded and integrated the industrial systems of Europe. It was thus a powerful force in the development of capitalism and nationalism.

The impact of European goods produced a change in neolithic America far more concentrated and rapid than anything in the history of white civilization. In 1500 Indian life north of Mexico was at a stage roughly equivalent to what we vaguely make out Mediterranean life to have been at, say, 6000 B.C. The first belt-knife given by a European to an Indian was a portent as great as the cloud that mushroomed over Hiroshima. The heir of the ages had thrust his culture into the era of polished stone. Instantly the man of 6000 B.C. was bound fast to a way of life that had developed seven and a half millennia beyond his own. He began to live better and he began to die.

Nicolas Perrot bringing a fire-lighter to the Mascoutens signalized a change far more revolutionary than that which the application of steam power to machinery produced in white culture. He was in Wisconsin; his predecessors had signalized the same change from tidemark on inland. A knife blade of the poorest steel, an axehead of worked iron, a needle, a file, a pair of scissors, any piece of steel or iron meant comfort, ease, and power not possible to an Indian without it. Fell a tree with sharpened stone or hollow out a log with fire, then with an axe; sew a dress with a bone awl and thread made of split animal sinew, then with a needle and silk or linen thread. A garment made of skins required the labor of hunting, skinning, curing, and tanning as well as tailoring, and it was in some weathers ineffective or unhealthy. Woolen cloth was immensely more versatile, comfortable, effective, and easier to work. The matchlock or firelock of the early seventeenth century, which as a trade gun was displaced by the flintlock musket in another generation, was an incredibly inept firearm; but in the conditions of forest hunting it was much better than the indigenous bow. (The Indians of the Northeast used a bow which was neither double-curved nor sinew-backed.) It outranged the bow and its heavy ball, from an ounce to an ounce and a half, made sure of the kill. It was an even greater advantage in war, in so much that tribes which had firearms invariably subjugated or drove out those which did not. Metal arrowheads, lance blades, knives, and hatchets had an equal superiority over flint weapons. Steel and brass and copper wire were not only stronger than sinews or buckskin thongs but infinitely versatile. An iron or copper kettle required no labor, could hardly be broken, and made cooking less arduous. It could be hung over a fire; earthenware pots could not be; watertight baskets could be but seldom were; pots and baskets were filled with heated stones to make the water boil. The white man had first got tobacco from the Indian but his curing processes made it so much more desirable than the home-grown kind that tribes which could buy tobacco stopped growing it except for religious rituals. That the pigments, paints, glass and porcelain beads, tin and brass ornaments, mirrors, burning glasses, and function-less novelties which formed so large a part of the trade had an equally revolutionary importance is not always obvious to modern students, but it always was to Indians. Finally there was alcohol, which the French traded in the form of brandy; it opened to the Indian experiences that, quite literally, were beyond all price.

The revolution affected every aspect of Indian life. The struggle for existence, for food and shelter to maintain life, became easier. Immemorial handicrafts grew obsolescent, then obsolete. Methods of hunting were transformed. So were methods — and the purposes — of war. As war became deadlier in purpose and armament a surplus of women developed, so that marriage customs changed and polygamy became common. The increased usefulness of women in the preparation of pelts worked to the same end.[2] These and related phenomena produced changes in social organization. The fur trade increased trade in general, so that there was more intercourse among tribes. This meant an acceleration of cultural change, for the tribes acquired arts, crafts, myths, and religious ceremonies from one another. Standards of wealth, prestige, and honor changed. The Indians acquired commercial values and developed business cults. They became more mobile. Hunts, raids, and trade excursions ranged farther. Tribal movements and shifts were speeded up.

In the sum it was cataclysmic. A culture was forced to change much faster than change could be adjusted to. All corruptions of culture produce breakdowns of morale, of communal integrity, and of personality, and this force was as strong as any other in the white man's subjugation of the red man. The wonder is that the Indians resisted decadence as well as they did, preserved as much as they did, and fought the whites off so obstinately and so long. For from 1500 on they were cultural prisoners.

The Fox chief who told Perrot, "You gave birth to us for you brought us the first iron," was telling the truth, and Perrot's threat to close the trade to the Ottawas was a threat that he would deprive them of the means of subsistence. To the last fragment of a broken axehead, the last half of a cracked awl, the last inch of strap iron, a better life depended on the trade. Down to just such items the goods traveled along the native trade routes, hundreds of miles beyond the white trader who had first sold them for furs. A tribe that traded directly with the whites was in the most favorable situation, fully supplied and better armed than its customers and possessing the power of any monopoly. Consequently there were trade rivalries, trade diplomacy, and trade wars. For most Indian tribes war had always been a sport, a cult, and a vocation. But the trade with industrial Europe made it for three centuries a fundamental condition of Indian society. Trade wars produced tribal displacements and migrations. The attacked fled before the attackers; those who had firearms or iron for points pushed back those armed with bone- or stone-tipped weapons. New frictions, tensions, and population pressures followed.

But it must be said with strong emphasis that these movements, though they were a supremely important force in our history, were only movements for a new reason. Too many historians have treated the Indians as if they were a static and even uniform society when the white men reached North America. Whereas sizable population movements were going on at that time and had always been going on through the prehistoric period as far back as archeological evidence extends. There was competition for hunting grounds and agricultural lands;

the less populous or the less warlike were forced somewhere
else. Exhaustion of food supplies produced some movements.
Climatic changes produced others, notably in the Southwest
and the Upper Missouri Valley. The massing or aggregation
of culture groups forced the displacement of some long-estab-
lished societies, as in the southward retreat of the Pueblos. For
some migrations, such as that of the Utes and Comanches, it
appears to be true, as it must have been true of many prehistoric peoples, that there was no reason beyond the common
American ones: an itching foot and a conviction that it's a
better country farther on.

When the French reached the Sault they met for the first
time a division of the Chippewas and named them the Saulteurs. This division was the rear guard of a migration, which
no evidence suggests was a retreat, westward from the Atlantic
coast. It had been going on for over a century and the bulk of
the Chippewa nation had reached Lake Superior. There migration became contention with the Sioux for possession of Wisconsin and Minnesota. The Sioux and their Siouan-speaking
relatives were latecomers to this region, though they long preceded the Chippewas there. The evidence is mixed but it
indicates that they reached the area by ascending the Mississippi and most archeologists believe that they had reached the
Mississippi from the southeast after a much earlier migration
eastward through the Deep South. In the sixteenth century the
Ohio Valley was almost empty of inhabitants though the
mounds show that it had previously had a large population.
North of it all the way to the Eskimo country stretched an almost continuous belt of Algonquin-speaking tribes.

These eastern Algonquins (there were many tribes of the
same family elsewhere) were the first Indians whom the French
and the English encountered. Most of them ranged widely
through the forest and some must be called seminomadic,
though all had permanent villages. They lived in individual
bark-covered huts or long loaflike communal houses built of
upright logs and covered with bark. White men found their
languages easy to learn and in the East they were so closely
related that a man who knew one dialect could usually make
himself understood in several others. They were usually
friendly toward the whites, as friendliness goes between savage

and civilized people. They were intelligent and shrewd and had a commercial sense and, in their own terms, an industrious enterprise that soon got them called "the bourgeois of the Indians." They had invented and at once contributed to white culture and the subjugation of the wilderness the birchbark canoe, one of the most remarkable boats mankind has ever used.

A group of Iroquoian tribes had followed the Ohio eastward or the Appalachians northward and had cut the Algonquin country in two. The Hurons, whose domain stretched eastward from Lake Huron south of and beyond Lake Simcoe in Ontario, were Iroquoian. So were their allied and dependent neighbors, the Neutrals and the Petuns. By the time Champlain reached Canada the fur trade had already produced an irreducible rivalry and hostility between these tribes and those whom history calls the Iroquois, the Five Nations of the Long House who occupied New York State.

Like the Hurons, the Five Nations had penetrated as far north as the St. Lawrence. By Champlain's time, however, they had been pushed back to the country with which they are permanently associated, the Hudson and Mohawk Rivers and the Finger Lakes. He found them on the defensive against the better-armed Algonquins but the war was still going on. He joined it on the Algonquin side. He could not have done otherwise; the fur trade had developed an organization that left him no choice. The great bulk of furs was west and northwest — it always would be, as far as the Rocky Mountains and Far North. This wealth reached the St. Lawrence from "the Far Nations" through the Huron country and with the Hurons as middlemen, and they controlled most of the great water route to the upper country.

But Champlain's decision was a fateful commitment; it meant the death of thousands of Indians, French, and English. It can hardly be said that New France was ever at peace with the Iroquois. There were short and long armistices; trade with the Iroquois was never completely cut off; sometimes there were only small and distant frontier incidents — but there was always war or readiness for war. The struggle became acute following 1624, when the Dutch established the trading post near the mouth of the Mohawk that was called Fort Orange and would become Albany. They gave the Iroquois access to firearms and all the manufactured goods, a market for furs that broke the French monopoly, and an irresistible incentive to secure the trade of the far Indians and the upper country.

In the 1640's these forces produced a series of actions that made the struggle for the trade part of the main stream of American history. Those actions eventually spread from Florida to the Mississippi and beyond, and brought the contending empires steadily nearer a showdown. The first phase was the struggle to control the St. Lawrence.

ల∾ల

No Indians of North America ever showed greater ferocity than the Iroquois; no tribe had more talent for military organization; few tribes fought better. As a military power they were able to destroy their neighbors and to spread terror for hundreds of miles beyond their country. And their country occupied one of the fundamental areas in the grand strategy of the continent.

They could control the upper St. Lawrence and Lake Ontario. So they could control the lower end of the route to the Great Lakes, the Mississippi River, the Far West, and Mexico. They were on the flank of the trade that came down from the upper country. They could also control the Hudson River–Lake Champlain–St. Lawrence route, the main highway of invasion, north and south, from prehistoric wars down to the September day in 1814 when Sir George Prevost's army turned in its tracks at Plattsburg and headed north again. And they could control another route to the interior of the continent which had the formidable advantage that its lower reach and its communication with the sea were not closed by ice in the winter, as the St. Lawrence was. It led up the Hudson and then. by way of the only gap in the Appalachian system, up the Mohawk to Lake Oneida and thence to Lake Ontario.

When Fort Orange was established the Iroquois confronted New France with the threat that they might divert from it the trade of the West and Northwest. Everyone who thought patriotically knew that the fort should be taken and the Dutch driven out. In 1664 the English did what the French should have done and captured it, thus bringing the frontiers of the two empires much nearer each other. But through all this time the significance for the Iroquois was that the slight resources in furs of their own country had been exhausted and they were forced to be a trading nation — to be middlemen for the tribes that had furs. And the Hurons had beaten them to it.

It was the Hurons who brought the furs to the French and traded them for goods, at first to Quebec, then to Three Rivers which was founded at the mouth of the St. Maurice River in 1634, finally to Montreal at the mouth of the Ottawa, which was founded in 1642 to defend New France against the Iroquois and which became the center of the trade. The Hurons had worked out a trade arrangement with the Nipissings who, as their retailers, collected furs all the way from the Sault to James Bay, the southern end of Hudson Bay. They had a similar arrangement with the Ottawas, who ranged among the Far Nations. The Hurons themselves traveled even more widely; their trading parties were out three-quarters of the year on routes almost as fixed as those of a milkshed. They had what amounted to a corner on all the merchantable beaver in Canada except that along the Saguenay and east of it, the least important field. Once a year their flotillas of big canoes brought the wealth of Canada down the St. Lawrence and took the goods of France back. This was the fur fair, the system of trade which had been established by royal decree. It was also the annual spectacle of New France: savage and Christian pageantry, high mass and dances in worship of the bears, Jesuit sermons and Indian oratory, breechclouts and velvet breeches, nuns, war chiefs, woods-runners, ceremonious banquets, tribal debauches, the nights crimson with fire, musical with river songs, hideous with the chanting and the yelping of the braves. The economy of New France received its lifegiving transfusion and the King's ships turned down the river with the furs. The continental and the world markets had met.[3]

The route west led up the Ottawa River instead of Lake Ontario in order to avoid the Iroquois. But the Hurons were vulnerable on the lower reaches of the Ottawa and on the St. Lawrence. For a quarter of a century the Iroquois ambushed and harried them there, and made repeated forays on the French settlements. But on the whole the French and the Hurons had the better of it. They held the Iroquois off, frequently beat them in force, violated treaties they made with them, and formed combinations against them. And the trade grew in value through the 1640's.

So, beginning in 1649, the Iroquois made a series of military campaigns that were among the most successful in our history. In that year they broke the Hurons and drove them out of their country. As many as eight thousand starved to death; the rest were dispersed. In the same year the Iroquois also broke and dispersed the Hurons' neighbors, the Petuns. In 1651 they did the same for the Neutrals. The first war was thus a complete triumph: Huronia, all but emptied of inhabitants, had become a private preserve of the Iroquois. But at once they learned the lesson of Napoleon. Though they had conquered the heartland the continent outflanked them, they could not hold in force an area so distant from their home, and they never did get the trade. France sent its colony nowhere near so much help as it should have done but what it sent sufficed. The colony and the Far Nations changed the trading system and there was nothing for the Iroquois to do but to extend hostilities a thousand miles farther west. The Eries stood between them and the trade: they destroyed the Eries as a tribe in 1654 and 1655. No use. They sent armies as far as Mackinac and the Mississippi and tried to make alliances with the Far Nations. The raids were magnificent military tours de force but the Far Nations, reinforced by the westward-drifting victims of the conquerors routed them, and the French broke the alliances like so many cobwebs.

Meanwhile a danger had opened to the south of the Iroquois, where they were as vulnerable as Lake Ontario had made the Hurons. Their kinsmen the Susquehannas had grown fat and aggressive on the English trade, and they were in a position to dominate the trade routes across Pennsylvania and down the northeastern tributaries of the Ohio. The Susquehannas therefore must be destroyed and after ten years of war they were. But they did the Iroquois more damage than all their other Indian enemies put together. The Iroquois were never again a dominant military power among the tribes and they could not match the growing strength of New France, which they had several times seemed likely to destroy.

English colonists, though against the general interest and for

reasons that made sense only locally, had joined with the
Iroquois against the Susquehannas. And New France had come
to a realization that Lahontan phrased, "I lay this down for an
uncontested Truth, that we are not able to destroy the Iroquese
by our selves" — so it had brought the Far Nations into the war
as allies. This is what it came to: the Iroquois wars to dominate
the fur trade (and to make it English) helped to precipitate the
imperial conflict that lasted just a century and was succeeded
by another one that lasted half a century. The lines of imperial
contention were all laid down by 1660; by 1670 it was under
way.

᙭

Pierre Esprit Radisson was captured by the Mohawks when
he was no more than fifteen years old but escaped from them.
He was recaptured and put to the torture, though the Indian
family who had adopted him were able to keep it mild.

The Mohawks let Radisson watch the slow roasting of his
fellow captives, and to personalize the drama pulled out four
of his fingernails. A squaw brought her four-year-old son to cut
off one of his fingers but the child was not strong enough to
perform the amputation and had to be content with swallow-
ing blood from the wound. When it was the old folks' turn, one
of them enjoyed holding Radisson's thumb in the bowl of his
pipe while he smoked. He could meanwhile watch the performance: "They burned a frenchwoman; they pulled out her
breasts and took a child out of her belly, wᶜʰ they broyled and
made the mother eat of it; so in short [she] died." That evening
"they bourned the soales of my feet and leggs. A [Mohawk]
souldier run through my foot a swoord red out of the fire and
plucked several of my nails." A little boy came and tried to
chew off one of his fingers.[4]

Radisson is talking about Mohawks but what he says describes all the tribes of the Northeast. They developed refine-
ments of torture which the zeal of religious Europeans during
the preceding centuries had never reached. Fire was the principal instrument. The torturers heated hatchets, swords, gun
barrels, any metal objects till they were incandescent, then put
them to work. Father Galinée has a striking phrase. A Shawnee

was being tortured and cried out when the searing iron was first applied, which meant a high score for the torturer. Galinée turned his head and saw an Iroquois drawing the red-hot gun barrel along the Shawnee's legs with, he says, a "grave and steady hand." . . . Radisson goes on. When a fingernail had been pulled out "they putt a redd coale of fire uppon it and when it is swolen bite it out w^{th} their teeth." From this stump the veins would be pulled out as far as possible and seared. Sinews exposed at the wrist would be wound round a stick and pulled out by windlass action. When a scalp was removed — the victim was still alive — it was enjoyable to dump a kettleful of glowing embers on the wound. Bullets were melted and the liquid lead poured into wounds, or for greater sport gunpowder which was then ignited. "They cut off yo^r stones and the women play w^{th} them." When the victim died, after being held from that release as long as possible, his corpse provided further sport.

This is an ecstasy of the primitive mind, though our generation has learned that the membrane with which civilization had covered it breaks easily. It became an instrument of international polity when the French first used Indians to fight the English. The English learned fast and the Americans inherited their learning. Torture became indispensable for diplomatic negotiations. You could always reanimate a slackening ally by giving him some prisoners to burn. If an alliance threatened between a French tribe and one of the English or Iroquois persuasion, a prisoner of either surrendered to the torture of the other amounted to handing the ambassador his passports: policy yielded to sensual esthetics. If no representative of the state were at hand, the clergy would assume the duty. La Potherie describes an occasion when peace threatened between some Huron *émigrés* and some Iroquois raiders. In compliance with a recent treaty the Hurons were about to liberate a captured Iroquois, but the resident Jesuit told them that France "peremptorily ordered them to put the Iroquois in the kettle." They did so; the danger of peace and commercial competition vanished.

∽

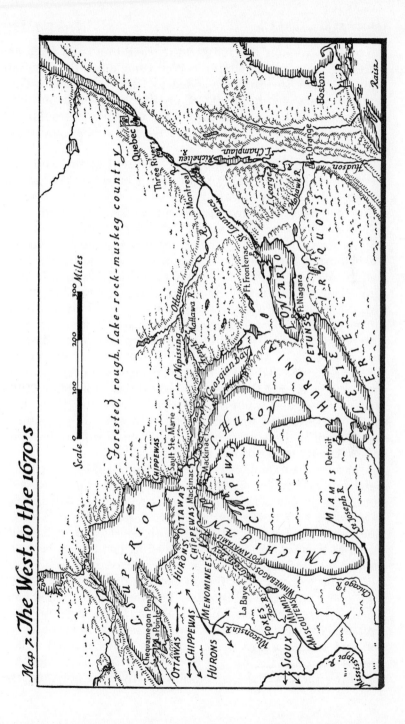

Map 7 *The West, to the 1670's*

Scale 0    100    200    300 Miles

Forested, rough, Lake-rock-muskeg country

Quebec
Three Rivers
Montreal
Richelieu
St. Lawrence R.
Ottawa
Nipissing
French R.
Mattawa R.
Georgian Bay
Sault Ste. Marie
Mackinac
Mackinac
L. HURON
Ft. Frontenac
L. ONTARIO
Ft. Niagara
PETUNS
L. ERIE S
ERIE S
HURON
IROQUOIS
L. Champlain
L. George
Mohawk R.
Ft. Orange
Hudson
Boston
Raisr

L. SUPERIOR
Chequamegon Pen.
La Pointe
OTTAWAS
CHIPPEWAS
HURONS
OTTAWAS
CHIPPEWAS
HURONS
MENOMINEES
WINNEBAGOS
POTAWATAMIS
CHIPPEWAS
CHIPPEWAS
L. MICHIGAN
MIAMIS
Detroit
St. Joseph R.
La Baye
Green Bay
Foxes R.
Wisconsin R.
MIAMIS
MASCOUTENS
Chicago R.
SIOUX
Mississippi R.

The first Iroquois war blocked the trade route and for four successive years no furs from the West came down the Ottawa. Fear of economic death was added to fear of massacre. The next few years saw the height of mystical rapture in the religious orders and the worst depression the colony ever knew. New France was saved by the market it had created and by a class of men it had outlawed.

The displaced Indians who were clustered at the Sault or wandering through Wisconsin and Michigan had to have goods and so did the more distant tribes. This commercial opportunity nerved the Ottawas to an unaccustomed daring and, helped out by some Hurons, they sent a delegation by a circuitous northern route to Quebec, offering to renew the trade. Their journey happened to occur at the same time that the first Iroquois raiding party to the upper Lakes was defeated, in fact annihilated; consequently the Long House would like peace with the French for a while and made a treaty. This was 1653 and next year a big fur fleet came down to Montreal. It was mainly Ottawa and Huron but intrepid explorers from some of the Far Nations — Marco Polos and da Gamas — accompanied it to gape at the slab-sided huts of Montreal and clasp their mouths in wonder of the Aladdin's cave at the trading fair. New France revived and when the canoes took back the goods that would revive the distributing system, two officially sanctioned traders went with them. Unsanctioned ones slid away in their wake. A new period in the trade had opened.

It had been a monopoly ever since Champlain organized the merchants and subjected them to government control to end an anarchy of competition and debauch. Under France it always would be a monopoly, whether of the crown, of a combine to which the crown farmed it on contract, or a limited system of licenses. And till now the effort had always been to confine the trade to the fairs at the three centers, to compel the Indians to bring their furs in and to prevent the *habitants* from going into the wilderness. The merchants wanted the protection from competition such a system afforded, the government favored it because it kept the citizenry under observation and built up agriculture, and the Jesuits favored it because it protected the Indians from civilization. But from the beginning there was the corruption of any monopoly and especially of those that were administered by the centralized bureaucracy of France, and the inevitable merchants within the ring who did illegal business outside it. So from the beginning there were illegal traders. In small parties they traveled the Indian country. They made the forests and rivers their country before the monopoly, the military, or even the missionaries began to follow their trails. They were the *coureurs de bois*. The term meant outlaw for a long time and it always carried with it the deadly opposition of the Jesuits but it meant too the admiration of those who stayed at home. And rightly so: those who escaped to the forest from the doubly dictatorial regulation of their lives by clergy and government were the best men in New France.

Europe in process of becoming America had to forge a blade to cut its way west. The *coureurs de bois* are the French variant of a new kind of man, an instrument fitted to the country beyond the frontier as the helve is fitted to the hand. Beyond all frontiers this new man was an Indian with a white man's mind and he lived free. The breed submitted to no constraints except their own. It is quite true that they fought, murdered, drank, whored, and wived as they pleased. Their morals were nothing the settlements could approve, their ethics except in relation to one another were even less admirable, economics could readily overcome their patriotism, but a cutting edge must be hard in order to be sharp. They lived by courage, resourcefulness, endurance, a skill hardly to be understood in a metropolitan world, and serenity of heart. They pulled the wilderness round them like a robe, they wore its beauty like a crest. For these Frenchmen it was the wilderness of inland lakes, white-water rivers, the forest that had no end, the muskeg, and the cold that grew more inconceivable as they traveled farther. "One of the Soldiers that accompany'd me," Lahontan says, "told me one Day that to withstand the cold one ought to have his Blood compos'd of Brandy, his Body of Brass, and his eyes of Glass." Like their kind everywhere they loved the unspoiled wonder of the world and they kissed the life of settled ways good-bye with a high and ribald heart. They were the forerunners. They pushed on, to live with Indians and feast or fight with them as the dice might fall, to lose their lives and find a continent — for others.

The two official traders who went out with the Ottawas in 1654 and came back in 1656 were of this breed. That it could produce an intelligence formidable in other fields as well is demonstrated by one of them, Médard Chouart, Sieur des Groseilliers. He will always be linked with his wife's half brother, Pierre Esprit Radisson. Radisson accompanied him on his second voyage and when he wrote a book describing both claimed to have been with him on the first one, though apparently he was not.[5] Groseilliers was the innovator and on these two journeys the commander, but both thought continentally and between them they changed the world.

Nicolet had reached Green Bay by way of the Strait of Mackinac in 1634. In 1654 Groseilliers reached it again, the first of record to follow him. But he and his companion took up the great adventure where the wars had broken it off and pushed the boundary of the known deeper into the continent. Apparently they did not go farther into Wisconsin than the foot of Green Bay. But they took their canoe and goods the entire length of Lake Michigan, the first white men who ever did so, and traveled part of Illinois and southern Michigan, among remnants of the displaced tribes that were wandering there.

Groseilliers' companion may have made an even more important journey. "By the persuasion of som of them we went into y[e] great river that divides itself in 2," Radisson says, "where the hurrons w[th] some Ottanake [Ottawas] & the wild men that had warrs w[th] them had retired. . . . This nation have warrs against those of [the] forked river. It is so called because it has 2 branches, the one towards the west, the other towards the South, w[ch] we believe runns toward Mexico, by the tokens they gave us. [Here for a sentence he perhaps slips into fable, for they told the ancient story that bearded Indians would be encountered. But, bearded or not, these Indians built big houses and had] such knives as we have had. Moreover they shewed a Decad of beads & guilded pearls that they have had from that people, w[ch] made us believe they weare Europeans. . . . "[6]

Historians have been divided over this passage.[7] But that Groseilliers' companion reached and crossed the Mississippi is more than possible. The Illinois, the Miamis, the Mascoutens, the Petuns and Eries, and perhaps other tribes which had lived farther east had by now crossed it, some of them in terror of the Iroquois. Possibly the bearded Indians are a faint rumor of the Spanish in New Mexico; if so, this is the first word of them to reach the French across the plains. The knives, though rumored, not seen, are firmer and the beads still more so. If this man did reach the Mississippi he was, of course, the first Frenchman who ever did so and the first white man after Moscoso to see it anywhere above the mouth. But whether he did or not it is hard to understand this in any other way than as a reference to the Mississippi River. If it is, then it is the first one in the documents of French America in which the Mississippi is recognized as a river. (Not, of course, as the river of De Soto and Moscoso — that was to wait for Jolliet.) It was the Big Water reported to Nicolet but he thought he was hearing about the Pacific. Since him the Jesuits and presumably traders too had been hearing about a great river farther west, but no one had understood the significance of the report, no one had even understood that he was hearing about a river.

Presumably, then, this is the Mississippi — reported to a Frenchman, if not indeed visited by him. What is equally important, here is mention of what Coronado had heard about in 1541, the Missouri River as a fork of the Mississippi. No name is given it, nothing is said about it except that it is a fork, and nothing is understood. But the geographical genius of Groseilliers, plus such map-ideas as there were, works out an idea of absolute importance: that here, somehow, by some means, in some direction is a route to Mexico.

With the country south of Green Bay, the Great Valley of the Mississippi comes into French awareness, and for the first time a white man understands its promise for mankind. "I can say that in [my] lifetime I never saw a more incomparable country, for all I have been in Italy," Radisson writes, speaking in his own person of this journey though his knowledge, if it is personal, must derive from the second one. "Whatever a man could desire was to be had in great plenty: viz staggs, fishes in abundance & all sort of meat, corne enough." The American heartland has found its first celebrant. And a vision of empire not based on fur, lacking in the French so far, decades from being born in the English colonies, vibrates in him on the shore of this, "the delightfullest lake in the world," the white sand leading from the cobalt-colored waters of Lake Michigan to the fat and waiting land:

> the country was so pleasant, so beautifull & fruitfull that it grieved me to see yᵗ yᵉ world could not discover such inticing countrys to live in. This I say because that the Europeans fight for a rock in the sea against one another, or for a sterill land and horrid country, that the people sent heere or there by the changement of the aire ingenders sickness and dies thereof. Contrarywise those kingdoms are so delicious & under so temperat a climat, plentifull of all things, the earth bringing foorth its fruit twice a yeare, the people live long & lusty & wise in their way. What conquest would that bee att litle or no cost; what laborinth of pleasure should millions of people have, instead that millions complaine of misery & poverty! What should not men reape out of the love of God in converting the souls heere, is more to be gained to heaven then what is by differences of nothing there, should not be so many dangers committed under the pretence of religion! Why so many thoesoever are hid from us by our owne faults, by our negligence, covetousnesse, & unbeliefe. It's true, I confesse, that the accesse is difficult, but must say that we are like the Cockscombs of Paris, when first they begin to have wings, imagining that the larks will fall in their mouths roasted: but we ought [to remember] that vertue is not acquired wᵗʰout labour & taking great paines

He was the wilderness spearhead of the adolescent capitalism of seventeenth-century France, and with his own hands he was to touch off a hundred years of imperial war. But he could think in continental terms and that he had profundity is shown here. The clouds opened to him and he foresaw the future of half the world.

Fully as important in Groseilliers' first voyage as his entering the Great Valley and hearing about the Mississippi and Missouri is the news he got, however distorted, that the tribes north of Lake Superior traveled to Hudson Bay — and traveled to it in canoes down rivers that flowed into it. Overland exploration had caught up with the great arctic voyages by the English, Dutch, and others that had been made to find the water route across the continent. Groseilliers would now work out the relationship between them.

He exhorted, ridiculed, shamed, and bullied the Indians till they dared to risk the Iroquois again. So in 1656 another fur flotilla reached Montreal and pumped more blood into the arteries of New France. There was enough fighting to terrify

the demoralized Hurons but there were enough miracles to
turn it the right way and to light ecstasy in the parish chapels.
This was the decisive year. From now on, though frequently
in danger and twice brought to the extremity, New France and
the trade it lived by would grow stronger.

Three years later, in 1659, Groseilliers went west again and
this time young Radisson went with him. At the very beginning, this still more important enterprise ran head-on into the
system. The heroic Jesuits had rallied from their failure among
the Iroquois and the dreadful shock of the Huron dispersion.
They prepared to follow the Hurons, whom they really had
converted, to the upper Lakes — and their plans and dreams
enlarged. The great interior was to be Jesuit country, inhabited
by safely Christianized and obedient Indian peons, free of the
impiety and debauchery and lechery of the traders who had
been invading it since Groseilliers broke the way — and with a
percentage of the traffic in furs to fill God's treasury. "Their
design," Radisson says, "is to further the Christian faith to the
greatest glory of God, and indeed [they] are charitable to all
those who are in distress and needy, especially — " the phrase
falls delicately — "to those that are worthy or industrious in
their way of honesty." The Jesuits ordered the explorers to
take the missionaries with them.[8]

Groseilliers' first journey had "maintained the countrey" —
saved it — but he had been assessed the monopoly's twenty-five
per cent tax on his furs nevertheless. Now the governor must
get his cut, must have personal representatives in the party and,
secretly for it was illegal though routine, must get half the
profits too. "We tould him . . . we knewed what we weare, Dis-
coverers before governors" — and they left between two days.
Not without the pang that even the most hardened could feel.
Radisson had already spoken of the wrench on forsaking one's
own "chimney smoak" and forsaking the life "when we can kiss
our own wives or kisse our neighbour's wife w^th ease and delight" for some years, to "worke whole nights & dayes, lye down
on the bare ground & not alwayes that hap, the breech in the
water, the fear in y^e buttocks, to have the belly empty, the
wearinesse in the bowels, and drowsinesse of y^e body."

They intended to find out about those rivers to Hudson Bay,
which might well be the most important discovery yet for the

trade. So, seeing no sea serpents in Lake Huron though Radisson had previously seen one in Lake Ontario, they reached the Sault and kept on north to Lake Superior, gun-barrel blue with the presage of winter. It was the farthest and largest of the Lakes, the end of the route that was to have taken Cartier and Champlain to the Pacific. Brulé had probably seen it and Radisson and Groseilliers must have had predecessors in entering it, but their names and dates are not known: history comes to Lake Superior with their canoes. (Radisson pays tribute to the dispossessed Hurons who had told them about it: 'those great lakes had not so soone comed to our knowledge if it had not ben for those brutish people [at the Sault]; two men had not found out yᵉ truth of these seas so cheape; the interest and the glorie could not doe what terror [of the Iroquois] doth at yᵉ end.") They coasted the beautiful southern shore in their canoes, rounded Keweenaw Point, and almost at the end of the lake came to Chequamegon Bay, protected from storms by Madeleine Island at its entrance. Radisson, the continental thinker, remembered the rival of France. "There are yett more countreys as fruitfull and as beautifull as yᵉ Spaniards to conquer, wᶜʰ may be done wᵗʰ as much ease & facility, and prove as rich, if not richer, for bread and wine." They had brought France to a new Farthest West. They were restoring the trade of tribes that had lost it and bringing it to others for the first time. And they were setting a bench mark, for victims of the wars followed them and settled here, the Jesuits came, and tribes moved up from the south. Chequamegon became the first of the rallying points where the Far Nations were brought within French influence and the great alliance against the Iroquois-English axis was welded.

When their Indians returned they went even farther, to Lake Court Oreille in Sawyer County, Wisconsin.[9] Here, among Indians who had seen no white men and no goods or only worn-out remnants, they were, like Perrot's company, spirits. "We were Cesars, being nobody to contradict us," and an Indian "durst not speake because we weare demigods," and again they were "the Gods of the earth among these people." As Caesars and gods they distributed bounty among the murderously humble — kettles, hatchets, knives, sword blades, awls, needles, combs, mirrors, paints, bells, rings, beads. As ambassadors of

France they told the tribes that the great father would maintain the trade which they had opened. The country was as rich
in game as in forests but the northern winter set in and there
came a snow so deep and soft that they could not use snowshoes.
Starving times began. They and the Indians ate dogs (normally
tabooed by these tribes), the stewed or powdered bark of trees,
boiled beaver hides. "Our gutts became very straight from our
long fasting," and "we mistook ourselves very often, taking the
living for the dead and yᵉ dead for the living."

They bound the Menominees to the French for good and the
Potawatomis almost as strongly. They traveled an undeterminable distance north of the lake and met the Crees, the far-
ranging people of the north. Now they learned unmistakably
that the region northwest, and north of Lake Superior was a
country far richer in beaver than any the French had yet
tapped. And they learned something about the rivers and the
configuration of the land. They learned enough to work out a
plan for mastering the Northwest, one that not only began the
long war when they acted on it but contained the answer to
which the British companies were to fight their way in the
nineteenth century.

Now when spring came on these instruments of empire made
another momentous journey. A delegation arrived from the
most formidable Indians hereabout, on the whole the most
formidable of all Indians, to do homage to the spirits who
owned the iron, to marvel at the goods which they now saw in
quantity for the first time, and to ask the Caesars to visit their
country and bring the trade. Nicolet may have heard of the
Sioux a year or so before his voyage of 1634. They are not mentioned in the *Jesuit Relations* until 1640, though the fathers
must have heard of them before that from the Ottawas. Radisson and Groseilliers now met them. Presently they were off to
visit in the country of the wild-rice lakes the nation who from
the earliest rumor had been known as the terrors of the West.
(In the *Relation* of 1669 Allouez called them "the Iroquois of
this country," Dablon gave them the same designation in that
of 1670–71, and many were to repeat it. It elevates the Iroquois
above their station.) So France reached a new people, as different from the Algonquin and Iroquoian tribes as Spaniards are
from Danes. They got more information about the Crees, the

great North, and the West which, obscurely, they began to see was also great. Stately ceremony, grandiloquent oratory — the first skin tipis Frenchmen had ever seen — a stone-age economy that meant a vast new market — beaver as common as dirt and the untouched Farther On to keep supplying it. Or so they were impelled to think. The Sioux said that the French were masters over them and the adventurers said, for Louis XIV, that indeed they were. They made abundant gifts to rivet this fealty, told the Sioux to make peace with their enemies the Crees so that the way to the farther country would be unencumbered, and as demigods departed.

They had worked out an idea so important that they based their whole plan on it, and in fact it is one of the two indexes to the entire history of the Canadian fur trade. They had no accurate idea of how far into the interior they had come, but they knew it was a long journey from Montreal. They thought that the continental distances were shorter than in fact they were, but they knew that they were very long. They had only a generalized conception of Hudson Bay, for the best maps showed only such a conception, but they knew enough about it for the purpose at hand. To reach the rich fur country northwest of Lake Superior by canoe from Montreal — by the Ottawa River route to Lake Huron, to the far end of Lake Superior, up whatever river route might lead from there — would be enormously expensive in time and labor. And there were always the Iroquois. What Radisson and Groseilliers proposed to do was to reach the rich fur country another way — ships to the western shore of Hudson Bay, and then by canoe up the rivers by which the Crees descended to it.

They went back to Chequamegon and picked up a group of Chippewas, inveterate enemies of the Sioux, but not till they had helped eighteen tribes (most of which must have been subtribes or mere bands) hold a Feast for the Dead and had thus further cultivated the soil that would germinate the great alliance. They returned to the settlements with beaver richer than any New France had ever seen — and again had saved its solvency. During their absence the Iroquois had come yelping north again, life was safe in no one's dooryard, and the Mohawks had raised an army in the high hope of wiping out the colony. There was a universal fear that they might be able to do just that. But the war was stopped short by an event which both the French and the Iroquois interpreted as a miracle. At the Long Sault of the Ottawa River seventeen French and five Indians under Adam Dollard held off the entire Mohawk army for five days, killed several score of them, and, though all twenty-two were killed, so terrified the invaders that they "went home dejected and amazed, to howl over their losses and nurse their dashed courage for a day of vengeance." [10] A few days later, while the mangled bodies of the heroes still lay unburied, Radisson and Groseilliers came down the Long Sault, and with them so many fur-laden canoes of the Far Nations "that [they] did almost cover yᵉ whole river." The great flotilla of 1660 reached Montreal. "By our meenes we made the country to subsist, that without us had been, I beleeve oftentimes, quite undone and ruined."

But it was not proper to save New France from ruin at the expense of anyone's graft. The governor had no share in this rich harvest and the monopoly, the leagued merchants, had not dared to risk the *pays en haut*. Now the governor (that "Bougre") levied 4000 livres — they may be translated as francs — ostensibly to build a fort at Three Rivers and fined them 6000 more for the dignity of government. This was his personal cut, and also he assessed the twenty-five per cent tax which was the monopoly's legal due but was always remitted to everyone who "did grease his chopps," which came to 14,000 livres more. Twenty-four thousand livres in all, more than forty per cent of the partners' profit in their two years which had given France hundreds of thousands of square miles. They had to pay it at the moment when the merchants who had risked nothing grew fat on the beaver which they had induced the Indians to bring to their doors.

It was a political mistake of the first magnitude. "Seeing ourselves so wronged, my brother did resolve to goe and demand Justice in France." Wasted effort. The systematized corruption of the bureaucracy and the blindness of the colonial office to the supreme Canadian opportunity made any thought of justice idle. Radisson and Groseilliers understood that opportunity; perhaps they could make the English colonies understand it, and if not there was the English court.

☙

Three months before Radisson and Groseilliers reached Montreal from the upper country in 1660 Charles II entered London in triumph, and when the Ottawas went back they took two Jesuits with them. Ripples from the first event would require fifteen years to reach the Great Valley but the second one made itself felt at once. The Jesuits closely interrogated the adventurers about the country to which their converts had fled. (It would have been graceful to acknowledge their source.) Adding what they heard to Nicolet's accounts and mixed information from other *coureurs de bois* they began to understand the vastness to which they might bring the worship of Jesus and the jurisdiction of His Society. And, mastering what the two had learned about its geography, they interpreted it in the light of the best knowledge of the time.

Father Lalemant reported the first results in the *Relation* of that year. From the western end of Lake Superior, he said, the bay of Espíritu Santo in the Gulf of Mexico lay seven hundred and fifty miles due south. He was short by more than five hundred miles; the error may be explained as his guess at the latitude of the lake, for the partners of course had taken no altitudes. His longitude was good: though maps of the time usually displaced the bay of Espíritu Santo (if we take that to be the mouth of the Mississippi) by at least ten degrees, Lalemant practically canceled this out by overestimating the distance Radisson and Groseilliers had traveled west. (He does not identify the bay as the mouth of the river, and of course had no idea that the big water they had heard about south of Superior and west of Green Bay was the river De Soto had crossed.) He went on to state three exceedingly important facts — what he took to be facts — about the geographical hub at the end of Lake Superior. First, eight or ten days' journey north would bring you to Hudson Bay at Port Nelson. Second, five hundred miles southwest of Chequamegon, there was a lake which emptied into the Gulf of California. (Into the Vermilion Sea, the Sea of Cortés; California had long since become a huge island again.) Finally, a tribe that lived only ten days west of Lake Superior — at most this would be a hundred and fifty miles and he was referring to the Sioux — said that ten more days of travel would bring you to the Western Sea.

Lalemant's estimated travel time to Port Nelson was grotesquely short of the reality, but the fact behind it was the Hayes and Nelson Rivers, the ones which Radisson and Groseilliers had heard about and decided would be the most economical route to the richest fur country. He calculated that a great circle course from Port Nelson to Japan was thirty-five hundred miles, two thousand short of the facts. (And he was assuming that there would be open water all the way.) He concluded too that further exploration of Hudson Bay would reveal a strait leading to the Pacific: the Northwest Passage. The Western Sea, which was at most twenty days from Superior, Wisconsin, also had an outlet to Hudson Bay, as well as a connection with the Gulf of California. The map in his mind is blurred but he has two interior passages to the Pacific and one by way of the Arctic.

Father René Ménard had been chosen to establish the first Jesuit mission in the *pays en haut,* St. Esprit at Madeleine Island in Chequamegon Bay. He was old and tired out. He died the next year, 1661. But in 1665 Father Claude Allouez took his place there, among the tribes that had by now given its once deserted shores a heavy population. Allouez was almost continuously attended by special providences and full-scale miracles, so that he could hardly step from a canoe dryshod except by the forethought of God, but he had a powerful intelligence, inexhaustible energy, and a love of wilderness travel that kept him forever on the move. He achieved a remarkable knowledge of the Indians, learned many languages, and served France and his Order well. In 1669 he was directed to establish a mission at Green Bay; it was named St. Francis Xavier but it is La Baye in the literature, as St. Esprit at Chequamegon is La Pointe. Claude Dablon was named Superior of the northwest missions, which he directed from the Sault. He was a trained and talented geographer and his was probably the best mind of all the Jesuits in New France. The frail saint Jacques Marquette succeeded Allouez at Chequamegon.

In the *Relation* of 1669–70 Dablon repeats Lalemant's information about Hudson Bay, the route to Japan, and the Western Sea and refines it in the light of what another decade has learned or heard. It is now understood that the Western Sea is to be reached by a big river (which of course must flow west), that it has remarkably high tides, and that ships have been seen on it. So Marquette had written him from La Pointe a year earlier, adding that the river is in the country of the Assiniboin Indians, only eight days out from his mission.

Suddenly there are a lot of big rivers, and this news coincides with a flurry of excitement about still another one, the Ohio. The one which Marquette is reporting leads to the Western Sea and this must be considered in relation to an item which Allouez had reported from the Sioux two years earlier, in the *Relation* of 1666–67. He quoted them as saying that there was only one more tribe west of their country and then "the earth is cut off, nothing is to be seen but a great Lake whose waters are ill-smelling, for so they designate the Sea." With these reports the area that has been called here the Northwestern Mystery rises for the first time to the awareness of white men. The mist has taken the shape of the Western Sea but behind it is Lake Winnipeg. The tides that enabled Dablon's critical intelligence to be sure it was the sea are a rise in the water sometimes produced by wind action at its shallow southern end; the name Winnipeg probably means a bad smell.

Dablon's summary of 1669–70, however, touches on something else that is new, and this is solid fact. It is now no longer possible to ignore Indian reports of a big river somewhere west of La Baye; it comes out of the north and flows south. The missionaries there are sure that it is not very far away and the Indians have said that their war parties have gone down it "a good many days' journey" without reaching its mouth. Dablon does not connect it with the Big Water that Nicolet had heard about but he concludes that it must flow to "the sea of Florida [the Gulf of Mexico] or that of California." The possibility that it might lead to the Pacific (which would make it the Great River of the West) is exceedingly important. So important that it must be explored at once and Dablon adds that the mission hopes to send an expedition to it the next year.

Even before this the big river had been named. In the report that mentioned the sea west of the Sioux, Allouez had said that they lived on a big river which they called "Messipi." Some students have believed that the tireless Allouez may even have visited it on one of his journeys with wandering bands of Indians.

The expedition to explore the Mississippi did not start till

1673, and then it was an enterprise of state. A good many reasons had combined to make it urgent. The great Colbert's plans to transform New France from a mission and a private proprietary of the fur business to a useful part of the expanding colonial empire were maturing. The privileges of the proprietary, whose current avatar was called the Company of the Indies, had been cut down so far that considerable progress in agriculture had been made. The Intendant Talon (the Intendancy was a device of absolutism to safeguard the king's power by paralyzing local administration) was investigating the colony's mineral wealth, especially the rumored copper and iron of the Great Lakes region. This meant increased governmental interest in the West. Since the only authority in the West was the Jesuits, this in turn stepped up the intensity of the contest between the clergy and the crown for control of New France, which Talon had been directed to win for the crown. In addition, the development of the colonial empire which Colbert envisioned would be greatly accelerated by the discovery of the water route across North America, which the Mississippi might turn out to be. And the imperial contest was sharpening. If the Mississippi should lead to the Pacific, which was the great hope, it would bring France out on the western flank of the Spanish Empire. If it should lead to the Gulf France would be weakening Spain in the East, where St. Augustine had been founded. Finally, and most pressing of all, the English had reached Hudson Bay.

For all these reasons Talon prepared to investigate the Mississippi. A new governor of New France, its greatest one, the Count de Frontenac, arrived in time to ratify the exploration and assume responsibility for it. He had secret royal instructions to curb the power of the Jesuits, instructions which his temperament and political connections predisposed him to carry out faithfully. That may be why he followed Talon's recommendation and named a layman to command the exploration, Louis Jolliet who had been born to the wilderness life and was a master of it. But there had to be a Jesuit too and the election fell on Marquette, indomitable, saintly, and a skillful linguist.

Just as Jolliet reached Montreal on his return his canoe capsized and his journal was lost. The narrative of the expedition

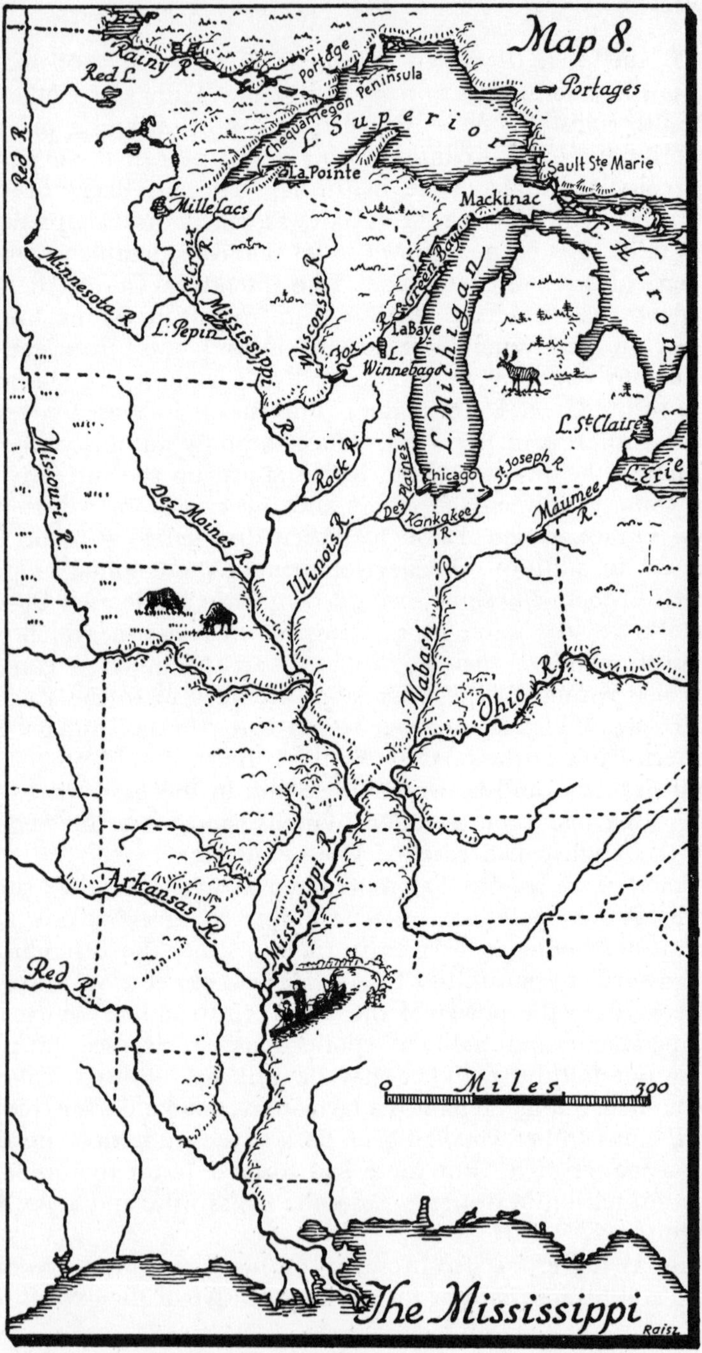

Map 8.

~ Portages

Red L.

Rainy R.

Portage

Chequamegon Peninsula

L. Superior

Red R.

La Pointe

Sault Ste Marie

L. MilleLacs

Mackinac

Minnesota R.

St. Croix R.

Mississippi R.

Wisconsin R.

Fox R.

L. Huron

L. Pepin

LaBaye

L. Winnebago

L. Michigan

L. St Claire

Missouri R.

Rock R.

Des Moines R.

Illinois R.

Des Plaines R.

Chicago

St Joseph R.

Erie

Kankakee R.

Maumee R.

Arkansas R.

Wabash R.

Ohio R.

Mississippi R.

Red R.

Miles

0                300

The Mississippi

Raisz.

that has survived is written in the first person, ostensibly by Marquette. Actually, however, it was written by Dablon after Marquette's death, from Marquette's notes and a copy of Jolliet's journal which he had made and kept.[11] The narrative says that the purpose of the expedition was to explore "the Great River called by the Savages Mississippi, which leads to New Mexico." This postulates the guess of Groseilliers, but is the hope, not a conviction. Everyone concerned clearly understood that the Mississippi might lead not to New Mexico but to the Gulf or through Virginia to the Atlantic.

Jolliet and Marquette collected all the information that the Indians had so far provided and drew a map. In May 1673 with five *voyageurs* and two canoes laden with goods they set out from the mission of St. Ignace, on the northern shore of the Strait of Mackinac. At Green Bay the Menominees, who clearly perceived the threat to their position as middlemen in the trade, invented fierce tribes who would murder the explorers if they went on, river monsters that would swallow them at a gulp, and a river devil whose behavior was even more appalling. The Frenchmen replied that they would take the risk and, reaching the end of Green Bay, paddled up the Fox River and went on to Lake Winnebago; all this stretch had been traveled by the missionaries of La Baye. Lake Winnebago is only a widening of the Fox; crossing it, they followed the upper river to a Mascouten village, where they also found refugee Miamis and Kickapoos. Nicolas Perrot and his companions had this stretch by heart but from here on everything was conjectural. They were traveling an important highway, the Indian route from Lake Michigan to the Mississippi which they now made a white man's road. Taking two Indians to guide them through the ponds and swamps of the upper Fox, they came to the portage across the height of land. It was a mile and a half long; beyond it was the Wisconsin River, whose waters had never known a white man's paddle. "Thus," the narrative says, "we left the waters flowing to Quebeq, four or five hundred leagues from here, to float on those that would thenceforward take us through strange lands."

The great day was June 17 [12] and Jolliet calculated the latitude of the place where the Wisconsin met "this so renowned river" within half a degree. They had reached the Father of

Waters. Jolliet named it the Buade (Frontenac's family name) and later the Colbert; Dablon says that Marquette called it the River Conception, to fulfill a vow he had made to the Virgin. They had been traveling country so beautiful that it was a hymn of praise to Her; from now on the landscape grew stranger. They saw a few monsters: big catfish, a spadefish, a panther. They did not see the devil or his demons but did see pictures of them painted on the rocks. Buffalo grazed on the banks but for a long time they met no Indians. Finally they encountered some at, probably, the mouth of the Des Moines River.[13] Since they were wearing wool, the explorers knew that they were in the sphere of French influence; they on their part recognized Marquette as a Black Robe. They turned out to be a band of the much traveled Illinois, a numerous, mixed, and now wandering people who had been displaced by the Iroquois wars. They were predisposed to like Frenchmen; some of them had asked the priests at La Baye to establish a mission in their own country, others had appeared at Chequamegon with the same request and it may be that some of these now encountered had seen Marquette there. The arrival of the explorers meant that the direct trade would now reach them, so they staged a delirious circus of welcome. But they saw a chance to become a most-favored nation and distribute goods to other tribes, so they too predicted catastrophe if the expedition should go farther.

The second great date is not given except that it was "at the end of June," that is, later than June 26. They were talking about the piasa birds (the demons painted on the rocks) when, "sailing quietly in clear and calm water, we heard the noise of a rapid, into which we were about to run. I have seen nothing more dreadful. An accumulation of large and entire trees, branches, and floating islands was issuing from the mouth of the Pekistanouï with such impetuosity that we could not without great danger risk passing through it. So great was the agitation that the water was very muddy and could not become clear."

They had closed an arc: in 1673 France had reached the river of whose existence Coronado, without knowing it, had heard in Quivira in 1541. For they had reached the arch of the continent and the river pouring through it was the Missouri, which at the moment when white men first saw it was the Big Muddy. (They got the name Pekistanouï on the return voyage.) Neither the Mascoutens nor the Illinois had prepared the explorers for it. Nothing had prepared them for it; chocolate brown, noisy, spinning logs like tops, it came boiling at them from behind the mist of man's ignorance. The mist curtained off everything except the mouth and would continue to for a long time.

They must have landed there for, after they learned where the mouth of the Mississippi was, Jolliet transferred the great hope of reaching the Pacific to the Big Muddy. In his letter to Frontenac he said that he had seen a village that was not more than five days' travel from a tribe that traded with the Indians of California. If he had got there two days earlier, he said, he would have seen some who had come from there and had brought four hatchets with them. (He meant New Mexico. The hatchets may have come from there but, if so, then they had passed through the hands of many tribes along the trade routes.) And Dablon concluded from the oral report he made on his return that by going up the Missouri "one will perhaps arrive at some lake which discharges toward the west, that which is sought and what is all the more to be hoped for as all those lands are covered with lakes and broken by rivers which afford wonderful means of communication between those countries one with the other." [14]

They went on to close the arc of De Soto. They reached the previously unseen Ohio, for it is not true that La Salle had discovered it in 1669. They had heard about it from the Indians, whose name for it sounded something like Wabash — so confusion about another river system, whose existence they did not suspect, may be said to begin here. The country grew tropical: canebrakes, live oaks, parroquets, lush undergrowth — and clouds of belligerent mosquitoes. They met a friendly tribe who had guns and goods, which could have come down the trade routes from the English in Carolina or from the Spanish in Florida. These Indians did not visit the sea but understood that it was ten days away.[15] So they went on farther and met a tribe that De Soto had fought, the Quapaws.

They were at first hostile but the calumet made them friends. In what the chiefs said the facts of geography got inextricably mixed with lies, guesses, and erroneous deductions. The explorers believed the Quapaws' yarn that the downriver tribes would be hostile and heavily armed. They took this as implying that there were Spanish on the Gulf coast. The Quapaws said it was still ten days to the sea; Jolliet reduced this to five and ended by believing that two or three days more would have taken them there. They had come far enough to be sure where the big river reached the sea. They were below the latitude of Virginia, so it could not flow to the Atlantic; they had come straight south whereas it would have trended west long since if it flowed to the Gulf of California. It must empty into the Gulf. They had, the narrative says, "obtained all the information that could be desired in regard to this discovery." That was true and there was therefore no reason to go on, though they would not have met heavily armed Indians or flung themselves "into the hands of the Spanish, who without doubt would at least have detained us as captives." How far south they had come is uncertain but at any rate they had passed De Soto's crossing.[16]

So toward the end of July they turned back, the first white men to breast the current of the Mississippi — and in the wrong kind of craft, for these were two- or three-fathom birchbark canoes, whereas the river demanded heavy dugouts. They had made a great discovery and had made it with ease, comfort, and security not usually accorded to explorers. There had been only two tense moments with Indians, both short, and only the mosquitoes were troublesome. Now they made the first traverse of another immensely important route, for the Illinois had told them of the river that now bears the tribal name and they turned up it. They learned what Perrot and Groseilliers had already learned, what Allouez was on the point of learning, that this country too was a paradise "as regards its fertility of soil, its prairies and woods, its cattle [buffalo], elk, deer, wildcats, bustards [Canada geese], swans, ducks, parroquets, and even beaver." It made so strong an impression on Jolliet that he made plans to come back and begin its occupation, preceding La Salle's vision, and Marquette was vouchsafed the happiness of saving a soul, for he baptized a dying child. They went up the Des Plaines, made the first crossing of the "Chicago portage," and reached Lake Michigan by the Chicago River.

(Jolliet suggested to Dablon just such a canal as was to be built in the nineteenth century, connecting the waters of the Great Lakes with those of the Mississippi.) They paddled up the western shore of the lake in the temperate northern air after the subtropics, beside meadows bending with grass and forested bluffs where the oaks were tawny and the maples red with autumn. October had come on, with the great flights riding the north wind, when they got back to the fellowship of their kind at the crude, momentous shacks of the mission at La Baye.

They had done a great deed: they had fixed the big river in geography from the mouth of the Wisconsin to the Gulf. In the *Relation* of the next year Dablon stated the paramount discovery: "It is certainly most probable that the river which geographers trace and call Saint Esprit [De Soto's river] is the Mississippi."

This determination of course removed the passage to India to other waters. Before the voyage they had understood that two of Coronado's kingdoms, Quivira and Tiguex (the upper valley of the Arkansas, the upper valley of the Rio Grande), bordered on Canada. They knew otherwise now. There remained the Pekistanouï, the Big Muddy, and Jolliet's idea that it would lead to New Mexico. Dablon had interpolated in Jolliet's report the suggestion that there might well prove to be a convenient lake — a species of Western Sea — whose western outlet flowed to the sea. In the *Relation* he enlarged on this with information which the Illinois had given the explorers. They said that when one went up the Missouri for five or six days he came to a prairie which was perhaps a hundred miles wide. A route led northwest across this prairie, to a river that was navigable by canoes. It was perhaps forty-five miles long and emptied into a small deep lake, "which flows to the west, where it falls into the sea." This would be the Vermilion Sea, the Gulf of California. And Dablon ascribed to Marquette a remark that may have been Jolliet's, "I do not despair of discovering it some day, if God grants me the grace and health to do so."

Thus in the year of its discovery the Missouri became, in French thinking, a highway to Mexico and its silver mines. This idea was to confute government, diplomacy, and military strategy till the Great Valley became American, and to confuse geographical thinking till Lewis and Clark got home. The interior lake, the short portage between watersheds (which would decrease from a hundred miles to less than ten), and the misconceptions about the Missouri all got tangled with the web of rumors about the Saskatchewan, the Assiniboine, and the Great River of the West. Though they had cleared up the Mississippi, Jolliet and Marquette had added to the confusion of the Northwestern Mystery.[17]

ぐ∞の

Jolliet and Marquette made their journey in the summer of 1673. Two years before that the increasing tension of both the international and the domestic situations had made it imperative to assert the sovereignty of France.

St. Mary's River, by which the waters of Lake Superior reach Lake Huron, has become the carotid artery of industrial North America. The Sault probably flows no more than a quarter of the water it once did, the rest being diverted to the manufacturing and military establishments of the two nations on its shores and to the three canals that in a season of seven months carry a greater tonnage than the Panama Canal does in twelve months. But it is still a hurricane of white and roaring water. Nothing has altered its geographical, political, and commercial importance, for it was a nerve center of all three in 1671. Its geographical relationships, however, are understood otherwise than they were when Talon wrote to the king that it was only seven hundred and fifty miles from the countries that bordered on the Western Sea and that "there seems to remain not more than fifteen hundred leagues [3750 miles] of navigation to Tartary, China, and Japan." [18]

The mission of Ste. Marie stood on what is today the American side, overlooking that rush of water, with the forest stretching away behind it. Near by, in June 1671, were the bark lodges of many tribes of the interior. Fifteen were there by delegation or by proxy and two others later ratified what their cousins did on June 4.[19] (The official count is seventeen but some of them were subtribes.) They were a congress from a vast area: Crees who lived north of Lake Nipigon and knew Hudson Bay and the Hayes River, Assiniboins from the vicinity of Lake

Winnipeg, Foxes from Lake Winnebago, Illinois who had drifted into Iowa and were now drifting east again. A wide country but the claim was wider: that the soil the mission stood on carried with it Lake Huron, Lake Superior, and "all other Countries, rivers, lakes and tributaries, contiguous and adjacent there unto, as well discovered as to be discovered, which are bounded on the one side by the Northern and Western Seas and on the other side by the South Sea, including all its length and breadth." The delegates assented and in witness thereof affixed their seals: "Some of them drew a beaver, others an otter, a sturgeon, a deer, or an elk."

Painted with their clan marks or the symbols of their private medicine, their hair greased and perfumed, dressed in their finest decorated deerskins or the colored woolens of the trade, brass rings and strings of shells or copper hanging from their ears, adorned with all the jewelry and amulets that carried the highest solemnity in Indian religions, reverent as Indians usually were to other religions, awed, their minds boiling with excitement but not in the least conceiving what this great medicine day portended — they gathered on a small hill overlooking the river, the mission, and the black wall of woods beyond. A cross of cedar poles had been prepared there, and a hole dug to receive it.

A procession of chanting men came slowly through the gate in the log palisade that inclosed the mission house and the chapel. There were Dablon as superior of the missions, Allouez, another Jesuit, all doubtless in their black habits, though probably Dablon wore a white stole for the ceremony. There were Nicolas Perrot, Louis Jolliet, a soldier from the Quebec garrison, and twelve or thirteen traders, bearded, as bespangled as the Indians, as gaudily dressed too in the greens, carmines, violets, and yellows of their vocational attire. A gentleman of France followed them, doubtless dressed in the tiring of the court, which would mean a plumed velvet hat held under his arm, a long curled wig, bright velvet skirted breeches to the knee, short jacket and long shirt, lace and ruffles at knee and elbow, a short ceremonial sword. Probably as he marched behind the white standard with the golden fleur-de-lis he held a perfumed handkerchief to his nose, for cause. He was the King's proxy, "Simon Francis Daumont, Esquire, Sieur de St.

Lusson, Commissioner subdelegate of my Lord the Intendant of New France."

Dablon blessed the cross. Then as it was raised and set in place, the trained voices of the Jesuits and the cruder ones of the *coureurs de bois* blended in the Hymn to the Cross. *"Vexilla Regis prodeunt"* — the standard of our King comes forth. But this was the act of an earthly king as well, so they prayed for him and then St. Lusson had the clerk read aloud the commission that gave his acts authority. Perrot translated it to the tribal delegations, who applauded with a soft, courteous hissing. Above the cross St. Lusson raised the arms of France, which were fixed to a cedar slab, and the voices chanted, *"Exaudiat te Domine in die tribulationis . . ."* In the breviary it is the Prayer for the King Going into Battle. Then:

With proper measure, "in a loud voice and with public outcry," he proclaimed, "In the name of the Most High, Most Mighty, and Most Redoubtable Monarch, Louis the Fourteenth of the Name, Most Christian King of France and Navarre, we take possession of the said place of Ste. Marie of the Falls, as well as of Lakes Huron and Superior," and all the land drained by their waters, and all the lands beyond, to the Seas of the North and the South and the West, to China and Tartary. He raised "a sod of earth" from the hilltop and cried "Vive le Roi!" and the priests and traders shouted "Vive le Roi!" after him. Three times: three proclamations, three plucked sods, three shouts of fealty to the King. And all Potentates, Princes, and Sovereigns, States and Republics were warned not to seize or settle on any place thus taken by the King of France, on pain of incurring his "hatred and the effects of his arms."

The Great Lakes were French now, the North to Hudson Bay and on to the Arctic Ocean about which they had so few ideas and those so marvelous, the Great Valley that was the richest possession in the world, the as yet unvisited river that led to the Gulf of Mexico, all its tributaries and streams and all the hillsides that fed them, all the countries adjacent and beyond, discovered or still to be discovered, Coronado's Quivira and Harahey, the lands that the mist shut off from view, all the unknown and unrumored lands, the lands of fantasy where the river led to the lake that led to the sea that led to China, the farthest springs on the highest peaks of mountain ranges no one

Map 9. Hudson Bay Region.

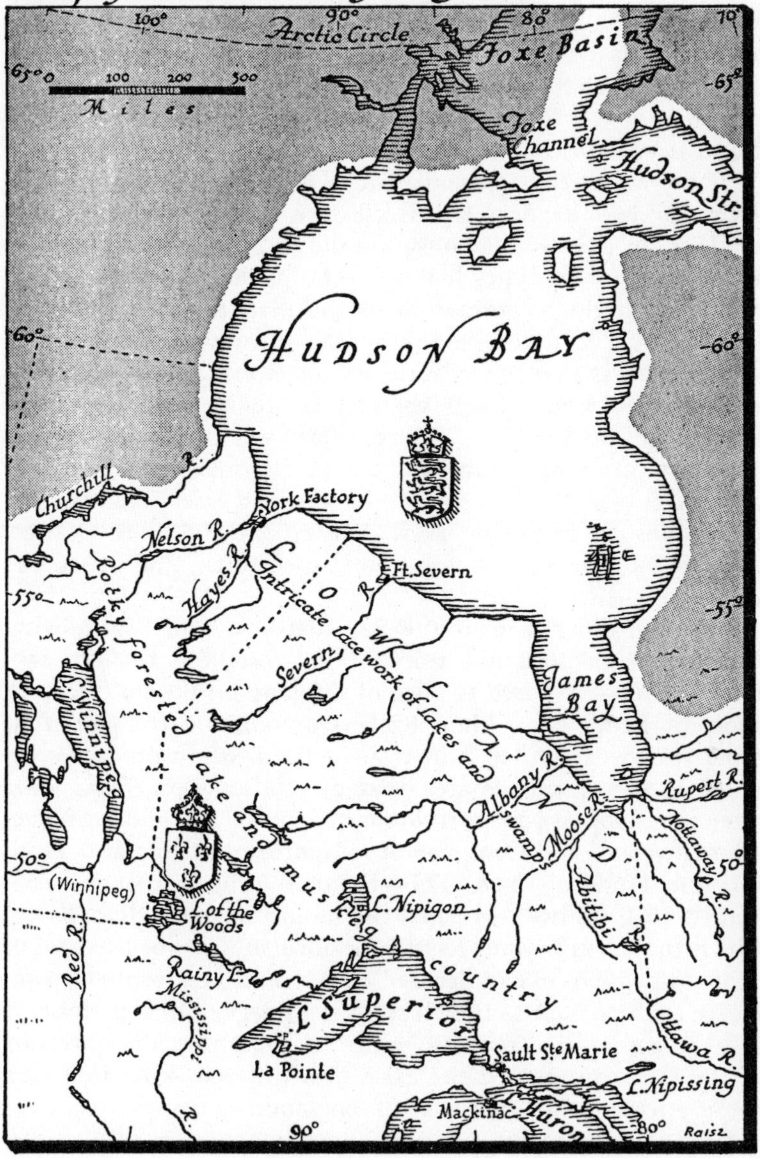
100°
90°
80°
70°
Arctic Circle
Foxe Basin
65°
0 100 200 300
Miles
Foxe Channel
Hudson Str.
Hudson Bay
60°
Churchill R.
York Factory
Nelson R.
Hayes R.
Ft. Severn
Intricate lacework of lakes and swamp
Lowland
55°
Severn R.
Rocky forested lake and muskeg country
L. Winnipeg
James Bay
Albany R.
Rupert R.
Moose R.
Nottaway R.
50°
Abitibi R.
(Winnipeg)
L. of the Woods
L. Nipigon
Red R.
Rainy L.
Mississippi R.
L. Superior
Sault Ste Marie
Ottawa R.
La Pointe
L. Nipissing
Mackinac
L. Huron
Raisz

had heard of yet, the mountains and valleys, plains and watersheds that the dreaming mind saw above the clouds at sunset. When the last sod fell and its loosened loam blew downwind and the last shout and musket shot came echoing back from the edge of the forest, louder than the frenzied rapids of the Sault, it was all French. . . . A brave deed in a brave setting.

There was already some dissent. Three years earlier, in June 1668, the ketches *Eaglet* and *Nonsuch* sailed from Gravesend for Hudson Bay. Radisson was on the first of them and Groseilliers on the other. They had found no justice in France and no one who would be interested in the plan to reach the great beaver country (and the South Sea) by way of Hudson Bay and the rivers that flowed into it from the west. So they had crossed over and offered themselves, their plan, and their knowledge in England, where there were rich speculators willing to attempt a continent and noble ones eager to follow the example of the Duke of York, who had lately enlarged his possessions by capturing New Amsterdam and Fort Orange. Among them was a general, admiral, scientist, and artist, the cousin of the king, Prince Rupert.

Radisson did not make the Bay that year, the *Eaglet* being forced back to port by a storm, but Groseilliers did. In September his party raised the flag of England on the southeastern shore of James Bay. Their backers organized a company that same year, with Prince Rupert at its head, who would give his name to the enormous area west of Hudson Bay. Two years later the company got a new charter and a resounding name, The Governor and Company of Adventurers of England Trading into Hudson's Bay. They began trading into it in 1668, and in 1670 Radisson reached the mouth of the Nelson River, which the Crees beyond Lake Superior had described to him in 1659 and which he had decided then would be a route of commerce as important as the Northwest Passage. On September 1 1670 in a shorter ceremony, with less pageantry, the governor of the Company nailed the arms of a different king to a tree there, the arms of Charles II of England.

Who granted to the Hudson's Bay Company

the sole Trade and Commerce of all those Seas Streightes Bayes Rivers Lakes Creeks and Soundes in whatsoever Latitude they shall bee that lye within the entrance of the Streightes commonly called

Hudsons Streightes together with all the Landes and Territoryes
upon the Countryes Coastes and confynes of the Seas Bayes Lakes
Rivers Creeks and Soundes aforesaid that are not already possessed
by or granted to any of our Subjectes or possessed by the Subjectes
of any other Christian Prince or State with the Fishing of all Sortes
of Fish Whales Sturgions and all other Royal Fishes in the Seas. . . .

That is, all the rivers of Canada east of the Continental Divide
that did not flow into the Atlantic, the Arctic, the St. Lawrence
or the Great Lakes, and all the land they drained, and all the
beaver in this vastness and all the Indians who would trade
beaver for goods. So England was on the northern flank of
New France, with a claim that almost crossed the continent,
and the struggle for the wealth of the North was joined.

That was one beginning, two years before St. Lusson's ceremony at the Sault. Three months after it Englishmen first
pushed over the crest of the Alleghanies from Virginia and saw
waters flowing westward into the domain St. Lusson had
claimed: another beginning. And in May of the next year,
1672, Louis the Fourteenth of that name launched the armies
of Turenne and Condé against the Dutch. France had become
the greatest nation in Europe and Colbert had made it the
richest one. But that was not enough: throwing away the scabbard, he went out, like the Iroquois, to conquer the world. His
incomparable armies thus opened an era of world wars that
would long outlast him, would make New France and its whole
continent stakes in the great game, and would draw it steadily
nearer the vortex that was to suck it down.

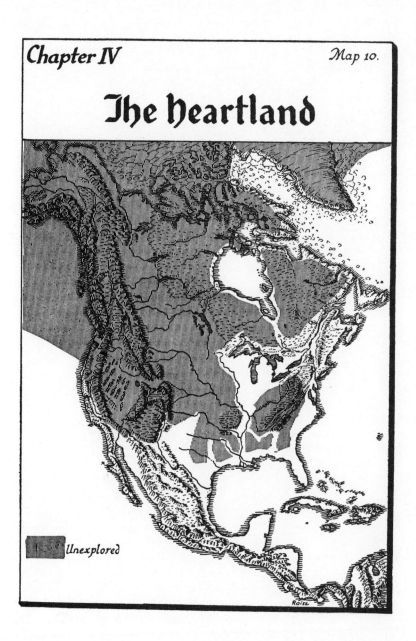

Chapter IV

Map 10.

# The heartland

Unexplored

Roise

# IV

# The Heartland

ON THURSDAY April 9 1682, the French Empire in North America got its final boundaries. Two months earlier Robert Cavelier, Sieur de la Salle, with his one-armed lieutenant Henry Tonty, his Recollect friar Zenobe Membré, twenty other Frenchmen, and thirty-one Eastern Indians, had floated out of the Illinois River and turned down the Mississippi. They passed De Soto's crossing and the farthest south of Jolliet and Marquette, followed the wake of Moscoso's retreat, and, reaching the Gulf, had now caught up with Pineda and Narváez. They had established where it was that the Mississippi reached salt water.

So on April 9 there was another ceremony, some distance upstream. Father Membré blessed the cross, they sang the solemn hymns, and La Salle took possession of the Great Valley for Louis XIV, the Sun King, who was now at his noon mark. When he claimed all the lands that the Mississippi drained, his ideas about its watershed were capacious and nebulous. He was able to divine that the Ohio River was the axis of its eastern half. Out of what the Iroquois and other Indians had told him about it he made at least four rivers; he could not give it sense but had the prescience to understand that it was the route between the Mississippi and the English colonies. About the western half of the watershed he had learned from the Indians along the Illinois River only the same fantasies that were reaching the Jesuits on the upper Lakes. The claim he made ran to the Rio Grande, which the map showed but showed with elastic longitudes and purely imaginary courses, and beyond it for a distance which he could not state in scale. The claim ran up the Mississippi to its unknown source and included its western drainage and whatever might lie westward "beyond the country of the Kious or Nadouessions," who were the Sioux. It embraced "all the nations, peoples, provinces, cities, towns, villages, mines, minerals, fisheries, streams and rivers comprised" in this vast watershed.

A month earlier, in the country of the Arkansas Indians, La Salle had made a preliminary declaration; now he repeated the name he had given then to the Great Valley. It was named for the King, it was Louisiana. The word is all that is left of Louis XIV on the mainland of North America.

La Salle was many men in one. The most powerful and corrupt energies in Canada and France assailed him, made use of him, knocked him about: rival economic interests, national and international religious rivalries, national and colonial and imperial party politics, the Jesuit International. Some of the heat they engendered survives still and he has never had an unimpassioned biographer. There was in him an imaginative splendor which shows that the Renaissance was not dead, and he was kin to Champlain. He is also seventeenth-century France, France as expanding capitalism with its primordial business ethics, and so kin to Colbert. But he was a man of the next century too, he is the first of a line that includes Pitt and Thomas Jefferson.

Thinking westward from Mount Royal, Champlain had had a premonition, an awareness: that what the mist of ignorance concealed might be a continent which was a coherent whole and must be French. By the time of Colbert's chosen instrument, the Intendant Talon, discovery had given the idea a kind of shape, and Talon worked methodically to add this continental wholeness to French power, as territory, subjects, known wealth, and wealth still to be discovered or created. Frontenac had a more powerful intelligence than Talon and more clearly understood what this continental wealth might be. But the idea comes complete with La Salle. He began in fantasy and hoped to follow the Great Lakes to the Northwest Passage and on to China. But he worked it out step by step till he foresaw the economy of the United States and till he understood the grand strategy of North America.

New France had the waterway to the heart of the continent, the St. Lawrence and the Great Lakes. Tributaries could be followed farther west; how far no one knew but it would be worked out. Louisiana had the highway to the Gulf of Mexico, the Mississippi River. East of its lowest reach lay Florida, a Spanish possession. West of it were Mexico and New Mexico, which in God's good time would be conquests of France. What else lay west of it no one knew, but the rivers that flowed into it would lead to whatever there might be. Two routes already known connected the Great Lakes with the Mississippi, one by way of the Fox and Wisconsin Rivers, the other by way of the Chicago, the Des Plaines, and the Illinois; and La Salle himself promptly established a third one. The St. Lawrence, the Great Lakes, and the Mississippi, then, were the access-route, for trade or conquest, to the Spanish possessions. And they encircled the British colonies on three sides, and so could keep them walled off east of the Appalachian mountain chain. Finally, whatever the Ohio might turn out to be, roads run two ways and if the French could take it toward the British colonies, the British, provided they were permitted to, could come down it to Louisiana.

Thus Louisiana was half of the hinge on which possession and mastery would turn. It was also a bastion necessary to the defense of New France, which sea power could throttle by blockading the mouth of the St. Lawrence. The St. Lawrence and the Mississippi were the compass of the New World.

This was geopolitical reality, the basis of a good deal more than half of continental strategy. But there was something else. More than four-fifths of the wealth of New France was furs, the rest was fish, and it had no agricultural wealth. One trouble was that whereas the crown's imperial policy required it to develop the colony's agriculture, the crown's economy required the colony's furs, an adverse interest. Another was that France could never find enough Frenchmen who wanted to go to New France and would not relax the rigid control that frustrated those who would go. At best New France could barely store enough extra food to supply a handful of the king's troops making a foray against the Iroquois. Apart from an occasional shipment of wheat to the Sugar Islands, it had no agricultural surplus for export at any time during the seventeenth century.

But Louisiana was Radisson's "more incomparable" country; Jolliet, Allouez, and Perrot had all found it a paradise of fertility. Its richness woke in La Salle the idea of an empire not founded on the fur trade.

Tonty called Louisiana "the most beautiful country in the world" and said that it contained "the finest lands ever seen." There was maize everywhere, wheat could be grown in the northern reaches and sugar in the southern ones — enough to feed New France and the Sugar Islands too. There was incalculable wealth of other kinds, known lead and copper, probably the richer ores, timber for all the navies of the world, few beaver (it was thought) but a big potential trade in other pelts and in hides, hemp, cotton, inexhaustible grazing, the buffalo, perhaps an industry from buffalo wool, perhaps another one from silk culture as the mulberries seemed to indicate, perhaps pearl fisheries. Louisiana must be a colony by itself. If it were, it would be richer than New France and would dominate North America.

It was, besides, beyond the reach of the fur merchants, who had sabotaged La Salle's earlier ventures, bribed and stolen his men, procured the deaths of some of them, dispossessed him, and used the government of New France to defeat him. It was beyond the present authority of the Jesuits too, who opposed him as vindictively as the merchants and only a little less violently. Only Allouez, faithfully accompanying the Illinois on their wanderings, was in Louisiana now but its future was as great a promise and a challenge to them as to La Salle. It was to be their preserve, protected from the Canadians who were civilization's carriers of vice and disobedience, and populated by baptized, peaceful tribes who would be subject only to the priests. But what La Salle saw, and what he made others see with his eyes all the way up to Colbert, who died in 1683, was the spread of France through the Great Valley — agriculture, trade, wealth, and power. It was not to be. But, hated, robbed, persecuted as if he were a common enemy, performing prodigies of travel and of managing the tribes, La Salle did what he could. When he died, a personal failure, he had erected jetties that were to divert the current of history.

He built a post at the mouth of the Chicago River; it commanded the highway that led from Lake Michigan over the

Chicago portage to the Des Plaines. He built another one where the St. Joseph River enters Lake Michigan; a short portage connected the St. Joseph with the Kankakee River. Both the Des Plaines and the Kankakee led to the Illinois River, which was a highway to the Mississippi and which became La Salle's river. He named it the St. Louis, for his king and for the patron saint of France, and then the Seignelay, for Colbert's son and successor. He found on it some villages of the tribe whose name it bears, the Illinois, and found them under threat of attack by the Iroquois. Where the river widens as Lake Peoria he built a post not badly named Fort Crèvecoeur, considering the disasters that had befallen him. Then upstream a little way at the natural stronghold later called Starved Rock he built Fort St. Louis; here he would make his stand against Indian and French enemies. It was from Fort St. Louis that he went down the Illinois to the Mississippi and down that to claim the land and name it Louisiana. The Mississippi was French now. So was its mouth, a geopolitical climax, a crux of trade and military power, a dominant point in the continental unity. But there was no end to desperation — by now he had been robbed of Fort St. Louis — or to the effort that must be made.

Frontenac had been his patron, protector, and partner. But the great governor's feuds with the clergy and spoliation of the fur merchants had finally brought about a combination of enemies more powerful than he, and he had been recalled to France. There he plunged into the intrigues of a dictatorship that was between two wars. They had divided the court, divided France, and stretched a network of treachery and conspiracy across Europe. Louis had brought France to a pinnacle from which, so he believed, in one more step he could master Europe. But Europe had foreshadowed — in the Triple Alliance — a device that would eventually defeat him. It had found — in Dutch independence — a cause that would be a core round which resistance to France could cohere. And in one of history's strangest heroes, William of Orange, mediocre, dauntless, immovable, untiring, incapable of retreat, it had found a man who could hold the coalition together and coerce it.

Louis had called a new turn in the Great Game. By the diplomacy of division and threat and by a series of *faits accomplis* too strong for a single opponent to resist but too slight for a coalition to act against, he intended to get the positions, territories, and satellites that would enable him to conquer Europe with one more war. This is a stage that all dictatorships reach; when they do the dictator can no longer afford mistakes and always makes them. Louis's mistakes began now.

Besides, there was another Great Game and its objectives diverged from his just enough. The Jesuit International covered the world from Lake Superior to the court of China and was now at the apex of its power. "The Church to rule the world, the Pope to rule the Church, the Jesuits to rule the Pope," so Parkman states the aim, and the last chance the Church had to rule the world came in the three years following the death of Charles II in 1685. France was the power-base of the Jesuit International but the key to the Great Game it was playing was England, whereas for Louis the key was the Low Countries. The deviation was enough to make Louis sometimes oppose the Jesuits, and like the other forces in the great power vortex of the 1680's it was felt across the world.

Between them the dictator and the International forced the Church to adhere to the enemies of both. To save its sovereignty, Innocent XI in the end supported the ultimate coalition. (It was called the League of Augsburg.) Louis made the mistake of strengthening William of Orange to bind James II of England more closely to himself. Innocent could have intervened but he spoke no word when the Protestant William sailed to dethrone the militant Catholic James. But what counts here is that in the power struggle at the French court which Frontenac had re-entered, and to which La Salle followed him, there was the focus of another world-wide conflict, between the Jesuits and the Franciscans. It affected North America from Hudson Bay to the Texas coast.

Frontenac had always belonged to the Franciscan party. His successor as governor of New France, La Barre, was allied with the Jesuits as well as the fur merchants. He had confiscated La Salle's estate, seized his forts, and sanctioned the Iroquois to plunder his men as much as they might be able to. Without hope of redress in Canada, La Salle went to France for justice or at least backing, for new supporters, for funds — and to give the French Empire so firm a hold on Louisiana that it could never be dislodged from North America. He had occupied the strategic centers at the north, the end of Lake Michigan and the Illinois River — thus guarding the routes between the Great Lakes and the Mississippi and establishing France in a position from which it could command the mouths of the Ohio and the Missouri. Now he would secure the most important site of all: he would found a colony at the mouth of the Mississippi.

But his star, always murky, had set. Passions as extreme as those he roused in others had driven him all his life; he had experienced repeated setbacks, frustrations, and defeats, each of them stunning enough in itself to destroy a lesser man. Clearly his mind had lost some of its power and some of its hold on reality; possibly he had gone mad.[1] But it is also true that in this project for Louisiana he was a man before his time. The enterprise was beyond the power of one man to effect: it was on too large a scale, demanded too much, cost too much. It proved in the end too great for the France that did colonize Louisiana later on. Whether or not it was too great for the France of Louis XIV before the Sun King began to bankrupt himself and his successors is an idle speculation, for in the event it disappeared under the first wave of the one last war that would conquer Europe and so the world.

The court greeted La Salle with one auspicious and one weeping eye. The onset of war with Spain and rumor that the English were encroaching on the Great Valley had focused the attention of Louis and Seignelay, Colbert's successor, on North American geopolitics, which La Salle was the first man to understand. For the moment the interests of France coincided with those of the Franciscan party. After an incomprehensibly tangled intrigue, which involved the entire trans-European underground conspiracy, La Barre was ordered to restore La Salle's estate. La Salle himself was commissioned Governor of Louisiana and directed to fulfill his dream by founding a colony on the Gulf coast. So at last there was something besides ministerial good wishes to support a French colonial enterprise: four hundred men (entirely unfit for wilderness life), something like adequate equipment, and as a culminating marvel something like adequate funds as well.

But they had a price: conditions that betrayed the vision, robbed the enterprise of meaning and use, and insured failure. The interest of the court in Louisiana colonization was to

secure a bridgehead for an attack on the silver mines of northern Mexico, not to develop the resources of the Great Valley.
Two powerful anti-Jesuit conspirators who came to direct La
Salle's fortunes succeeded in blending his enterprise, the court's
interest, and the wild scheme of an adventurer who had become one of their protégés. This was Diego de Peñalosa, a
former governor of New Mexico, who had been condemned by
the Inquisition and banished from New Spain. In revenge he
went to England, like Radisson and Groseilliers whose interests the two religious were also serving in the underground, and
offered the ministry a plan to conquer Mexico and New Mexico
and to seize the mysterious lands of Quivira and Thayguayo,
which he claimed to have explored. Failing there, he went to
France, where the conspirators took him up. The plan was for
La Salle to found his colony at some convenient place on the
coast of Texas, at Galveston Bay or near it, and for Peñalosa
thereafter to join him, for a conquest of the mines, with an
army which he would somehow raise among the buccaneers
and freebooters of the West Indies.

In order to get even this oblique and catastrophic support,
La Salle was forced to forfeit the integrity of his plan and to
falsify the geography of the Mississippi, moving its mouth much
farther west and taking advantage of European ignorance to
represent it as very much like the Rio Grande. So he was a for-
sworn man when his expedition started in 1684, and a doomed
man too. The court had already abandoned and forgotten the
enterprise it had begun by sanctioning; no one had any interest
in it except the investors and the two religious. Peñalosa raised
no pirates and presently died. La Salle himself wanted to give
up before they made their landfall. The sea voyage was a
chapter of errors and accidents. La Salle was monstrously despotic, he was often sick, he quarreled with his naval commander. Agents of the opposition had been planted in the
company, there was always treachery or betrayal to beat down.
Less than half the original company remained when they
landed at Matagorda Bay, a hundred miles from where Cabeza
de Vaca had ended his voyage a hundred and fifty-six years
earlier. There has never been an adequate or even a logical explanation of the story of the colony: a chronicle of disease,
starvation, aimlessness, treachery, mutiny, and murder. It had
dwindled to forty-five when, in 1689, La Salle began his last

expedition, a desperate overland attempt to reach Fort St. Louis at Starved Rock and the incorruptible, heroic Tonty who held it for him. He took half the survivors, leaving the other half to die after his failure but eventually to give France and the Republic of Texas a claim to Texas, which later on the United States both asserted and disavowed. On the way some mutineers murdered him.[2]

La Salle's plans failed altogether — as his plans. But his forts on the Illinois and St. Joseph altered the alignment of forces in the West. They and the Indians he induced to settle in quantity near them helped to keep the Far Nations French and to frustrate the Iroquois. His determination of the Mississippi and his effort to colonize its mouth created part of the pattern of destiny. Spain reacted at once to his landing at Matagorda Bay. A series of expeditions from Mexico explored the lower Rio Grande, parts of interior Texas, the Texas coast, the eastern Gulf coast. A Texas mission was established in 1690, to serve as an outpost against the French. In 1698 Pensacola was founded on the west coast of Florida, to flank the mouth of the Mississippi and to support the stubbornly surviving settlement on the east coast. Further posts and settlements on both coasts followed.

War and an at last aroused sense of urgency impelled France to act on La Salle's vision. In 1699 a settlement was made on the Mississippi, at Cahokia, across the river from St. Louis; for some years *voyageurs* from the upper Lakes, in trouble with their leaders or the priests, had been slipping away to live there with the friendly Illinois. In the same year the French moved to safeguard the key site, the mouth of the Mississippi, reconnoitering it just in time to turn back two English ships on the same errand. The establishment of 1699 was to the eastward, at Biloxi. It was moved farther east, to Mobile, in 1702. At last the proper site, New Orleans, was occupied in 1718. In 1713 a frontier outpost was established at Natchitoches on the Red River, a focus for the Indian trade and (the hope was) for trade with Mexico too, and a counterweight to the Mexican outposts. Trade and, slowly, settlement spread downriver from Cahokia and upriver from New Orleans. Traders ventured up the Arkansas River and a little way up the Missouri. Fort Chartres, between Cahokia and Kaskaskia, was built in 1720 and Fort Orleans, actually in the Missouri, in 1722. In 1722 also a per-

manent post, replacing temporary ones, was established at Vincennes, to protect another route between the Lakes and the
Mississippi that had become important, by way of the Maumee
and the Wabash Rivers, and to resist encroachment by the British trade.

By this time La Salle's prophecy had become fact. Louisiana
had been added to the Great Lakes as part of the French con-
tainment of British imperial expansion.

৶

*Coureurs de bois* had been going to the upper country ever
since Radisson and Groseilliers led the way there in the 1650's.
Those who had licenses were legal, the rest were not, and they
forced a change in the fur trade system, steadily diminishing
the effectiveness of the monopoly till its corporate embodiment,
the Company of the Indies, had to give up in 1674. By then
there were few families in New France which the trade and its
graft did not touch and few, if any, officials who could not be
bought on shares. Even the missionaries, who did their utmost
to keep the trade out of the West, had to connive at the illicit
participation in it by mission employes, in order to keep them.
The vigorous youth of the colony turned to the Western woods
and waters as the best escape from regulation of their lives and
the best hope for advancement. When the crisis came in the
1680's there had been a full generation of them in the West,
and they had mastered the skills of managing Indians.

By now nearly four-fifths of the furs came from the *pays en
haut:* the upper Lakes, the country west of Lake Michigan, and
that north and west of Lake Superior. The Hurons and the
Ottawas, the established carriers, still sent their canoe flotillas
to the annual fair at Montreal and carried back goods for the
internal market.[3] Sometimes parties from various lake tribes,
many of which were not canoe Indians, joined them to gape
among the marvels, sign treaties of eternal loyalty, and take
home impressions of wonderland and French power. This was
the trade system which the government, the clergy, and the big
merchants favored, to bring the Indians to the easily controlled
market at Montreal. But the *coureurs de bois* in increasing
numbers took goods to the Indians in their own country. The
Indians greatly preferred this system, so did the small merchants

who could enter it with a smaller investment and could capitalize on opportunity and chance, and there was always the left hand of a big merchant to finance a trader without the knowledge of his right hand. So by now traders were going everywhere in the West, following the market. They tended to base on the missions, to the resentment of the missionaries, till they developed their own posts — usually near by. They worked from Chequamegon till the Hurons and Ottawas fled it in fear of the Sioux, from the Sault, Mackinac, and Green Bay — these first, then the Chicago River, the St. Joseph, interior Wisconsin, finally Minnesota.

This trade developed such men as Nicolas Perrot, who roamed Wisconsin, Illinois, Michigan, and the North as opportunity and the service of New France might call him. Another like him was Tonty. It was he who rallied the Illinois at Fort St. Louis during La Salle's absence. Eventually the fort and La Salle's authority passed into his keeping — it cannot be said into his hands for he had but one. Of unlimited fortitude and hardihood, with the same genius for managing Indians that La Salle had, he was one of the rocks to which the new alignment in the West was moored. Another was his cousin, Daniel Greysolon, Sieur du Lhut, whose name has taken a different spelling in the American.

Duluth began his westering in 1678, the year when La Salle, having finished his wilderness apprenticeship, moved on the Mississippi Valley by way of Lake Michigan. The next summer he tackled the job of making a peace between the Chippewas and the Sioux. Nearer the trade goods and better armed, the Chippewas were slowly forcing the Sioux westward in this ancient war. But now the Sioux were getting arms not only from the French tribes in Wisconsin but also from those that were trading with the British at Hudson Bay, the Assiniboins and the Crees, with both of whom they were intermittently at war.

The measure of Duluth is that he did procure a truce. Then he went on to the heart of the Sioux country, Lake Mille Lacs in eastern Minnesota, the farthest west that France had so far gone and not far from a great secret, the sources of the Mississippi. He sent some of his men still deeper into the unseen. They met a band of Sioux who had some salt and said that they had got it on a raid twenty days' journey to the westward. When he learned this Duluth's response was inevitable: he knew that he was only a short distance from the Western Sea.[4]

To reach the Western Sea became the fixed star of his ambition. But first there was another peace to make, this time between the Sioux and their even less-known relatives and enemies, the Assiniboins. He made it, easily maintaining his ascendancy over the feral Sioux, and spent the winter of 1679–80 at Kaministiquia, on the northwestern shore of Lake Superior (Fort William, Ontario). It was an immensely important site. Later on the canoe route to the Far West and the Northwest would begin here — the route to Lake Winnipeg, which was the bad-smelling, bad-tasting Big Water that they kept identifying as the Western Sea. The route to the Northwestern Mystery. When Duluth began his discovery the following summer, however, it was not by this route, whose existence he did not suspect, but by one which he pioneered — unless Radisson and Groseilliers had traveled part of it. It led to the Mississippi from Lake Superior, up Brulé River and down the St. Croix. But now he learned that the Sioux had some French prisoners and, knowing that French rule depended on French prestige, postponed the discovery of China and made his lightning journey to Lake Mille Lacs where he found the three whom La Salle had sent to investigate the northwestern corner of his caliphate. One of them was the Belgian Recollect, Louis Hennepin, and so chance had brought together deep in the forest wilderness two veterans of the Dutch wars. Both Duluth and Hennepin had been on the field six years before at the great Condé's last battle, the bloody, indecisive one he fought with William of Orange at Seneffe.

Hennepin was a fascinating man: a devout Franciscan with Jansenist leanings who became a heretic, jovial, adventurous, intrepid, tireless, a sound geographer, a good observer, a braggart, a liar, and an indefatigable admirer of Louis Hennepin. One of the chaplains assigned to La Salle's great enterprise of 1678, he visited Niagara Falls and his description — freshened with a pinch of Munchausen — is the first ever printed. He made the voyage to Green Bay on the *Griffon,* the first ship that ever sailed the farther Lakes, whose subsequent disappearance was one of La Salle's heaviest losses. He took part in the canoe trip to the St. Joseph River, the building of the post there, the exploration of the Illinois River, and the winter (1679–80) at

Fort Crèvecoeur. Then La Salle, before beginning his tremendous journey to Fort Frontenac in search of the *Griffon* and his suborned deserters, directed him to accompany two *voyageurs* whom he was sending to explore the upper Mississippi. They had a canoeful of trade goods and were to build a post at the mouth of the Wisconsin River. (Another of La Salle's sagacious moves.) Somewhere short of it a war party of Sioux captured them. Their boisterous jailers distributed the goods and, not daring to kill Frenchmen, took them wandering about the wild-rice country. Hennepin saw the Chippewa and Minnesota Rivers, the widening of the Mississippi that is called Lake Pepin, and Lake Mille Lacs. Permitted to make a probationary journey on his own, he saw and named St. Anthony's Falls, the site of the Twin Cities. He was back at Lake Mille Lacs when Duluth arrived and cowed the Sioux into releasing the prisoners. Duluth saw them out of the country to Mackinac — in Hennepin's book the rescue is just a happy meeting of fellow countrymen. After wintering at Mackinac, Hennepin went back to Montreal and, the following year, 1682, to France. In 1683 he published his *Description de la Louisiane*. It is a valuable book for he had seen much action, territory, and Indian life, and had learned more about all of them from La Salle, Duluth, and the winterers at Mackinac. It was a tremendous novelty in Europe and acquired a wide fame and popularity. So Hennepin took up the literary life, creating staggering and entirely fictitious achievements for himself and trying to pass off La Salle's explorations as his own.

Duluth had to go to France too. La Barre, the governor, had charged him with illegal trade — none of La Barre's own, illegal, traders had come within hundreds of miles of him but the principle was that he was entitled to Duluth's furs for they might have done so. He had to clear his name and he tried to get help for the discovery of the Western Sea. No chance. From the beginning you had been supposed to open up the New World for little more than love of country and the chance to pay a heavy tax on the furs from whatever new beaver resources you might discover. So you explored for the King's glory under mortgage to the fur merchants, and, like both Duluth and Perrot, lost every dime you made. From now on there would be no help at all except a brief monopoly on the trade in the new region you had explored, a monopoly which bore the same tax and was usually taken away from you forthwith. For Louis was conquering the world now and could spare no funds for discovery.

Returning to Lake Superior in 1683, Duluth had so many jobs to do that he never did get started toward the Western Sea. He was always near it — so near that there remained only the Sioux and that single tribe beyond them, and the Sioux themselves had a Chinese accent.[5] One job was to build a trading post at Kaministiquia. That was part of the energetic reaction to the Hudson's Bay Company's establishments. As part of the same effort to encircle them and cut off their trade Duluth's brother, Claude Greysolon, Sieur de la Tourette, ranged north to Lake Nipigon and built a post there. It became enormously successful with the Assiniboins and the Crees. But Duluth himself was everywhere, disciplining rebellious tribes, breaking up Iroquois or British conspiracies, maintaining the trade through periodic scarcities of goods and the mountainous chaos of regulations. When the wars came he commanded a fort at one of the strategic centers of the Great Lakes, the St. Clair River, the passage from Lake Erie to Lake Huron. Later he commanded another stronghold of this fundamental line of defense, Fort Frontenac at the outlet of Lake Ontario. He led the life of audacity for the King till he was old and crippled. But the great deed was that he gave the French their precarious but long-lived ascendancy over the Sioux. Perrot supported him in maintaining it, as he supported Perrot in maintaining an equally precarious ascendancy over the belligerent Foxes. The two tribes were the fiercest and the most problematical in the West.

These wilderness leaders and the men they led — the *voyageurs,* the rivermen — were a remarkable specialization. A text speaks easily enough of a canoe voyage from Montreal to Lake Winnebago, Kaministiquia, or the Chicago River. The words have no overtone of the skill that took a birchbark fragility up furious streams and along the treacherous Great Lakes water, the labor of carrying its freight across scores of portages, or the extremity of effort that was wilderness life. That life imposed the maximum tax on strength and resourcefulness and for avoidance of the tax imposed the simplest penalty, death. Such journeys as La Salle's winter crossing from Fort Crèvecoeur to Fort Frontenac and Tonty's dash from Fort

St. Louis to the Red River to succour the survivors of La Salle's last failure are prodigies of endurance and resolution. They rouse wonder that a mortal body could stand so much or the human will hold it to such an ordeal. But all wilderness journeys were routine in prodigy.

And there were the Indians. Friendly, hospitable, genial (most of them), generous, amusing, they were also children with tantrums and deadly weapons. At any moment and without warning friendship could become murder: murder on impulse, in spite, in remembered grievance, for honor, for appeasement of the supernaturals, as a courtesy to an ally, for no reason, and always for the trade goods. Wilderness man living with neolithic man had to live with him as with a jaguar in its den.

Hundreds of Frenchmen died in the wilderness, murdered on caprice or for an ounce of iron, drowned in the rapids, smothered by blizzards, snuffed out by a momentary lapse in skill, or starved. But they held the West. New France numbered some ten thousand inhabitants in 1680: there were twice as many people in Connecticut, six times as many in Massachusetts. The ten thousand held the continent against English colonies that were populous, powerful, vigilantly backed by the crown, secure in agriculture and commerce — everything that New France was not. They held it because they had mastered the wilderness. In that mastery a decisive part was an invention of just these men, La Salle, Tonty, Perrot, and Duluth: the confederacy of the Far Nations.

৩∾৩

In 1682 Radisson, who had been lured back to the French interest for a while, seized a Hudson's Bay Company ship and a recently erected fort at the mouth of the Nelson River. It was the Company he had helped to found, and the site was the one which he knew would dominate the trade of the richest beaver country. Till now not much had been done about it and the Hudson's Bay Company had frustrated the designs of Radisson and Groseilliers, the men who knew, as stupidly as the French had done. Its returns had been fantastically rich, though it was managed ineptly, corruptly, and with not only ignorance but fear of the land wilderness to whose Bay coast it clung. Its three permanent establishments were at James Bay, the southern arm of Hudson Bay, and till now it had merely sent ships to meet the tribes that came down all the long way from the Northwest. All the wilderness men were French; the maritime English were afraid of the dark and frozen land. The Company's policy was what it would be for another full century, to make the Indians come down the rivers to its posts on the seacoast. The two French experts, despised as Frogs, suspected of treachery, denied a share in the profits and incredibly underpaid for the only skill with Indians the Company could engage, in vain urged penetration of the West in order to get more customers, spread the word farther, and circumvent the traders of New France. Even so, the response of the distant tribes had been so great that a French riposte was imperative.

The piratical seizure of Port Nelson was part of the reaction to the arrival of the English in Hudson Bay. It was an extensive reaction, for it must be seen as including St. Lusson's declaration of sovereignty, La Salle's expeditions, and the augmentation of religious and commercial activity and of exploration in the West. To reject the foresight of Radisson and Groseilliers, to let the English act on it, had been an immeasurable mistake. The Bay must be got back and Port Nelson was got back with Radisson's minute foray of 1682. But the complexities of French-English diplomacy forced its return and again there was no future in New France for Radisson and Groseilliers. They were trebly defeated, by the crown's diplomacy, by the religious rivalry that was almost a civil war, and by the commercial corruption; when they returned to the English service it was for good. Meanwhile in 1687, while the two nations were still formally at peace, an amazing foray by a handful of men on snowshoes captured all three Company posts on James Bay. These could be held longer but in the end they too had to be given up.

The war came in 1689, ten years of it, King William's War in our texts, with raids that terrorized New England, Sir William Phips's capture of Port Royal, his failure to capture Quebec, and the scalpings and burnings along the New England border. In broader texts it is the War of the League of Augsburg, with the Grand Alliance forged against Louis, William of Orange defeated but never beaten, the battles of fortresses such as Mons and Namur, frightful but indecisive field engagements such as Steinkirk, the slaughter at the Boyne, and the reduction of the great French navy as the Dutch and British hewed away at it. Part of it, microscopic but momentous, was the seesaw fight of tiny ships and little groups of men for the posts on Hudson Bay.

e∾ͽ

But before this, in fact from 1680 on, there were the Iroquois again.

There was the permanent danger that they might be able to divert the trade to the British. The British could not begin to match the French in wilderness skills or the skill of managing Indians but they could always outsell them. British woolens were better than any made in France or any that could be bought elsewhere in the world market. The manufacturing system of Great Britain had developed so far beyond that of France that other goods, hardware, firearms, powder, ornaments, the whole miscellany, could be delivered to the customer so cheaply as to force the French out of competition. A more modern credit system, more efficient government, far freer competition, price flexibility, commercial ethics sufficiently hair-raising but much less sodden with corruption — all these worked to the same end. Beaver would always buy twice as much at Albany as from any French trader, and frequently three or four times as much. From its establishment as Fort Orange by the Dutch, the Iroquois had been competing with New France for furs from the West. Yet they were customers of New France too, even during the crises of the wars. And there were always Canadians who, shut out by the merchants' combine or merely following the trade balance, traded with Albany direct or through the Iroquois. The merchants themselves and even the government sometimes had to do the same, in order to get goods for the West and keep the trade of New France alive. There was always, that is, a black market.

The Iroquois had fought their wars of extermination in vain: they could not get the Western trade. The Hurons and Ottawas maintained their position as distributors and the *coureurs de bois* did a steadily growing business in the West. The Iroquois were furiously active. They ambushed the river routes and harassed the Lakes tribes with raids. They kept sending embassies west to form combinations: trade agreements, treaties of nonaggression or nonintervention, coalitions against the French. But they could not themselves keep the peace they made nor could one nation of the Long House force another to respect its peace.[6] French diplomacy in the West invariably defeated theirs and the Western tribes too knew enough economic theory to realize that British goods would not be cheap if the Iroquois should be able to corner the market. A people poor in natural resources, in beaver, the Iroquois eventually had no recourse but to try conquest again. After they had destroyed the Susquehannas, they put the general staff to work on the problem. The plan that resulted was to increase trade and diplomatic pressure on the Western tribes, and to send expeditionary forces to their country. This was realistic thinking but the strategy broke against a new French invention, the confederation of the West.

It was primarily La Salle's invention but it was implicit in what had been happening out west. The establishment of missions and trading posts changed tribal movements and brought about concentrations of tribes. The eastern half of Wisconsin, southern Michigan, and Illinois grew in population after the direct French trade reached them. The tribes that settled there were a big market in themselves and began to forward goods to more distant ones, some well beyond the Mississippi. The Iroquois plan was designed to squeeze the French out of this trade, with the backing of Albany. Its biggest objective was the two most populous groups, the Miamis and the Illinois.[7]

La Salle found a growing settlement of the Illinois on the river that is named for them. He built Fort Crèvecoeur near them and spent the winter of 1679–80 there. He sent Hennepin to the upper Mississippi and, leaving Tonty in charge, made his tremendous journey to Montreal. In the summer of 1680, while he was gone, the first Iroquois army arrived. Its coming terrified the Illinois — Indians were usually stampeded by an attack and the Iroquois by now had the reputation of any Battalion of Death — but they rallied. Tonty, who was wounded in the first skirmish, was wise and heroic and, after all, to the Iroquois as to the Illinois, he represented the mysterious power of the Great Father. They did not dare kill him and when they tried to bribe and intimidate him into neutrality he kicked the symbolic gifts away. Fighting a successful rear-guard action, the Illinois moved downriver. The Iroquois cut to pieces a group who stayed behind, massacred women and children wholesale, burned towns and cornfields, pictured themselves as fury on the loose, and went home with a lot of scalps and a fine feeling of accomplishment. On the way they treated themselves to a small but tasty massacre of Miamis, of the tribe who were most necessary to their plan of conquest and with whom at that moment they were trying to make a treaty against the French. The Illinois moved back to their towns and the Iroquois expeditionary force had won a battle but lost the war.

For the Illinois country was at the heart of La Salle's plans for the Great Valley, the base of operations and the stronghold that commanded the routes of transportation. The Illinois themselves were a market, a reservoir of military power, and an instrument to be used for expansion. If either they or their country should be lost to the Iroquois his design would be wrecked. So when he got back he assumed the function of William of Orange and began to organize a grand alliance of Western tribes against the Iroquois. He had posts on the Chicago and the St. Joseph Rivers, and he built Fort St. Louis to replace Crèvecoeur on the Illinois. Here, he pointed out to the Indians, goods would come direct to the customers and here the soldiery of New France would be garrisoned to protect them. The customers would be wise to settle permanently near these centers. He and his ambassadors, especially Tonty, succeeded in shaping a confederacy and the Iroquois embassies that came to argue alliance got nowhere. Additional bands of Illinois settled near Fort St. Louis; bands of other tribes began to locate there too. The idea spread to the Northern tribes, among whom it was vigorously seconded by the Jesuits and by Perrot, Duluth, and their kind. The tribes wavered, backslid, quarreled, intrigued with the Iroquois underground, blew hot and cold, sometimes were on the verge of shifting their allegiance to the English. But they held, the Iroquois were unable to form a combination in the upper country, and they had to try the military arm again. In 1683, contemptuous of the new governor, they sent a flying column in the direction of Mackinac. Duluth was in command, the confederation held, and the Iroquois commandos turned homeward without trying a fight.

This coincided with the new governor's effort to divide the spoils of office. La Barre, a puppet of the anti-Frontenac merchants and clergy, confiscated La Salle's forts and furs and cut himself in on the exultant return of the merchants to the upper country. Frontenac had exacted from the Iroquois ten years of better behavior than they had ever shown before, but they sized up La Barre as a weakling and a braggart. Under his nose and against his solicitation (Frontenac would have made it a command) they dispatched their raiding party to the Lakes. The next year, 1684, they captured a large consignment — seven canoes — of goods which La Barre himself was sending to the trans-Mississippi trade that La Salle and Tonty had organized.

This contemptuous piracy was incidental to the last all-out military effort the Iroquois made in the West, another campaign against the Illinois. They and Tonty, who was again at Fort St. Louis, got plenty of advance warning. "We repulsed them with loss," Tonty says imperturbably. "After six days' siege they retired with some slaves which they had made in the neighborhood, who afterward escaped and came back to the fort." The conquerors went home without a conquest, in fact with their whole plan for the West shattered forever. But La Barre opened a way for them to regain prestige and win victories in the East.

He mounted an expedition against the Iroquois, assembling an army of regular troops, militia, and the friendly Eastern tribes. But he did not know how to lead it, he was unable to provide it with supplies, and at Fort Frontenac, La Salle's post at the eastern end of Lake Ontario which he had confiscated, an epidemic broke out and seriously weakened it. Not daring to order the Iroquois to meet him in council there, he took his dwindling force on to the Iroquois side of the lake. More of them sickened there and in their own good time a delegation of Iroquois chiefs appeared. La Barre treated them with a mixture of threat, bluff, and anxiety which they understood at once. With stately insolence they ridiculed and defied him.[8] They told the governor that he would be wise not to uproot or endanger the tree of peace which they and his predecessor had planted. As for them, they would go where they wanted to and do what they pleased. La Barre, they helpfully suggested, must be talking in his sleep — these stories of Iroquois belligerence and these odd threats. They had no present intention of making war on New France but if they should make war on the Illinois and Miamis, why that would be none of the governor's business. . . . This reassurance about the colony could sound to a scared man like a diplomatic victory and La Barre marched his army home. But when the Western tribes heard the news it sounded to them like the double cross. It sounded just as bad in France and La Barre was dismissed.

In a year and a half he had lost New France the strength it had built up under Frontenac. And he had come close to losing it the West, for he had produced a momentous innovation and then had failed to use it. That is, he had called on the Western tribes to make their alliance offensive as well as defensive, asking for war parties to join his expedition and share his victory. That Perrot, Duluth, and La Durantaye (a first-rate army officer who commanded at Mackinac) were able to persuade them to try it shows how great an ascendency they had achieved. Some tribes refused the hatchet but Duluth swung the Hurons, Perrot the Ottawas, and the Winnebagos, Sauks, Foxes, and Menominees sent token forces. Over four hundred warriors all told made the long canoe journey from the rendezvous at Mackinac, and they managed their expedition better than La Barre managed his. The idea of large-scale co-operative action was new to these Indians, the farther they traveled the more heavily fear of the Iroquois oppressed them, and spectacular omens of misfortune scared them worse. But the *coureurs de bois* kept them going, fought down the panics, composed the quarrels, fanned the war spirit, and got them as far as Niagara nerved to invade the Iroquois lands. But there couriers from La Barre reached them and told them to go home; the new weapon that had been forged was not to be used. The Indians understood that the governor, the vicar of the Great Father, did not dare to attack the Iroquois and had been frightened into abandoning to them his allies of the Far Nations whom Frontenac, La Salle, and all the others had said he would protect.

They did go home and "French prestige in the West now touched its lowest point." [9] Disaffection and pro-British sentiment ran through the tribes and La Barre's weakness was giving the Iroquois the victory they had been unable to win. There were plunderings and much talk of massacre; intertribal feuds were renewed. And at this moment British trade reached the upper Lakes. There had always been a trickle of *coureurs de bois* who crossed over to the freer, more profitable Albany trade, and in 1685 one of them guided a party to Mackinac. It was a rich convoy, captained by an Albany Dutchman and financed by British merchants.

Its arrival at Mackinac dramatized for New France the grand strategy that La Salle had blocked out. From now on the Ohio River must be a military frontier: the British must be stopped south of it, to keep them away from the Great Lakes.

Jacques de Denonville, the army colonel who had replaced La Barre as governor, called Duluth back from another start toward the Western Sea, to build a fort on the vital access-route between Lake Erie and Lake Huron, the St. Clair River. But it did not control all the routes and the next year, 1686, a larger flotilla from Albany reached Mackinac, sold the cheap British goods to delighted French Indians, and took a small fortune in beaver home. . . . It could not be permitted. The British had returned to Port Nelson and from Charles Town, Carolina, the southern outpost of the British colonies, traders had reached the Cherokees and would soon get to the Chickasaws. It was now that the amazing winter foray set out for James Bay and seized the three Hudson's Bay Company posts there. Denonville also prepared to attack the Senecas, who had mounted the raids on the Illinois, had done most of the intriguing in the West, and were soliciting the rest of the Long House to join them in a total attack on New France itself.

The situation was so desperate and the colony was so weak that he felt only a miracle could save it but he seized the initiative. The ministry had sent more troops, not many all told but enough to double his small command. He left the new regulars to guard the home base, organized the militia round his veteran regulars, and drafted the Eastern Indians. Then he called on the Western tribes to make a second venture. Everything depended on them, on their joining him. But they were festering from La Barre's betrayal, the arrival of English goods was a powerful argument, and the incitation of the Iroquois to transfer their allegiance had more convincing logic than ever before. It was up to the *coureurs de bois* to save the French Empire.

Perrot was in the Sioux country when Denonville's summons reached him. He had spent the winter of 1685–86 north of La Crosse and in 1686 built the first post for the Sioux trade, on

Lake Pepin. They were a big market and here and at a later
post at Prairie du Chien he did a rich trade. His success outraged the Foxes, Mascoutens, and Kickapoos, who were already
rebellious, for the trade should have gone through them. Perrot
was superb, defending his posts with a handful of men, ranging everywhere, and purifying the hearts that had turned bad.
He beat the rebels into submission by sheer moral force, and
in a climactic council told the Foxes that he had closed the
road to the Sioux. "A few days afterward, I set out with two
Frenchmen to go across the country to [Green] Bay; and at
every turn I encountered some of these savages, who showed
me the best roads and entertained me very hospitably." He enlisted some of the chastened Foxes for Denonville's campaign,
rounded up the Potawatamis and the Menominees, and in person led them to Mackinac.

When he got there he found that a big victory had already
been won. Albany merchants had sent out two more trading
expeditions. As the first of them approached Mackinac, unaware that Denonville was making a war and that his allies were
mustering, La Durantaye captured it — fifty men, the Iroquois
guides, and twenty trade canoes laden with, Lahontan says,
goods to the value of fifty thousand crowns. The wavering Ottawas now saw a tremendous thing: Iroquois warriors forced to
stand in the bows of canoes paddled by bearded *voyageurs* and
to sing as ignominiously as any of their white or red victims.
The sight counted, and so did the exultant bullying of Perrot
and his men. Detachments of Ottawas and Hurons followed
Perrot and La Durantaye to Duluth's fort on the St. Clair.

The alliance was working. Tonty led to the St. Clair a hundred and fifty Illinois, a third as many Shawnees, and a scattering of Miamis. François de la Forest, once a lieutenant of La
Salle's with Tonty and now his partner, arrived with additional
bands. In their birchbark troop-carriers the confederated force,
unstable, jealous, likely to explode at any moment and make
war on one another, likely to be turned back by any omen in a
bird's flight or the shape of a cloud, but held firm by the masters
of Indian psychology — the Army of the West, more than five
hundred all told, moved off down the blue Lake Erie water, to
the silver flash of paddles and the bass grunts and falsetto ulula-
tion of the war songs. North America has seen no more striking
spectacle than this procession, the Far Nations off to fight for

New France, and the white masters who chivied and folded them like a shepherd's dog nosing his flock along.

Almost at once they met and captured the other band of Albany traders, with their renegade Frenchmen and their contingent of Iroquois. Tonty calls the two successes a great stroke of good fortune and says that "from the quantity of brandy and merchandise which they had with them [they] would have gained over our allies and thus we should have had all the savages and English upon us at once."

Lahontan, off to the wars again with his company, says that the Indians — meaning chiefly the Western ones — not only forced Denonville to continue the campaign when he wanted to call it off but saved it from disaster. That is Lahontan's bile, for the single battle was fought sagaciously by all divisions of the invaders except the Ottawas, whose customary discretion led them to withdraw from it. But it is true that Tonty, Duluth, Perrot, La Forest, La Durantaye, their *voyageurs,* and their Indians kept an ambush from turning into a rout and held the battle firm till the regulars and militia could come up and win it. It was a bad defeat for the Senecas and a sorely needed victory for Denonville. He marched on to the big Seneca town which its owners had already burned, destroyed the cornfields and the stored corn, and burned such forts and small villages as he could reach. Then he went back to Montreal, a victor but not enough of one.

The Western alliance had succeeded — so well that the one later attempt which the Iroquois made to break it up was easily repulsed. The West was saved — but for a time it seemed as if New France might be lost. For the Iroquois reaction to the invasion of 1687 was an outpost war that through the next two years harried the colony more terribly than anything since the crisis of a generation before. Trade came to a standstill, west of Quebec the whole colony was in panic, raiding parties annihilated villages, and no life was safe. Denonville crumpled, finding neither the wit nor the nerve to make advantageous use of such forces as he had. Finally, in August of 1689 at the village that had grown up at La Salle's seignory, La Chine, only six miles from Montreal, the Iroquois perpetrated what Parkman calls "the most frightful massacre in Canadian history." Denonville forbade a counterattack that might have butchered the drunken Iroquois in turn and, with two hundred dead and a hundred and twenty taken prisoner for the torture, all New France fell into a panic as desperate as his.

But by now the whole world was in flames. In 1688 the Glorious Revolution had put William of Orange on the throne of England, in 1689 the League of Augsburg had become the Grand Alliance. Louis XIV had no ally in all Europe, and everything that France had in men and money was needed for the war. There was but one step he could take to save the North American continent: he could send Frontenac there again. He did so and it sufficed.

New France was at the extremity when Frontenac landed, except that his presence reawakened hope, sobered the Iroquois, and flung the threat of his name as far as Mackinac, where the Hurons and the Ottawas were once more conspiring with the Iroquois to overwhelm the French. To hold Mackinac, the center of the upper Lakes, was his most urgent task. He depleted his meager strength to send a small garrison there. This reinforcement, the authority of his name, and the labors of the Jesuits and the wilderness leaders stopped the uprising in its tracks.

Frontenac had to make war with the scanty resources of a small, impoverished colony. It was a dire and fearful war, a bloody stalemate which for all its random killing got nowhere and had to be fought over again a few years after it ended. It so solidified the mutual hatred between the French and British colonies that there could never be a compromise but only a total victory for one or the other. It developed to a far greater fiendishness the raiding parties of Eastern Indians led by a French officer and stiffened by a few militia that the last war had used. Such commandos burst on the border settlements of New England and New York, killed, scalped, burned, and slipped away beyond pursuit. With the Iroquois Frontenac fought a war that fluctuated like the other one, while many men and women died, but it was more decisive. In the end the Iroquois had taken all the licking they could stand. They gave up their dream of empire and made peace, content to be a second-class power.

By the end of King William's War everyone understood the continental strategy of imperialism. It was the geopolitics that Talon, Frontenac, and La Salle had learned and declared, based on the St. Lawrence, the several crossroads of the Great Lakes, the Mississippi, and the Ohio. The energies of New France increased a great deal now but nowhere near enough to keep pace with the steady growth of the British colonies. They were expended mainly in the development of the lower Lakes frontier, the strong point of the lower Mississippi, and the strong point on its middle reach that came to be called "the Illinois." The effort was to pen the English colonies in their walled province east of the mountains, to keep their trade away from the Ohio and the Lakes. It had to be almost exclusively a military effort, though it had originated as a dream of colonization. And it was the military talent shaped and perfected by the wilderness that enabled the French to postpone their certain doom.

The military resource was paired with, in fact consisted of, the Indian trade. The West was the great market and its business was the first business of French America. Such energy as was left for expansion was mainly focused west of Lake Michigan and Lake Superior. This meant a new set of conditions, of which the controlling one now became a nation of Indians more important in the history of North America than the Iroquois, the Sioux.

# Chapter V

Map II.

# Converging Frontiers

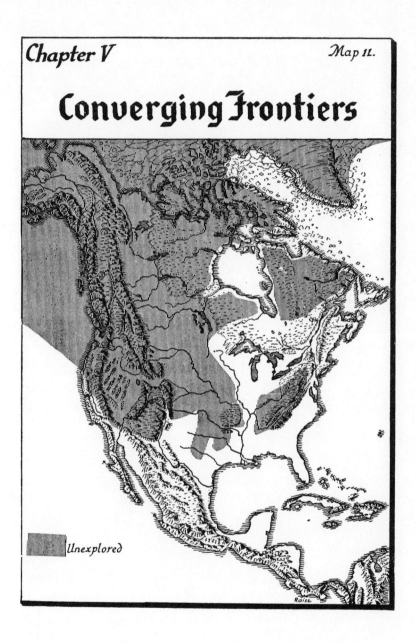

Unexplored

# V

# Converging Frontiers

THE POST which La Tourette had built at the northern end of Lake Nipigon was on the flank of the Hudson's Bay Company trade and had been established to attract Indians who made the long canoe voyage to the Bay. Many of its customers were Crees, who roamed in bands of little more than family size through the desolation of rivers, lakes, and forests west of the Bay. The Crees were always friendly to white men, they were a people of great intelligence and skill, and the arrival of manufactured goods had launched them on a wide expansion, north and west. Radisson and Groseilliers were the first white men who met them, and they had later traded at Chequamegon and the Sault, but the British were bringing them goods in much greater quantity than they had had before. Assiniboins also came to the Nipigon post. Those stodgy relatives of the Sioux were a Plains tribe, the first the French had ever seen,[1] but they were also canoe-users. The French called them Stone Indians, picking up an Algonquin designation that may have referred to a characteristic way of using hot stones in cooking, and they come into English records as the Stoneys. They were nomadic and though when the French first met them their country extended well to the east of Lake Michigan, most of it was west of there. So they came from the Northwestern Mystery, though as yet the French had heard so little that they hardly realized it was mysterious. Some of the western Crees raided into this area and many knew the country east of it, the black-spruce forest broken by many lakes.

Constant questioning of their visitors gave the French at

Lake Nipigon lively if vague ideas of this country, where obviously there were many furs and therefore a big market — and
where one would find the Western Sea and, at last, the route to
China. By 1688 an officer of the Nipigon post, Jacques de
Noyon, had acquired some understanding of the canoe route
the Indians took from it to Lake Superior, which they reached
at Kaministiquia, where Duluth had had a winter fort eight
years before. Heading inland from there, de Noyon paddled
up the Kaministiquia River, portaged the height of land, came
down to Lake des Mille Lacs of Ontario (which must not be
confused with Lake Mille Lacs of Minnesota), and went down
the watery labyrinth to Rainy Lake. He had thus traveled the
first stage of what was to become the great highway to the Far
West, though for a long time this stage would be abandoned
for a less troublesome route to Rainy Lake from Grand Portage. He built a post and spent the winter there and heard from
the Indians about the near-by sea and about a race of dwarfs
they were going to fight next year on the way to it. (Probably
Chinese and certainly out of place, for the true home of this
myth was on the Missouri below the Grand Detour.)   Then in
1689 he pioneered the next stage too, going down the eighty-
five-mile stretch of Rainy River to Lake of the Woods.[2] He
had made an extremely important exploration. At 94° W. —
the meridian bisects Lake of the Woods — he was farther west
than any Frenchman before him had gone and had almost
reached the eastern edge of the mystery. But the wars, white
and red, absorbed the energies of New France, for a time the
whole West was abandoned, and because of the dislocation
that resulted nothing came from his venture for twenty-five
years and not much for forty. Eventually it was followed up,
however, whereas nothing whatever came from the much more
remarkable exploit of a young "servant" of the Hudson's Bay
Company.

This was Henry Kelsey, who in 1689, after five years in the
North, was nineteen years old and who then and from then on
was a tremendous figure. The overland expedition that captured the three Company posts on James Bay in 1689 had griev-
ously shrunk the Company's rich trade. Only one post remained, York Factory, high on the western shore at the mouth
of Hayes River. Still refusing to venture away from salt water,

the Company built two posts out of reach of the French — out of reach, that is, of forays by land — one at the mouth of the Severn River, the other at the mouth of the Churchill. The latter was almost at 59° N.: hoarfrost four inches thick on the walls of rooms with big log fires in them, nine months a year on snowshoes, only a two months summer and then the soil, where there was soil, thawing only a few inches down with five feet of frost beneath. The crude post burned in 1689 but meanwhile the Company had sent Kelsey into the barrens to drum up customers among the Northern tribes.

He is unique among the English on the Bay: he was interested in the Indians and in the country beyond and back of beyond. He had traveled to the Bay in the ship that brought Radisson there on his second trip after he rejoined the Company, when the French had once more frustrated and defrauded him. He knew of the hundred-mile voyage up Hayes River Radisson had made in 1682 to bring in Indians. It would be impossible to know that master of the wilderness, who had worked out the orientation of the interior by sheer intelligence, without catching fire from him. Kelsey caught fire: he had the wilderness in his heart. He was the first poet known to have engaged in the fur trade; it can never have had a worse one, but he managed to express the need:

> *For many times I have often been oppresst*
> *With fears & Cares y$^t$ I could not take my rest*
> *Because I was alone & no friend could find*
> *And once y$^t$ in my travels I was left behind*
> *Which struck fear & terror into me*
> *But still I was resolved this same Country for to see*
> *Although through many dangers I did pass*
> *Hoped still to undergo y$^m$ at the Last.*

In the caste-bound, quintessentially British Company his liking for the "wild men," the red "niggers," must have made him despised and his desire to travel inland, though an Englishman, must have got him suspected of insanity. But the oddities proved useful, and he had already displayed an aptitude for wilderness life that no other Company man even faintly approached. When he traveled the barrens in 1689 he traveled more expertly than the Indian who was his only companion. The Indian, in fact, was scared most of the time, kept refusing to go any farther, made Kelsey speak in whispers lest the terrible Eskimos hear him, and told him he was a fool "for I was not sensable of dangers." The journey was fruitless but Kelsey describes a big animal he shot: he appears to have been the first white man who saw a musk ox.

The next year, 1690, the governor of York Factory took what was for the Hudson's Bay Company a revolutionary measure: he sent Kelsey inland to find the country of the Assiniboins, bid them keep the peace with the Crees, and solicit them to come in greater numbers to the Bay. Kelsey's long journey cannot be mapped in detail — his report mentions almost no identifiable landmarks — but there is no doubt about his general course. The two highways to the vicinity of York Factory from the Winnipeg country were the Nelson and Hayes Rivers. The Indians sometimes used the Nelson; later on when the trade developed, they usually and the Company always used the Hayes, since it was easier traveling, especially for the big York boats which replaced canoes, and led more directly to Lake Winnipeg. With a homeward-bound party of Crees he went up one or the other, probably the Hayes, since the factory was on its bank and Radisson had gone that way. They may have crossed over to the Nelson farther west — a much used later route did — but eventually they came to the web of tributaries and intercommunicating lakes that is spun round the big lakes of southern Manitoba. The long voyage brought them at last to Lake Winnipeg, probably, or perhaps the route may have gone north of it and on to Cedar Lake or Lake Winnipegosis, which are directly west of it.[3] They went on to enter the Saskatchewan, probably, or the Red Deer River that reaches Cedar Lake from the west. It does not matter: Kelsey must have been on the Saskatchewan this year or the next, and the Saskatchewan is the highway and focus of this area. He was the first white man to see it and would have to be called its discoverer except that he was unable to make it known.

This river too they traveled for a space and then they cached their canoes and started overland, southwestward. Kelsey was deep in the Northwestern Mystery now and he went on an indeterminable distance west of the hundredth meridian, traveling what no Frenchman and no Englishman had ever seen, the Great Plains. A very few small parties of New Mexicans had reached the plains east of the Rockies in Colorado and perhaps had gone on to the vicinity of the forks of the Platte, in Nebraska. If so, traveling east they had reached about the longitude where Kelsey now was, something over seven hundred miles south of him.

Constantly counseling peace and trade, Kelsey had encountered another branch of the Crees and eventually he met the Assiniboins.[4] With one or the other tribe he spent the winter deep in the plains, somewhere north of the Assiniboine River and west of Lake Winnipegosis. The next year he went back with his hosts to the place where he had left the river and sent word to York Factory, by a party who were going there to trade, that he needed more goods for presents. This summer too he spent wandering with the Indians about the plains, on his errand of pacification and advertisement, and somewhere in the plains he passed another winter. In the spring of 1692 he joined another party bound for York Factory and went down the long waterway again and reported to his chief, having been gone two years.

Kelsey had covered a sizable part of the most important area in fur trade history, though no one can be sure just what part. Almost certainly he had traveled both the Nelson and the Hayes and their tangle of related waters, Lake Winnipeg, Lake Winnipegosis, and the Saskatchewan. He must have seen the upper waters of the Assiniboine and may have gone as far south as $51°$, perhaps even farther. The place where he left his canoe, to which he twice returned, must have been at or near The Pas, a great staging point — a crossroads of the prehistoric intertribal trade, a depot on the route to the Company's posts on the Bay, and eventually a transfer center on the route to the far North and Northwest. He was the first white man who saw all this, and who saw the northern plains, the great plains buffalo herds, and any tribe of Plains Indians in their own country. His notes contain the first description of the Plains way of life that anyone not a Spaniard had written. And the poet speaks of

> an outgrown Bear w$^{ch}$ is good meat
> His skin to gett I have used all y$^e$ ways I can
> He is man's food & he makes food of man
> His hide they would not me it preserve
> But said it was a god & they should Starve

In his report this becomes "a great sort of a Bear w^ch is bigger than any white [polar] Bear & is neither White nor Black But silver hair'd like our English Rabbit." This is the earliest mention of grizzlies, which would be "grizzled bears" till about 1800.[5]

Kelsey could not have consorted with Assiniboins for so long without hearing about mountain ranges to the west, about the Rockies, and he speaks of the farther tribe he tried so hard to find as the "mountain poets." This early, then, the most important of all the facts about the unknown Far West, the dike that dammed off the water route to the Pacific, came to the attention of the British. Whatever idea of the mountains he may have formed, his report says nothing about them, and whatever he may have told the Company, it made no deposit on anyone's thinking. The Rockies disappeared in men's ignorance again.

In fact Kelsey's great feat was without issue or consequence. It shines like a shooting star in the annals of discovery and then the sky is dark again. At York Factory he "had my labour for my travell" and it was as if the first penetration of the northern plains had never been. The Company saw no reason to follow it up, no reason to send anyone after him — let the wild men come to the Bay. The trace he had made across the void vanished completely from awareness. Back of the again impenetrable curtain the Rocky Mountains ceased to exist and the great Saskatchewan River, which he had seen, reversed its direction and once more flowed to the Pacific Ocean.

ᘉᗧᓂ

The imperial frontiers in North America were captive to the forces Louis XIV had loosed. The one last war that would master Europe exhausted Europe but settled nothing. For two years after it ended diplomacy tried to create a stable alignment of the powers. The best hope of peace lay in the fact that for half a century Spain had been falling like Lucifer son of the morning and was now prostrate. Its possessions spread across Europe without logic of geography or nationality. If they could be satisfactorily distributed among the powers peace might follow like the well-being of a man who has dined well.

The solution depended on finding the right successor to

Charles II of Spain, who was dying but took an unconscionable time about it. He was the last of the Spanish Hapsburgs and childless; whoever succeeded him must have dynastic logic and must be willing to acquiesce in the partition of his estates. Louis had found it wise to join his late enemies, England and Holland, in working out a balance of power — which would put him in a position to master Europe with one more war. And the ambassadors ended by deciding that the Spanish crown should go to a German Hapsburg. He was nine years old and he died before Charles was able to, so that everything had to be done over. Another scaffolding of peace was laboriously erected; now the heir would be the Archduke Charles of Austria. Everything, however, must turn on the consent of the dying king, who destroyed the scaffolding. Remarking that only God could dispose of kingdoms for only to God did they belong, he willed the territories of Spain to a Bourbon duke, Philip of Anjou, and three weeks later died at last. It was November 1700. Philip was a grandson of Louis XIV.

The treaty which allied France with the two maritime powers had notably advanced Louis's Grand Design. But now the God to whom kingdoms belonged had offered Louis the chance to complete the Design with a single step, at the cost of no more than his honor and that of France. If he were to repudiate the treaty he had worked furiously to shape, Spain and France would be united under a single house, there would be no more Pyrenees, and at last just one more war, and that one a war against an unstable coalition, would stand between him and the mastery of Europe. The good faith of nations lasted a few days. On November 16 at Versailles Louis received the court of France with his grandson on his arm and spoke to three-quarters of the world: "*Messieurs, voici le roi d'Espagne.*"

The words began a war that lasted fourteen years. In Europe it was the War of the Spanish Succession. England had never had such a general as the one it now gave the Allies, Marlborough. Blenheim in 1704, Ramillies in 1706, Oudenarde in 1708 brought France down bleeding from wounds which a full century would not heal. But the victors demanded too much in reparation, Louis raised new armies, and the war went on. Malplaquet in 1709 was followed by a spate of small campaigns which spread pillage, rapine, spoliation, and disease till half of Europe was devastated. Peace parties formed in the fog of despair, Marlborough was betrayed and deposed, and in 1712 the plenipotentiaries met at Utrecht to restore logic to the world.

In North America it was Queen Anne's War, a border war, the New England border. The Iroquois had been tamed to discretion and neutrality; of the Long House only the Mohawks would fight and they only perfunctorily. But the Christian Indians — Caunawaughas, Abenakis, Micmacs — gave the Yankees ten winters of horror. The commandos raided across Maine, New Hampshire, and Western Massachusetts, and even Boston smelled the smoke of barns burning just twenty miles away. The garrisons and farmers died and the raiders slipped into the winter forest with a file of women and children who might be worth ransom money — and who gave New England a new literary form, the narrative of Indian captivity. Lack of organization, local politics, universal thrift, and the great north woods prevented effective retaliation. But following the massacre at Deerfield in 1704, the Yankees made an abortive attempt by sea on Acadia. A more carefully prepared one followed in 1707. It was all Yankee, it failed, and the next year the Canadians countered by breaking the English hold on Newfoundland. Now, however, the British ministry sent ships and troops to help defend the colonial possessions and this innovation marked a decisive turning point. For it made the border war an imperial war; all colonial wars would be imperial wars from now on.

The plan, that is, was not only to capture Acadia, a wasp's nest of privateering and a center of illicit trade, but to conquer Canada and add it to the British Empire. Never again would the awareness that Canada was an imperial possession weaken. At once New York, which had remained passive while the Yankee war went on and so had enabled its own borders to sleep sound, prepared for war. For if the conquest of Canada would give the West to the Empire it would give the Western fur trade to New York.

The Bavarian Corridor of the East, the predestined route of invasion north and south, was the Richelieu River, Lake Champlain, and the Hudson River. It had been the Algonquin-Iroquois corridor before the white man came, from Champlain on it was the French-Iroquois corridor, it was the intercolonial corridor as soon as there were hostilities between colonies, and now it became imperial. An army worked north to Fort Edward, south of Lake Sacrament which the English were to call Lake George. It stalled there for lack of direction and support, the first of several military failures which the site was to see. The invasion of Canada by land ended here but the sea-borne expedition started the next year, 1710, and the Yankees captured Port Royal for themselves and the British Empire. They had captured it twice before, only to find that they had spent their blood for nothing when it was restored to France, but this conquest was for good. Port Royal became Annapolis Royal, Acadia was to become Nova Scotia, and the British colonies and Empire were on the Gulf of St. Lawrence, the Atlantic portal to the continent, and would stay there. Now a single victory would complete the conquest of Canada. In 1711 the New England colonies raised another huge levy of men, money, and supplies, and the ministry sent transports and war-ships. It was to be Quebec itself now, and it might well have been for there was greater force than New France could have withstood. But the convoy made a late start and off the Gaspé Peninsula a disastrous storm struck it. Bad navigation and timid leadership cost a thousand lives, whereupon the com-manders turned craven with fear of the Canadian winter and sailed back to Boston. So God had wrought another of the miracles with which He was used to save New France. "Phar-aoh's chosen captains are drowned in the Red Sea," they sang at the high altar.

There was no more fighting — there could be none for Europe had to have peace. So the Congress met at Utrecht to bring the seventeenth century to an end. The sun of Louis XIV and of his people had set. He had begun by raising them to a pinnacle high above Europe; he ended by bankrupting their economy and their society, dooming them to a series of steadily more disastrous wars, and making the Revolution inevitable. France had to accept at Utrecht such terms as might be vouch-safed. The terms were not so harsh as they would have been if the great Whig ministry of England had not been overthrown by the Tories under Harley and St. John, who were almost as desperate for peace as Louis was. But those that related to North America were sufficiently drastic. The Iroquois were made British subjects. Newfoundland and Nova Scotia became British. But the most important article was the one by which France acknowledged British sovereignty over Hudson Bay.

The Peace of Utrecht, 1713, marks the true beginning of British imperial expansion. It gave Great Britain, which was already the greatest sea power, the most favorable position for colonial development. In North America the British held the southern bastion of the St. Lawrence access-route to the Great Lakes and the interior of the continent. Their new subjects, the Iroquois, held the southern shore of Lake Ontario, which threatened Canada and the West and by means of which one might be severed from the other. The way was cleared for the occupation of the Ohio River and the encirclement of New France.

The French promptly built the greatest fortress they ever had in Canada, Louisburg on Cape Breton Island, to counterbalance Annapolis Royal and what would presently be Halifax. Then they set about bolstering continental strategy. It remained a limpid simplicity and an iron solidity: the British must be held east of the Alleghanies. This meant in turn, as it always had, that the West must be developed. But more urgently than ever before it also meant that the unknown Western lands must be explored.

๛

The war had concentrated the energies of New France in the East. The Iroquois problem had been settled before the war began but this gain fell far short of offsetting a blunder that had preceded it. In 1696 fur trade licenses were revoked and another attempt was made to control the trade of the entire colony from Montreal, and to control Montreal from the colonial office at court, which could get news from Lake Superior, with good luck and at the right time of year, in six months. The revocation of licenses was unworkable and came close to being catastrophic, for it resulted in almost complete abandonment of the West.

The outlawed trader, gouged by his illegal but protection-paying principal in Montreal, was under pressure to desert to the British merchants at Albany, who charged less for goods and paid more for beaver. So were the Western Indians, who would not come to Montreal in force and were now vastly undersupplied. With the French coming west unsystematically and in small numbers, the tribes lost much of the discipline they had learned. Grown dependent on European industry, they would perish without iron for tools and weapons, cloth for clothes, powder, guns to shoot it in, and mechanics to repair the guns. The only occasionally broken peace which the French had imposed on them became anarchy as they resumed their old feuds and made new wars for what goods there were. Pursuit and flight set up new migrations and the population map of the upper country changed again.

Within five years the British trade and the menace of conquest had compelled the reoccupation of Mackinac, the lower Illinois River, the Chicago River, and the Detroit highway between Lake Erie and Lake Huron. These were of course dominant centers in continental strategy, and it was from the necessities of strategy that a new system of trade developed. As old posts were reoccupied and new ones opened, the commercial and the military functions blended. The commandant, usually a veteran of the border war or the campaigns in Europe or both, tended to get the license and become the resident trader; his garrison, always small and composed of militia, were also business hands.

The war had offered new careers to young men. There were both freedom and opportunity in ambushing farmers along the Piscataqua, or if one crossed the sea to the great war there might be a seignory for him when he got back. So the daring and adventurous took to the armies, or to the smuggling by which besieged and blockaded Canada supplied itself from the New England enemy, and the West could not be adequately reoccupied till the war ended in 1713. There had been nearly twenty years of vacillation, neglect, and blunders; the harm to New France proved permanent.

In 1715 the garrisons at Mackinac and on the Illinois River were strengthened, a post was built on the Maumee River to take care of the Miamis and to guard one route from the Lakes to the Ohio, and an old highway was protected by the construction of a new post on the St. Joseph River to cover the portage to the Kankakee. To abandon La Baye, the site that commanded Green Bay, had been a staggering mistake, for it not only controlled the route to the Mississippi by way of the Fox and Wisconsin Rivers but curbed the impetuosity of the belligerent Fox Indians. A new post was built and the first military garrison stationed there in 1717.

Now that Hudson Bay was English, the company named for it must be contained at any cost. The access-route to Lake Superior was secured by Mackinac but there remained two key sites. Chequamegon, near the western end, had been unoccupied since 1671, when the Ottawas and Hurons fled it in fear of the Sioux and the Jesuits had to follow them. A fort was built there in 1719, for the Sioux whose importance as customers grew steadily greater and if possible to end their Hundred Years' War with the Chippewas. Kaministiquia on the northwest shore of the lake, which was even more important, had been reoccupied in 1717. No one since de Noyon had gone up the canoe route to the Western Sea; now the French prepared to follow him and Duluth into the great unknown.

This took care of the lake frontier except for Lake Erie, which was anchored at the western end by Detroit and the Maumee post and was soon to be protected by others but remained vulnerable, and Lake Ontario, which could never be secured because of the Iroquois. A new trading post was built at Niagara in 1720 and it had to become a fort when the British trade reached the lake at last with the establishment of Oswego in 1724. One was built in 1726 — aimed square at the English trade, the English army, and the English settlements. It lasted as long as New France did and was fought for in all wars. The frontier outposts of the two empires were nearer each other here than anywhere else.

Though La Salle had recognized the importance of the middle Mississippi country, "the Illinois," the beginning of its development had been almost inadvertent. It attracted old wilderness men grown weary of snow. It was so far from Canada that, though the settlements formed round the mission at Cahokia, it was a refuge from the espionage and prohibitions of the Church. And since many routes of travel met and crossed here, the trader in advance of history who can never be identified or followed knew it before the settlements were made, and after they were made tarried here on his unchartable ventures nowhither. The tribes were a miscellany, Illinois, others who had settled here on solicitation from priests or traders, late arrivals from the population shifts following the Iroquois wars and the abandonment of the Western posts. The important ones were the Shawnees, the Miamis, the Mascoutens, the Tamaroas, and the Kickapoos. It was visited by tribes from the south and southeast and with more important results by others from the West, Missouris, Osages, and Iowas.

Though its population was indolent, the Illinois was an agricultural paradise. But its great importance was strategic — geopolitical. The Ohio, the Missouri, the Lower Mississippi, and the Upper Mississippi met here. Since it controlled the mouth of the Ohio, it could bar the English from the interior and might be decisive in keeping them from the Great Lakes. Since it stretched across the Mississippi, it held the north-south highway that made French North America a unity. And since it controlled the mouth of the Missouri, it was the entrance to the unknown West.

The last corner of the empire was Louisiana. The worst failure of the French colonial system, a victim of the world-war finance, straitjacketed in mercantilist and absolutist regulations, sodden with corruption, enervated by disease, it nevertheless remained a hinge of the continent and a stronghold of the empire. La Salle had not said, "Whoever rules the mouth of the Mississippi rules whatever people may live in an area of three million square miles," but only because he did not know how many square miles there were. The crown drew abreast of his principle in 1699 and founded Biloxi while English ships were actually entering the Mississippi. To dissuade others Iberville built a fort above the mouth of the river in 1700, and in that same year his brother Bienville went some distance up the Red River of Louisiana. (They were the most celebrated of five Le Moyne brothers, the most considerable family in Canada. Iberville had been second in command of the expedition that seized the three posts on James Bay; then, turning admiral, he had commanded the seesaw war for domination of Hudson Bay.)

In 1700 too Pierre Le Sueur, who had joined the colony at Biloxi, made an amazing journey. He had previously reached the Sioux country along the Great Lakes route and had maintained trading posts there. Now he sailed *"un felouque et deux canots"* up the Mississippi to the Minnesota River and up that to the Blue Earth. Here in southern Minnesota he built another trading post for the Sioux and began mining operations. Since his ore was the colored clay that gives the stream its name, the enterprise foundered, but it displays one of the fantasies that had mingled with realism in the founding of the colony. An afterglow of the vision which the gold of the Spanish Conquest had lighted all over Europe lingered on, and there was a conviction, which three-quarters of a century did not shake, that the wilderness beyond the Mississippi must be full to the brim with precious minerals.

For Louisiana was not only the southern bastion of a fortress whose purpose was to keep the British Empire east of the Alleghanies. And it was not only the southern anchor of the continental axis. It was also the imperial frontier beyond which to the west and southwest — somewhere — lay the Spanish Empire. Through all its stages — crown colony like Canada, leased private monopoly, corporate monopoly of the Mississippi Company which was an iridescent part of John Law's plan to rebuild the finance of France — through all three stages it was warped to the court's belief that the West was crosshatched with open veins of gold and silver. And, thinly populated and all but bankrupt, it was always to be the base for expansion toward the Spanish lands.

Spectacularly little was known about those lands. Spain had communicated almost no information about them to the world, the French had had no experience here, and the maps cartographers drew were almost entirely speculative. The maps, for instance, commonly give Santa Fe a latitude north of the mouth of the Illinois River, though it is more than three degrees to the south; this affected ideas about the routes to New Mexico. The interior topography was invented. Errors in longitude were enormous and they were increased by everyone's inability to conceive how wide the country was. Mexico and New Mexico were but foggily differentiated and the French in Louisiana were not always sure which of them they were trying to reach. They knew, however, that silver had displaced gold as Mexico's greatest treasure and that the mineral frontier had been moving northward. (By 1700 it had reached the rich deposits of Sonora and Coahuila, and shortly afterward those of Chihuahua were discovered, but there were no mines in New Mexico and so far as the Spanish had learned no deposits.) Louisiana was to complete the errand of De Soto by making contact with these treasures, absorbing as much of the output as possible, and discovering equal or greater wealth on the way.

Perhaps the Red River was a route to Mexico. Its lower reaches were explored for twelve years after Bienville's entrance and in 1713 Juchereau de St. Denis founded a permanent outpost and watchtower at Natchitoches in western Louisiana. The next year he tried a southwestern bearing, entered Texas which had been thinly garrisoned since La Salle's fatal venture alarmed the Spanish, and was captured and taken all the way to Mexico City. When he was courteously escorted back to Mobile (with a wife he had married while a prisoner) he had little information that was useful to Louisiana.

There was a fair trade with the Indians at Natchitoches but Spanish authorities outlawed the trade with Texas that had begun promisingly enough. The northern bearing of the Red River suggested that it might be a route by which the authorities could be avoided and the supposedly rich mines of New Mexico could be reached. In 1718 Bénard La Harpe was dispatched to find out. . . . On the Red and other rivers that came out of the West the French waterman entered a new phase. These were a new kind of river, turbid, choked with sandbars, sown with snags and frequently blocked with "rafts" of matted logs and trees, forever shifting their channels, now sluggish and too shallow for any craft, now deep and furious, subject to floods that took them far out of their banks. The canoe of the northern waters, which had opened a third of the continent to the French, was useless here and they had to take to the dugout. It was heavy, hard to control, and above all slow but serviceably tough, and this ungainly, snubnosed craft took them up the western watercourses — rowed, poled, pushed over the bars and through the shallows, pulled up the rapids with handlines.

La Harpe ascended the Red River some distance into Oklahoma, then crossed to the Arkansas, which he reached somewhere between Muskogee and Tulsa. He hoped to find the sources of both the Red and the Arkansas but was far short of both. (The Indians told him that the Spanish settlements were far away but he thought himself much nearer to them than he ever got. So, when they heard about him, did the Spanish.) He hoped to reach the Pawnees, about whom the French had been hearing for forty years, and the Comanches. The French understood that the Comanches, the Tartars of the plains, traded with the Spanish and they kept getting rumors about them in so many places, spread over so wide a territory, that they took them to be the most important Indians of the West. La Harpe met neither Comanches nor Pawnees but he did meet a number of tribes new to the French, find new potential markets, and learn about the Plains culture at first hand.

The great culture revolution of the West had spread widely: the revolution made possible by the horse. Tribes along the eastern and western base of the Rocky Mountains as far north as Canada had horses. All the southern Plains tribes had them; if one could speak of a "horse frontier" it could be seen as working northeastward toward the Missouri River in a wide arc. A tribe that had horses could follow the buffalo herds or make war for several hundred miles; it could go on horse-stealing raids, as the Comanches did, several times as far. A tribe that had horses could always defeat one that did not, unless the latter had plenty of guns, and one that had both horses and guns was a first-class power. Indeed the Plains Indians as light cavalry were more effective than any military force the white man could use against them till the middle of the nineteenth century. Part of the alarm which La Harpe and his immediate successors caused in New Mexican settlements was the fear that the French would sell the Indians guns, as the Spanish would not. It was justified; even without guns, raiding tribes, especially the Comanches, were terrorizing the Spanish borders.[6]

Like all later Louisianans who got far enough, La Harpe was ecstatic about the prairies and the short-grass plains. The country was wonderfully beautiful, with its "savannas," its river bottoms choked with berries and fruit trees and wild roses, its inexhaustible forage, its salt springs, its endlessly receding horizon, and the wind forever running across the bending grass. It was richer by far than all the mines of all the Americas. It begot the dream of a *pays* inhabited by peasants happily working a soil more fecund than that of France and incomparably finer than that of Canada, a land of Norman barns stretching to infinity and lasting to eternity. . . . It gave this dream to the briefly wandering representatives of a colony so poor that the colonists fought for preference when a bull, the regal gift of Versailles, was driven down the gangplank from one of the king's ships. France would never possess this fat land, would only pass over it like the shadow of a cloud.

La Harpe brought back a mass of information about the geography of the country toward which the French were moving. There was no way of orienting what he himself had seen. Much of what the Indians told him was true but practically all of it was misunderstood. The French had no experience by which to interpret what they heard — and they did have preconceptions, which derived less from the mysteries and marvels told them by the northern Indians than from the logic of cartographers. Lacking data, cartography and formal geography had to proceed by deduction and extrapolation. Estimates of longitude and even some determinations of latitude were grossly erroneous, the interior orientation was worked out from fictitious landmarks, and most of all there were fantastic assumptions. Of the last, the lively inventions that Lahontan had contributed to knowledge in one short chapter now began to exert their powerful influence. As a result the West, beyond the French penetration, was constricted, bent out of shape, and allowed but little relief.

The Missouri River was the master key to all the geography beyond its mouth. The French frontier had totally erroneous ideas about it, never corrected them, and bequeathed them in full flower to the Spanish when they took sovereignty over Louisiana. Of the thickly clustering misconceptions three were lethal: persistent underestimate of distances, the assumption that the Missouri's upper reaches (wherever they might be) were far south of where they actually were, almost as far south as Santa Fe, and ignorance of the nature and even the existence of the Rocky Mountains.[7]

No doubt traders from the Illinois River posts were on the Missouri before 1700: no doubt, but also no evidence. A few years after 1700 they were trading on such lower tributaries as the Gasconade and the Osage and at the mouth of the Grand. They may have reached the first big elbow, at the mouth of the Kansas River, before 1712 but this is unlikely. In that year,

## Map 12. The Heartland

L. of the Woods

Rainy R.

Kaministiquia

L. Superior

Sault Ste Marie

Mille Lacs

Brûle St Croix

Chequamegon Bay

La Pointe

Mackinac

Minnesota R.

Ft. Beauharnois

L. Pepin

Wisconsin R.

La Baye

Winnebago

L. Huron

Niobrara R.

SAND HILLS

Des Moines R.

Mississippi R.

Chicago

L. St Clair

Detroit

L. Erie

SAND HILLS

Villasur Massacre

Platte R.

Republican

Missouri R.

Ft. St Louis

Starved Rock

Ft St Joseph

Maumee R.

Ft. Miami

Smoky Hill

SMOKY HILLS

de Bourgmond

Kansas R.

Ft. Orleans

Illinois R.

L. Peoria

Ft. Crèvecoeur

Ft. Ouiatenon

Wabash R.

Arkansas

du Tisné, 1719

Osage R.

Cahokia

Vincennes

Ohio R.

Ft. Charles

Kaskaskia

Ste Geneviève

Cimarron R.

Canadian R.

la H.

OZARK PLATEAU

APPALACHIAN PLATEAU

Red R.

OUACHITA MTS

la Harpe

Ft. Assumption

Tennessee R.

Brazos R.

Trinity R.

Arkansas Post

San Juan Bautista

St Denis, 1715

Natchitoches

Sabine R.

Ft. St Pierre

Ft. Rosalie

Mobile

Rio Grande

Colorado R.

Matagorda Bay

Baton Rouge

Biloxi

Pensacola

New Orleans

Portage

Grass Forest

0   100   200   300

Miles

Raisz

however, a major figure arrived in the Illinois, Etienne de Bourgmond, who had in the greatest measure the skill with the Indians that the French in the Lake country were beginning to lose. As a young ensign he had commanded Detroit, succeeding the great Tonty; he had fought in the Fox wars and had lived the life of a *coureur de bois* to the full. Now he spent five years among the Missouri Indians. During that period he went up the river at least as far as the Platte and acquired a basic knowledge of all the tribes on the way, together with much information, necessarily of mixed quality, about more distant ones. The Missouris and Osages had already been attached to the French Indians; so had the Iowas, whom de Bourgmond met at some unidentifiable place near the Platte. He met the Kansas and at last the Pawnees, the most important tribe in these parts. He also met the Otos, probably the Omahas, and perhaps the Poncas — all of them displaced tribes. He got more news of the Comanches, whose country he estimated to be not far away — and in this he was not wholly wrong for though they now lived in the plains of the Southwest they raided northeastward almost as far as the Black Hills. Like La Harpe he had heard about the Arikaras and heard that they traded directly with the Comanches and the Spanish, which was not true. He won the friendship of all the tribes he met; and that the French never had Indian enemies on the Missouri, in sharp contrast to their experience on the lower Mississippi, was due primarily to him.

De Bourgmond's achievement has the élan of the great age, which by now was over. How far he went above the Platte is disputed and indeed he may have gone no farther, but that in itself was a great deed.[8] It does not matter: the mouth of the Platte was the equator of the Missouri, and it was to remain the farthest upstream reach of the French for many years. He appears not to have gone inland. He remained convinced that the Missouri would lead to New Mexico and there is no evidence that he heard of a route to it by way of the Platte.

Following de Bourgmond's exploit, French traders may oc·casionally have visited the Pawnees and the neighboring tribes but the direct trade usually got no farther than the Osages and the Missouris, who had the position of middlemen and were determined to keep it. Rumors of the Comanche commerce with the Spanish settlements kept on reaching Louisiana, however, and the need of finding out about it continued to be urgent. One of La Harpe's objectives was to find the Comanches, and in the year when he left the back country, 1719, Claude du Tisné was dispatched on the same errand. The Osages turned him back and when he tried again the Missouris did likewise, but he circled them and went on till he met a band of Pawnees on the Arkansas River, which was south of their own country. He also heard much talk about the Comanches but learned from the Pawnees, who had some Spanish objects that may have come from them, that though they did trade with the Spanish they were not allies. There for a time both exploration and information rested.[9]

∽

The Spanish reached the Platte from the west, it may be more than a hundred years before the French reached it from the east. In 1594, fifty-three years after Coronado, a small party made the second journey from New Mexico to Quivira, reached the Arkansas River in eastern Kansas, and then traveled north for twelve days to a bigger river, which would be the Platte.[10] They revisited Quivira in 1604 when Juan de Oñate, the colonizer of New Mexico, led an impressive company to about Wichita. Occasionally thereafter small parties of New Mexicans on scouting or punitive expeditions took the Pecos, Canadian River, or Cimarron exits from their northern piedmont and ranged to the Llano Estacado of Texas or the plains of southern Colorado. But they were not a mountain people: they did not venture into the Rockies until 1761.

Most of the New Mexican settlements were in Coronado's Tiguex — the valley of Rio Grande in the vicinity of Albuquerque — and around Santa Fe, which became the capital in 1610. This northern frontier was the farthest reach of Spain's imperial energy till a brief renaissance just before the American Revolution carried the Franciscans into California. The province of New Mexico was tiny, precariously held, and almost at the end of the world. A typical settlement would be a small congeries of ranches, a small presidio with a captain and a handful of musketeers, a church and a school for the Indians.

In 1700 there may have been five or six hundred Spaniards in Santa Fe and perhaps four thousand in all New Mexico. They had worked out an adaptation to the high plateaus that made the haciendas rich in at least horses, herds, and flocks.[11] The poor, which meant nearly everyone, were very poor and yet not wretchedly so for the soil was fertile and they had learned much from the Indians. Political and religious direction had to come all the way from Mexico City, in fact all the way from Madrid; the weary distance and the brittleness of the bureaucracy left them effectively free. They had individuality, courage, and toughness, and needed them for their society floated on an ocean of Indian hostility. But they had lost the pioneering spirit.

The subject population of New Mexico may have been seventy-five thousand Indians, who theoretically were Christian and obedient. The Spanish had learned to deal more gently with Indians, but no conqueror's hand is soft and the pueblo dwellers held to their own culture more obstinately than any other Indians. Even today the American industrial culture, which dissolves others like strong acid, has affected it only slightly; in the seventeenth-century Catholicism and the mission peonage made only a thin film over the tribal life. An unwarlike people, the Pueblos accepted conquest, hated the conquerors, and harkened to what the gods said at night in the kivas. In 1680 the gods said the hour had come. The Pueblos rose all across New Mexico, killing several hundred of their oppressors and driving the rest out of the land. For twelve years New Mexico was theirs again; they sprinkled the sacred cornmeal and sang the chants in the churches of the Christian God. Then in 1692 Diego de Vargas brought the army back, and the small image of Mary called La Conquistadora, which gave him victory and is still carried in procession through the streets of Santa Fe every June.

The reconquest was comparatively mild but the Pueblos had learned no fondness for being ruled. Some of them slipped away to canyons in the New Mexican distances and some went even farther, to the plains of eastern Colorado. There they built their adobe dwellings in the country of Apaches who were friendly to them and to the Spanish. These were semisedentary bands who had set up relations with the settlements. They came in to trade, had built up large horse herds, and gave promise, or so the Spanish thought, of becoming Christians. The hope was that they would be a buffer against more hostile Apache tribes which periodically raided the settlements.[12] The need of such a border defense had been greatly increased by the Comanches, who through most of the century had been moving down from the northern Rockies. Now they had permanently occupied the southwestern plains as their country and were too numerous and too fierce to be shut out.

In 1695 a band of Apaches coming in to trade at a village near Taos announced that their people were retreating on the settlements because they had been repeatedly attacked by — amazingly — Frenchmen. They said furthermore that "a large number of French" were marching in this direction, "towards the plains of Cíbola." [13] They were not entirely sure that the invaders were French but they were "white and ruddy," which could mean no one else, and they were most certainly headed this way. Thus, four years before Cahokia and Biloxi, word of the French on the Mississippi traveled the thousand miles to New Mexico for the first time. This stupefying news was forwarded at once to the Governor at Santa Fe, who dispatched a commission to find out all the Apaches knew. For this was an incident on the imperial frontier, it corresponded to the alarming landing of La Salle on the Texas coast eleven years earlier, and no one in Mexico or New Mexico had a realistic idea of how wide a space lay between Santa Fe and the Mississippi. (In fact the Spanish did not know where on the distorted map of North America, apart from the St. Lawrence and the Great Lakes, the French were.) When pinned down to facts, the Apaches admitted that they had not seen these white, ruddy men and had not been attacked by them. They had heard about them by the Indian underground, by the grapevine telegraph which, they said, had passed through seven nations on the way to them. It factored down to a story that "certain white men came to the bank of the water and made war on the people of Quivira and other parts, and presently they go away and return and make war and go away, and that it is very far off." The rumor was, then, that beside some water somewhere the French were fighting the Wichitas, whom no Frenchman had yet seen and none would see till La Harpe, twenty years later. There were no invaders, no white men were making war on the Wichitas or any other tribe, the border Apaches were not being forced west by white pressure. That is how the Indian news service worked, and the story is probably a roving correspondent's conception of the posts on the Illinois River. At most it could mean only a few traders crossing the Mississippi from those posts and traveling inland a little way with some of their customers.[14]

From that time on the frontier was always anxious about the threat from the east. Though the French had not yet made contact with the Pawnees, New Mexico heard in 1697 that they had joined with them to massacre four thousand Navahos who had tried to raid them. (Probably there were not four thousand Navaho fighting men and certainly they had never fought the Pawnees.) Two years later the Navahos said that they had some goods to show for it. They reported that the French, whom they could not possibly have seen, were magnificent fighters.[15]

Out on the Colorado plain the Pueblo *émigrés* grew disenchanted with freedom as Comanche raids increased and the almost equally vicious Utes began to harry them from the west. So in 1706 they sent an embassy to Santa Fe, making submission and asking to be escorted home. (They were Picuríes.) The governor sent a company of soldiers and Indians under his sergeant major Juan de Ulibarri. With them went a Frenchman who had deserted from La Salle's Texas colony twenty years before and had settled in New Mexico. From the divide above Taos they went down to the interior plateau drained by the Canadian River (they called it the Colorado), turned north across the Raton Mountains to the Purgatoire (their Las Animas), and sticking to the foothills of the Front Range reached the Arkansas River at about Pueblo. From ripening plums, cherries, and grapes of the piedmont they turned east into the plains, the trail marked with sods by the Apaches "who lose even themselves here." Ulibarri found the scared Picuríes in their adobe town, perhaps fifty miles down the trail, and before organizing them for the return he once more formally annexed eastern Colorado to the Spanish domain as Santo Domingo of El Cuartelejo.[16]

The Apaches who were co-tenants of El Cuartelejo invited

Ulibarri to go campaigning and help them kill some Pawnees, who they said lived seven days' journey away. (Not much more than a hundred miles, or less than half the distance to the nearest Pawnees.) They said that the Pawnees were allied with the French and that the alliance had been raiding them. They claimed that just a few days since they had killed a Frenchman and his Indian wife and that they had her scalp — he, alas, was bald — and his guns and personal equipment. But when pressed they changed their story: he wasn't a Frenchman, he was a Pawnee. They produced a gun, an antique model conceivably French, but nothing else. On the strength of it they asked Ulibarri to leave one of his guns, even a useless one, with them when he left. Evidently that was the purpose of the yarn: to bring the Pawnee menace closer to the Spanish so that they would sell some firearms. But Ulibarri accepted the whole story as true and reported it to Santa Fe, where it was added to the now sizable but entirely fictitious evidence of French pressure on the frontier.[17]

Through the next ten years the Spanish heard that that pressure was growing steadily. In actual fact there were only the trade on the lower Red River and that on the lower Missouri which did not reach beyond the Osage and the Grand. De Bourgmond did not reach the Platte till 1714.[18] Then in 1717 there was war between Spain and France (Philip V had seized Sardinia and Sicily) and all the frontiers must be fortified. In New Mexico the explorations of La Harpe and du Tisné, magnified by the Indian news service, looked like military invasions. New Mexico had to be armed against them and the loyalty of the Indians must be secured. In 1719 the governor of New Mexico led another expedition to the Arkansas River in Colorado, to intimidate the Comanches and reconnoiter the French. On the Arkansas he fell in with some Apaches who reported that they had recently — so recently that a wound had not yet healed — been shot up by another flying column of Pawnees and Frenchmen. The skirmish had occurred, they said, in northern Colorado on the South Platte River — within sight of the Rocky Mountains and five hundred miles farther west than any Frenchman had yet traveled. What was worse, the Apaches said that the French had built in that region two "large pueblos" each as big as Taos. There they lived with their beautiful white women and there they were building up a stock of munitions. They were training the Pawnees and the Jumanos (a hostile tribe far to the south whom the French knew only as a name on maps drawn in Europe) to use firearms and act as their hussars. It was another fantasy, told idly or in hope of presents; the facts behind it, if there were any, could have been only a brush with some Pawnees who had guns. But it convinced the governor and alarmed Mexico City when he reported it. The messenger who carried his report picked up on the way another bulletin for the viceroy: that six thousand Frenchmen (more than the total population of Louisiana and the Illinois in 1719) were already within a hundred and fifty miles of Santa Fe.[19]

To defend the frontier the viceroy ordered the construction of a fort — it would have a garrison of twenty-five — in Santo Domingo of El Cuartelejo, east-central Colorado. More than two hundred miles from Santa Fe, such a fort could not be held against anyone and the Comanches would cut it off. The governor of New Mexico told the viceroy so but sent another expedition to locate the hostile French. He should have led it himself for his lieutenant-governor, Pedro de Villasur, botched it disastrously. With a command of a hundred soldiers and Indians Villasur took the familiar trail to El Cuartelejo. Finding neither Pawnees nor Frenchmen there, he went looking for them on a course that took him farther into the interior than anyone from Spanish America had ever gone before. He found the Pawnees on the Platte River, in Nebraska.[20] They killed him, most of his troops, and many of his Indians. Survivors reported that there were Frenchmen among them and the governor made it two hundred, besides "an endless number" of Pawnees. But nobody claimed to have seen any Frenchmen; none could have seen them for there weren't any. Villasur had reached the country of the Pawnees and a band of them, helped by some Otos, rubbed him out.

They made their attack at dawn, August 13 1720. By October 5 tidings of it had reached the Illinois.[21] Now the French were alarmed, for the news reached them as a bulletin that the Spanish were marching on the Illinois, that there were two hundred of them, and that a large number of the terrible Comanches were marching with them. A year later the Indians told them that the Spanish were preparing to return, avenge their defeat, and build a fort on the Kansas River. Since such a fort would threaten all Louisiana and would probably alienate the lower Missouri tribes, it must be prevented. Bienville ordered the construction of a fort at some appropriate site on the Missouri, but that would not be enough — the Comanches must be detached from the Spanish and made allies of the French. But the lower Missouri tribes would object to such an alliance: they were all hostile to the Comanches and they would want to monopolize any trade that might be opened farther west. The only hope was a general pacification of the Indians all the way to New Mexico, a Pax Gallica of the plains. It was a grandiose idea and there was only one man who could possibly carry it out, de Bourgmond. He had gone to France after taking part in the capture of Pensacola by which, when the war broke out, Bienville had briefly driven the Spanish from the eastern flank of Louisiana and established the French on the southern flank of the British colonies. They sent for him.

In November 1723 with the small force that was all Louisiana could raise for him he reached the villages of his adopted people, the Missouris, at the mouth of Grand River. There he built a post which he named Fort Orléans and there during the winter he learned of a danger much more serious than any threat from New Mexico. The Otos and Iowas, mutinous because they had not been getting goods, were singing war songs with the now disaffected Sioux and the implacable Foxes, who had been raiding the French in Wisconsin and Illinois. This alliance, which foreshadowed the confederacy which the Foxes were almost able to put together a few years later, would have brought in the Pawnees and the Omahas, as de Bourgmond at once realized, and if that happened the French might be unable to retain any part of the Illinois, even its stronghold, Fort Chartres (some distance down the Mississippi from the mouth of the Illinois River, built in 1719 to replace Fort St. Louis). He was able to bring the wavering tribes back to their allegiance, by personal authority and plenty of presents. Then in June 1724 he set out on what he considered a futile mission, to make a treaty with the Comanches.

His summer and autumn journey took the French farther west than they had ever been before and they made a pageant of it. They pushed the heavily laden dugouts up the Missouri to the mouth of the Kansas River and some distance up that, then took to horse. Their banners, music, and military forma-tions delighted a large village of the Kansas Indians, with whom de Bourgmond held a stately council. The Kansas de-cided to combine their summer hunt with this opportunity to escort French cavalry to the territory of their toughest enemies. Three hundred warriors, three hundred squaws, three hundred travois-pulling dogs lengthened the column that pushed on through prairie heat and ferocious prairie storms. An epidemic traveled with it and eventually de Bourgmond fell so ill that he had to turn back, all the way to his new fort. His deputies pushed on, with letters of state for the Spanish and ransomed Comanche captives to be returned to their people. Word that they had reached the terrors of the plains and found them friendly brought de Bourgmond back posthaste from Fort Orléans. On his way he managed to bring together big delega-tions of Iowas, Otos, Pawnees, and Kansas for a council, by far the most impressive ceremony the Indians of the region had ever seen — rapturous oratory, presents that strained the solvency of Louisiana, presentation of royal banners, resolu-tions of universal peace and brotherhood.

He hurried on across Kansas, reached the short-grass country, and on the hazy October horizon finally saw the signal smokes of the Comanches. He met them in Quivira but the Wichitas of Coronado and Oñate weren't there any more; they had drifted into Oklahoma to get away from the tribe which de Bourgmond was now making an ally of France.[22] His envoys had already found out that the Comanches had never seen fire-arms and now it was clear that they had never before seen Frenchmen. They were impressed. The presents de Bourg-mond gave them were modest compared with the minimum lagniappe necessary in the upper country, but the chiefs had nothing to judge them by but what they had got from the im-pecunious and stingy Spanish. They promised to love the French and never to make war on any of their Indians. From now on, they said, the Spanish would be as the dirt and the French as the sun, but see to it that plenty of goods are sent our way. So de Bourgmond, his orders carried out, went back to New Orleans and on to France, where the letters of nobility he had so thoroughly earned were awarded to him.

Thus the Spanish farthest east and the French farthest west reached and a little passed each other. But the imperial frontiers approached no closer than they had been. As raids by the Comanches, Utes, and Apaches became more frequent the Spanish had to abandon the outposts they had maintained east of the valley of the Rio Grande and no frontier protection remained except a fringe of friendly Indian pueblos, which eventually had to be left undefended. They remained apprehensive of the French and believed that their province was open to military expeditions from the east. They resolutely refused to understand that the distance itself was the strongest possible military safeguard. When at last an army crossed their open frontier a hundred and twenty-two years had passed and the army was American, Stephen Watts Kearny's Army of the West in 1846.

And de Bourgmond's final effort had been wasted. Louisiana was not able to support even the outpost of Fort Orléans, halfway across the state of Missouri. In 1726 the garrison had to be reduced to eight men and two years later the fort was abandoned.

Trade on the lower Missouri increased but slowly, as the French had to fight the Foxes, the lower Mississippi River tribes, and even the Chickasaw from the southeast, who were British Indians. The economic structure of Louisiana was changing and the principal French effort reverted to the north. The Illinois did begin to prosper a little at last, and the trade with it. The long voyage to the upper Missouri could not be attempted and that to the Platte remained too expensive to be attempted very often, but in 1739, fifteen years after de Bourgmond, the dream of opening trade with New Mexico got at least a little fulfillment, when Pierre and Paul Mallet and six others succeeded in reaching Santa Fe.[23]

Like Coronado, de Bourgmond had been on the best route to Santa Fe from Louisiana, up the Missouri to the Kansas River and up that to a place convenient for crossing to the valley of the Arkansas — eventually the route of the Santa Fe Trail. The Mallets acted on the established, the unshakable conviction that the Missouri River would lead to New Mexico and they went up it as far as the mouth of the Niobrara in northern Nebraska, 275 miles above the mouth of the Platte. Here they were induced by geographical reasoning, topograph-

ical intuition, or advice from the Pawnees to abandon the convention they had acted on. They turned up the Niobrara for
some distance and then struck southwestward across lots. By
this tremendous detour, which now took them across the Platte
again, the Republican, and Smoky Hill Fork, they reached the
valley of the Arkansas in western Kansas. They followed the
river upstream to El Cuartelejo, where they found an Indian
who guided them to the Raton Mountains and down to Taos,
and they spent the winter in Santa Fe. Their presence here
was illegal; Spanish law permitted neither trade with nor visits
by foreigners. But they were welcomed as pioneers of a commerce which, for all the distance it would have to travel, would
not have to travel so far and would not cost so much as the
commerce New Mexico now had. The officials began to organize the formal smuggling system that would make the illegal
Santa Fe trade possible through a century only occasionally
interrupted by fits of patriotic honesty. In the spring one, perhaps two, of the Mallet party stayed on but the others started
back to Louisiana. Rightly dissatisfied with their outbound
route, they undertook to find a better one and, with help from
the Indians, they found one. Taking the Pecos corridor, they
reached the river which the New Mexicans called the Colorado
— from now on it would be the Canadian — and followed it
to familiar territory on the lower Arkansas. The commercial
contact had been made but the commercial dream turned out to
be an illusion. Though occasional parties got to Santa Fe again
there was practically no trade till long after Louisiana became
Spanish.

ॐ

The Platte was the equator and may be taken as the northern
boundary of empirical knowledge of the Missouri. How far
beyond it the French got is problematical but they did not
reach the Arikaras, whose villages were in the vicinity of the
White River, just north of Pine Ridge. Somewhere between
the Platte and the Arikaras knowledge ends and the mist closes
in.

Louisiana, to whose jurisdiction the Illinois had been added,
could not cross the equator. Mismanaged by the monopolies
that controlled its trade, impoverished by wildcat finance, ad-
ministratively paralyzed by the collapse of the French colonial

system, it was simply too weak. The French there were all but smothered by the vastness of the West. Add the Missouri River: swift, hurling its matted debris at any craft that entered it, perilous with snags and boils and sandbars and crumbling banks, the channels to be found only as they were come upon and always changing. The distance to the mouth of the Platte was a little more than six hundred miles, not a formidable journey for a people who had traveled all the North, but infinitely slow and laborious by dugout.

The Indians too slowed and eventually stopped the French advance. Beginning with the Kansas, at the mouth of the river named for them, they grew capricious and no one after de Bourgmond had authority over them. The Omahas were volatile and treacherous and could determine policy for the Otos and the Poncas. Like all Indians who saw a chance, they wanted to control the power and wealth of the trade in European goods; it must pass through their country and they could harry it like Barbary pirates. Eventually they were able to close the river.

The distant Foxes were even more important. Shortly after 1700 their truculence and pro-Iroquois, pro-British activity convinced the government of Canada that they could be tolerated no longer, they must be exterminated. Nearly forty years of war followed, during which they sometimes seemed likely to drive the French out of the West. They were the central problem from La Baye to Fort Chartres, slaughtering Frenchmen, disrupting the trade, driving the friendly tribes in on one another. While the war lasted they were a barrier stretched across Wisconsin; penetration of the West from Lake Michigan was impossible. It was primarily a Canadian war but the Foxes and the allies whom they got from time to time raided into the Illinois and so absorbed energy that Louisiana might otherwise have been able to put into a western thrust. But another result was more harmful: the Foxes changed the Indian conception of the French. They proved abundantly that Frenchmen were neither spirits nor supermen; they could be terrified and defeated. By mid-century the tribes between the Kansas and the Platte had learned to treat the occasional small parties of traders who came their way with an arrogance and contempt that no Frenchmen in North America had ever been exposed to during the seventeenth century.

Imperial expansion had to lie farther north. As the irrepressible conflict with Great Britain for the possession of North America drew nearer, the French were under the increasingly critical necessity of breaking through to the farther West and whatever it might contain. They had to reach the Pacific by some route, whether the Northwest Passage, the Strait of Anian, or the Sea of the West. The effort would have to be made from Canada. The older colony was much sounder and stronger than Louisiana and, after the Peace of Utrecht, had entered the period of its greatest growth. And the great route west was Canada's. Chequamegon, at 91°, was actually farther west than Fort Chartres; Kaministiquia was not quite so far but from it a water route, of which a long stretch was already familiar, led west into the unknown. From Montreal to these strong points the route passed tribes who had been known since 1660 or before and all of them except those occasionally seduced by the Foxes were friends. The Chippewas, whose loyalty never wavered, were at the western terminus. Just beyond them were their enemies the Sioux, who stretched between the French and the farther West just such a barrier as the Foxes did in Wisconsin. They were much more numerous than the Foxes, they were fully as warlike, and they thought more intelligently. They had periods of hostility and had sometimes joined the Foxes, whom normally they hated almost as much as they did the Chippewas. But they were frequently friendly and there was the economic bond, for they desperately wanted trade goods, especially guns. From now on the Sioux were to be a decisive influence on every move in the search for the West. Their great cultural separation was already in progress. The Chippewas had been slowly forcing them westward and southwestward from Wisconsin, northern Minnesota, and Lake Superior. For those at the western fringe of this forced migration, the Minnesota River had become a political and cultural road fork. At its elbow, which is where the Blue Earth River reaches it from the south, some turned north up the river toward the great Minnesota forest and others kept on moving slowly west, toward the plains and the Missouri. These last would become horse Indians and embrace the Plains culture.

As the French abandoned their efforts to ascend the Missouri, it remained thickly encrusted with misconceptions. The most durable and costly of these was the belief that its headwaters lay somewhere near the Spanish towns. The French began by misconceiving the latitude (as well as the longitude) of New Mexico and then, when they came to understand it better, concluded somehow that Spanish settlements must extend north of those which they knew about. Or if not Spanish, then some other civilization. The Jesuit Pierre Charlevoix, a trained and sagacious scholar who was sent from France to find out all that was known here about routes to the West, phrased this idea clearly: "I have good reason to think that there are in this continent either Spanish or some other European colonies much more to the north than what we know of New Mexico and California, and that after sailing up the Missouri as far as it is navigable, you come to a great river which runs westward and discharges into the Sea." There are no mountains in between, and it is striking that, this late, he not only can believe that there are colonies in California and so demonstrate that the name is still a mere word to Europe outside of Spain but can seriously assume the possible existence of colonies entirely unknown to anyone except the nation that founded them. That was seven years after de Bourgmond reached the Platte, three years before his journey to central Kansas.

Another basic misconception resulted from the persistent failure to understand how wide the continent was. In 1717 the Sieur Hubert of the Council of Louisiana reported to the colonial office on the desirability of exploring the Missouri. One sentence of his report has sometimes been interpreted as showing that this early the French had already learned from Indians the master reality of their geographical search, a river flowing to the Pacific from the height of land. It is not that, for they had not, but is instead a deduction from the assumptions of symmetrical geography. Hubert says that the colony has heard "of a large river which is understood to issue from the same mountain [range] that holds the source of the Missouri. It is believed, even, that a branch of it falls into the Sea of the West." (The Pacific is the Sea of the South: this river would fall into the interior sea.) He says that the Canadians whom he proposes be sent west would discover it "very soon" and that they would be able to open commerce with China and Japan, "the way to which would be short." Hubert proposes that the Canadians be sent up the Missouri not only because China is just a little farther on but because the river is the best route to

the Spanish lands and mines. "They would have to go *only* to
the upper Missouri," he remarks, and at least the mines "are
said to be *northwest* of that river."

Here, with everything just a little farther on, the mist closes
off the Missouri. Behind the curtain were China, the Western
Sea, the River of the West, the Northwest Passage, and people
of great interest. In 1718 someone heard that de Bourgmond
was going to make discoveries a thousand or twelve hundred
miles upriver from his favorite Missouris, and was going to
open the trade with a populous nation who lived there. They
were little men with big eyes set an inch out from the nose.
They dressed like Europeans and wore boots studded with
gold. They lived beside a large lake (which would be the Western Sea or perhaps Lahontan's invention), devoted themselves
to fine workmanship, and had much gold and many rubies.
These people had been reported to de Noyon much earlier
and far to the north, where the Crees described them as three
and a half or four feet tall, heavy-set, and kinky-haired. The
Crees said that they lived in fortified towns beside the Western
Sea and added that they were white men and wore beards. The
Chevalier de Beaurain, who annotated La Harpe's report,
located them more precisely: they were fifteen days beyond the
Arikaras, who he thought lived just above the mouth of the
Kansas River. They were white, he said, they worked metals,
and they traded with the Comanches. De Bourgmond found
that they had withdrawn farther up the Missouri; they were
beyond his highest ascent, up near the river's forks. (It had no
forks.) They were called the White Omahas but they weren't
dwarfs, they were the handsomest Indians on the continent.
They were nomads and wandered along both sides of the river,
and they did not make war on anyone.

The myth had developed an intricate structure by now:
dwarf Indians, white Indians, bearded Indians, of an advanced
civilization, in cities beside an interior lake or sea. All Europeans had always heard rumors about such Indians just a little
farther on, and these particular variants were shaped somewhat
in French thinking by Lahontan's bearded Mozeemleks and
splendid Tahuglauks, and by the romancer Sagean's more magnificent Acaanibas. But now there was some fact too, from the
very center of the Northwestern Mystery. The French had
begun to hear about the Mandans.

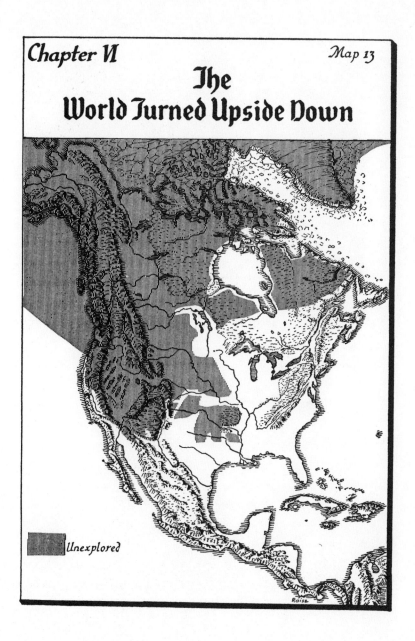

# The
# World Turned Upside Down

Unexplored

Raisz

# VI

# The World Turned Upside Down

AT MALPLAQUET, the Duke of Marlborough's "very murdering battle" — many such battles have been fought near Mons — a Canadian who was a lieutenant in the Bretagne Regiment received a bullet wound and four saber cuts. At twenty-four he had already served two years in the European war and exactly half his life had been spent under arms: he was twelve when he joined the Canadian forces and nineteen when he went on the long raid through the winter forest that ended with the massacre at Deerfield. After four years more in Europe, he carried back to his birthplace at Three Rivers the scars of nine wounds suffered in the King's service. He was Pierre Gaultier de Varennes, Sieur de la Vérendrye, the last of the great line that began with Champlain.

To any resident of Three Rivers, from childhood on Indians from the upper country were as familiar a sight as priests from Quebec and stories of the farthest frontiers as familiar as the legends of the saints. One grew up with the forest and the white water coloring one's dreams and to go west was the noblest ambition. Vérendrye joined the family fur business at Three Rivers, where four sons were born to him. He needed advancement, he wanted to make discoveries, and he came to understand how Canada could checkmate the Hudson's Bay Company. He applied for the post at Nipigon; the Marquis de Beauharnois, who was now Governor of Canada and one of the best it ever had, got it for him. This was in 1728; Vérendrye was forty-three.

The Nipigon post carried with it control of the one at Kaministiquia, which had been re-established by Zacharie de la Nouë. (He had gone all the way to Rainy Lake but had been

forced back to Lake Superior by the Sioux-Cree hostilities.)
The two posts were a valuable holding; with Chequamegon
they got most of the fine furs of the Northwest. And their customers were a source of information about the farther West.
By the time Vérendrye went to Lake Nipigon the whole route
to Lake Winnipeg was understood and some fairly accurate
ideas about Lake Winnipeg had been worked out, though inaccurate and fantastic ones remained.

Two hundred and fifty miles from Chequamegon and more
than three hundred from Mackinac in an air line, farther by
canoe, the Nipigon post was on the edge of beyond. What else
was it on the edge of? In a six months' winter when the temperature might not rise to twenty below zero for a month on
end, there was time to inquire. Dressed in skins or Hudson's
Bay Company woolens but most welcome when they wore the
robes of stitched beaver skins that in a year or two would become *castor gras,* the most prized of all furs, the Indians gladly
bartered for his goods and gladly told him about beyond. Made
affectionate by brandy and warmed by the big log fires, while
auroras flamed beyond the stockade or the fort shook to gales
out of the Arctic, they told him what they knew, what they
had heard, what they guessed, and what it amused them to invent. They scratched maps on pieces of bark or hide or in the
ashes of the hearth. In a year Vérendrye had heard all that his
predecessors had heard and a great deal more.

There were many lakes. There was a treeless country. There
was a little mountain which glowed night and day. (*"La
montagne dont la pierre luit jour et nuit."* It was only one
of the five-hundred-foot-high hills that lie on the Manitoba —
North Dakota boundary and are called the Turtle Mountains.
It was not as yet the Rockies, which remained west of rumor,
but it would end by giving them one of their earliest names, the
Shining Mountains.) There were copper, lead, many minerals,
many ores. There was a nation of little men, three feet tall and
very brave. But the tremendous news was that beyond this
maze of lakes was a great river which flowed west, apparently
out of Lake Winnipeg, and at its mouth was the sea. Here then,
Vérendrye decided, many decades of rumors came true. This
must be the route to the Western Sea and the journey that
Jacques Cartier had begun for France in 1534 could be brought
to triumph in 1730.

*Map 14. The Canoe Route L. Superior~L. Winnipeg*

Vérendrye had a soldier's direct mind. His entire career shows that he was telling the truth when he wrote to the supercilious Minister of Marine, the Comte de Maurepas, of "the zeal by which I have always been animated for the King's service and especially for the discovery" — of the Western Sea — and added, "Money, Monseigneur, was moreover always a secondary consideration with me." If it had not been, he could not have endured for so many years the gouging and fraud of the partners who were forced on him and of the merchants to whom his business and his time were always mortgaged. He was charged with making the great discovery for the King, ordered to finance it by the fur trade, forbidden to make a profit or even recoup his expenses from it, and constantly rebuked and censured for not more rapidly destroying his own business in order to be about the King's which it financed. He was the pre-eminent thinker and most skillful leader of the fur trade of his time. He worked out the only possible check to the system that had been based on Radisson's ideas. He established the system by which the fur trade, the fundamental business of

Canada was conducted down to the merger of the Hudson's Bay Company and the North West Company in 1821. He opened the richest fur area. He understood the political economy of the trade far better than the merchants, the government of Canada, or the Ministry of Marine and Colonies. But he could not deal with its finances or its graft — if there was a difference. He never had a chance.

He summarized his information, sent it to the governor, Beauharnois, proposed to make the inestimable discovery, and asked for help. What Vérendrye reported to him could be rationalized to agree with and extend the total sum of what the French had heard about the West. What counted more, it agreed with the theorems of geography in which the governor and the scholars at Quebec were adept. They worked it out that the great river which Vérendrye had heard about was the fourth, and so far missing, one that the continent required for symmetry. From a fundamental if as yet unvisited height of land the Mississippi flowed south, the St. Lawrence east, and the Colorado to the Vermilion Sea. (To the Gulf of California, and the Colorado had at last been given its name by the great Spanish Jesuit Eusebio Kino.) From that same height of land a river must flow west, else nature would be unsymmetrical, which was against reason. Here is a fundamental postulate fully expressed: a pyramidal divide from which four rivers took the drainage of North America to the oceans. The height of land would have various locations and the rivers would flow in various directions but from the early eighteenth century on the conception, which contained implicit contradictions, always influenced thinking about the West. And now thinking about the Sea of the West became remarkably ambivalent. Sometimes it was thought of as it had been originally conceived, a big interior sea or the widening of an arm of the Pacific that began as a strait and penetrated deep into the interior. Sometimes it was thought of as the Pacific itself, the Western Ocean, the South Sea. Sometimes it was both.[1] With similar ambivalence Beauharnois was able to hold mutually contradictory ideas, for if the River of the West flowed from this cardinal height of land it could not also flow from the Western Sea.

He knew there was empirical proof that this logically derived river existed, for lately — *"nouvellement,"* but actually 127 years earlier, in 1603 — the Spaniard Martin Aguilar had seen the mouth of a big river on the California coast. (This was one of several supposed river mouths on the coast that either did or did not get assimilated to the Strait of Anian.)² And he saw further merit in Vérendrye's news in that it supported the conclusions of Father Charlevoix. And since Charlevoix, Canada had been taught the extreme desirability of finding a route to the sea that would avoid the Sioux.

In 1720 the learned Jesuit Charlevoix had been sent to Canada, where he had previously spent some years, to collect all that was known about the far country, appraise it, and determine the most promising route to the Western Sea.³ On a sweeping tour of the West he talked to every Frenchman and Indian who presumed to know anything. He arrived at absolute conclusions. The Sea of the West will be found to the southwest of the Lake of the Assiniboins — by which he probably understood Lake of the Woods.⁴ It is therefore west of the Sioux but so near their country that there is only one tribe in between. Europeans have been seen on its coast but it is an inland sea which communicates (somehow) with the Northwest Passage and so (somehow) with California. Having thus compressed geography by bringing the Western Sea close to the western frontier, Charlevoix then gave it logical neatness. The Mississippi, the Minnesota River, and the Missouri, he decided, all have their sources in the same height of land. . . . When he wrote the Mississippi was familiar for some distance above the Falls of St. Anthony, the Minnesota to its elbow, and the Missouri to the mouth of the Platte. Cartographers had fluently imagined their upper reaches, which logically could not be very long. The proximity of the sources of the Missouri and the Mississippi was an obstinate and long-lived idea, and their location on the same height of land was an elastic error easy to build upon, as Beauharnois built on it by adding the River of the West.

On the basis of his study Charlevoix proposed alternative ways of reaching the Western Sea. One was to ascend the Missouri, "whose source is certainly not far from the Sea." The other would be slower but more thrifty since it would make discovery self-supporting: to re-establish the Sioux post, found a mission there, and work west from that base. The Sioux would occasionally bring in captured Assiniboins, and the Assiniboins were understood to trade with the Iowas, "who live near the

Missouri and know all about its upper reaches." [5] (The Iowas did not know those upper reaches and the Assiniboins did not trade with them.) Charlevoix's information was, of course, entirely erroneous and his reasoning therefore was fantasy, but nevertheless the Missouri was one of two routes that would have led to the Pacific. The weakness of Louisiana and the dislocation produced by the Fox wars had made the ascent of the Missouri impossible, however, and the alternative plan was adopted. In 1727 a post was built and a mission founded at Lake Pepin in the Sioux country. That was three years before Vérendrye made his proposal and though the Sioux had remained friendly, their friendship was not affectionate enough to inspire confidence.

Beauharnois decided that it all fitted together. So he and the Intendant Hoquart forwarded to the colonial office Vérendrye's request that he be authorized to open trade in the Lake Winnipeg country. The Sea of the West, they told Maurepas, was from ten to twenty days' journey beyond Lake Winnipeg, though with serene ambivalence it was also seven hundred leagues, 1750 miles. The post there would get furs that the Crees were now taking to Hudson Bay. The enterprise would be only a slight expense to the crown. They did not allude to the important additional fact that this route west would be outside the Sioux country. It was a two-edged fact, for the Sioux would not like the opening of trade in the country of the farther Crees and the Assiniboins.

However slight the expense, it was more than France would assume, as Maurepas at once made clear. He would appropriate no money to win the West and Asia: let Vérendrye pay his own way, and let him not try to turn a profit. He was granted a trading monopoly of the country he was to open up. This tied him to the fur merchants, which was equivalent to commercial peonage. From now on the Post of the Sea of the West — the collective name of the posts that Vérendrye established and the ones that were tied in with them as time went on — composed the richest holding in the trade. The merchants got the profits, Vérendrye was always in debt. They broke their contracts with him, bribed his men, kept him half-impotent for lack of supplies. Because the licenses and privileges which he controlled as commander of the West were valuable, they filled the governor's ears with lies about him. The same lies went to Maurepas, Louis XV, and the circle of wits, gamblers, and

courtesans who governed France. Beauharnois knew Vérendrye's value and supported him. But Maurepas sat in Versailles,
polishing epigrams and moving dividers across fantastic maps.
He perceived that Vérendrye would have reached China long
since if greed had not led him to disregard the crown's purpose in order to steal the crown's furs. This witty mediocrity
had seen nothing at the court he had grown up in to suggest to
him that honor or patriotism could move a man. Beauharnois
answered his criticism with daring sharpness but Vérendrye was
ordered to give up trading for himself, was repeatedly made to
feel the minister's contempt, and was repeatedly forced to make
the three months' canoe trip to Montreal to defend himself and
even to get supplies for the posts that sent the merchants six
hundred packs of beaver a year. Thirty thousand beaver pelts.

He spent his full income from licenses and ran hopelessly
into debt. He had a few men of his own, a penurious and un-
dependable amount of Indian presents from the governor, and
the traders and *voyageurs* of the merchants. In 1731 he began
the great enterprise by pioneering a new route through the
clotted lakes, swamps, bogs, and creeks to Rainy Lake. His
predecessors had gone by way of the Kaministiquia River; his
route, much easier though it had more portages, was by way of
Pigeon River and what was to be permanently known as the
Grand Portage. A bought mutiny kept him on Pigeon River
this year but he sent one of his men and his oldest son, Jean-
Baptiste, on to Rainy Lake, at the outlet of which they presently built a post named Fort St. Pierre. Rainy Lake, which
was named for the mist of a large waterfall, was 225 miles west
of Lake Superior: the thread of communication with Montreal
now stretched across twenty degrees of longitude.

The next year, 1732, Vérendrye pushed on down the beautiful Rainy River a hundred miles by canoe, to Lake of the
Woods, and there built the post that was to be his headquarters, Fort St. Charles. It was on the west shore of this
multiple lake, in the Northwest Angle, the detached part of
Minnesota that has only a water connection with the United
States. In March 1734 he sent Jean-Baptiste down the furious
water of the Winnipeg River to Lake Winnipeg, and in the
fall sent him there again to build a third post, upstream from
the mouth of Red River which empties into it. It was called
Fort Maurepas.

This was momentous. Fort Maurepas was on one principal

route of the Hudson's Bay Company's customers, close to even more important routes, and close to the crossroads of the richest beaver country. All three posts were in the area, between Lake Superior and the Red River, of which Alexander Mackenzie said there was not "a finer country in the world for the residence of uncivilized man." And when they reached Lake Winnipeg and the Red River, farther west than they had ever had posts before, the French had come out of the northern forest. They were on the edge of the plains, where the conditions of life were fundamentally different from those that had governed the Canadian trade up to now and where the Indians had a different culture. The Assiniboins were a Plains tribe, so were the Western Crees, beyond them to the west were the Gros Ventres or Fall Indians and the Blackfeet, and the native trade routes extended to many other tribes.

For three years Vérendrye developed the trade at these posts and did not explore beyond. Jean-Baptiste had gone down the Winnipeg River at least once but its "shocking rapids" were desperate going and he had looked for and found an easier route. It led through the morass west of Lake of the Woods to the Roseau River, which empties into the Red well upstream from the mouth of the Assiniboine. There had been no reason, then, to survey the confused marshy delta of the Red. As late as the spring of 1737 the eastern shore of Lake Winnipeg as far as the mouth of the Winnipeg River was but dimly understood, and beyond the Winnipeg it was unexplored. At that time none of Vérendrye's men appear to have gone up the western shore of the lake or to Manitoba and Winnipegosis — there was no need to, the customers were coming in.

Vérendrye kept getting information but it was confused, and though some cardinal items had been cleared up, he projected farther west the same confusion he had brought with him. He learned now that the great river which flowed from Lake Winnipeg to the sea flowed northeast, not west — it was the Nelson. Since it led to Hudson's Bay Company territory he had no concern with it except to intercept the furs. But he kept on hearing about a great river that flowed west, which must be the one he had started out to find. The trouble was that in the accounts he got, which he interpreted as referring to a single river, at least three rivers were being reported, the Missouri, the Saskatchewan which flows east, and an imaginary one.

With too small a company and too thin a supply of goods, limed not only in the treachery of the merchants but in their fears as well, Vérendrye managed to keep in business — and to send rich yields to Montreal. Maurepas fumed at what seemed his greed, dishonesty, and inefficiency — surely all the man had to do was to travel ten or twenty days and find the Western Sea. Meanwhile Louis XV had got France involved in a war that was the prelude to a prologue, the War of the Polish Succession. The aged Cardinal Fleury kept it from becoming a general war and hostilities did not extend to North America, but war was always bad for the fur trade. Goods reached the traders in smaller quantities and they cost more. The Fox wars that Beauharnois had inherited went on spreading blood and terror across the West. Disaffection increased among the tribes and the Ohio Valley grew ripe for the English traders. They were coming into its eastern margin now from the Carolinas, from Virginia, and in greater numbers from Pennsylvania. And on the Champlain corridor the French had fortified Crown Point, where the lake narrows to half a musket shot.

Vérendrye proved himself a master of Indians: he had the temper of command that is so marked in Duluth and Perrot, La Salle and Tonty. His tribes were the Assiniboins, the Crees, and their near relatives the Monsonis. They were Hudson's Bay Company customers but he got their best furs and as many of the rest as the scarcity of goods permitted. The trouble was that all three tribes periodically fought the Chippewas, who were located on the Lake Superior abutment of the long route to the West, and all three were inveterate enemies of the Sioux, who were within raiding distance of that route throughout its whole length. The endless war between the Chippewas and the Sioux also ranged through much of the country in between. They raided each other and both raided Vérendrye's tribes, who raided both in turn. All these tribes wanted guns and powder and iron, all resented the sale of munitions to their enemies, and all were a powder keg of jealousy and vengefulness that any spark could explode. The trade, the colony, the Empire, and especially the great discovery depended on keeping all this wide country peaceful.

The re-established post at Lake Pepin maintained a precarious ascendancy over the Sioux and his presence at Lake of the Woods was a deterrent as well as an irritant. He kept his own

## Map 15.  *Vérendrye's Progress*

Indians pretty well in hand. He traveled constantly from post to post and harried the merchants to maintain a flow of goods. He had a brother's affection and a father's authority, he excelled in the ceremony and oratory they admired, and he made good his will, repeatedly closing to them the road that led to the Sioux. But misfortune dogged him. In 1736 his nephew, the skillful and sagacious La Jemeraye, died. In the same year he paid tragically for a blunder he had made two years before, though the balance of necessities was so even that he must be said to have been forced into it.

Four hundred Monsonis had arrived at Fort St. Charles burning to avenge a defeat by the Sioux. When they had danced themselves to the required frenzy, they demanded as a token of alliance that he send Jean-Baptiste down the war road with them. It was a cruel, inescapable dilemma. Not only might he send his son out to be killed if he agreed, he might also upset the equilibrium of the Sioux peace. If he refused, his allies and customers might become enemies. The trade of the Northwest might be lost, the way to the Western Sea might be closed,

and the prestige of the French might be gravely damaged —
they might "take the French for cowards." He made his decision. He added Jean-Baptiste to the war party, which numbered nearly a thousand when the Crees joined it, and contributed a lot of munitions. After seizing the opportunity for
some sales promotion, he joined the war dance, sang the war
songs, and "let them see the wounds I got at Malplaquet, which
astonished them." When his report reached France Maurepas
approved this measure of international co-operation, but after
the fact Beauharnois remembered that he had foreseen the
disastrous outcome.

The war party got its scalps and glory but the Sioux it ambushed learned that Jean-Baptiste had been with their enemies.
They deposited him among their grievances and the chance
came in 1736. Vérendrye sent a party from Fort St. Charles to
Kaministiquia to hurry the year's goods, for they were always
late as well as short. Jean-Baptiste and the resident missionary
went with it. Twenty-one all told, they were camped on an
island in Lake of the Woods when the Sioux who had gone
looking for them found and killed them. They cut off their
heads — in the sign language a gesture which pantomimed
decapitation meant "Sioux" — and wrapped them neatly in
beaver skins, which was Indian irony. They had won a great
victory, by far the biggest victory over white men the Sioux had
ever won. They grew overbearing and began to menace the
Mississippi River post. The next year, 1737, the commander,
Jacques de St. Pierre, had no choice but to burn it and get the
missionaries and his handful of men away while he was still
able to. Except for the route to Vérendrye's posts, Minnesota
was now lost to the French.

The massacre seriously cut down Vérendrye's strength and
postponed for another year the discovery which he kept promising Beauharnois and Maurepas he would soon make. All this
time, while sending to Montreal a wealth of furs whose sale
netted him no margin for exploration, he had been amassing
information about the farther country. By now he had heard
so much about the Mandans and the river they lived on that
he was sure they had the answers to all the problems.

What he heard ran through the whole scale of fact and
fantasy; he had no way of differentiating between them and
little awareness that fantasy entered in. Some of his informa-

tion came from Crees and more from Assiniboins, who made regular trade visits to the Mandans.[6] Everyone agreed that the Mandans were a markedly superior people who had an astonishing civilization. They lived in fortified cities, raised crops and horses and goats, hunted the buffalo, and excelled at arts and handicrafts but had neither iron nor firearms though they did have rich outcrops of ore. Theirs was a plains country and its flora and fauna were fabulous, including two-headed snakes and others *"d'une grosseur prodigieuse."* Their seven — or nine — cities were on the banks of the River of the West. Presently it came out that they were white men, men just like the French, and wore beards.

The river of the Mandans was the River of the West and just what the name says: it flowed west. Vérendrye learned that near its mouth there were tribes which had iron tools and European cloth; it was a long way from the Mandans to those tribes, with seven nations in between. (This was true: the misunderstood reference was to the mouth of the Missouri and the more than seven Missouri River tribes.) At its mouth were Frenchmen. (The settlements of the Illinois.) Soon, however, it transpired that the river of the Mandans did not flow west. It flowed south and (Vérendrye judged) southwest and emptied (he concluded) into the Pacific. Confusion was now chaotic and very intricate. Vérendrye had no way of knowing it but stories out of New Mexico which came over the Indian news service had begun to mingle with those that came up the Missouri from the Illinois. The white Indians, the Mandans, had blended with both the French and the Spanish. And, without realizing it, he had three different conceptions of the River of the West. Sometimes it was the Missouri, sometimes a river that flowed west to empty into the Pacific or possibly the Western Sea, and sometimes both. Desire and logic had combined to produce this amalgamation, but also there entered into it Indian allusions to the Saskatchewan, which flows east.

Two misconceptions are paramount, the usual contraction of distances and a postulated height of land where imaginary watersheds came together. At Fort St. Charles Vérendrye was a little more than three hundred miles in an air line from the Mandan villages; Fort Maurepas above the mouth of Red River was about the same distance from them. He judged that at most the Europeans on the seacoast could be no more than three hundred leagues away. Seven hundred and fifty miles would take him to the fulfillment of the dream.[7]

And still he had been unable to build up the margin of time, manpower, and goods for presents that he needed to visit the white Indians who could direct him to the sea. He sent them word that he intended to come, invited them to visit him, and once learned that one of them made up as an Assiniboin had sought shelter at Fort Maurepas but had been refused it by a sentry who did not perceive the opportunity. He fretted within the arc of his holdings, soothed Indian haughtiness and fed Indian pride and listened to Indian brag, cut into the British trade, bore the censure of Beauharnois and the denunciations of Maurepas, and waited for the light. He had turned fifty and his body would not always answer to his will in the plains heat and Arctic cold, when he traveled hundreds of miles by canoe and hundreds of miles on snowshoes. "I stayed on for seven days more to recover from the fatigue of the journey, my old wounds giving me so much pain that they threatened to stop me on the road."

Once, it is good to know, the Governor of Canada had had enough irony from the Minister of Marine, who was currently outraged by the abandonment of the Sioux post and Vérendrye's willful failure to reach the Western Sea. It is indeed annoying, Monseigneur, Beauharnois wrote (in effect) to Maurepas, that the Indian country does not in some ways correspond to the ideas held about it at Versailles and that Indians sometimes depart from your conception of appropriate docility. Sometimes we encounter obstacles which the foreign office had failed to foresee for us, sometimes risks must be run, and if we are forced to gamble we must sometimes lose the stake. Would your excellency be more content if I had ordered the garrison of the Sioux post to stay and be killed? Consider too that this misadventure may have added to the difficulty of Vérendrye's project. And observe that in spite of the Sioux this bungler has maintained his posts and has kept his plans for discovery intact. Truly this is at Versailles an insignificant achievement, but in the view of those who must solve in work and fortitude the problems you set for us in ink it is not wholly without virtue.

⚭

But at last, in 1738, Vérendrye had enough goods to make the venture and had found a trader who was willing to help. In late September he took his canoes up Red River from Fort Maurepas. At the mouth of the Assiniboine River they built a small post; his sense of commercial geography was always true, and the city of Winnipeg stands on the site today. They turned up the Assiniboine, heading west at last. They halted at the place where a main Indian highway crossed it, an ancient trade route between the Missouri and Lake Manitoba. A busy trade now moved along it to the Hudson's Bay Company and so they built a post at this focal point, where fortunately there was timber. It was called Fort La Reine and the job lasted through the plains autumn with its long gold and lavender twilights, to the middle of October. From here they had to travel overland — and on foot, which meant that they could take but few goods for trade or presents. The route followed the well-worn trail due south to the escarpment called the Pembina Hills, across them and west and west-southwest to the higher hills known as the Turtle Mountains (the ones which he had heard at Lake Nipigon shone night and day), and from them south-southwest for a hundred and fifty miles to the Mandan villages on the Missouri.[8]

They had crossed the transition province into the Great Plains. The great north woods ended east of Lake Winnipeg and Red River. Along the Red there were marshy meadows, large groves of poplars and willows and plums, smaller groves of ash, oak, and cottonwood, and much brush. Similar groves were not uncommon south of the lake as far as the Assiniboine, but thereafter they were to be found only on the hillsides and along the occasional streams.

An Assiniboin village with which they had been forced to waste much time decided to escort them to the Mandans.[9] Its chief sent messengers ahead to invite them to send out a welcoming committee. Three or four hundred strong now, they marched across the frozen plain. The committee of greeters met them on November 28 . . . and though they were Mandans they were not white men. They were just another tribe of Indians in buffalo robes.

The greeters ceremoniously welcomed the French and began to trade with the Assiniboins — and, in spite of what Vérendrye had heard, they did have iron and guns. After thoroughly outwitting the Assiniboins they tried to dissuade them from coming on as guests by saying that they had arrived opportunely, just in time to help fight the Sioux. Shocked and bewildered by the Mandan reality, angered by remembering the years of lying he had had from the Assiniboins, Vérendrye pushed ahead to the first Mandan village, the King's banner going on before. The one to which he had been invited was smaller than the five he counted on the Missouri, and less than a day's travel inland from them. They reached it on December 13 and the Mandans, who had prepared a great celebration, rushed out and hoisted all twenty-seven white men to their shoulders. The French had now met the tribe about whom they had been hearing for decades. As soon as Vérendrye's sons went on to the river they reached the master key to the real and imaginary geography of the West. And still the Mandans were not white.

Everybody, except the younger Alexander Henry who hated all Indians, always liked the Mandans. At this period the first of the two great smallpox epidemics in their history had not halved their numbers — the second one, just a century later, was to extinguish them as a tribe. Nor had the westward migration of the Sioux reached them except for occasional raids. It is a measure of the Mandans that, unlike the surrounding tribes, they were not afraid of the Sioux. Though less well supplied with guns, they either met the raiders on the plains and battled or scared them into peaceful behavior or else retired into their impregnable villages and laughed at them. They were handsome and robust, sunny, humorous, hospitable. The cellar pits of the big domed earth lodges and the caches near the fields were affluent with corn, beans, pumpkin meal, dried melons, jerky, pemmican, and buffalo tallow, and there was game everywhere. As Vérendrye watched them negligently gyp his escorting Assiniboins (and then send them hurrying home by repeating the rumor that the Sioux were coming) he noted the superiority of their leather and quill work and the variety of artifacts from farther tribes that reached them in the trade. He was fascinated by their folkways, which differed from those of the forest tribes he had known. He saw light-skinned and fair-haired and redheaded Mandans. Still bemused by the stories he had heard, he came to the odd conclusion that they must be of mixed black and white blood. But the question still remained, whence the white blood?

These villages were on both sides of the Missouri, at or near the mouth of the silt-choked Heart River. The Mandans said that their country was at the very center of the world and in a way they were right,[10] but by Vérendrye's calculations he had not covered half the distance to the Western Sea. He had, however, suffered two misfortunes. The box containing most of the goods he had brought for presents was stolen during the uproar of the welcome, so that he had little currency, and when the Assiniboins decamped his interpreter went with them. The second loss was disastrous for now he could get information only by the sign language of the plains, an excellent means of communication but not capable of geographical exactness. He did what he could and so did his son Louis-Joseph, called "the Chevalier," at the river village, but it was very little. He decided to leave two men among the Mandans for the winter to learn the language and gather information, go back to Fort La Reine, and make the discovery the next year.

He had learned some facts, picked up additional rumors, and misunderstood much — but just how much cannot be determined from his report. The river of the Mandans flowed southwest by south — this westering was a purely local series of bends — but he heard that a little farther on it took a more westerly course to the sea. This seems to have convinced him that the sea it finally reached was the Pacific — in which he was not much interested, since it would mean New Mexico and the Spanish. He charged the men he left with the Mandans to find out what tribes lived *farther up* the Missouri ("*quel nation au desue*") and whether a height of land lay *upstream*. This is revealing and important. Though he had not recognized the River of the Mandans as the Missouri he had concluded that its source, not its mouth, lay near the Western Sea. As a matter of fact, an ascent of the river would have taken him to the Pacific: it was the route the Americans eventually followed.

He learned about the Arikaras, south of the Mandans and known by hearsay to the French of the Illinois, and got some trustworthy information about Plains Indians in general. He kept hearing about white men who lived at the mouth of the river, and vague hints of the New Mexicans mingled with what the Mandans had heard about the Illinois. The rest is misreported or misunderstood.[11]

It was mid-February 1739 when Vérendrye got back to his

Fort La Reine on the Assiniboine from the Mandan villages
after a brutal winter crossing — "never in my life did I endure
so much misery, pain, and fatigue as in that journey." As soon
as the ice was out of the rivers he sent the Chevalier to find a
site for a post on Lake Winnipegosis — whose importance had
become clear to him since 1737 — and then to go on to Rivière
Blanche, which he now knew flowed east, not west. It was the
Saskatchewan, the key to the North and the second great key
to the West, a highway to the Hudson's Bay Company posts;
it drained part of the richest fur country and led to rivers that
drained the rest. The Chevalier must be called its discoverer,
for Kelsey's visit to it forty-nine years before had left no im-
print on knowledge. He went up it to the forks and when he
asked the Crees there where its source was they "replied with
one voice that it came from very far, from a height of land
where there were very lofty mountains [and] that they knew
of a great lake on the other side of the mountains, the water of
which was undrinkable."

At last there are mountains and a lake of bitter water
which cannot be mistaken. The Chevalier was hearing about
the Rocky Mountains, about the Pacific, and about the Saskatchewan as a route by which it could be reached. All the
clues were now in and in less than twelve months the Vérendryes had reached the two dominant features of the Northwestern Mystery, the Missouri River and the Saskatchewan.

When the men who spent the winter among the Mandans rejoined Vérendrye at Fort La Reine, they had additional information. They had learned about the far-reaching trade network of the Plains Indians, in which the Mandans had a
conspicuous part because of their craftsmanship, agricultural
surplus, and relations with the Assiniboins who got European
goods. Several bands of horse Indians — Vérendrye already
knew that the Arikaras had horses — had arrived from the
West and stayed for long visits. There is no reason to suppose
that these were not the same tribes who in the 1790's came regularly from the West to trade with the Mandans and who can be
positively identified: Cheyennes from north of the Black Hills
and Crows from the Powder River country. From these Indians they heard about white men who must now be identified
as New Mexicans. There was tangible evidence about them at
last too, for Vérendrye's men brought him a bit and bridle that

were unmistakably Spanish and *"une couverture"* (possibly a rebozo or serape) and a shirt made of cotton. It may be confidently assumed that the Indians to whom the men talked — in halting Mandan — had not themselves been to the Pueblo villages and Santa Fe: they were repeating what they had learned from other tribes with whom they traded.

There was little enough here to clear up confusion. All that Vérendrye learned certainly was that the Far West and its sea were more distant than thus far he had believed. The insistence of the horse Indians that there were cities of white men on the seacoast supported the misconception he had already formed. (These cities, however, must have been villages of New Mexico, whereas the earlier ones had been Cahokia and other settlements in the Illinois.) And though he now heard that there were "high mountains" in the country of the white men, he did not relate them to the ones the Chevalier heard about on the Saskatchewan and he formed no conception of the Rockies. It does not appear, in fact, that the Chevalier made anything of the mountains the Crees described.[12]

Through the next three years the trade was so active and his Indians and the Sioux were so constantly at war that Vérendrye could not renew the search. In 1741 he sent the oldest of his remaining sons, Pierre, to the Mandans to join the Indians of the West when they returned to their country and so find the Western Sea. The trading tribes did not make their annual visit this year, however, and Pierre went back to Fort La Reine. He took with him two horses that he had bought from the Mandans, so it may be that the Mandans had entered the horse age in these years, 1737–41.[13] He also brought an embroidered counterpane and some specimens of "porcelain," possibly the tubular or oblong beads which the Mandans made from pulverized European glass beads. Vérendrye sent these to Beauharnois, who forwarded them to Maurepas as evidence that contact with the Spanish settlements had been made at last. And Pierre went on to build a post named Fort Dauphin on the western shore of Lake Winnipegosis. Presently, but just when is not clear, the Vérendryes built another post on the Saskatchewan itself, near the great crossroads called The Pas, where the Pasquia River empties into it. Here or near here Henry Kelsey had abandoned his canoe for overland travel and it was a central meeting place of routes. Vérendrye had now established a semicircle of inland trading posts that cut across the routes by which the Hudson's Bay Company's customers reached its Fort Albany, Fort Severn, and York Factory.

But Maurepas, in a France which had by now joined the War of the Austrian Succession, had had enough of Vérendrye's failure to find the Western Sea. He ordered Beauharnois to send out someone qualified to help him — better qualified than he for the job, presumably — and to get one of his sons out of the West now and the others soon. Beauharnois protested but Maurepas would hear nothing in Vérendrye's favor. Humiliated, mired in debt, and now in bad health, Vérendrye gave up hope and resigned his office. Beauharnois appointed to succeed him Nicolas de Noyelles, a veteran who had commanded at Detroit and fought in the Fox wars. Meanwhile, in 1742 and 1743 Louis-Joseph and François Vérendrye had made a supreme effort to go on to the end of the quest.

ᘒᕈᗀ

The Vérendryes have been chiefly known for this heartbreaking effort, though it is less important than their having reached the Missouri and the Saskatchewan and having established posts on the flank of the British trade. Unfortunately the Chevalier's report is one of the vaguest documents of exploration in American history. Only one place he went to can be positively identified, and that not from his text. For the Indians he saw and the distances he traveled there is only inference, within the limits of what was possible.

The two young men with two *voyageurs* reached the Mandan villages, opposite Bismarck, in May 1742. At first on foot but on horseback after they found the tribe they called the Horse People, they wandered about North and South Dakota during the rest of the year, passing from band to band of Indians and stopping when their hosts did, sometimes for weeks at a stretch. They called the first bands they traveled with the Beaux Hommes and the Little Foxes and it would be idle to conjecture who they were. The Chevalier believed that these bands knew the route to the sea, but gradually he became convinced that this must be a sea which was already known. The commonest direction of travel, southwest, seemed to imply the South Sea or the Vermilion Sea — the Pacific or the Gulf of California. His guess vividly demonstrates how unreal ideas of continental distances were; he was east of the Dakota badlands.

The Horse People, when the Vérendryes finally reached them, had recently been raided by the *Gens du Serpent*, the Snake People, who were hated and feared throughout this country. And the Snake People, so they had been hearing for three years, barred the way to the Western Sea. They cannot have been Shoshones, the tribe called Snakes later on, who are not known to have come so far east, even on their forays against the Blackfeet in the north. They may have been Kiowas, who were much like the Comanches and shared their ferocious reputation, and the Horse People were probably Cheyennes but possibly Crows. One tribe had an illustrious reputation, they were said not to be afraid of the Snake People; these were called the *Gens de l'Arc,* the Bow People. These may have been Arikaras; some divisions of the Pawnees are perhaps more likely; it is possible that they were Cheyennes or Crows. The story was, too, that the Bow People traded with tribes that went to the sea.[14]

The Chevalier reached the Bow People in late November and found them mounting a big invasion of the Snake People's country. Village after village joined up and when the campaign began it resembled a mass migration. (But there cannot possibly have been the two thousand fighting men the Chevalier counts.) Slowly and with infinite circuitousness they traveled the winter plains toward the Snake People. As they traveled — tipis, travois, great horse herds, brawling squaws, howling dogs, innumerable false alarms, pantomimes and parades asserting ferocious courage — various chiefs told him about that country and the sea beyond it. They had not seen either the country or the sea (and they understood that the Snake People had not been to the sea) but they had heard a great deal. What they told him was some factual information and much fantasy about New Mexico, and much fantasy and perhaps a little fact about the Pacific north of California. Unable to distinguish the last, he was able to identify the first, including an Indian account of the Villasur expedition of 1720. He was now sadly sure that he was traveling toward "a known sea." He meant the South Sea in the latitude of New Mexico and, as he thought, the seacoast of New Mexico. Thus at the moment when he was at last hearing about something unquestionably new and perhaps to be relied on, he was constrained to believe that failure lay ahead. Nevertheless no Frenchman had ever reached the South Sea overland and though gloom settled on him, "still I greatly wanted to go there."

The country of the Snake People, so the Bow People said, lay at the eastern base of a range of very high mountains. On the western side of that range was a strip of coast inhabited by the white men about whom the French had been hearing for so long, and their great cities, herds, soldiers, priests, and slaves. It was a narrow strip, for from the ridge of those mountains you could see the sea. The Bow People had not climbed to the ridge, they had not even seen the mountains, but Snake People whom they had captured had told them what it was like.

On January 1 1743 mountains stood up on the western horizon. They were the Black Hills of South Dakota but to the Chevalier they were the mountains of rumor on the edge of the South Sea. . . . Or just possibly, since the hope would not die, on the edge of the Western Sea men had so long yearned to behold. . . . Preparing for hostilities, the Indians stopped and made a camp for the women. The Chevalier left his brother there with their outfit and took his two *voyageurs* on with the war party. On January 21 they reached the mountains, which were timbered and very high. Scouts learned that a village of the Snake People, the objective of all this panoply, marching, and incantation, was near by.

What followed was typical of Indian warfare. The Snake People had been terrified by the approach of the Bow People and had abandoned their village. Finding it abandoned, the Bow People were terrified in turn, for it must mean that the enemy were at this moment massacring their women and children at the undefended camp. Instantly the Bow People turned back. Fear became panic and panic became a stampede. The Chevalier and his men went with them. There was nothing else they could do for they had so boxed the compass, had wandered so unintelligibly about the midwinter plains, that only the Indians could get them back.

But there was that high ridge. As he looked back over his shoulder at it, his horse at the gallop, the Chevalier understood that if he had been able to climb those mountains he would, at last, have seen the sea where the West ended.

The stampede began on February 6 1743. The return journey was shorter but equally reckless, and now they were delayed by heavy snows. They stayed with the head chief of the Bow People till the middle of March, when they joined a band of Indians who were friendly to his tribe. The Chevalier called them the *Gens de la Petite Cerise*, the Little Cherry People. Now a positive identification is possible: these were Arikaras who lived near where the Teton River flows into the Missouri, opposite Pierre, South Dakota. They reached the Arikara village on March 19 and on March 30 the Chevalier buried on a hill above it the lead plate that claimed his farthest west for France. On February 16 1913 a fourteen-year-old schoolgirl named Hattie May Foster saw a corner of it sticking out of the ground and kicked it free.

It was July when they got back to Fort La Reine on the Assiniboine, "to the great delight of my father, who was very anxious about us," having heard nothing of them in fourteen months. But the admission must be made, that the supreme effort had come to nothing, and Vérendrye's resignation would have to stand. The French never again tried to find the West or its sea by way of the Missouri River.

სი

"Consider, Monseigneur," Beauharnois wrote to Maurepas, "six years of service in France, thirty-two in this colony without reproach . . . nine wounds in his body . . ." This record seemed to justify his asking for the impoverished Vérendrye the captaincy of a company of Canadian troops — for he had served all these years in the rank of lieutenant. He got the captaincy at last in 1745 and was sixty years old. Four years later his crowning reward came, the pretty ribbon and cross which certified that he had been made a Chevalier of the Order of St. Louis and thus was acknowledged to have worthily served his King. He remained impoverished.

De Noyelles, his successor in command of the Post of the Western Sea, did nothing about the discovery and after two years asked to be relieved. Beauharnois wrote to Maurepas that he intended to appoint the man best qualified to continue the great quest, the man who had established all the holdings of the Post of the Western Sea and had lived with its Indians for fourteen years. Vérendrye, expecting service again at sixty-two, expressed his deep gratitude to Maurepas, who directed the governor to deprive him of all command unless he straightway did better than before. It was, however, an interim governor, a naval man not in awe of ministerial thinking, who spoke out to Maurepas. "What has been reported to you with reference to the Sieur de la Vérendrye . . . is entirely false." Furthermore, officers who are asked to make explorations "will be under the necessity of giving a part of their attention to commerce as long as the king shall not furnish them with other means of subsistence. . . . These explorations cause heavy expense and expose a man to greater fatigue and greater danger than regular wars. The Sieur de la Salle and the son of the Sieur de la Vérendrye furnish the proof of what I say." Nothing would convince Maurepas that the last great French explorer of the West was not a grafter and bungler, but presently he polished a phrase too wittily and, aiming it at the king's mistress who was ruling France, got himself thrown out of office. It was a new Minister of Marine who sent the veteran his pretty cross.

In the fall of 1749 Vérendrye, now sixty-four years old, prepared to resume in the following spring the discovery of which he had first dreamed twenty years before at Lake Nipigon. In the years since then he had groped through frustration to great achievement and to some solid knowledge which he could take with him into the unknown country, surely this time with success. He knew, as the fur merchants were to find out after Montreal became English, that the schedule of business and travel in the far Northwest must be on a two-year basis, that it was so far away one year would not do. So he would go to Fort Bourbon which his sons had built, and which had now been moved to Cedar Lake, and spend the winter there. The next spring he would take his canoes up the Saskatchewan, which his sons had discovered, and go on to the lofty mountains about which the Indians had told his sons, and cross them and so at last reach salt water. He was sure now that the Saskatchewan would prove to be a route to the sea. It was one — but Vérendrye did not live to make his start. He died that winter and his sons, who knew more about the Northwest than any other white men, were not permitted to succeed him.

A veteran of the Sioux country, Jacques Legardeur de St. Pierre, was named to succeed Vérendrye. The trade that centered at Lake Winnipeg and the lower Saskatchewan remained rich but prodigious skill was required to keep it going in the face of the increasing difficulties of transport and finance. St.

Pierre was charged, as everyone was, with finding out just where the Western Sea was, but his hands were full and he may not have been interested. In 1751 his lieutenant Boucher de Niverville went far up the Saskatchewan and built a miserable little temporary post. How far he went cannot be determined but to the forks and far up one of them, perhaps to about the western boundary of the province of Saskatchewan, perhaps almost across Alberta to Calgary. The latter is almost at the foot of the Rocky Mountains, and Niverville said that he saw them — if he did he was the first white man who had seen them anywhere north of Colorado — and it is almost certain that he talked with Indians who had crossed them. Apparently he was to have gone on and crossed them himself but he fell desperately sick — or perhaps it was just the universal sickness of Canada.

Wherever he was, he and St. Pierre at Fort Maurepas and La Reine continued to hear about strange beings beyond the mountains — about white men and black men, and also about strange waterways off somewhere in the still longer nights and thicker ice of the North. Just possibly they were getting rumors of a people who had come an even longer way than the French, who had already touched the Aleutians and the Rat Islands, and who with a ruthless and barbarous vigor would bring a fourth imperialism to North America. The Russians.

In 1753 St. Pierre was called east and the Chevalier La Corne replaced him. La Corne built a post on the Saskatchewan well above The Pas which intercepted Indians coming down from the north to trade at Hudson's Bay as the one at The Pas intercepted those coming from the southwestern part of the great fur country. The trade routes to York Factory were pretty thoroughly choked off now and the lethargic Company was roused by the decline in its dividends. Forced to do something at last, it sent into the interior the first of its servants who had gone there since Henry Kelsey in 1690, a young man named Anthony Henday.[15]

Before Henday an adventurer had made an odd circuit of the Northwest, one Joseph La France, a French halfbreed who had had a considerable experience in the Great Lakes trade and had been on the upper Mississippi. Like many another trader he ran afoul of the graft. Determining to join the British trade, he started out to reach York Factory by the canoe route west from Lake Superior. La France, an illiterate, is only some twenty pages of recorded talk in a book by Arthur Dobbs, the Irish engineer and civil servant who undertook to compel the Hudson's Bay Company to find the Northwest Passage, who not only was a fanatic and a beholder of visionary lands but had, perhaps, the greatest talent the eighteenth century ever knew for misconstruing geography. But it is clear that from 1739 to 1742 La France did go down Vérendrye's route by way of Rainy Lake and Lake of the Woods to Lake Winnipeg, traveled much of the interlaced waterways around the lower Saskatchewan, and eventually reached Fox River and went on down the Hayes to Hudson Bay. Nothing came of his three-year adventure but he made a fine bludgeon for Dobbs. He was in the Lake Winnipeg country during the most active years of the Vérendryes but if he heard of them or they of him neither made a record of it.[16]

The Company sent Henday inland in 1754 not to trade with the Western Indians but to induce them to give up their trade at the French posts and resume the interminable journey to Hudson Bay. Except for the Crees with whom he originally set out and who were already Company customers he had no success. He was surcharged with contempt for the grimy foreigners whom he met at posts so squalid that it seemed impossible they could have damaged the lordly Company's business, but he found that the foreigners had the Indians' liking and loyalty as well as their trade. The Western Crees would continue to trade with the French, they said, and so would band after band of Assiniboins he met. He kept on and, like Kelsey, had to abandon canoe travel and go overland with various bands of nomads. Alone among Plains Indians, with only a frugal store of presents, he made a tremendous circuit. He reached the North Saskatchewan, the South Saskatchewan, its far fork the Red Deer, the Rocky Mountains. He was the first Englishman who ever saw them, the first who ever saw horse Indians, and the first white man who ever saw the Blackfeet, of whom Kelsey and presumably the French had heard. He met the Fall Indians (the Atsina, who were affiliated with them) and then the Bloods and other proper Blackfeet. He even wintered with them, for the Blackfeet first come to European knowledge as an amiable and hospitable folk — they who were throughout the history of the fur trade to be nearly always hostile and always suspicious, sour, bilious, and vicious. They too declined to visit the Bay. They were a Plains people, unable to use canoes and afraid of the lakes and rivers; much as they needed guns for their wars with the marauding Snakes, they would continue to get them at high prices from the Assiniboins, who would continue to get them from the French.

Henday therefore took no customers back to the Bay, though he did take back a vast store of information entirely new to the Company. He was received with considerable skepticism — Indians who used horses, egad! — and complete disregard. The moral of what he had seen was that the Company was going to have to abandon its policy of making the Indians come to its seaboard posts. (And of regarding the customer when he got there as a red bounder who had to cover his nakedness, must not be given rum, and would please be so good as to like the treatment.) It was going to have to go into the interior and compete with the French, who at the posts Vérendrye had located were strangling its trade. That was the moral but as it turned out the Company didn't have to go inland for quite a while: the French were finished.

That was why St. Pierre had been called east: in the governor's opinion he was the best officer in Canada and the one most likely to make himself feared and respected by Eastern Indians. In the part of Pennsylvania that reaches Lake Erie between New York and Ohio, at Fort Le Boeuf on French Creek, St. Pierre had an appointment with destiny. For the final war was at hand. At Fort Le Boeuf on December 11 1753 he met a young Virginia major named George Washington and rang down the curtain on the prologue.

℮∾◦

The prologue had begun as a war: five years of it here as King George's War, eight years of it abroad as the War of the Austrian Succession, 1740–48. All it settled in either hemisphere was that another war must come. New England, New York, and Pennsylvania had five years of burning, pillage, massacre, and lonely murder. The English colonies made an agonizing effort to abort jealousies and commercial rivalries, to postpone their rebellion against royal and parliamentary and proprietary control of societies that were becoming states, to extemporize some kind of common defense against the French.

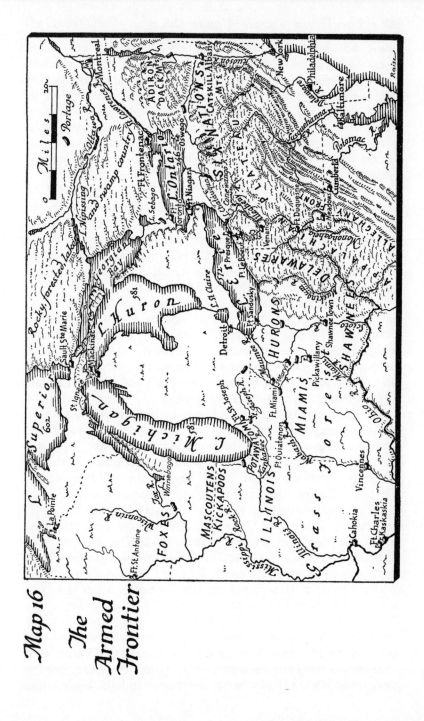

Map 16

*The Armed Frontier*

There was guerrilla warfare across Cape Breton and Nova Scotia and a futile effort to invade Canada up the Champlain corridor. In an unorthodox, heroic campaign a British fleet and a New England army captured the greatest fortress in North America, Louisburg. Then just to make inconclusiveness entirely symmetrical, the peace treaty which was signed in 1748 restored Louisburg to France. . . . It may be that Great Britain lost New England when the royal seal went on that treaty.

The five years that followed were only an armistice in Europe. They were not even an armistice here but a period in which military conflict was not called war. The five-year preparation for what would have to be called war centered where the irrepressible conflict would have to center: the eastern frontiers, the eastern end of the Great Lakes frontier, the Champlain corridor, and the Ohio Valley.

The semi-strangulation of trade produced by the war years still further alienated from their loyalty to the French the Indian tribes along the Great Lakes. A prime fact of world history as the eighteenth century approached the halfway mark was that a very large Indian population, the tribes east of the Mississippi River, had become incapable of supporting life without European manufactured goods.

The English could do much better by them than the French. Though French shipments to America increased greatly, the English easily outmatched them in quantity, quality, and especially price. They would have won easily except for two things: the greater military talent of the French and the fact that, at just this moment, the English colonies matured to the point where they were ready to cross the Appalachians not for trade only but for land. The Ohio Company was chartered the year before the peace treaty was signed at Aix-la-Chapelle, received its first grant two years later, in 1749, and sent its first land-scout over the mountains in 1750. The Indians understood its significance as well as the French did. Some tribes that had become British in fact or predisposition became French again. Nearly every tribe in the Ohio Valley had pro-British and pro-French factions. Besides being scalped, English colonists who were murdered might now have their mouths filled with dirt.

Louisiana was too poor and feeble to meet the trade, but the great Ohio Valley might be saved. Some Indians had shifted eastward: Shawnees up the Ohio, Miamis out of Illinois into Indiana. Some had shifted westward: many Delawares out of eastern Pennsylvania and many nominal Iroquois out of New York, to the forks of the Ohio and northwest of them. For a long time the Tuscaroras had been migrating north into the southern and western Iroquois company; they were received into the Iroquois confederacy and at some time after 1715 it becomes proper to speak of the Six Nations. Mixed tribes and bands of uprooted refugees also helped to fill up the Valley. British traders in ever increasing numbers, with ever increasing amounts of goods, packed over the mountains to supply it.

By the peace, 1748, they were not far short of controlling it. They had reached as far west as Sandusky, where they would have affected all the upper Lakes, but were presently thrown out of there. They had regular routes of call below the forks of the Ohio, and semipermanent establishments near or on the Muskingum and the Scioto. The most dangerous of their concentrations was even farther west and north. A large group of Miamis renounced even nominal allegiance to the French and moved east to attach themselves to the English, settling on the Miami River at the site of Piqua, Ohio. Their town there, the famous Pickawillany, became a center of British trade, a focus from which sedition could spread everywhere.

The only answer possible was to drive the British traders east of the mountains and keep them out. It was a military answer; Canada became a colony in arms. The line of the Wabash and Maumee was stiffened. Vincennes and Ouiatenon (Lafayette), Detroit at the passage between Lake Erie and Lake Huron, Fort Niagara at the western end, and Fort Frontenac at the eastern end of Lake Ontario were all either rebuilt or at least strengthened. All the resources of intrigue, flattery, and intimidation that the French used so expertly were put to work among the hesitating tribes. Strenuous efforts were made to increase the flow of goods. The Church redoubled its efforts and something like the fervor of the early seventeenth century again animated the missionary-diplomats.

In 1749 Céleron de Bienville with an insufficient force was ordered to sweep the Valley free of British traders and intimidate the Indians. He made a difficult circuit, Chautauqua Lake to the Allegheny, the Ohio, the Scioto, the Miami, and at last the Maumee. It was a failure; the traders were not scattered nor the Indians impressed. But Canada had an instrument that had been tempered by a full century, the Western officer. In 1752 Charles Langlade led a force of Ottawas and Chippewas — Western Indians — to Pickawillany. This was not war, it was only hostilities, but it was decisive. The Miamis and some other tribes returned to their allegiance; all tribes were reminded that the French were mighty. British trade west of the mountains was obliterated and now it would be a contest for the forks of the Ohio, the Monongahela and the Allegheny. Whoever held them would rule the entire valley.

That contest was not war in 1753 when the new governor of Canada, the Marquis Duquesne, called on another veteran of the West, the elder Marin, to open a new route to the Allegheny by way of French Creek and to build the post called Fort Le Boeuf. Marin brought Western Indians with him and built the post but fell mortally ill. St. Pierre was called from the West to succeed him, and it was still not war when on that December 11 Major Washington brought him an order from the governor of Virginia, directing the French to get out of the lands of the English king. . . . For if Pennsylvania had dominated the commercial expansion, the speculation in Western land was primarily Virginian. The Ohio Company itself was a Virginia enterprise and had two Washingtons among its proprietors. The two forks became the Ohio River in Virginia, the claim was, and Virginia extended all the way to the sea. . . . It was not war when St. Pierre refused to lead the French back to Canada or when in 1754 Contrecoeur went on to the forks, captured the little British force that was trying to build a fort there, and himself built Fort Duquesne.

Major Washington led an expedition against it and at the Great Meadows he ordered the volley that set the world on fire, and still it was not, officially, war. Braddock had been defeated, the Acadians had been subjected to a forced migration, and probing campaigns had been made in three different areas when at last, in May 1756, it was decreed to be war. It was the war that changed the world.

The colonial possessions had precipitated the war: neither North America nor Europe nor yet the world was large enough to contain both the British and the French empires. Europe shattered at the touch — France was leagued with the ancient enemy Austria. It was France, Austria, Russia, Sweden, and

Saxony against Prussia, and therefore by force of necessity Great Britain was allied with Prussia. The first stage might be called the first of two successive wars, if not indeed of three. It was the war of Prussia as a military state, Frederick the Great as a military genius, the "damnable money" that Great Britain financed him with, the British fleet gradually destroying the French navy and sweeping French merchant shipping from the seas, and in the nick of time enough British troops released for service in Europe. Pitt had wondered whether North America was to be won in Germany. As it turned out, no; but while the European war spread slaughter everywhere, there were times when both Great Britain and North America came close to being lost there.

As the agony went on Great Britain appeared to be bankrupt — but France truly was. And whereas the art of government had stagnated for a generation in Great Britain, it was a generation only of mediocrities, not of men who drifted beyond cynicism into political paralysis, which is what had happened to those who were supposed to govern France. England raised up Pitt and Pitt was the father of victory. He organized a second war, the global one. He roused the British people to the highest pitch of patriotism they had ever known and made them — by that time they had repudiated him, which is what happens to British geniuses who win wars — the greatest commercial nation and the greatest colonial power. India was theirs, most of the West Indies was theirs, the west coast of Africa was theirs — but most of all, Canada was theirs.

That Canada sustained the war from 1754 through 1760 was an achievement so great that in retrospect it is hardly credible. The English colonies were, though they were not yet called, dominions; the French had no thought of colonies except as dominion — and the single *s* means much. The more than paternalistic, the in fact dictatorial management of every human enterprise could not work in the New World. It could seem to work in the French West Indies, where there were a true aristocracy, slave labor, and a staple crop so rich that vast profits were left over after the graft had been harvested — but even there it had begun to collapse. In Canada dictatorship could not withstand the strains of colonization. The little town of Quebec, perhaps eight thousand people when the war came, was a provincial capital in some ways luminous and charming,

but it was fully as corrupt as Versailles. At the beginning of
the war the fur trade that was more than four-fifths of Canada
was mortally sick with graft. And before the British fleet cut
off Montcalm's army from reinforcement and supplies, the
corruption that was the government of Canada had already
cut its power in half.

What Canada had was men raised from childhood to be
soldiers, to be wilderness soldiers. They had the maximum of
military skill and military virtue; they were incomparable
fighters. If the war they fought was cruel and vengeful and
revolting, no war is ever otherwise and theirs was also in the
highest degree heroic. They are to be remembered in the
chronicles of war as having held off the British Empire through
seven years.

Concern here is with their skill in using the weapon they had
forged, the Indian auxiliaries. Indians from as far west as the
Des Moines River fought along Lake Champlain. Under Montcalm and his staff, the French officers who fought the war were
mostly those who had first been known at Detroit, Mackinac,
La Baye, La Pointe, the Post of the Sea of the West. Fighters,
cajolers, and commanders of Indians. The Western Indians
attached to the French army at Lake George appalled its general. Chippewas, Ottawas, Potawatomis, Winnebagos, Hurons,
Miamis came east to fight Braddock. They and other tribes
came east repeatedly to harry the borders, raid the communications, raid the raiders, and do the outpost work of every campaign. They were with the horrified Montcalm when he took
Oswego. If Christian Indians started the massacre of British
prisoners after the surrender at Fort William Henry, the Western Indians finished it. They were as unstable as their dreams
but they were always bloody. As the war turned against the
French, so did they, but they had been of immeasurable help.

And when the war ended, the Indians of the West found
that for them as well as the French the world was turned upside down. They had bet on the wrong power. But they had
made no worse mistake than the Indians who bet on the British.
No one may say what the perpetuation of France in the New
World would have meant to the Indians north of the Spanish
lands, but at worst it would have meant a far kindlier fate than
they got. The English wanted land: their mouths were choked

with dirt. Victory for them could mean only one thing for the Indians, destruction.

The closing-in began in 1758 with the fall of Louisburg, that great fortress, of Fort Frontenac which meant that the West was cut off, and of Fort Duquesne which as Pittsburgh would hold the Ohio. By 1759 Montcalm was thinking of the solution that suggested itself to Jefferson Davis at a similar time, escape beyond the frontier to rally and hold the West. A year earlier he had won at Ticonderoga by defeating a big army under a stupid general. But now he faced not only an overwhelming superiority in men and arms but perfect specimens of the two types of general who have won most of England's victories and who have in common only distrust of the intellectual process. Amherst was indomitable, slow, thorough, and by the book. Wolfe was intuitive and touched with the masochistic sentimentality that can be counted on to rouse the British heart. Montcalm had to pull in to the citadel of Canada, abandoning Ticonderoga, where Amherst began to build with enormous labor the unneeded fortress whose guns would eventually drive the British army out of Boston. Oswego was reoccupied. Niagara fell. Rogers was off to St. Francis. The last Western Indians gave up the fight and went home. In September Montcalm and Wolfe met on the Plains of Abraham in the battle that killed both of them.

There was sporadic fighting the next year, 1760, but the war truly ended and New France fell when Quebec surrendered. The river of Cartier was English. Champlain, Frontenac, La Salle, Jolliet, the martyred Jesuits, Perrot, Duluth, Vérendrye, the shining company of great, far-seeing men moved beyond the sunset into the dark, leaving North America only a memory of their dream.

ᐁᣠᐁ

At the beginning of his climactic volume Francis Parkman says, "The most momentous and far-reaching question ever brought to issue on this continent was: Shall France remain here or shall she not?" No, not quite. The most momentous and far-reaching question ever brought to issue on this continent was, "whether that nation or any nation so conceived and so dedicated can long endure." Yet there was another question that linked these two together, a question which the continent itself asked — and was to answer. It was: Are there geographical units here to which political units must correspond? It was asked so quietly that down to today many have never realized that it was asked at all, though by the time Canada surrendered many had heard it. None knew the answer then for more than half the continent remained unknown.

Men had to get about answering it in conditions that had changed altogether: the issue that Parkman phrased had turned the world upside down. Unquestionably, of the changed conditions the most important was that the English colonies had become, though not yet. a nation, a people of themselves, an imperial people. Whether the Americans had completed or were just completing their separateness counts less than that they had entered on an imperial expansion. As both a dream and a fact the American Empire was born before the United States.

It was important too that the colonies no longer had an enemy at the north to fear, to fight, or in any way to take into account. But the thing more immediately important was one part of the peace settlement. In the distribution and rearrangement of lands conquered during the war Cuba, which had become British, was restored to Spain, who paid Florida for that redemption. Canada became British except for two minute specks off the coast of Newfoundland, St. Pierre and Miquelon. So did everything that lay east of the Mississippi River, except some 2800 square miles which included the mouth of the river and the town of New Orleans. This master key to the continent, this portal to the heartland, had been secretly ceded to Spain before the final peace treaty was written. With it Spain had also acquired everything west of the Mississippi and south of Canada that had been French.

From then on that was Louisiana: North America south of Canada, west of the Mississippi as far as its drainage might extend, but not including what had been Spanish territory before the cession. That was Louisiana; Louisiana was now Spanish. It comprised an entirely new pattern of forces, which would exert a constantly increasing torque. And nobody knew how far west Louisiana extended.

Chapter VII

Map 17.

# Twelve-Year Armistice

Unexplored

# VII

# Twelve-Year Armistice

ENGLISHMAN, although you have conquered the French, you have not yet conquered us." Thus a Chippewa chief at Mackinac in 1761, to the Alexander Henry who is called "the elder." A native of New Jersey who had been a sutler in the war, he was one of the first English traders to reach the upper country after the surrender. He traveled the whole distance through tribes still hostile and bewildered by the surrender. In order to go at all he had had to take one of the conquered Frenchmen as a partner.

The chief explained that the English had been able to win the war only because the French king, who was old and worn out with fighting, had fallen asleep. But now his sleep was nearly ended; when he should awake, "What must become of you? He will destroy you utterly." The Chippewas had sent many of their young men to fight for the French. Many had been killed and that their deaths remained unpaid for was a charge on those they had left behind. The debt could be settled with blood of their enemies, a fee easy to collect at Mackinac. Or their spirits would stop wandering in the between-world and go to their long home if their bodies were covered, covered with such presents as had been healing the sorrow of mourners for a century and a half. But the English had made no move to dry the mourners' tears, and the chief came to his exordium. "Englishman, your king has never sent us any presents nor entered into any treaty with us, wherefore he and we are still at war; and until he does these things we must consider that we have no other father nor friend among the white men than the king of France." Furthermore, in order to keep the international situation from degenerating, it would be a good idea for Henry to set out some English milk right now. Rum: it would replace brandy as the medium of diplomacy.

Few Indians west of Iroquoia had dreamed that their Father over the water could be defeated. They therefore assumed that the news out of Montreal and Quebec meant a merely temporary setback. The French worked hard to encourage the illusion, hoping that they could use the Indians to get back something of what they had lost. Parts of the Western garrisons and many traders had fled to Kaskaskia, beyond the Mississippi, or all the way to New Orleans. Along the Western border from then on there were always intrigue, conspiracy, and preparations for the Return. The dream of crushing the conquerors was secret but efforts to undermine them and to limit their expansion was overt.

The British heartily co-operated in their own frustration. In the wilds of that wild country a servant of the king who was not yet Lord but only General Jeffrey Amherst was forced to extemporize policies for dealing with the conquered and their Indians. No libel ever accused him of intelligence and a general officer could not stoop to accept advice from such men as William Johnson and George Croghan who, though the most skillful leaders of Indians the English had developed, were colonials — that is, hicks. As for the Indians who had tortured, killed, and scalped British soldiers, let them learn discipline and let them feel the King's hand. To give them presents — a necessity in all dealings with Indians — why, damme, sir, it would be like paying tribute to pirates. He instructed agents and commandants, who dared not obey him so long as they still had goods, that there would be no presents. Especially there would be no gunpowder, which the Indians had to have in order to live and which the French had always given them in adequate amounts without expecting any return, for to make them earn gunpowder would be the shortest way to teach them discipline. Amherst expected some unrest to follow the adoption of his policy, but Indians were cowards and if they should make any trouble a handful of lobsterbacks could bring them to heel.

Amherst was abetted by others of similar ability and by the

renewed pressure for the sale of Indian land. So that by the time news of the peace settlement reached America, Great Britain had come close to losing the conquest. Before the Western tribes could organize an uprising of their own, the Iroquois were sending embassies west to foment one. The Iroquois knew that the new empire would be territorial and that territory would cease to be Indian. For a century they had been trying to organize a Western confederation against white men. They had never succeeded and neither had the Foxes, who had tried to effect a pro-British coalition, but now conditions were favorable. The Western tribes were French and they believed that Canada would rise again. They refused to see themselves otherwise than as sovereign peoples who owned the West and who tolerated the white man there only on a commercial basis. They had not been defeated, they were nobody's allies now, and they set about preparing a war. They had learned a good deal about working together, so that something like a general movement got under way. Ambassadors bringing war belts from the Iroquois, the Delawares, and other Eastern tribes were welcome.

What followed was the most formidable Indian uprising in history. It has been called a confederacy but it was rather a series of spontaneous wars and there was no general staff. But it was a coalition war and it coordinated some campaigns, especially in the West, where it was animated by a remarkable strategist, the Ottawa chief Pontiac. His objective was simple: throw the English out of the West (or as earlier continental thinkers had put it, confine them east of the mountains) and bring the French back. The war aim made an intense appeal to such tribes as his own Ottawas, the Hurons, Chippewas, Illinois, Miamis, Winnebagos, Kickapoos, Menominees, and Potawatomis. He found an effective propaganda for it in the teachings of the Delaware thaumaturge who is known only as the Prophet.

The Prophet was a millenialist, the first of a numerous line whose vision was to restore the ancient purity of the red man, as of Eden before the Fall. He preached the abandonment of everything the white man had brought — iron, guns, clothing, tools, alcohol — and the restoration of the old-time religion and the primitive crafts and skills. And the white man must be expelled: America for the Americans. When the red man should forswear foreign ways, deer and buffalo would return to the salt licks, the forest would grow back over the clearings, streams troubled only by Indian paddles would fill with fish, the lakes would again be covered with wildfowl more numerous than the eye could take in. The supernaturals would walk in beauty and the Indian would recover his lost wholeness. This not ignoble vision of turning the clock back was an evangelical racism. Pontiac made it pro-French.

So in 1763 General Amherst had a war on his hands, the bloodiest war the Indians ever fought. At least five hundred of the king's soldiers were killed — many of them carved up, boiled, and eaten in religious ceremonies. There is no knowing how many colonists were killed but the lowest estimate would have to be well above two thousand. Almost all the tribes south of the St. Lawrence and east of the farthest frontier came in, including the Civilized Tribes of the South who had been incorruptibly British. The demobilized militia had to be re-enlisted to join the army in full-scale campaigns and the local rangers were under arms everywhere. Mackinac, Sandusky, Venango, Ouiatenon, Le Boeuf, Presque Isle, St. Joseph all fell to the Indians. Fort Pitt and Detroit held out. They were attacked repeatedly, and the measure of Pontiac is that he kept Detroit under siege for six months.[1]

Only one outcome was possible and a principal force in bringing it about was the economic need of the farthest tribes, the Crees, Chippewas, Sioux, and their neighbors. The vision of a re-arisen France was all very well but the Far Nations needed traders. Peace broadened down with a series of councils, pageants, and treaties. A rash of new land companies broke out in the East and the routes west filled with trade canoes that were changed in no particular except that their owners were English.

Great Britain had just begun to organize a government for the enormous territory that the French and Indian War had won for it. To create a workable imperial system for half a continent was a task so difficult that perhaps there was not enough political intelligence and administrative experience to perform it well, even if there had been no exigency. But there was the most desperate exigency. In the sense with which the United States was to be familiar from 1776 on, Canada had never had an Indian Problem. In exactly that sense Great Britain had one from the moment when Quebec surrendered. The universal uprising, and especially the attempt of the Western tribes to assert their independence, confronted the ministry with the entire Indian Problem in a condition of crisis before the peace treaty was signed. It responded with one of the *ad hoc* expedients which governments make up as they go. Quickly redrawing its plans for organizing the conquest, it issued what is known as the Proclamation of 1763. Twelve years later the man who had precipitated one world war at the Great Meadows assumed command of an American army at Cambridge Common and another one began. The Proclamation hastened the maturing of the American consciousness. It increased the momentum and accelerated the velocity of Independence. It set conditions and focused energies that would determine the territorial shape of the United States.

It was intended as a temporary makeshift but it hardened into permanence. It was intended to make the settlement of the Western lands an orderly economic and political process. (To all governments Western settlement always seemed susceptible to orderly direction but the people who made the settlements could never follow the blueprints.) It created the first Indian Country, a domain reserved to Indians as their own, where they were assured protection from white exploitation and chicanery. From then on there would always be an Indian Country, a legal fiction, till Oklahoma became a state in 1907.

Besides setting up governments for East and West Florida, the Proclamation erected one solid abutment for Independence to rest on in the kind of government it organized for Quebec. More directly to the point, however, it set aside the area south of Canada, west of the mountains, and east of the Mississippi for "the several nations of Indians who live under our protection," who "should not be molested or disturbed in the possession of such parts of our dominions and territories as, not having been ceded or purchased by us are reserved to them . . . as their hunting grounds." It forbade surveys, ownership, and settlement in this area. It prohibited private treaty or purchase. And it commanded everyone who had settled in the area to get out forthwith.

What, the stunned colonials wondered, what did the King's

ministers think that a war which brought the whole world in had been fought for? Not only did the Proclamation undertake to put a stop to the process of settlement, which nothing could interrupt, and to that of speculating in the future value of land, which nobody would abandon; it also extinguished all the claims of thirteen colonies — chartered, conquered, inherited, theoretical, and imaginary claims — to the West. . . . "It is time," Jefferson wrote in August 1774, "for us to lay this matter before his majesty and to declare that he has no right to grant lands of himself." [2] And among the facts which Congress submitted to a candid world in July 1776 were the King's "raising the Conditions of the new Appropriations of lands," and keeping "among us in Times of Peace Standing Armies without the consent of our Legislatures," armies whose original purpose had been to assure compliance with the land policy and the Indian policy.

In the welter of conflicting economic interests, one issue was sharply and urgently defined. The result may be almost, though not quite, generalized as a victory for Montreal that eventually made the canoe route to the fur country the international boundary between Canada and the United States. Settlement would destroy both fur production and the Indian market. The value of the trade had steadily increased; far greater wealth in both furs and markets was known to exist in areas almost or quite untouched. Existing investments were large, potential profits seemed incalculable. The ministry, bewildered by imperial immensity, as distant as Versailles from the New World, required to make a way as needs must through problems to which its experience gave no clue — the ministry supported furs and the Indian trade as against land. Just long enough to make the Revolution, for still another reason, inevitable.

ॐ

The ending of the war precipitated a vigorous contest between Albany and Montreal for the wilderness business, a contest in which Montreal rapidly drew ahead. And in 1763, the year in which the peace treaty was signed, Pierre Laclede Liguest went up the Mississippi, now the boundary between the British and Spanish Empires, from New Orleans, one of the hinges of the continent and now a Spanish town and a center of the French underground. New Orleans had grown rich enough to support moderately large enterprises but the war had disrupted business and Laclede represented one of the resulting combinations. The French had held the continental arch, the mouth of the Missouri, from Fort Chartres and Kaskaskia and the lower Illinois River — from the east bank of the Mississippi. Laclede would hold it from the Spanish side. There early in 1764 he built the trading post that grew into the city of St. Louis. His adjutant was his stepson who was not yet fourteen years old, René Auguste Chouteau. He was founding the greatest dynasty in the fur trade of the United States. He was also seizing the Indian trade of the lower Missouri, which might have gone to the British. Soon St. Louis had engrossed the trade of the Illinois as well and much of that which victory on the Ohio had seemed to secure. Its competition was felt all the way up the Mississippi and to the center of British activity on the Great Lakes, the increasingly populous depot at Mackinac.

The war had completely cut off the Western trade for only a short time and even in the far West it was partially renewed as early as 1761, when British traders got as far as Rainy Lake. But if not destroyed, it was shattered and the whole market was starved for goods. A near-vacuum had been created; colonials were rushing to fill it before the small seditions left over from Pontiac's war were stilled.

The trade began to develop a new structure, though necessarily its ground plans remained those which the Indians, distances, and routes had imposed on the French, and at first only the top level was changed. Direction, goods, and financing became British, everything else had to be French. Only Frenchmen knew the fur country, the routes, and the Indians. The standard methods and procedures, the skills, the transport, barter, and credit system were all French. The indispensables, the *voyageurs* who took the canoes up the rivers and carried them and the goods over the portages, were French. Alexander Henry had to have a French partner not merely for personal safety but in order to do business at all — though he was an experienced Albany trader. A considerable time passed before anyone could do without French clerks and guides and nobody was ever able to do without the *voyageurs*. One reason why the Hudson's Bay Company delayed coming down into the Saskatchewan plains, and why it at first fared so badly when it did, was the impossibility of getting anyone to compete with them. It could not hire enough Indians to transport goods by canoe nor get Englishmen who could learn how. Finally it brought Orkneymen to the West, a people bred to life on the water but not the Western waters. By the time the large York boats they built came to compete in inland transport, the last era of the fur trade had begun. Long before that, Montreal as a British town had trained its own clerks, guides, and bourgeois. They were an expert and enterprising class but prosaic. Dash and élan did not return to the fur trade till the career open to talents oddly lifted highland Scots to leadership.

The French trade had been straitjacketed with regulation. The English trade was all but unregulated: intercolonial rivalries and the welter of private competitions progressively broke down the system of licenses plus military supervision that was instituted. Even worse evils than those of monopoly control resulted. The goods went west on an always rising flood of rum; debauchery of the customer became systematic. A business that had never been immaculate dipped close to brigandage. "At the Grand Portage," Henry wrote of his arrival there in 1775, "I found the traders in a state of extreme reciprocal hostility, each pursuing his interests in such a manner as might most injure his neighbor. The consequences were very hurtful to the morals of the Indians." Not only to the Indians. The traders hijacked one another's Indians and even one another's furs; chicanery, intimidation, personal violence became common hazards and from now on there would be stresses that resembled guerrilla warfare.

This was the first North American business that required large-scale organization. In the course of the next two decades problems of control, supply, finance, routing, and traffic control produced a system remarkably like that of any big corporation today. It was an elaborate and exquisitely interdependent system, so delicately adjusted that only experts could keep any part of it going and so expensive in the overhead that combinations were inevitable.

Even the physical basis of the system, transportation, was

marvelously complex. It had evolved during a hundred and fifty years of French experience and had elaborated as the distances lengthened west. The goods of the trade — iron and fabricated hardware, woolen and cotton cloth, garments and blankets, muskets and other weapons and powder, trinkets, paints, beads, spices, and the illimitable rum — the goods came by ship to Montreal, where they were packed in *pièces* intended to weigh exactly ninety pounds. At La Chine the canoe voyage began. The entire international trade rested on the properties of birch bark and on the craft which the white man had developed from the Indian models. The *voyageurs* had to know how to make canoes but the best manufacturers were Indians, and guilds of specialists had long since gone permanently into the business at Three Rivers, St. Joseph Island near Mackinac, Rainy Lake, and elsewhere.

The canoe frame was best made of white cedar, the lightest wood of the northern forest and one of the toughest. The bark was that of *Betula papyrifera,* which is variously called the yellow, white, silver, paper, and canoe birch. Strips of its bark, sometimes up to thirty feet long, taken from the trees preferably in hot weather and pressed flat, were fitted over the cedar frame. They were sewed together with watape, a cord made from the roots of various evergreens, commonly the juniper or red cedar. The bark was usually not much thicker than a coin but sometimes (from trees of the farther north) up to a quarter of an inch. All seams were made watertight with pine gum, which had to be renewed at least once a day, frequently oftener. A supply of it went in every canoe, together with spare watape and a big sponge for bailing.

The Indian's personal canoe, ten or twelve feet long, could not be used for freight; the trade canoe was modeled on the larger war canoes of the Far Nations but had undergone a beautiful specialization and refinement. As far as Grand Portage, whether by the route up the St. Lawrence or the more laborious but faster one up the Ottawa River and down the French River to Lake Huron — as far as Grand Portage the carriage was by big *canots du maître*. These were up to six feet broad and up to thirty-five or even forty feet long. They could carry four tons of lading and a crew of fourteen, though eight or ten were more usual. After the packs had been taken across the nine-mile Grand Portage — a ten-day carry — they were loaded in *canots du nord,* which averaged twenty-five feet long, had a crew of five to eight, and carried a ton and a half of lading. An intermediate size, not feasible in the far country, was the *canot bâtard* with a crew of up to ten; a "half canoe" was smaller than the *canot du nord,* say twenty feet. Canoes were usually painted and nearly always had designs in bright color at the bow.

The two places of greatest responsibility and skill were those of the bowman and the steersman, who got higher pay than the "middlemen." In calm water the *voyageurs* could drive their craft as fast as six miles an hour. If there was a following wind they could rig a sail from the oilcloth that covered the freight and make perhaps nine miles an hour. A headwind, however, risked swamping and put a full stop to navigation; it was the commonest and most maddening reason for delay and might last for days. Between Montreal and Lake of the Woods, by the Ottawa River route, there were more than ninety portages, from a few yards to nine miles long and of every degree of cussedness — through bogs, up or down steep inclines, along the knife-edge of cliffs, through crooked rock crevices, or knee-deep in mud or waist-deep in flood water. Over each of these portages the *voyageur* trotted with two ninety-pound *pièces* on his back and a tumpline round his forehead, in stages that averaged six or eight hundred yards. The river-horse type shouldered more than two packs, and legendry has folk heroes toting up to eight. Since portaging was both slow and exhausting, and therefore expensive, it was kept to the minimum that seemed, in a way, safe. The skill of the *voyageurs* was to take through the boiling white water of the *saults* a craft which a moccasined foot could puncture and a second's contact with a snag or rock would rip apart. In big rapids few who were upset lived to swim ashore or be pulled out by their companions. Small wooden crosses clustered on the shores at such places, voices cried out of the water at night, and the rivermen appeased the saints and the Indian deities alike by many kinds of amulet and incantation.

Like the mouth of the Platte, Grand Portage marked the equator. The big canoes turned back from here, now laden with packs of fur. Their crews were the humble, the *mangeurs du lard.* The élite were the *hivernants,* the winterers or north-men who went on to the high country, after such ceremonies as have marked the decisive crossing on all our waterways. One who made the crossing for the first time had to undergo initiation. "I was instituted a *North man* by Batême," John Macdonnell says, "performed by sprinkling water in my face with a small cedar Bow dipped in a ditch of water, and accepting certain conditions, such as not to let any new hand pass by that road without experiencing the same ceremony, which stipulates particularly never to kiss a *voyageur's* wife against her own free will, the whole being accompanied by a dozen Gun shots fired one after another in an Indian manner. The intention of this Batême being only to drain a glass. I complied with the custom and gave the men, between Mr. Neil McKay and self a two gallon keg as my worthy *Bourgeois* Mr. Cuthbert Grant directed me."

Englishmen and Scots found the *voyageurs* hard to bear. They were dirty, profane, excitable, noisy, forever talking, forever singing, of volatile anger and quick delight, comradely with Indians who were even grimier than they. They were everything their new masters were not, and there was no damned sense in the superstitions and rituals with which they surrounded every detail of their job. David Thompson, who was Welsh and evangelical and may have missed the more esoteric allusions in Indian small talk, says that their swearing shocked even the Indians. "The fact is," he added, "Jean Baptiste will not think, he is not paid for it; when he has a moment's respite he smokes his pipe, his constant companion and all goes well; he will go through hardships but requires a belly full at least once a day, good Tobacco to smoke, a warm blanket and a kind Master who will take his share of hard times and be first in danger. Naval and Military men are not fit to command them in distant countries, neither do they place confidence in one of themselves." He should have said, British naval and military men.

It was well that they had gaiety and pride, for only men reared in the poverty of French Canada would have done work of such skill and danger for such insignificant pay. A middleman might make the trip between Montreal and Mackinac, a full season, a year's work, for a hundred and fifty livres worth four-fifths of a provincial shilling each, or the longer one to Grand Portage for two hundred and fifty. (Frequently paid in goods, which encouraged him to think he was making a profit when he traded them, though his employer had taken a big discount.) He could have a blanket, shirt, and trousers for his outfit — winterers got more — and could find the rest of his clothes himself. He would not live high for canoe-space was too valuable to be wasted on anything but concentrated food and there was seldom time to fish or hunt or even pick berries on the way. Leached corn and buffalo tallow would suffice him as far as Grand Portage or Rainy Lake, wild rice and tallow from there to Lake Winnipeg, and the hazard of the commissary beyond, with a diet of frozen fish in the winter time without salt or any variation except that supplied by the annual failure of supplies. But the rum was good and wind and rain were part of the job. A friend died of an arrow in his guts if the Sioux were up or you dragged his body from the calm water below the *sault,* but the Virgin watched over you and when the autumn came you would be home. Ahead of you were the forest and the white water, the shadow and the silence, the evening fire, the stories and the singing and a high heart.

⎙

Not till 1768 did the British traders out of Montreal get as far into the high country as their French predecessors had reached. First the intervening country occupied them, the upper Lakes and the upper Mississippi, whose Indians were half naked and half starved. The tribes blotted up enormous quantities of goods and English milk, and paid for them with enormous quantities of furs. Henry notes that on one day, July 1 1767, a hundred canoes from the Northwest, a single area, arrived at Mackinac. Each one would be carrying nearly three thousand pounds of beaver, fifteen hundred skins. Though St. Louis and New Orleans were draining the trade of the Illinois and making inroads along the upper Mississippi watershed, the value of the Canadian fur trade in that year, exclusive of the Hudson's Bay Company, must have been about 150,000 pounds sterling.

From St. Louis the French-Spanish combinations pushed up

the big tributaries such as the Des Moines and Minnesota and down Lake Michigan to Mackinac. The two trade frontiers met squarely at the mouth of the Wisconsin River, where the lush flat called Prairie du Chien became a common rendezvous. In this nearest West beliefs but little younger than the white man's awareness of North America found a new focus. It was here that the British began the search by land for the Northwest Passage that they had not found in two centuries of heroic search by sea.

After the fall of Montreal in 1760, the celebrated partisan leader Robert Rogers was dispatched with two companies of his Rangers and one of provincial infantry to receive the surrender of the Western forts. Since Detroit was the symbol, he held there the rude, momentous ceremony that made the West British. Presumably his interest in the far country was first roused by Indians and French veterans whom he met on this expedition. The next year he was ordered south on the campaign that was to subdue the Cherokees and here he caught fire from the Irish zealot who was now the Governor of North Carolina, Arthur Dobbs.

More than twenty years earlier, Dobbs had convinced himself that, notwithstanding the negative reports of many mariners, a water passage led to the Pacific from Hudson Bay. Though in time he embraced every conception of the Northwest Passage extant, the basis of his *idée fixe* was the imaginary voyage in the high latitudes of a probably fictitious Spanish admiral named De Fonte who was supposed to have proved a hundred years before that ships could sail from ocean to ocean. Dobbs assimilated the imaginary geography of De Fonte to the Hudson Bay region and concluded that the Hudson's Bay Company, whose charter required it to search for the Northwest Passage, had developed an institutional interest in concealing the truth. He headed a formidable attack on the great Company and, a member of Parliament, sponsored bills which would have canceled its charter for failing to explore its domain. Mastering (in his way) all the available information, he achieved new theories and in 1741 sent Christopher Middleton, a navigator with experience of the Bay, to make a careful examination of its western and northwestern coasts. Middleton proved that there was no channel that led toward the Pacific. But he did not prove it to Dobbs, who accused him of taking bribes from the Company to falsify his records. He sent out another expedition to prove Middleton a suborned man — the last one that ever tried to find the Northwest Passage from Hudson Bay — and remained entirely unconvinced when it led to the same dead end. His hatred of the Hudson's Bay Company, though of high voltage, was incidental; he was a carrier of the expansionist energies that intended to make the Pacific a British instead of a Spanish lake.

Dobbs continued his agitation and pamphleteering for a water route to India. In 1744 he published *An Account of the Countries Adjoining to Hudson's Bay,* a vigorous, absorbing book which assembled everything that was known, rumored, guessed, logically deduced, and imagined about the Northwest. It is a visionary's argument and perhaps the most shining eighteenth-century example of what the imagination can do when it has a blank map to work on and is handicapped by no empirical knowledge whatever. With its factual data (which are confined to the Bay fringe and the minute specks that the Pennsylvanians and Virginians had acquired along the Appalachians) mingles a hash of Lahontan and his kind, of Dobbs's grotesque deductions from the illiterate La France's account of his travels, and of Daniel Coxe. Coxe, whose father had sent the first British ships into the Mississippi, had published in 1727 a book called *A Description of the English Province of Carolana.* Its Carolana was roughly Louisiana, its information and thinking were precisely the same quality as Dobbs's, and it exerted a baneful influence throughout the century. All this and additional bits gathered wherever there were bits to gather proved to Dobbs that there was indeed a passage to the Pacific, if not De Fonte's then the Strait of Anian, which usually lay south of De Fonte's route, or something connecting a crescent of lakes and rivers pinned centrally on Lake Winnipeg — or something else. He never did leave off crusading. He was an old man when Rogers went south in 1761 but, the royal Governor of North Carolina, he was still so intent on the passage to India that the colonists accused him of ignoring the Indian wars to further its discovery. By this time he had forced the Admiralty to offer a reward of twenty thousand pounds for the discovery of the Northwest Passage.

In 1763, ordered to Detroit again, Rogers fought his way into it through Pontiac's besieging Indians. Two years later he was in London, a fashionably successful author and a candidate for the discovery of the Northwest Passage. . . . He published his campaign diaries, a handbook on the colonies, and, anonymously, a drama called *Ponteach,* full of stately rhetoric and one of the first plays ever written by an American. Like his lieutenant Jonathan Carver, he may have forfeited the respect of scholars because he sold well. . . . He had the backing of influential men, introduced to them by Dobbs who died early that year. His plan for the discovery of the Northwest Passage postulated much of Dobbs's thinking. He intended to carry it out from Mackinac, to the governorship of which he also aspired. His petition for a subsidy to explore the West was rejected but he got the governorship and decided to make the exploration as a private enterprise.

Since Rogers would have to administer Mackinac, he had to delegate the exploration to others. He selected to lead it James Tute who had captained one of his Ranger companies and was like him a Yankee. (Rogers was from Methuen, Tute from Deerfield.) On returning to America he met and hired another Ranger officer, Jonathan Carver of Weymouth. That Carver was a mapmaker appears to explain why Rogers employed him. And that by the time Carver wrote his book Rogers's enemies had got him into the furious trouble which dogged him all the rest of his life appears to explain why Carver did not mention him in the book but presented the journey it describes as his own enterprise.

On the assumption that the western opening of the Northwest Passage could be more easily discovered than the one in Hudson Bay, Rogers projected an overland journey to the Pacific. This was therefore the first effort to cross the continent that Englishmen had made since the earliest years of the seventeenth century, when the continent ended at the Alleghanies.[3] What was in Rogers's mind cannot be stated in detail, for there was little concrete detail, but his petition of 1765 makes a lordly simplification: "The Rout Major Rogers proposes to take is from the Great Lakes towards the Head of the Mississippi, and from thence to the River called by the Indians Ouragan which flows into a Bay that projects North-Eastwardly into the [country?] from the Pacific Ocean and there to Explore the said Bay and its Outletts, and also the Western Margin of the Continent to such a Northern Latitude as shall be thought necessary." [4]

A great name has been found. Rogers was to make it Ourigan in his second proposal, 1772, and in 1778 Carver was to use the spelling that endured, Oregon. No one knows its provenance. And no one can mistake its reference: this was no actual river, it was that product of pure thought, the Great River of the West.[5] In Rogers's mind, presumably, it was associated with the Strait of Anian, the entrance of De Fonte, or something else he had got from Dobbs. When after seven years of turmoil, calumny, and disgrace Rogers again proposed his exploration, he had refined his ideas in the light of Carver's. He now intended to go to the source of the Minnesota River, whose latitude he missed by less than one degree, and thence "to cross a twenty-mile Portage into a branch of the Missouri, and to stem that north-westerly to the Source: To cross thence a Portage of about thirty Miles into the great River Ourigan; to follow this great River through a vast and most populous Tract of Indian Country to the Straits of Anian. . . ." This later proposal would be of only speculative interest, since Tute and Carver acted on the first and vaguer one, except for the positive statement that only a thirty-mile portage separates the Missouri waterway from the Oregon. As a generalized concept this idea was almost immortal, dating back to Verrazano, and it had had concrete embodiment since Jolliet, but this is the form that was to be a fixture till 1805. Presumably it is due here to Dobbs, who probably got it from Coxe. Coxe made the "land carriage" even shorter, half a day "between the River Mechsebe [the Mississippi, not the Missouri] and the South Sea Stretching from America to Japan and China." [6]

The Missouri played no part in Rogers's original plan. He directed Tute to avoid it, in fact, and to travel northwest from the Sioux country to "Fort La Prairie." Presumably he meant the Assiniboine River, and even so was moving it a long way south. From there Tute was to bear northwest of north to the "forks of the great River Ourigan." He would find that the Oregon emptied into an arm of the sea between 48° and 50° and that, not far from its mouth, a channel led to Hudson Bay

at about 59°. It must be remembered that the Oregon flowed
west and that nothing Rogers had heard from French wilderness men could have given him these ideas, which are from
theoretical geography. The latitude of the Hudson Bay end of
the Northwest Passage corresponds to a favorite conception of
the Strait of Anian but that given for the Oregon's mouth seems
to have no mooring even in mythology, though it corresponds
to Juan de Fuca's imaginary strait. And Tute is not going to
have any trouble with mountains.

Carver entered the wilderness before Tute, wintered among
the Sioux before joining him, and got farther than he into the
interior, beyond the forks of the Minnesota River. Rogers was
unable to forward supplies, however, having been overtaken
by the troubles that culminated in a trumped-up charge of
treason. So the discovery of the Northwest Passage petered out
on the western shore of Lake Superior.

Carver made his journey in 1766 and 1767. He did not publish his *Travels in the Interior Parts of North America* till 1778.
The book profits from what had been learned in the intervening years but in the main it represents Carver's own knowledge.
The caustic Yankee with the flamboyant spelling, Peter Pond,
says that Carver "Gave a good a Count of the Small Part of the
Western Countrey he saw But when he a Leudes to Hearsase he
flies from facts in two Maney Instances." [7] But his book de-
serves the trust and popularity which a fashion in scholarship
undertook to destroy. It is the work of an intelligent man, a
sharp observer, a shrewd analyst of rumor, and a good writer.
It is the first detailed description of Minnesota and even of
Wisconsin — Perrot and the Jesuits had described the latter
only briefly — and it is not only a reliable account of the Eastern Sioux but literally the first account that had anything useful to say about them. The first English report on the far West,
it is better than anything the French had produced. If it dis-
covers a deserving hero in its author and if that author finds no
reason to inclose in quotation marks some passages where a
niggardly caution would have used them, there were distinguished precedents.

Knowledge of the nearer West had increased — knowledge
of many parts of North America grew rapidly during the 1760's
and 1770's. Carver ably analyzed this nearer geography, but

Beyond proved resistant. He fell victim to the symmetry of the neat pyramidal watershed that had enabled Charlevoix, Vérendrye, and Beauharnois to rationalize their vision, and probably did so on their authority.

He works it out that the Missouri and the Minnesota River rise in the near-by height of land, within a mile of each other. The same tableland — it is the Shining Mountains, "the highest lands of North America" — also holds the sources of "the four most capital rivers on the Continent . . . the St. Lawrence, the Mississippi, the River Bourbon [here probably the Nelson, sometimes the Hayes], and the Oregon or the River of the West." (Note that the Missouri is not a capital river.) The first three of these rise within thirty miles of one another; the Oregon, which Carver takes to be a river of equal size, is somewhat farther west, but not much farther. It is especially noteworthy that in these mountains, "a capital branch of the River Bourbon, which runs into Hudson's Bay, has its sources."

The composite shows that some facts have been winnowed since Vérendrye and the actual geography must be brought to bear on it. The Shining Mountains *here mentioned* are still Vérendrye's mountains whose stones shone night and day. They are the Turtle Hills on the Manitoba–North Dakota boundary, nearly a hundred miles west of the Red River, traversed east and west by the 49th parallel and north and south by the 100th meridian. Geographical thinking had long made the Great Lakes an extension of the St. Lawrence and therefore Lake Superior is its proximate source. (The streams that fall into it from the west, coming down the Superior Upland, are short.) The headwaters of the Red River are more than a hundred and fifty miles from the height of land that divides it from Lake Superior drainage, though a hundred and fifty miles downstream the watersheds approach within a few miles. But it is true that Red River drainage and that of the Minnesota are less than two miles apart — in the neck of land between Lake Traverse and Big Stone Lake — and in this sense, which is by no means what Carver meant, the Hudson Bay and Mississippi watersheds are even closer than he thought. At this point, too, small streams of the Missouri River drainage are only half a dozen miles away. The *source* of the Mississippi, however, is a hundred miles east of the Red River and a hundred and sixty miles northeast of its source.[8]

## Map 18. Jonathan Carver's North America

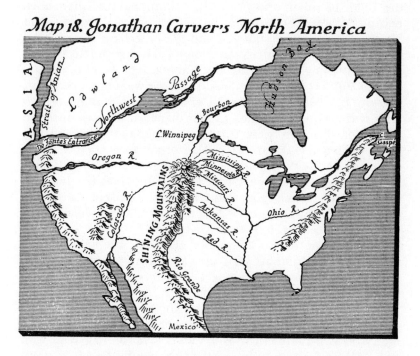

On the whole, rather than shrinking, the map has expanded. From his prime center, Carver says, each of the four capital rivers flows "upwards of two thousand miles" to the sea. This more than doubles the actual distance from his height of land to the mouth of the Nelson, and a little less than doubles the distance to the Gulf of Mexico. His east and west distances are good, though one of them is accidental. The distance from the western end of Lake Superior to Cape Gaspé is almost exactly thirty degrees of longitude; from the same place to the Pacific, along the 46th parallel, is thirty-two degrees. Carver understands that the Oregon, the River of the West, flows "upwards of two thousand miles" west from the Turtle Hills, and empties in "the bay at the Straights of Annian."

The core of his betrayal is the Shining Mountains. They were "the highest lands in North America," and he made them just a little more than thirty miles west of the Minnesota River.

He made a fix on these, the Turtle Hills, which are actually on the 100th meridian and nearly three hundred and fifty miles beyond the westernmost point he himself had reached. For the Sioux with whom he wintered and the miscellany of tribes he met at Prairie du Chien and along Lake Superior told him many true things. He heard much about New Mexico, including the Villasur massacre. (Spanish horses were now shod with silver.) But what counted most of all was the Shining Mountains, a conception of which Carver managed to work out. He concluded that they were three thousand miles long and ran from Mexico northward, east of California, and ended at about 47° or 48° N. (To provide comfortable space for the sea-level Strait of Anian.) They separated the waters that fell into the Gulf of Mexico from those that fell into the Gulf of California, and the Mississippi watershed from that of the Pacific. They were, in short, the Continental Divide.

The concept of a Continental Divide had been clearly phrased by Sir Humphrey Gilbert. Most (but not all) formulations of theoretical geography had to assume some kind of fundamental divide and it was especially necessary to conceptions of the River of the West and to the teleology by which the eighteenth century's right reason imposed a beautiful symmetry on North America. But Carver was the first man who ever expressed a conception of the Continental Divide that approximately corresponded to the actual one.

The trouble was that he had two sets of Shining Mountains which he understood as one. One supplied the actual parting of waters, he heard about it in reports of Indians who had heard about it, and he did exceeding well to analyze their reports as he did: these were the Rocky Mountains. The other set was the Turtle Hills, which he took to be only a few miles away and which for him not only separate the Mississippi drainage from that of the Pacific but also separate it from the rivers that "fall into Hudson's Bay or the waters that communicate between two seas." The waters that communicate between the Pacific and Hudson Bay were the Northwest Passage, which he identified as the Strait of Anian. The same bay whence it leads eastward holds the mouth of the two-thousand-mile-long Oregon, which has rounded the northern end of the Shining Mountains after flowing from the country of the Sioux through "plains that according to their account are unbounded and probably terminate on the coast of the Pacific Ocean."

That Carver was able to get so much right which Vérendrye did not shows that the accretion of knowledge had been considerable, but his cardinal misconception diverts him into apocalypse. The mountains are called Shining because of an "infinite number of chrystal stones of an amazing size with which they are covered and which, when the sun shines full on them, sparkle so as to be seen at a very great distance." Beyond them lies the future, waiting for new Columbuses and Raleighs to open their incalculable wealth to mankind. Carver is clear that it will be American mankind: La Salle's Garden of the World has been reborn.

Others after Carver worked out almost identical answers and European and American cartographers adopted them. But far beyond this still fogbound part of the world map, out in the Pacific Ocean and along its bounds, a new great age of discovery dissolved away huge areas of ignorance as the 1760's ran out and the '70's came on. Western man learned much about the earth he intended to organize when the British Empire challenged Spain for the Spanish ocean, and when Spain found in Charles III a king who gave promise of meeting the challenge and restoring the glory of the great reigns. Fundamental in this harvesting was the fact that various sciences had achieved finer accuracies. In 1744 Cassini de Thury presented to the French Academy the first eighteen sheets of an outline map of France, a triangulated, checked, dependable map such as there had never before been of any moderately large area anywhere. Various prime meridians were fixed and their land relationships examined. A century of effort by gifted men to make possible the determination of longitude at sea reached a climax with John Harrison's first chronometer in 1735 and a more brilliant one with his great Number Four in 1759. Such advances as these are the bench mark of a new era.

Though unknown areas of great extent remained, cartographers were mapping the Pacific Ocean far more accurately than before, and now they would begin to get the western coast of the United States located. But they would not refine the Far West beyond Carver's Shining Mountains and River Oregon.

Even before he wrote his book his fellow colonials who were

in the fur trade had amassed empirical knowledge that revealed the falsity of his logical derivations. But their knowledge was of the Northwest, not as yet of the West.

∽

By 1766 Canadian traders had reached the southern part of Vérendrye's domain, Lake Winnipeg, the Red River, the Assiniboine, Lake Winnipegosis. They proved the soundness of his commercial strategy and, like him, cut deeply into the Hudson's Bay Company trade in this plains area. They went on to the Saskatchewan country, as Vérendrye had done, and learned the great importance of its furs, routes, crossroads, and strategic portages. One "Franceway" was at The Pas in 1769, James Finlay at Nipawin (150 miles upstream from it) in 1769, and Thomas Corry at The Pas and Cedar Lake as advance agent for the formidable Frobisher brothers in 1771. In this short time they deprived the Company of so great a bulk of the furs it had been getting that it was at last coerced into changing its hallowed and by now solidified system. In spite of very great difficulties of supply and transport, they were taking the trade to the interior of the great fur country, as Vérendrye and his successors had done. They were getting such returns that a couple of seasons would make a man rich and the total return of what was so far only a few traders was notably increasing the capital wealth of Montreal. Since Henday the Company had sent out many of its servants as public relations officers, to travel with the Indians, create goodwill, and encourage them to come to the Bay. Now, though it sneered at the Canadians as "Pedlars," it had to do what the great Radisson had wanted it to do from the beginning. In 1773 it sent its Matthew Cocking to make a circuit of the interior and report on the activities of the Pedlars and the next year it took the revolutionary step at last and packed off Samuel Hearne to build the first post it had ever had far inland.

On the Saskatchewan the Pedlars learned about a country even wealthier in furs and prepared to move into it. So they set up still another deadly competition with the Company, passed the farthest French penetrations, and arrived at new things. North of the Saskatchewan the dry plains ended and the forest

# Map 19. The Canadian West, 1763-1793

Miles
0  300

Raisz

began again, a forest thickly sown with streams, swamps, and lakes. North (and northeast) of the forest belt stretched the tundras, whose map looks like the Milky Way, all lakes, all streams, all muskeg, all barrens. Winters lengthened to seven, eight, nine months. The summer nights were only twilight. If beaver fur grew finer as you traveled north, the swarms of mosquitoes grew thicker and deadlier till you longed for the terrible cold that would put an end to them. There were new Indians, a medley of Athapascan tribes, the *babiche* people who made from deer and caribou hide long skeins of infinitely versatile rawhide cord and who were masters of travel by snowshoe and sledge. The first whom the Pedlars met were those whom the Hudson's Bay Company called "the Northern Indians," the Chipewyans. Beyond them were tribes and subtribes whose names are so confused that they can seldom be confidently identified, such names as Red [copper] Knives, Hares, Dog Ribs, Caribous, Quarrelers, Beavers, Slaveys.

They were a shy, baffling people. They seemed timorous and seemed to grow more timorous farther north. Yet they were violent and murderous (though never toward white men) and their wars were revolting. The squalor of their barren country made them seem revolting in many ways, but their adaptation to that land was superb. If they were primitive among primitives they solved the problem of existence with complete success, and today they live much the same kind of life they lived before the white man in a country which he has never been able to occupy. The snowshoed Bedouins of the West, the Crees, had been raiding them for a century, irresistible with Hudson's Bay Company iron and guns, shoving them farther into the drowned lands and selling them goods at a ferocious markup. The Chipewyans and others on the southern fringe had gradually become direct customers of the Company, paddling the endless distance through the maze of waters to the Churchill River and down it to Fort Churchill at its mouth. So they became middlemen for the still higher tribes. The goods they took back with them worked farther into the desolation, to Lake Athabaska, Slave Lake, beyond the 65th parallel, and on toward the Arctic Ocean.

From the base of the lower Saskatchewan the Canadians began to work toward this country. They were so far from

Montreal that now it took two seasons to get the goods to the Indians, so far from Grand Portage that the round trip could not be made in a single season and they had to work out a way of forwarding goods from an advance depot. On the plains no one ever lacked food for long but game was scarce in the northern forest and at the posts there was no food but fish all the long winter and but little else at any time. The posts had to be built on lakes where fish were abundant and the garrisons had to spend the autumn seining them, mostly whitefish but whatever else there might be, carp, pike, lake trout, and the huge sturgeon of the region. "The freshwater hog," David Thompson called the sturgeon, explaining that it was abundant only in muddy waters. "Whatever is not required for the day is frozen and laid by in a hoard, and with all care is seldom more than enough for the winter." There was seldom quite enough, in fact, and every winter at every post beyond the buffalo country there were periods of very short rations and some when starvation seemed inevitable.

It was a winter country, for the short season when the streams were open was a frenzy of transporting furs, goods, and supplies between the depots and the posts. (A new kind of food for the canoe fleets had to be found and a system of distributing it developed.) Travel was by snowshoe and dog sledge, under threat of blizzards and sudden drops of temperature from the normal thirty below to sixty or more. But one welcomed this necessary business of intercommunication for life at the posts was monotonous to the verge of the wilderness anxiety for which the American language has found such names as snow fever, cabin fever, moonwalking, and the like. A hundred miles through gales and snowstorms was a small price for the sight of a different log hut, the sound of English words, and a chance to get drunk with a man whose thought ran in the same images. "I took leave of Mr. Frobisher," Alexander Henry writes, "who is certainly the first man that ever went the same [so great a] distance in such a climate and upon snowshoes to convoy a friend." Books were beyond price. Rivals who would bully, beat up, or take potshots at one another would also carefully forward through the snow whatever London magazine of three years ago might have turned up in a bundle from York Factory or Grand Portage. Indeed the frequency of conversion to the

Christian life among these casehardened men is to be explained by the escape from boredom, world-emptiness, and drunken Indians provided by the King James Bible.

They came to hate their *voyageurs* not only for their slovenliness but for their acceptance of the wilds, their high spirits, and their cordial if verminous brotherhood with the Indians. They themselves had no trouble with Indians this side of the Gros Ventres and Blackfeet (so long, that is, as they were sober) for good customer-relations had to be preserved and the Indians were cordial to the providers of goods. But most of them despised the customers and treated even the women they bought like foul animals. But they bought the women: a man's nature asserted itself when the wildfowl had gone south, the elk had withdrawn deep into the woods, and the clouds piled up ever higher lead-colored mountains that would soon loose snow.

> *Thursday October 10.* [This is the Vermonter Daniel Harmon, who for a long time postponed the convention, from qualms which few besides him ever felt.] This day a Canadian's daughter, a girl of about fourteen years of age, was offered to me; and after mature consideration concerning the step which I ought to take, I have finally concluded to accept her, as it is customary for all gentlemen who remain for any length of time in this part of the world to have a female companion, with whom they can pass their time more socially and agreeably than to live a lonely life, as they must do if single. If we can live in harmony together my intention now is to keep her as long as I remain in this uncivilized part of the world, and when I return to my native land I shall endeavor to place her under the protection of some honest man, with whom she can pass the remainder of her days much more agreeably than it would be possible for her to do were she to be taken down into the civilized world, to the manners, customs and language of which she would be an entire stranger. Her mother is of the tribe of Snare Indians, whose country lies along the Rocky Mountains. The girl is said to have a mild disposition and an even temper, which are qualities very necessary to make an agreeable woman and an affectionate partner.

They were a tough breed in a tough business. The trade was hardening into the most brutal practices, but if they could stand theft, hijacking, and occasional murder, their finances could not. The new conditions required concentrations of capital that could underwrite credits of three or even four years, investment in sloops for the Great Lakes which lowered the overhead on freight, larger payrolls, and the storage of goods and food in depots for the high country. But the price wars of competitors were threatening them with bankruptcy. South of Mackinac small operators could do a profitable business but the Northwest demanded the organization and methods of big business. The Canadians stumbled toward the solution, the first American trust, in great part by extemporizing small co-operations against the Hudson's Bay Company.

The Company was both reluctant and ill prepared to meet the competition from Montreal that in the early 1770's was getting its best customers and its best furs. It was staid, handcuffed by tradition, imprisoned in a rigorous caste-slavery that gave its servants no incentive to be enterprising, with few wilderness men and entirely without rivermen. Its first reaction to the Canadian invasion of the Saskatchewan was to increase its advertising there, to send more agents to persuade the Indians to scorn the Pedlars and resume trading at York Factory for auld lang syne. It was wasted effort and the Company had to imitate the Pedlars. The remarkable explorations of Samuel Hearne were part of the new adaptation but at first a temporizing part. The first effort was to try to increase the trade in the far North and Hearne was ordered to travel far beyond any penetration that had yet been made. His principal objective was to drum up more business but he was also to reach the Northwest Passage if it existed there, and to find a big river about which the Company had long been hearing, a river of unknown direction but apparently rich in virgin copper.

After two failures Hearne made a third journey which in 1771 took him to the rumored river, the Coppermine, and across the Arctic Circle, down to the Arctic Ocean, and on the return trip to Great Slave Lake. It was a tremendous feat. It practically proved that there was no Northwest Passage but some time passed before anyone who was interested understood, even, that it narrowed the possibilities. It accomplished little for the Company's business and the Pedlars had to be met. In 1774 the Company sent Hearne to meet them, in the heart of the country they had taken over. On Pine Island Lake, between the Saskatchewan (which, having changed its channel now flows through it) and Cumberland Lake, he built a post which he named Cumberland House. With the Pedlars all about it, the Company was for a long time still outmaneuvered, outguessed, and out-traded but the establishment of Cumberland House marks the beginning of what was to become the most violent competition in fur trade history. It also hastened the big combination that would be called the North West Company.

But something intervened.

Thomas and Joseph Frobisher, who had already done a rich business in the Northwest, had solved some of the problems of further advance, and were beginning to solve others, spent the winter of 1773–74 on Cumberland Lake, not far from the site where Hearne was to build Cumberland House. The next spring, having arranged to be supplied by associates in the plains country, Joseph Frobisher went on north up the Sturgeon-Weir River and the long crooked lakes at its upper end to the carrying place which led to the Churchill River. Here he met Indians in large numbers coming down from the North to descend the Churchill for their annual trade at Fort Churchill. They were valuable Company customers and Frobisher did so profitable a business with them that from then on the carrying place was called Portage du Traite, Trade Portage.

In 1775 two Americans, Alexander Henry and Peter Pond, abandoned the Great Lakes trade where they had both been eminently successful, and joined the increasing company on the Saskatchewan. They were first-rate men and Pond was rather more than that, a man who was to cast a long shadow and initiate great actions. They met at Lake Winnipeg, traveled together toward Saskatchewan, and presently fell in with the Frobishers. Cumberland House had been built, the Hudson's Bay Company was in the Pedlar's best territory, and it would not do to cut each other's throats — hence the first combination. (Pond did not join it this year but went to Lake Dauphin on his own.) Henry and the Frobishers pooled their goods and agreed not to cut prices. . . . Henry lists some of those prices. A beaver skin would buy ten musket balls or half a pint of gunpowder. A one-pound axehead cost three beavers, a colored stroud blanket ten, a trade musket twenty. Compared with Mackinac prices these are, except for the musket, unholy.

Henry set about mastering the conditions of the plains and the northern forest. He had an adventurous winter, with mass drunks for the customers and a couple of grueling journeys. In the spring he visited some Assiniboins on the plains and then accompanied the Frobishers to the vital Portage du Traite, between Saskatchewan and Churchill waters. They went on past it to Lake Isle-à-la-Crosse and presently met some northern Indians. They had come from a country far beyond any the white men had heard of, beyond the reach of rumor and guess, all the way from Lake Athabaska. They talked about a previously unheard of river which came down from the Stony Mountains, the Rockies, and they had a name for it; it was the great Peace River. They said that beyond the mountains was a big salt lake, the Pacific. They went on to say that from the lake into which Peace River emptied another great river issued, flowing north to a sea, but whether the same sea as the one beyond the mountains they did not know.

They were decorous people, and a little sad. They asked for rum but wanted it diluted, lest the young men get too drunk. It had already been diluted to an appalling profit but Henry could bring himself to meet their wishes. First appointing guards to stay sober and protect this lordly bringer of iron and red milk, they had a quiet, rather melancholy two-day drunk. Then they traded their skins and slipped away into their brown, barren land.

Thomas Frobisher stayed here, preparing to go on to the distant Lake Athabaska. Joseph Frobisher and Henry turned back, heading for Grand Portage and next year's goods. The day they started out was July 4 1776.

Beaver Lake, the Saskatchewan, Lake Winnipeg, Lake of the Woods — all that long way east. Their start was late but they reached Grand Portage on October 15, a tremendous trip whose speed was made possible by unusually favorable wind and water. But it was at Lake of the Woods that they heard the world had changed again.

The Indians there told Henry "that some strange nation had entered Montreal, taken Quebec, killed all the English, and would certainly be at the Grand Portage before we arrived there. On my remarking to Mr. Frobisher that I suspected the *Bastonnais* had been doing some mischief in Canada, the Indians directly exclaimed, 'Yes, that is the name, *Bastonnais*.' "

The American people had become the Americans and had

made a nation. As a nation they were fighting a war to establish in politics the independence they had achieved in fact. As imperialists they were breaking up the British Empire in North America to organize an empire of their own from its fragments. What Henry heard about at Lake of the Woods was their capture of Montreal, the center of the fur trade, and their first attempt to conquer Canada.

# Prime Meridian

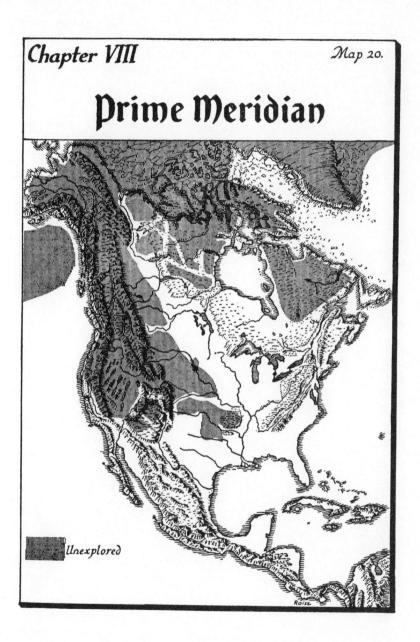

Unexplored

Roist

# VIII

## Prime Meridian

THE LORDS COMMISSIONERS for Trade and Plantations formally represented to the ministry: that "the great
object of colonizing upon the continent of North America has
been to improve and extend the commerce, navigation, and
manufactures of this kingdom." They pointed out that His
Majesty's Proclamation of 1763, to which certain moot questions might properly be referred, was intended to keep the
colonies in subordination to and dependence on the mother
country and to make sure that settlement should extend no
farther west than the kingdom's trade should reach. No one
could have more completely or more justly summarized the
mercantilist principle that had opened a cleavage line across
the British Empire in North America. The Commissioners
proceeded to exemplify the insensitiveness to colonial realities
that characterized the government of Great Britain in good
King George's glorious days. They took judicial notice that
there was important capital in Montreal and a rich trade in the
Indian country. So, "the extension of the fur trade depends
entirely upon the Indians being undisturbed in the possession
of their hunting grounds and . . . all colonizing does in its
nature, and must in its consequences, operate to the prejudice
of that branch of commerce."

The colonial agent in London, Benjamin Franklin, preparing to take the scalp of the Commissioner who wrote the report,
must have felt that the Lord had delivered his adversary into
his hands when he read the further admonition, "Were they
[the Indians] driven from their forests, the peltry trade would

decrease and it is not impossible that worse savages would take refuge in them." Colonial savages.

The Commissioners' report was of 1772. Two years later, following hard upon the "Intolerable Acts" that undertook to punish an already rebellious Massachusetts, the policy thus formulated became law in the Quebec Act. In the provisions that secured to seventy thousand Canadians the Catholic establishment and their inherited legal institutions, this Act was as enlightened a measure as any that a British government has ever adopted. (Even so, it gave the rebels of Massachusetts, whose Protestantism had the highest voltage, a propaganda beyond all price.) But the rest of it was consummate folly. Ratifying the Proclamation of 1763, it decreed the persistence of Yesterday in an era already warm with Tomorrow's sun. It committed the British Empire to the support of furs as opposed to land, it triply sealed the West to settlement and to speculation, and it legislated mercantilism as the reef on which colonial expansion would be shattered. Great Britain undertook to confine the colonies east of the mountains. That was precisely the policy that had forced Great Britain to erase the French Empire from the map of North America.

The Quebec Act was a decree that the motion of the stars should cease. In its final form it passed the Lords in June 1774. In that same month a Pennsylvanian named James Harrod at the head of thirty men went up the Kentucky River, crossed over toward the head of Salt River, and began to build the stockade and cabins of the first settlement west of the mountains. The ghost of La Salle need no longer revisit the glimpses of the moon: the people who were to fulfill his imperial dream had reached the Great Valley. All the way across the continent there would be an advance screen of long hunters and mountain men, but the wealth the Americans were to create would not come from furs. These were settlers, farmers, builders of communities. Throughout the war that now broke out they kept following where Harrod (and Henderson, Boone, and Logan) had led. The war splashed their wilderness stations with their blood but they kept coming, mostly to Kentucky, and at least twenty-five thousand of them were settled west of the mountains when it ended.

From the beginning it was a world war. And from the beginning the West was the stake of four empires. Great Britain, which had risen to world dominance in 1763, might be reduced to a second-class power, and if it lost North America might lose India too. The France of Louis XVI hastened its own downfall when it lined up beside the United States, made official the war it was already fighting unofficially, and in February 1778 signed a treaty of alliance. The aid which Spain was supplying to the United States was almost as vital as that which came from France. But the establishment of the United States put on the eastern flank of Spain, which was historically the great colonial power of the Americas, a colonial revolution and the threat of an expanding nation. Spain could tolerate neither, but first things first. The United States could be used as a weapon to break British colonial power — and in the process it would conveniently exhaust itself.

So there were Spanish harbors for American privateers and credit for munitions. Arms and especially powder went up the Mississippi from New Orleans, up the Ohio, on to Fort Pitt where it was invaluable to the border defense, and some went over the mountains to the Continental Army. At the arch of the continent the village of St. Louis dominated the Indian trade south of the Great Lakes. Spain had a lieutenant-governor there, it had Laclede and Chouteau, and it had the sense to adopt French methods of dealing with Indians and to let the *émigré* French manage the frontiers. If British traders illegally (though with the approval of Spanish officials when money was short) reached toward the Missouri by way of the Minnesota and Des Moines Rivers, the Spanish went up the Mississippi and traded to the eastward with British tribes. In 1774 Peter Pond had noted at Prairie du Chien "a Large Colection from Eavery Part of the Misseppey. . . . Even from Orleans Eight Hundred Leagues Below us." [1] When the war came activity on this commercial frontier was intensified and Americans soon arrived to help out. Meanwhile there were southern Indians to incite against scattered loyalist settlements, and when groups of American rangers on the same errand came within reach they could count on the Spanish for sanctuary and supplies. In 1779 Spain entered the war against Great Britain — but not as an ally of the United States — and the attack on the British Empire became overt.

Both France and Spain intervened in the War of the Revolution for realistic reasons of imperial policy: not to help the Americans make the United States but to halt British expansion. There was no intention that the United States should long survive the war: politics hath bubbles as the water has and this temporary aggregation of rebellious colonials was to be of them. If it should remain an independent nation at the end of the war, it would be a weak one too and it could not last. It would break up into its component parts and disappear, or it would attach itself to a stronger power, or when the next war came it could be picked off. And imperial strategy committed the French and Spanish to the same principle as that of the British: the United States must not extend into the heartland, the West must not be American.

The military actions in the West were microscopic and exceedingly important. They were not decisive in American survival but the very nature of American nationalism turned on them. So did the area of the American empire; so did the greatest national wealth the United States was to have. Here was an action in delicate equipoises, constantly oscillating. Extemporized organizations of frontiersmen, ranging across great distances and seldom acting for very long at a time, provided just enough weight to turn and keep the balance American.

It was an Indian war. Both the British and the Americans undertook to make allies of such Western tribes as they could get and to buy the neutrality of the others. The British had many advantages: the posts and forts all the way to Lake Superior and beyond, possession of the entire lake frontier, unmatchable quantities of trade goods and munitions, and what counted fully as much, the policy of the Quebec Act which guaranteed to the Indians the country between the Ohio and the Lakes. The United States was now the threat that the British had been, and British Canada the promise of security that the French had been. The tribes clearly understood the Quebec Act and had learned that the Americans meant land companies and land companies meant extinction.

So the tribes north of the Ohio gave American legendry the belief that the word "Kentucky" means "dark and bloody ground." Before the war ended most of them were raiding the Kentucky stations and reaching on to the Tennessee country and the Wilderness Road. The most tireless were the Shawnees. They had begun as friends but the unmotivated violence that was always to characterize the American frontier murdered their chief Cornstalk, and from then on they were never to be at peace with the United States till Wayne broke the last confederacy in Washington's second term. The Delawares also chose the losing side as was their custom, though they too began as allies of the Americans. Piankeshaws, Miamis, and Illinois raided with both tribes. They were directed by the British commandants of the Western posts, especially Henry Hamilton at Detroit, whom the Americans probably hated more than any other British officer. The British decision to use Indian auxiliaries in the general war — a foolish blunder for it meant a net loss to them and much worse to the Indians — at once involved the far tribes too. Ottawas, Chippewas, Hurons fought in the border campaigns and took the ancient war road down the Lakes and the St. Lawrence to fight in the East. So, though in smaller numbers and as pro-British factions of divided tribes, did Sauks, Foxes, Potawatomis, Menominees, and others. The Western Indians were only a minor weapon in the East, though a terrible one, but they became the principal British force in the effort to hold, subdue, and extend the West.

The decisive stay to that effort was provided by one of the suddenly disclosed geniuses shaped to the need and the hour whom the American Revolution was able to supply in such amazing numbers. He was George Rogers Clark, a Virginian, a neighbor and friend of Thomas Jefferson's. He acted as a military officer of Virginia, not of the United States. In fact he was working to bring Richard Henderson's Kentucky government, Transylvania, under Virginia jurisdiction when he perceived that the West was in danger of being lost and that there was a way to save it.

It was 1778, the year following Burgoyne's surrender and the dolorous winter of Valley Forge. Twenty-six years old, a lieutenant colonel of Virginia, promised more money than was ever appropriated for him and authorized to raise more troops than would join him, Clark set out on a campaign as romantic as it proved to be successful. He had four companies, one hundred and seventy-five frontiersmen. News of the French alliance reached them just before they started and they shot the falls of the Ohio during a total eclipse of the sun — it was just that histrionic. On the night of July 4 they captured the small post of Kaskaskia without a fight, Cahokia followed, and they had won the Illinois. Its French inhabitants, who had been but indifferent subjects of Great Britain, gladly became Americans. Moreover, Clark had made contact with the Spanish at St. Louis, not yet declared enemies of Great Britain but at his service. Munitions and supplies came from them only, and the expenses of his occupation were thereafter mostly paid by the Spanish and mostly with private capital. . . . The trader and ex-soldier Francis Vigo, who spent a sizable fortune to maintain the conquest, has been memorialized with a number of statues, and heirs of the speculators who bought up his claim against the United States were paid off ninety-eight years after he saved Clark.

Clark summoned Indians from all about to a council. In spite of frantic British efforts to make it futile some four thousand of them gathered at Cahokia in August. Most of them were factions of tribes that also had pro-British factions but they represented formidable power. Clark proved to have genius at dealing with Indians. Not Perrot, Duluth, or Frontenac himself had achieved a more brilliant success than that won by Clark's speech of martial flattery, scorn, defiance, and invitation to ally themselves with the new-risen star, the victorious Americans. "Hear," his own note of the peroration runs: all the Clarks were inspired spellers, "hear is a Blody belt and a white one    take which you please    behave like men." They took the white belt. That he was able to make American or at least Spanish-American Indians out of the Sauks and Foxes (who may be thought of as by now a single tribe) proved to have decisive importance, and it was only less important that the majority of Potawatomis stood with them. Ottawas, Chippewas, Miamis, and delegations of other tribes came to the council too, but he did not win many of them over. Nevertheless his success was so impressive that minorities along the vital route of the Wabash River, where the British had abandoned the commanding position at Vincennes, changed their allegiance.

In two months Clark had permanently changed the alignment of forces in the West. He had also forestalled a grandiose

British plan of invading it from several directions at once. At Detroit Hamilton, "the hair-buyer," had been sending Indians on raid after raid against the Kentucky border.[2] (It was now that the three renegade Girty brothers, of whom Simon became the most notorious, began to earn their reputation for savagery.) He had planned a campaign against Fort Pitt, the way to which had been opened by the failure of an American foray in the direction of Detroit. He himself did not have enough force for it but he hoped to bring to his support the loyalist bands that were scouring the Tennessee country, such Southern Indians as the Cherokees, and even British regulars from the fluctuating war in the Carolina piedmont. The loss of the Illinois, which established the Americans on the right flank of his invasion, could not be tolerated. Summoning the Northern Indians, Hamilton moved with negligent slowness on the key post, Vincennes. He occupied it December 17 — Clark had been too weak to garrison it — and then dispersed most of his force, planning a spring campaign to clean out the countryside and confident that he was invulnerable during the winter.

Hamilton was thus poised on the frail and vital line that connected the United States with its imperial frontier. The winter proved less than a safeguard for him. A hundred and eighty miles stretched between Kaskaskia and Vincennes, a third of it "drowned" with the winter floods of prairie rivers. In as resolute a feat of leadership as our military annals contain, through two and a half weeks of February 1779 Clark led across this wilderness of mire and ice a small force that was half American and half Illinois French. Never warm, seldom dry, getting their first full meal from the loyal inhabitants of the town they had come to recapture, they completely surprised Hamilton and captured him and his whole force.

That did it. On the far frontier American power and prestige could now withstand anything brought against them. When Spain entered the war, the British undertook to invade Louisiana and capture St. Louis. The movement was based on Mackinac and utilized the Sioux, whose firm attachment to the British was to be a conditioning influence in the West for the next quarter-century. Besides Sioux there were Ottawas and Chippewas in the force that was assembled; it was so impressive that the Sauks and Foxes were pressured into neutrality and a few of them persuaded, with a scattering of Menominees and Potawatomis, to join up. Moving south, the Indians and the small British contingent captured various Spanish detachments, took a few scalps, and appeared before St. Louis, where the commandant had dug trenches and built some crude fortifications. On the morning of May 26 1780 they made their attack but with so little conviction that it broke on the trenches, ebbed back, and became a retreat. This failure of a well-planned effort, which had sufficient strength to overwhelm the town, has never quite been explained but may have been due to the Sauks and Foxes. Except for brief periods they had always been enemies of the Sioux, they had become partisans of the Spanish, and they had been impressed by the Americans. They simply omitted to fight and this appears to have deterred the Sioux, who had any Indians' reluctance to storm fortifications. Across the Mississippi a parallel move on Cahokia was stopped short by the arrival from downriver of Clark and some of his men. The invaders hurried northward, burning what they could and taking some scalps. This outcome so discredited the British that their influence collapsed. Upper Louisiana remained Spanish, which is to say permeable by the United States. The commandant at Mackinac, anticipating a counterattack, withdrew the fort from the mainland to the island, where it remained thereafter.[3]

Thus the British effort to encircle the United States from the west failed. It had been based on sound strategy, on the geopolitical ground plan, and the Spanish now undertook to effect the same strategy. Their riposte to the attack on St. Louis, a brilliantly executed raid up the Illinois River, was in consideration not of the military present but of the imperial future. They were asserting Spanish sovereignty east of the Mississippi, that is eastward from Louisiana. Spanish policy was necessarily committed to the theorem that the United States did not extend west of the mountains. If the British colonies had ever had even a shadowy claim to the Mississippi Valley, the theorem ran, the Proclamation of 1763 and the Quebec had extinguished it. To support the policy the government of Louisiana seized a number of places on the east bank of the Mississippi; then it made a conclusive demonstration by capturing West Florida. The full extent of the claim thus affirmed, sovereignty over the entire Mississippi Valley, might not be made good during the war or at the peace settlement when it ended. But Spain was nevertheless established on both banks of the Mississippi and control of the river, so every imperial thinker assumed, implied control of the American empire and would very likely destroy the United States. A friction bound to be deadly was thus insured. When hostilities ended, the Spanish policy was continued by other means: the incitement of Indian trouble along the southern border to keep the West weak and conspiratorial activity to detach it from the United States.

The remainder of the war in the West was an episodic seesaw, a stalemate determined by Canadian and American geography. The Americans were never able to make the sustained attempt on Detroit which would have split Canada in half, might have given them all of it, and would certainly have changed the conditions of the western and northwestern fur trade. The British were never able to make the reciprocal attempt on Fort Pitt that would have detached the West and enabled them to combine with the loyalists of the South and East. The Indian fighting that continued nearly everywhere had no effect on the outcome.

After the victory at St. Louis saved the Illinois, the Americans practically abandoned it. Almost at once Clark was back in Kentucky, in time to take a strenuous part in turning back another invasion by the British and their Indians. A year after the surrender of Cornwallis, he led an expedition against the Shawnees at Chillicothe that held the central frontier against the last raids boiling out of Detroit. He had built a fort at the falls of the Ohio, the future site of Louisville, and there he was, in command of a small body of troops which fluctuated in size and was nearly impotent of itself but constituted a force in being, a nucleus round which the frontier militia could always form. There he was, a curious resultant of many energies that all but neutralized one another and yet were just threat enough to the Canadian West. Nothing much was said about him at the peace conference but everybody there and elsewhere knew where he was.

◦⌥◦

## The Kentucky Merchant

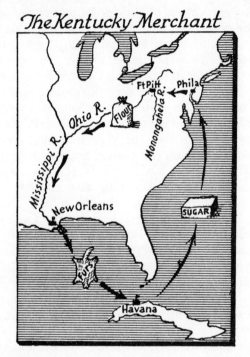

Lord North's "Oh God, it is all over!" when word of the surrender at Yorktown reached him was right. The end had come in 1781 only a few miles from Jamestown where so much had begun in 1607.

Immediately the West exerted its imperial leverage. Washington began to plan a campaign west of the mountains. As the commissioners for peace prepared to meet at Paris the common interest of three enemies was manifest to everyone. The United States had proved that a nation existed but the three empires were resolved that there must not be a fourth one. Washington might cross the mountains but the nation must not. And a Kentuckian went to Pennsylvania, loaded a boat with flour, floated it off down the Monongahela, passed Fort Pitt with the shades of Contrecoeur and Braddock watching impassively, passed the falls of the Ohio, drifted into the river of De Soto and Jolliet and La Salle, reached the New Orleans Iberville had founded, and exchanged his flour there for furs. He took the furs to Havana, exchanged them for sugar, sold the sugar in Philadelphia, and crossed the mountains to Kentucky.

That was why there would be a fourth empire: geography said so. It was also why, whatever treaty might come out of Paris, there would be no settlement of the continental strategy till 1804. But the Virginia roads converged on Cumberland Gap and the village of Wheeling, and the Pennsylvania roads on Pittsburgh. The buffalo began to leave Kentucky.

John Jay, straight from the frustrate humiliation of trying to negotiate with Spain; John Adams, feverish with suspicion of the British that only his distrust of the French kept short of explosion; Franklin, far subtler than either and thinking of more extensive territory than they dared to, thinking indeed of all Canada as the great prize of the war — the American commissioners were a strange and striking trio. But however diverse they were, all three had beyond the margin of consciousness the unconscious feel of the continent, the continental will, that shaped a demand. And they set a precedent that has held for American diplomacy ever since: they got more than there was any realistic expectation the United States would be able to get.

Their triumph was possible because they recognized the moment when the growing British realization that a stronger United States meant a weaker France intersected the belated British recognition that the United States, even though independent, remained the principal market for British manufactures. They got the whole West — all the way to the Mississippi. With the West the United States got the most cohesive force of nationalism and got its empire. The French and Spanish Empires were left with no weapon but conspiracy to use against it till there should be another war. Nobody doubted that another war would come tolerably soon.

There was a period during the negotiations when the 45th parallel might have been made the boundary. . . . That would have made the Michigan peninsula Canadian and the Ontario peninsula American, and would have satisfied the powerful interest of Montreal by keeping the canoe route west from Lake Superior entirely free from our control. . . . Many considerations made it instead the line of the Lakes and from the end of Lake Superior the established route, "the water communication," to Lake of the Woods. So far accurate geographical knowledge and a just balance of commercial realism had agreed on the "natural channel" — though the United States got a valuable, and to Montreal very disturbing, dividend in that Grand Portage lay south of this line. Here, however, ignorance of geography took over. The terms agreed on were that the United States was to extend to Louisiana. Louisiana was undefined except that its northeast corner must be the source of the Mississippi. How, then, to fix the western boundary of the United States? David Thompson ascribes to Peter Pond, whom he calls "an unprincipled man of a violent character," an intention to injure his former associates in the fur business by advising the American commissioners to insist that the Mississippi lay due west of Lake of the Woods.[4] He appears to have remembered something that never occurred and the true explanation was the insufficiency of the maps, though in fact no one knew where the source of the Mississippi was. John Adams said that "it was the Mitchell's map upon which was marked out the whole boundary lines of the United States." This was John Mitchell's "Map of the British and French Dominions in North America," originally published in 1755 but almost entirely unchanged in the edition used at the peace conference. On the whole it was the best map that had yet been drawn but Mitchell showed the Mississippi more than two hundred and fifty miles west of Lake of the Woods. He did not show its source — he inserted a detail map in the upper left-hand corner where it would have come — but made a note saying, "It is supposed to arise about the 50th degree of Latitude." So they decided it: "Thence through the said lake to the most northwestern point thereof, and from thence on a due west course to the river Mississippi." Since the source of the Mississippi is well to the south of Lake of the Woods, the northwest corner of the United States was to be at an intersection that did not exist.

The French minister at Philadelphia had clear sight. He wrote to his chief, Vergennes, "The Americans, in pushing their possessions as far as the Lake of the Woods, are preparing for their remote posterity a communication with the Pacific."

The treaty made the West as far as Louisiana American, and the states with claims to the intervening lands had already relinquished them to Congress, to the Congress which governed the United States under the Articles of Confederation. The possession of this area was the most powerful centripetal force the nation had. It proved to be just strong enough to hold the United States together under the Confederation, while every year increasing thousands of Americans crossed the mountains to develop the empire, which in turn strongly conditioned the nation that made the Constitution. Of several final determinants established under the Confederation one was a momentous provision first embodied in the resolutions of 1784 "for a temporary government of the western territory" and fully expressed in the Northwest Ordinance of 1787. The latter stipulated that from three to five states were to be created from the lands north of the Ohio and that these states should be admitted into the union "on an equal footing with the original states in all respects whatever." The American empire would not be mercantilist but in still another respect something new under the sun; the West was to be not colonies but states.

The peace settlement gave the Floridas to Spain, which thus flanked the United States at the south. But it fulfilled the most ominous threat to Spain by bringing to the east bank of the Mississippi, separated from Louisiana by only that traversable water, a nation that was both a democratic republic and an expansionist empire. Necessarily Spain had to assert the thesis that Americans had no right to navigate the river.

The settlement also appeared to cut off at least half of the Canadian fur trade. The routes west remained secure and the route from the West to the Northwest had been guaranteed. But Detroit, St. Joseph, the Sault, Mackinac, La Pointe, La Baye, Prairie du Chien, and Grand Portage were American. The economic shock was severe and was increased by the promptness with which American competition appeared in areas where Montreal had previously had to meet only Spanish competition or none at all. The Montreal traders, who had the bulk of Canadian capital and were the energy of Canadian expansion, seemed likely to be bankrupted forthwith. There was also a sincere concern about the Western Indians. The period of transition to American control would at best inflict scarcity and therefore anarchy on them, and of course they must slowly die, for whereas Canada was Indian furs the United States was Indian land.

Montreal, then, which had been unable to influence a ministry that was determined to make peace, succeeded in asserting its influence as soon as the peace was made. It made powerful representations: there would be a general Indian war, the British Empire would have a bankrupt Canada on its hands, it would lose half its imperial wealth or more than half. The reasoning sufficed.

The treaty had bound Great Britain to evacuate the still garrisoned military posts it held on American soil, various unimportant ones but mainly the strategic centers at Oswego, Niagara, Detroit, the Miami, and Mackinac. It bound the United States to enforce the payment in sterling of private debts owed to British merchants and to do what it could to make the states restore to the loyalists the estates that had been confiscated from them. The United States had done nothing to carry out either provision and clearly was going to do nothing. The breach of faith made a sufficient rationalization for another one: Montreal got its way and the garrisons remained where they were. As long as they did so the trading posts and the fur trade seemed safe.

৽

Peace treaty to peace treaty, 1763 to 1783. Through the two decades of turmoil the accumulation of knowledge had powerfully accelerated. The Age of Reason rocked societies as much as the wars did and the dynamics of expanding knowledge were preparing additional upheavals. . . . As minister plenipotentiary to France, Benjamin Franklin was commissioner of American privateers in European waters. One of his first officials acts, March 10 1779, was to issue a "passport" for a man who he could not know had been murdered on the island of Hawaii a month earlier. It commanded all Americans on the high seas to consider Captain James Cook and his two ships' companies "common friends to all mankind," to afford them all possible assistance, and to speed their return to England without hindrance or delay.[5]

Cook's *Resolution* is one of history's great ships and she was following the wake, two hundred years old, of a greater one, the *Golden Hind*. At daybreak on March 7 1778 she made a landfall on the coast of Oregon somewhat to the north of Coos Bay, stood out a little, and began to coast northward along the shore that Cook recognized as Francis Drake's New Albion. One hundred and ninety-nine years before, June 17 1579, standing to the south from about the same vicinity, Drake had anchored the *Golden Hind* in a California bay.[6] Among the secret purposes of Drake's voyage, though he made no effort to carry it out, was to discover the great Southern Continent. This was one of Western man's oldest myths, born centuries before Ptolemy, in whose version it was a continent that extended to the pole from southern Africa and stretched a hypotenuse to join with Asia and make the Indian Ocean a lake. As Terra Australis Incognita it was among the richest of the horizon lands and though the centuries shrank it inward a little, it lay gigantic and fair as a bride to be possessed just behind the curtain of mist to the southward, in the unimaginable distances of the Pacific. The great mariner who now followed Drake to the Northwest coast was on his third voyage of discovery in the Pacific. His second voyage there had ended in 1775 and in the course of it he had laid the myth forever: there was no Southern Continent, no Terra Australis.[7]

Another purpose of Drake's voyage had been to find the Pacific gateway of the Northwest Passage. About this too he had done little but, circumnavigating the globe, he had kept secret the route of his homeward passage. It had been widely believed ever since, especially by officials of Spain and New Spain, that with the help of secret Portuguese charts he had found and sailed the Northwest Passage, probably the shape of it that was called the Strait of Anian. To settle once and for all whether that strait or any other Northwest Passage existed was the primary purpose of Cook's third voyage. Belief that one led across Canada — or taking off to the south of Canada in the hazy and extensible New Albion — would not down. Maps of North America not dry from the press when he sailed showed such a continental passage in various latitudes, by various courses, among various archipelagoes, though most often as the Strait of Anian and in the high north.[8] Maps showing its eastern portal at Hudson Bay were drawn in all countries in spite of the fact that there were people with open and certain knowledge that they were wrong. The Royal Society and the Hudson's Bay Company — at least some members of each — knew that they were wrong. If surveys before that of Middleton in 1741 had not proved that no passage led to the Pacific from the western shore of Hudson Bay, Middleton had proved it to all minds except those drugged with fantasy. In 1746 the expedition to the Bay of William Moor and Francis Smith had proved it again, and Samuel Hearne's magnificent descent of the Coppermine River to the Arctic Ocean in 1771 had clinched it once more.

The believing mind cannot be dissuaded, however, and from various conjectured entrances imaginary straits still continued to reach Hudson Bay. Moreover, one logical possibility remained and it had lately been made attractive by a theory that the Arctic Ocean was an open, ice-free sea. It might be that the strait which Bering had almost discovered and which had been named for him led to the Arctic. (It does.) Or it might be that some other passage led from the Pacific to the Arctic in the unknown area west of the Coppermine. If there were such an entrance to the ice-free northern ocean, then it would be possible to sail near the Pole to Baffin Bay, to Davis Strait and on to the Atlantic. England could at last end Sir Humphrey Gilbert's search and possess the short strategic route to China. Or rather to India, for India counted more than China since the Seven Years' War had made it almost wholly British. The clearest minds foresaw that the British Empire was going to be oriented on the Pacific.

Was the Arctic in fact open in winter? Was the land west of the Coppermine River continental and how far did it extend? Was it an archipelago? Was there, indeed, any land there? Was there anywhere in the high latitudes any kind of passage from the Pacific to Hudson Bay or Baffin Bay? Was there one to the Arctic? Such questions could now be so asked as to be capable of being answered, whereas previously there had not been enough knowledge to give the phrasing reality. Cook was directed by the Admiralty and the Royal Society to answer them, and the terms of the standing offer of twenty thousand pounds to anyone who would find the Northwest Passage were changed to permit a naval officer to receive the award. Cook sailed on his third voyage July 12 1776, when no one in England knew what had happened in Philadelphia eight days before.

As the *Resolution* worked northward from her Oregon landfall Cook noted that there was no opening where Martin Aguilar, of the Vizcaíno expedition of 1602, was believed to have seen one. "In the very latitude where we now were geographers have been pleased to place a large entrance or strait, the discovery of which they take upon them to ascribe to the same navigator, whereas nothing more is mentioned in the account of his voyage than his having seen in this situation a large river, which he would have entered but was prevented by the currents." (And Aguilar had seen his river only with the eyes of faith.) Fog and weather that kept him well out from shore prevented him from seeing the river mouth which, unknown to him, another Spaniard, Bruno Heceta, had seen at 46° 17′. The same weather made him miss the important entrance to what would eventually be called Puget Sound. On March 29 he found an anchorage in an inlet which he named King George's Sound. Juan Perez, for the viceroy of New Spain, had preceded him there five years before, probably anchoring in the same strait but not landing, and had called it San Lorenzo. It was to come down as Nootka Sound. At first Cook supposed that the shore less than a hawser's length away was the mainland but later learned that Nootka Sound and a series of similar narrow inlets made an island of it. He never suspected that the far shore was that of a greater island, which was eventually to be named for one of his young gentlemen, George Vancouver.

Cook had been a warrant officer in the fleet that supported Wolfe in the campaign against Quebec, and so he thus became the first Englishman since Drake who had seen both coasts of North America. And with him on the *Resolution* was a corporal of marines, part genius and part moongazer, John Ledyard, a native of Groton, Connecticut, twenty-six years old. "It was the first time I had been so near the shores of that continent which gave me birth from the time I at first left it," he wrote, "and though more than two thousand miles [this, though Cook's determinations of longitude were exact] distant from the nearest part of New England, I felt myself plainly affected. All the affectionate passions incident to natural attachments and early prejudices played round my heart, and [I] indulged them because they were prejudices. I was harmonized by it. It soothed a homesick heart and rendered me very tolerably happy." This was the continental consciousness speaking. Thirty-three hundred miles from Connecticut, Ledyard had touched home.

As soon as Cook anchored, big canoes put out from shore and the English met a new kind of Indians. They were the tribe who gave their name to the island, the Nootkas, of the culture that extended from Oregon to Alaska west of the coastal mountains. Seafaring Indians. The big canoes were made from the big cedars of the forest that came down to the water's edge, and their navigators had such consummate skill that when Alexander Mackenzie saw them, fifteen years after Cook, he said that his *voyageurs,* of the breed that had mastered the Western waters, were tyros in comparison. They lived in houses made of split planks and were master craftsmen in wood, the only Indians north of the Aztecs who ever were. They were also fine artists; many artifacts in museums today are instantly recognizable from Cook's descriptions. They lived in a country of gentle climate, rich with easily procurable food. Fishermen, sealers, even whalers, they had an economy that centered on the regional fish, especially the salmon, just as that of the Plains tribes centered on the buffalo. They were affluent and Cook said they were lazy — he thought all primitives were. They had an intricate ceremonial system and a hierarchy of social castes from slaves to plutocracy. Cook noted that though phlegmatic they had violent tempers. He may be granted authority about tempers, being himself a perfected specimen of the type which the naval victories of the eighteenth century had created, God's and England's sea dog. (He would presently die of it.) He noted too that they were unawed and unfrightened. Right: they were a tough race and it was just as well, for from now on they were going to have to deal with sea dogs and with the equally basaltic American equivalent.[9]

They swarmed around the ship, haranguing the white man, singing their chants of welcome and exorcism, and disinfecting him with the sacred pollen. Some of them wore grotesque masks, ceremonial, totemic, or plutocratic, of carved and painted wood. Some had garments made from the skins of sea otters and these were momentous. Cook had an idea of their value — the Admiralty and the Royal Society had had them in mind — but it fell far short of the reality. "The fur of these animals, as mentioned in the Russian accounts, is certainly softer and finer than that of any others we know of; and therefore the discovery of this part of the continent of North America, where so valuable an article of commerce may be met with, cannot be a matter of indifference." He was very right.

Since the Indians wanted metal, preferring brass, the crew bought sea otter skins at bargain rates. "Whole suits of clothes were stripped of every button; bureaus of their furniture, and copper kettles, tin canisters, candlesticks, and the like went to wreck." Linen hung up to dry could be left unguarded but everything metallic disappeared — twenty-pound hooks cut off the tackles, all the fittings of small boats. And the local chamber of commerce kept visitors away from the English, in order to turn a profit on the metal. Within a few days Cook perceived that the native trading system he had touched here was intricately ramified and extended far into the interior — eastward. "The most probable way . . . by which we can suppose that they get their iron is by trading for it with other Indian tribes who either have immediate connection with European settlements . . . or receive it, perhaps, through several intermediate nations. The same might be said of the brass and copper found among them." Ledyard, making the same observation, decided that they must come from a great distance and were "not unlikely from the Hudson's Bay Company." Cook guessed as much too but allowed for the possibility that the stuff might come up from Mexico and added "Canada." The last was right: the Pedlars, out of Montreal.

But wherever the iron originated, a circle had been closed. On the Pacific coast an Englishman and an American saw trade goods that had crossed the continent.

His ships watered and new masts set, Cook got about his business. He made for 60° N., where Vitus Bering had fixed a point on the American mainland. It was the peak called Mount St. Elias, at the place where the modern boundary of the Province of Yukon turns eastward in order to keep the coast Alaskan. From here on Bering had established practically nothing about the American shore, and the maps by his polyglot chroniclers and commentators were so full of guesses that Cook repeatedly denounces them. Mount St. Elias might be the northwest corner of North America; at any point one might perhaps sail northeast to the Arctic Ocean. All of Alaska, that is, might be islands and most of it might be open water.

So from here on Cook carefully explored the coast. Once he thought he had found something and sailed two hundred miles up a big estuary that might be "a strait communicating with the northern seas." But no; it narrowed, "the marks of a river displayed themselves," and the water was "very considerably fresher than any we had hitherto tasted." Thick and muddy water, "with large trees and all manner of dirt and rubbish." He sent out two boats to make sure. "We were now convinced that the continent of North America extended farther to the west than from the most reputable charts we had reason to expect." This was a fact of basic importance and he was glad he had explored the estuary, for otherwise "it would have been assumed by speculative fabricators of geography as a fact that it communicated with the sea to the north or with Baffin's or Hudson's Bay to the east, and been marked perhaps on future maps of the world with greater precision, and more certain signs of reality, than the invisible, because imaginary, Straits of de Fuca and de Fonte."

By thus ending one hope, however, he created another. The estuary was Cook's Inlet, which the 60th parallel crosses at 152° W. At its eastern end a complex of unimportant rivers flow into it. Cook assumed that their sum meant there was "a great river" there, and when his maps were published this suggested to the Northwesters at Lake Athabaska that they could reach it from their far country and go down it to the Pacific.

He went on, turned the Alaskan peninsula and the long scythe blade of the Aleutians, and entered Bering Strait, having established that there was no other route to the Arctic. Hard and sudden storms struck, vast mountain ranges rose from the shore, glaciers crumbled to the sea from their slopes, volcanos exhaled their thin smoke, the natives showed a strong tincture of Eskimo culture and then were true Eskimos — he was in the terrifying, desolately beautiful Far North. On August 9 he passed the narrows of the Strait, and later Ledyard expressed the awe of seeing a different continent on either side. Bright ice was everywhere. On August 18 it was "compact as a wall" and rose ten or twelve feet out of the water. He was half a degree north of the Arctic Circle, he could go no farther, winter was closing in. He turned south, calling the nearest point of land Icy Cape.

He had proved to the hilt that Alaska was mainland. Reaching the Aleutians again, on Unalaska he twice got messages carried by amiable natives and written in a language no one could read. It was certainly Russian and he sent Ledyard to find the authors. The born cosmopolite found them, fainted in their steam bath, was revolted by the stench of their stewed seahorse and bear meat (to which, however, Cook paid a gourmet's tribute), and had a fine time. He helped them drink the liquor Cook had sent with him, for of many novelties the most amazing was that they were Russians and yet had no alcohol. He took three of them back to the ships and the leader followed with most of his outfit. They were fur traders of course, commanded by one "Erasim Gregorioff Ismyloff," whose Arabian Nights autobiography Cook could hardly bring himself to believe. He found them a genial, boisterous, naïve folk and got only hints of the incredible life they led and the incredible massacres of Aleuts they had perpetrated. Considering that no one could speak anyone else's language, he got an astonishing amount of geographical information from them, copying their charts and correcting his. What was almost as important, he learned where their knowledge ended.

He had proved that neither the Strait of Anian nor any other form of Northwest Passage existed. But the ice had prevented him from determining whether it was possible to sail eastward, or westward, through the Arctic and, as his instructions had foreseen, he would have to try again next year.[10] He was supposed to winter at Kamchatka but did not care to spend the long months doing nothing, so he sailed from Unalaska for the Sandwich Islands, intending to explore them. He had discovered and named them earlier this same year, 1778. (The Spanish had once glimpsed the archipelago from afar, more than two centuries before, and had long forgotten it.) Cook had landed on several islands but not the Island of Hawaii itself. He did so now and was received as a god. But it was as an Englishman teaching natives their place that he got himself killed there in February 1779.

Cook was, James A. Williamson says, "the greatest explorer of his age, the greatest maritime explorer of his country in any age." He forever destroyed two shining myths, erasing the Southern Continent and the Northwest Passage from the map of human ignorance. He added great stores of knowledge to the intellectual estate of mankind. Not the least item among them were the calculations that made him the first man who had ever known for certain how wide North America is.

❧

Cook's appearance on the Northwest coast was the first fulfillment of a threat against which the Spanish had begun to take action nine years before. The threat was Russian as soon as Bering directed the attention of his employers, who had expanded across Siberia with amazing speed, to the Aleutian furs. At almost the same time it became a British threat, with a naval expedition appearing in the Spanish ocean in 1742 and capturing the Manila galleon. When the Seven Years' War focused the imperial energy of the greatest sea power on the Pacific, the threat became acute. So at last Spain and New Spain turned to California.

For generations Spain had lacked both money and energy for frontier expansion, and the west coast was fenced off by the peninsula of Lower California and the deserts of Arizona. The pioneering spirit seemed dead, though a renewal of zeal among the Jesuits just before their expulsion had established a firm base of missions in Lower California. Otherwise it was enough to hold the New Mexico settlements and to develop the silver mines. In fact it was to protect the mines that action was taken at last. They were the greatest concern, the constant anxiety; they seemed always to be threatened from somewhere, from New Orleans, from St. Louis, from Canada, and now from Alaska or the Strait of Anian.

That was the impetus. And Charles III of Spain found in José de Galvéz, whom he sent to reorganize the finances of New Spain, a man of vigor and vision much like his own. Galvéz, who had some of the imagination of Cortés as well as the cynicism of the wearier age, conceived and launched the last Spanish expansion in North America. For a few splendid years the sixteenth century seemed to live again.

Since the Vizcaíno expedition of 1602 the Spanish had had a

reliable chart of the California coast as far north as Cape Mendocino. Almost no one had touched the coast in all the years
since it was made, though the Manila galleon sometimes had it
in sight and sometimes desperately needed a harbor there. All
other knowledge of California belonged to a time even earlier
than Vizcaíno and, such as it was, had been lost under encrus-
tations of legend, myth, and conjectural geography. But it was
known that the Northwest Passage took off from somewhere on
this coast.

In fact, a wide choice of Northwest Passages. The Strait of
Anian was already a century old when a rumor that Portuguese
mariners were using it provided the chief stimulus for Vizcaíno's voyage. Its entrance was then supposed to be at about
48° N. (Just below the actual entrance to Puget Sound.) The
latitude of 60° N. (that of Mount St. Elias) which appears so
often in speculations about it was the one commonly attributed
to a certain Maldonado, who in 1609 reported that he had
sailed through it in 1588. The imaginary Admiral de Fonte
was supposed to have entered it — or a river, or something —
at 53° (halfway up the coast of Queen Charlotte Islands) and
to have taken a labyrinthine passage to high above the Arctic
Circle. Drake was believed to have taken the *Golden Hind* and
the Spanish gold through either it or still another passage, "the
Englishman's strait," and no one could say where, except that
it must be above 43° or above 48°. (The southern boundary
of Oregon is 42°.) Juan de Fuca's supposed entrance was between 47° and 48°.

Any of these entrances could be moved to any latitude the
dream required. The Pacific end of the fantasy whose Atlantic
portal had been any inlet north of Florida, they constituted the
"Northern Mystery" which must be dissipated if New Spain
were to erect California into an outpost-stronghold to defend
the silver mines. For the entrance of the Northwest Passage, or
rather the exit, was one site that must be held at any cost, however great.

Galvéz worked out a plan and in 1769 ordered it carried
out. He was using the threefold instrument with which Spain
had always approached the wilderness, the soldier, the priest,
and the colonist. Because the Jesuits had been expelled from
New Spain his religious were the Franciscans. His good for-

tune was to find at the head of them in Lower California, where the effort must be based, the American saint who has not yet been canonized, Junípero Serra. And in Gaspar de Portolá he found a first-rate captain.

The first objectives were the known harbors of San Diego and Monterey. The expedition moved on the former in four divisions, two by land and two by sea. Despite the mountains and desert the colonization should have been no more than difficult, but all the breaks were bad. The land parties were in good health but almost starved. Those on the ships ate well but were decimated by scurvy — accepted as inevitable by everyone except Cook, who found out how to prevent it — and a mysterious additional epidemic.[11] Late in 1769, which made it a hundred and sixty-six years after Vizcaíno, the first California mission was founded; it was San Diego. Going north from there, neither the army nor the navy recognized the bay at Monterey, though the ancient description of it was exact. The search for it was what brought the first white men to the Golden Gate, a scouting patrol under a sergeant named Francisco de Ortega. Eventually Monterey was located and there on June 3 1770 the second mission, San Carlos de Borromeo, was founded and with it the first presidio. California had begun.

Two necessities remained: to connect it with Sonora by a trail and to push exploration up the coast. Portolá's naval commander was Juan Perez. In 1773 he was ordered to sail to 60° N., the Strait of Anian. Scurvy and the storms common in these parts kept him from quite making it. He coasted Queen Charlotte Islands, without recognizing them as islands, and then was forced to turn back. On August 7 1773 he anchored in Nootka Sound, San Lorenzo to him, and gave Spain the claim to it that antedated Cook by five years. The regional fogs kept him from seeing any of the actual entrances to the interior and from checking the imaginary ones.

Two years later Perez was back as chief pilot of an expedition dispatched with renewed urgency and headed by Bruno Heceta, the corvette *Santiago* and the schooner *Sonora*. They proved that there was no inlet at the latitude which Juan de Fuca had given as that of his passage inland. Somewhere on the Washington coast north of Gray's Harbor they anchored and, as Bancroft says, "on July 14 Europeans set foot for the

first time on the soil of the Northwest coast." They made other
landings and found the Indians viciously hostile. The stormy
season was coming on, the inevitable scurvy had broken out.
The captains decided to make as much northing as they could
before weather and sickness drove them back to California.
The schooner, captained by Bodega y Cuadra, who was to have
a further career in these waters, made a very high one. He
reached 58°, landed, and claimed Alaska for Spain.

In the *Santiago* Heceta and Perez reached about the latitude
of Nootka and then turned south, the ship leaky, the crew very
sick. They ran along the shore when they could but again fog
hid the entrances. But on August 17 — this is 1775, two months
to the day after Howe's regulars won the battle of Bunker Hill
— Heceta "discovered a large bay, to which I gave the name of
Assumption Bay." He could not land; "if we let go the anchor,
we should not have men enough to get it up." He lay to, currents swept the ship out to sea during the night, and in the
morning he was so "far to leeward" that he could not enter his
bay. But he had fixed and described the two bold headlands
that define it. He named the northern one Cape San Roque
and his 46° 17′ is exact to the mile; the other one he called
Cape Frondoso. They are Disappointment and Adams now.

> Having arrived opposite this bay at six in the evening and placed
> the ship midway between the two capes, I sounded and found bottom
> in twenty-four *brazas* [twenty-three fathoms]; the currents and eddies
> were so strong that, notwithstanding a press of sail, it was difficult
> to get out clear of the northern cape, towards which the current ran,
> though its direction was eastward, in consequence of the tide being
> at flood.
>
> These currents and eddies of water cause me to believe that the
> place is the mouth of some great river, or of some passage to an-
> other sea . . .[12]

At last someone was right about a great river on this coast.
This was the river that flows more water than any other in the
United States except the Mississippi. There never was the
Great River of the West but Heceta had found the reality that
corresponded to the fantasies men had made up about it. To
the pure fiction of Le Page du Pratz and many others, to the
logic and desire of Vérendrye and Carver and many others, to
the felt necessity of Champlain at Georgian Bay thinking west-

ward toward the Sea of China, to nearly two and a half centuries of the western lodestone behind the mist of thought and dream. For seventeen years after Heceta no one would recognize it again or enter it or give it a name. Then it would be the Columbia.

৵৹

Meanwhile the route from Sonora to California had been worked out by a series of superb explorations. The hinge must obviously be the long-known junction of the Gila and Colorado Rivers. In 1771 another great Franciscan, the saintly, desertloving Francisco Garcés, set out from there, went down the Colorado almost to the mouth, and crossed the chaos of Lower California and its most northern mountains. He won the friendship of the Yuma Indians, a powerful tribe at the river junction. The next year, 1772, the governor of Upper California, pursuing deserters eastward across the San Joaquin Valley, which was thus entered for the first time, pushed on to Cajón Pass and the Mohave Desert. In 1773 the same rumors of British and Russian activity that sent Perez toward the 60th parallel instigated an effort to break the overland trail from the Gila to Monterey. Juan Bautista de Anza who made it was one of the greatest of all Spanish frontiersmen, and he had Garcés with him. In 1774 with magnificent skill and endurance they took a large party to the Gila and across the deserts, and forced the crossings of the Cocopas and the Sierra. Anza came out at San Gabriel, the mission that had been founded in 1771 just north of Los Angeles. The next year he took a larger company of colonists to Monterey and on to San Francisco Bay and settled them there. The Arizona-California trail was marked and there was a watchtower on the Golden Gate.

The final chapter must be called poetry. In 1761 the New Mexicans had at last entered the Rocky Mountains, more than two centuries after Coronado and with the same motive, to find gold and silver and precious stones. The pioneer was named Juan de Rivera and he made at least three journeys. They were all northward from Taos to Colorado west of the Continental Divide, and his farthest penetration was the junction of the Uncompahgre River with the Gunnison. Occasional traders and perhaps punitive expeditions against the Indians entered the same region. Some knowledge of the country and of the various Ute bands that lived in it was amassed.

In Coronado's land of Cíbola, at the pueblo of Zuñi, the Franciscan missionary at Our Lady of Guadelupe was named Silvestre Velez de Escalante. He was zealous, energetic, restless, insatiably curious. He collected all the stories and rumors he could, a great bulk of them, about the deserts and the tribes who lived in them. Off to the westward, across the chromatic wasteland of bare rock, were the Hopis, who were mysterious, sullen, suspicious, and hostile. They must be converted or, as he at last came to believe, conquered. North of them and west of the area in Colorado that the New Mexicans had visited was an expanse about which it was hardly possible even to make guesses. The usual stories of horizon-land floated out of it; among them the bearded Indians make their last appearance on the Spanish stage. Hearing about them, Escalante decided that they must be descendants of Spaniards who had been wrecked on the California coast and worked their way inland. Deep in the rock desert there really were some Indians who wore mustaches and a kind of beard and eventually he found them.

The Gila-Monterey trail was long, difficult, and exposed to raiding Apaches. Perhaps a shorter or a better route to Monterey could be developed from New Mexico. The idea was logical and naïve; it succinctly expresses the ignorance of the interior West that the Spanish had maintained since 1541. In 1775 (Daniel Boone was founding his station on the Kentucky River) the authorities determined to find out about this postulated route through the unknown void. Instructions went to Garcés, who was with Anza, and to Escalante, who forthwith made a spirited journey to the Hopis. They remained recalcitrant and heathen — apostate rather, for long ago they were supposed to have professed Christianity — and would have nothing to do with him. He picked up further rumors about the deserts but got some information too. A Havasupai was visiting the Hopis at the time and from him Escalante heard about the big river on which his people lived. Escalante correctly identified it as the Colorado and conceived that it might have some bearing on the new route to Monterey. At any rate, that route must avoid the Hopis until they were disciplined.

It must detour them — to the north. Perhaps it could circle round to the Havasupais and certainly it must strike into the unknown land. So Escalante was convinced; he was also convinced that God intended him to make the great *entrada.* The poem, dramatized against the backdrop of the rock deserts, is of the summer of 1776. Liberty fires marked the routes of couriers from Philadelphia bringing word that the colonies were and of right ought to be free and independent states, and General Howe followed the defeated Washington out of New York so sluggishly as to suggest that he agreed with the Declaration. The first fantasy in colored stone was by Garcés, who set out to find the Hopis from the West and try his hand at reconverting them. Alone except for the guides who took him from tribe to tribe across the chromatic spectacle, he went from the mouth of the Gila to the Mohaves near Needles and from there across northern Arizona to the Grand Canyon country. So he found the Havasupais. That he could reach Cataract Canyon where they lived and descend its vertical wall is against reason but he did so. He spent five days with this remote tribe, who even today are no part of the known world. A gentle people, they responded to his gentleness and invited him to stay with them and forsake his foolish, heretical ideas. When he would not, they guided him to the south rim of the Grand Canyon; no other white man had seen it since Cárdenas two hundred and thirty-six years before. Gradually deserted by all his Indians except a child and an old man, he went on to the Little Colorado region and eastward across what is now the Hopi Reservation till he found the apostates he was looking for, at Oraibi. They would not listen to God's word but told him they would kill him unless he went away. (This was Independence Day.) His poem ends with a three weeks' canto of thirst and weariness that took him back to the Mohaves. It is a great work of art, this traverse of the unbelievable country, but a greater one began immediately.[18]

It will always bear Escalante's name, though he was outranked by the civilian leader, a veteran frontier captain with a distinguished record named Miera y Pacheco, who was a good man with an astrolabe and a good cartographer, and at least technically by his superior in the New Mexican missions, Father Francisco Atanasio Dominguez, who was as active as he

on the journey and did most of the preaching to Indians. Escalante would earn that reward of authorship anyway, for he kept
the journal, but there can be no question that he was the principal person of the expedition and its principal sustaining
force.

The reasons for the expedition were mixed and curiously
vague. The route to Monterey was the principal one, but it is
not clear why the fathers traveled so far north nor how, from
that northing, they expected to reach the Havasupais and their
river. The journal occasionally seems to have in mind another
great river too but it was not one they ever found; perhaps
there was some idea they might reach the inland shore of the
Strait of Anian, which of course would make the northern
route to Monterey simpler and more valuable. There was the
constant desire to take the Cross to the Indians who were
known to be in the region they would cross and to others who
would probably be found there. Among these last may have
been a tribe which the rumors said lived by a lake, for when
Escalante first meets the ones he calls the Lagunas he seems to
have heard of them before.

Besides Escalante, Dominguez, and Miera, there were seven
others, several of whom had been to the western Colorado
country before and one of whom could speak Ute. They were
well supplied with horses and mules for mounts and with presents for the Indians and had a small herd for beef. They left
Santa Fe July 29 1776 and got back there January 2 1777. They
traveled more than fifteen hundred miles; most of the route
lay through country as difficult as any in the United States and
some of it may fairly be called the most difficult.

From Abiquiu, the New Mexican outpost west of Taos, they
crossed northwestward into Colorado. Sometimes in the Southern Rocky Mountain Province of geographers, sometimes in
the Canyon Lands, they worked north across a landscape that is
shattered like the landscape of the moon. Much of it is high
plateaus rimmed by higher peaks, much is naked rock gashed
by canyons or the smaller arroyos that are heartbreakingly
beautiful with their explosive opulence of trees and flowers.
They reached the Uncompahgre River and the Gunnison,
which is where the known country ended. For a considerable
time they met no Indians but finally bands of lethargic Utes

showed up and pointed the way for them when they did not know it themselves. And the way was begemmed with the names of holy saints, scores of them, names they gave to creeks, canyons, patches of shade from stunted junipers where they nooned. The journal is all arroyos, gulches, canyons, mesas, mountainsides, ridges, labyrinths, mazes, rock slides. It is all untroubled too, certainly the most serene document in the annals of American exploration. He gave his angels charge over them lest they dash their feet against a stone.

Now they heard that some of the Laguna Indians were visiting a near-by band of Utes. They heard something about the lake these Indians lived beside too, and the story was that they had pueblos there. They sought them out — the Utes they were visiting were on the Gunnison in the vicinity of the Grand Mesa — and, determining to visit their country, hired one for a guide. The journal gives small reason for the decision; everything is tranquil, sunny, and of God's favor but has little speculation about how to get to Monterey.

They crossed White River to the Yampa Plateau and reached the southern edge of what is now Dinosaur National Monument. God's architecture had here taken the Egyptian style: pyramidal, terraced, painted with brushes of comet's hair. The Uinta Mountains, the one great range of the Rockies that has an east-west axis, stretched a barrier across the north; the trail to Monterey would have had to turn west here even if the Indians at the lake had been left out of account. Less than a hundred and fifty miles farther north is the abrupt end of an even mightier range, the Wind Rivers; at its base is South Pass, the portal through which the American odyssey would reach the interior west toward which Escalante now turned. A river comes down to the northern base of the Uintas, skirts them eastward for a space, and then cuts a way through them by a series of spectacular gorges, the last of which is called Split Mountain Canyon. This too is a great river and the fathers reached it where it hurtles through the scarred gateway of Split Mountain. They named it San Buenaventura. It would be called the Spanish River as soon as American fur traders reached it, and the Siskadee, and eventually Green River. It should be called the Colorado for it is the parent stream but governmental edict has bestowed that name on the Grand, which they

had crossed on the way to the White. Escalante had no awareness that either was the river of the Havasupais which he was
seeking.

Beyond the Green they traveled the eroded Utah plateaus,
easier going than most country they had crossed so far but with
lung-straining slopes and terrifying gulches nevertheless. It got
more fertile and more beautiful as they traveled west, reached
the Wasatch Mountains, and on September 21 crossed the
divide into the Great Basin. It was already autumn in the
Wasatch, with nipping winds and a crystalline sky. At the
shoulders of the peaks the fading silver of aspen would make
the evergreens black and on the lower slopes oak brush would
be scarlet fire; the lavender of canyon shade would be purple
hours before sunset. They had reached the valley of Utah Lake
at the season of its greatest beauty; small wonder that Escalante's heart went out to it. As they travel with the smokes of
signal fires alerting the timid bands who lived there, his journal
describes the promising soil, the water and timber, the grazing,
and what he says about it could be smoothly fitted into the first
official American report on Utah Valley, which is Frémont's.
On September 23, they reached both its lake and its Indians.
It is not clear whether the Lagunas were a band of Utes whose
culture had decayed or of the humbler Paiutes. Escalante rendered their native name in several spellings and ethnologists
make it Timpaiavats, but the variant that has come down is
Timpanogos.

The Valley of Our Lady of Mercy of Timpanogos — that is
the name Escalante gave to the plateau which a more prosaic
religion would cover with harsher names. The Indians were
kindly, timorous, and feckless, and they did not live in pueblos
as the Colorado Utes had said they did. They seemed eager
for conversion, however — in such a promising country any
Indians not actually on the warpath would have seemed excellent prospects — and the fathers promised to come back and
establish a mission. (In a burst of optimism Escalante said that
this single valley was so rich it could support more pueblos than
all New Mexico.) They agreeably consented to supply guides
for the trip to Monterey, though they had no notion where it
was or how to get there.

They explored the eastern shore of Utah Lake. The Lagunas

told them about a river that flowed out of it northward to another, larger lake, whose waters were salty and harmful, in so much that anyone who washed with them would find his skin inflamed.[14] They were strangely uninterested in this news. The other lake was nature's loyal realization of Lahontan's fantasy, Great Salt Lake, though there were none of his magnificent Tahuglauks rowing galleys down canals between houses thatched with gold. Escalante could have reached it in less than two days' travel and that he did not make the journey is inexplicable. It is all the more so because, as Miera's report makes certain, the Indians said, or were understood to say, that a very large and navigable river flowed west from it, and surely their first duty was to explore any such river as a possible route to Monterey. Miera thought it must be the Tízon, which he believed Oñate had discovered and named; but Tízon was Melchior Diaz's name for the Colorado. When he drew his map he showed it flowing west from the larger lake — and so created a cartographer's myth, for later maps would show just such a river flowing out of Great Salt Lake to the Pacific and would give it the name that Escalante had given Green River, the San Buenaventura.

They knew that Monterey was somewhere to the west, and knew little or nothing more. They made off southwestward, the direction forced on them by the mountains, and straightway were in the Utah deserts. For a time it was merely desert travel, requiring the minutely vigilant skill that they took to be God's favor but not such a nightmare as the Canyon Lands had been. But their horses were faltering, their supplies had reached the vanishing point, snowstorms blew out of the mountains, and the two priests had to acknowledge that, God having sent no guidance, they knew neither where Monterey was nor where they were. They were afraid of being caught for the winter in such deserts as they were now traveling or in the passes of ranges farther west, without food or Indians to get it for them. (A brawl in camp had scared their guide into deserting.) Their decision to head toward Santa Fe across the desert was the right one but it angered the laymen, who were for sticking to their orders and going on to Monterey, which was just over yonder a little way. Miera and his men, in fact, objected so violently that the priests determined to leave it to God. Bidding every-

# Map 21 Escalante's Journey

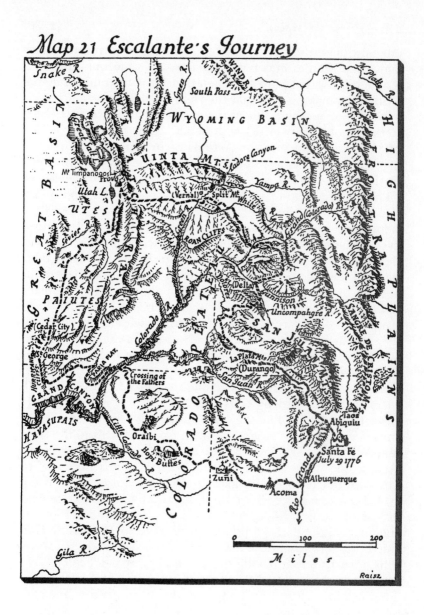

Snake R. · Great Salt L. · Mt Timpanogos · Provo · Utah L. · Sevier R. · WYOMING BASIN · South Pass · WIND RIVERS · N. Platte R. · HIGH FRONTIER · UINTA Mt S. · Isadore Canyon · Yampa R. · (Vernal) · Split Mt. · White R. · Green R. · Grand (Colorado) R. · ROAN CLIFFS · GREAT BASIN · UTES · PAIUTES · Cedar City · St. George · Colorado R. · KAIBAB PLAT. · Crossing of the Fathers · GRAND CANYON · HAVASUPAIS · Little Colorado R. · Oraibi · Hopi Buttes · Zuni · Acoma · Gila R. · COLORADO PLATEAU · Gunnison · (Delta) · Uncompahgre R. · SAN JUAN · La Plata Mt. · (Durango) · San Juan R. · SANGRE DE CRISTO MTS · Taos · Abiquiu · Santa Fe July 29 1776 · Albuquerque · Rio Grande

0   100   200

Miles

Raisz

one put himself in a reverent state of mind, they said the rosary and some litanies and then cast lots. It came out Santa Fe. There has always been a suspicion that Escalante nudged divine wisdom by making sure that the right lot fell.

Now the journey is out of the apochryphal books, incredible to those not greatly gifted with belief. They passed to the east of St. George and into the Arizona Strip along the Hurricane Cliffs, then northeastward through a gigantic lacework of stone, to the Paria Plateau and on to the Grand Canyon wilderness. "The descent to the river is very long, high, rough, and rocky, and has such bad ledges of rock that two pack animals which went down to the first one were unable to climb up in return, even without pack saddles." Many days were like that — many descents, many ascents, and no sure trail. They found their bearded Indians, who were frightened as all tribes hereabout seemed to be, and others as well; but though all were fluent with advice none would guide them very far. Frequently in danger of death from thirst, sometimes with only sickening dried seeds or cakes of "sunburned" cactus to eat, they kept on going with serene assurance, and when there were no Indians to find crevices for them to travel they went alone. There was only one place where the Colorado could be crossed, an immemorial ford used by Utes and Navahos. It was frequently described to them but they had no idea where it was. They missed it when they at last got down from the cliffs to the river and so had to go back for a time into the apochryphal geology. At last one more narrow canyon, where "it was necessary to cut steps in the rock with axes" so that the horses could make it, led out to a place where the Colorado looked slack enough. It was the ford. "About five o'clock in the afternoon [the last members of the party] finished crossing the river, praising God our Lord and firing off a few muskets as a sign of the great joy which we all felt at having overcome so great a difficulty." No white man had crossed it before them. They named the ford for the Most Holy Conception of the Virgin Mary but ever since Americans got to this country it has been the Crossing of the Fathers.

After that it was an anticlimax in fortitude that they had to cross similar country till they reached the sullen Hopis and could go on from there to Santa Fe in something like comfort.

... Nothing came of this sunnily stupendous journey. Thirty-odd years later, traders out of Santa Fe and Taos took to using some of their route in Colorado, and on the way to California crossed their trace in the Parowan country. Little is known about these shadowy figures, whose pathway came to be called the Old Spanish Trail. But no one followed Escalante into the Great Basin now, to baptize the Utes and Timpanogos whom he had promised salvation. The brilliant promise of the Spanish reawakening turned out to be a sunset color after all and it had already begun to fade. That power and energy and desire alike had failed, events at Nootka Sound would soon demonstrate. California itself became a dreamy anarchy. The Spanish made no effort to explore beyond Escalante's pathfinding or to colonize the Valley of Our Lady of Mercy of Timpanogos. That had to wait for the next century, William Ashley and the Rocky Mountain Fur Company, and Brigham Young and the Latter-Day Saints.

❧

After the islanders killed Cook, Captain James Clerke succeeded to the command. In the summer of 1779 he tried Bering Strait again but the ice stopped him and he concluded that the assigned task was hopeless. Presently he died and the American-born John Gore took the ships to England, ending the great voyage in October 1780. But in December 1779 he touched at Canton, and the sea otter furs that had been so casually and cheaply acquired at Nootka and up the coast sold for a small fortune. At least two-thirds of them had been spoiled or given away; the remnants, which were certainly disposed of at much less than their value, fetched two thousand pounds sterling in China goods. "The rage with which our seamen were possessed to return to Cook's River and, by another cargo of furs to make their fortune at one time, was not far short of mutiny." So wrote James King, who was now keeping the expedition's journal. Forthwith in two and a half pages he sketched out, practically to the last detail, a trading plan that adventurers of England were to put into effect, a plan for the Northwest Trade. The first British trading ship went to Nootka in 1785.

The American marine John Ledyard also saw the opportunity in full. In December 1782, a year after Yorktown but a year short of the peace, his ship anchored off Huntington, Long Island, and he deserted the King's service. He wrote his book and turned with terrible intensity to the ambition that dominated the rest of his life. It was twofold: to put the United States into the Northwest trade, whose triangularity he foresaw, and to cross North America eastward from the Pacific coast.[15]

Ledyard was the first American to understand the possibilities of the Northwest maritime trade and the first to propose it. The time was propitious. Shipping interests had been all but ruined by the peace treaty. The fisheries barely hung on and were losing money. Trade with the British West Indies was closed and the development of a satisfactory smuggling system would require years. The United States was being forced to the high seas and especially to the Pacific. Ledyard, like King before him, conceived just such a system as the merchants presently worked out. But he was misfortune's godchild: many times he seemed launched on his way but each time something beyond his power to affect stopped him short.

Shipping interests in New London were impressed and stimulated but would not invest. At Philadelphia Ledyard convinced Robert Morris, who undertook to execute his whole plan and make him its director — but then withdrew and instead backed the *Empress of China,* which in 1784 actually opened the China, but not the Northwest, trade for the United States. He could get no American backing and went to France, where Cook's discoveries had produced furious activity. Merchants of Lorient prepared to send him to the Northwest coast but this enterprise too failed, being blocked (probably) by the official voyage of Admiral La Pérouse to California and the Northwest coast. He went on to Paris and there formed a partnership for the great endeavor with a man whose career was even more romantic than his own, John Paul Jones. His star would not rise: this too failed. Meanwhile Ledyard had met Franklin, who was retiring at last from the American ministership, and his successor Jefferson.

Ledyard told Jefferson his plan for exploration. All other ways of getting to Nootka having failed, he now proposed to go to Russia, cross Siberia, and at Kamchatka or elsewhere find a ship engaged in the Aleutian trade. When he reached the

North American coast, he would cross the continent to New York on foot. He would do this with no outfit and no white companions: he would be Moncacht-Apé in reverse. Apart from his effort to establish the Northwest trade, this intention is the principal reason for his celebrity.[16]

Jefferson warmed to the idea, though he understood how much of it was fantasy, and they saw much of each other. Suddenly there was a chance in England and Ledyard was off to London. This time he actually got to sea on a ship bound to Nootka Sound but its papers were irregular (it was infringing on the chartered monopoly) and it was overhauled and forced to return. This was in 1786, the British trade to the coast had begun, and Ledyard came under the highly important patronage of Sir Joseph Banks, Cook's original backer, who was about to become head of the Royal Society. He could have got Ledyard a place with the Colnett expedition but gave it to another favorite instead. Eventually there was nothing left but the trans-Russian plan. In December 1786 he set out without a passport and with only a few pounds to travel on — ever since leaving the United States he had lived by the penurious favor of rich men — and when Jefferson heard from him from St. Petersburg he had "only two shirts and yet more shirts than shillings."

Ledyard's frenzied, tragic, and the truth is half-mad, journey does not come within the purpose of this narrative. He crossed the Urals to the Ob and at last reached Irkutsk. He had had various companions and he was being watched. Now he came under the scrutiny of Russian fur traders who were developing the Aleutian field and at that moment were preparing to expand southward along the coast to California. They could not let him carry out his plan and they didn't. They let him make a slow descent of the Lena to Yakutsk but forced him to spend the winter there. In February 1788 he was arrested on orders from St. Petersburg — technically for espionage, actually to keep him from reaching Okhotsk or Kamchatka — and was hurried all that long way back to Europe. Less than a year later, while on an expedition to Africa for Banks, he died in Cairo.

☙❧

Alexander Henry and the Frobishers had taken a momentous step, economically and geographically, when they moved from Saskatchewan waters to those of the Churchill basin, but they learned at once that an even more important one was indicated. In 1775 they had heard about distant Lake Athabaska, and the Chipewyans whom they intercepted at Portage du Traite in 1776 told them still more about it, and about Peace River and another river that ran northward to some sea. There was word about the Rocky Mountains too (which would sometimes be called the Chipewyan Mountains because of these same Indians) and the people who lived west of them. Some of those people were with the Chipewyans as captives, and their owners said that the Pacific could be seen from their country. But the main thing was the big lake farther north, whose name everyone would spell differently from now on, and its wealth in furs, which evidently was greater even than that which the Pedlars had found on the Saskatchewan and the Churchill. When Joseph Frobisher and Alexander Henry turned back to Grand Portage in 1776, it was with the purpose of reaching Lake Athabaska that Thomas Frobisher stayed behind. He did not make it; he had to return to Lake Isle-à-la-Crosse and spend the winter there.

For not only the furs grew more numerous as you went north: so did the difficulties. At Lake Isle-à-la-Crosse they were pretty far north, 56°, but it had taken them eight days to get there from their temporary post at Beaver Lake — and Lake Athabaska was much farther north, so much that the route to it from Grand Portage was 1850 miles long.[17] They had crossed into another geographical province, the subarctics. God had not sufficiently divided the land from the waters there. In a ten-mile square there may be a hundred lakes big enough to be mapped and a hundred smaller ones making the spaces between them impassable. Rivers widen into lakes for miles at a stretch but the channel through them is narrow and twists like the threads of a screw, with a gashed canoe if you miss a bend in it. Infrequent hills are naked granite, except for an occasional slope of dolomite which has a welcome covering of grass. But mostly the shores rise only a few inches above the water. Near the water they are sandy and so, like patches farther inland that have been burned, they are heavily timbered with

jack pine. Back from the shore is the muskeg, neither land nor
water nor mud but a scrofulous compound of them all, patch
after patch of it to infinity, each patch ringed with black spruce.
The country is four-fifths drowned and when not frozen is half-
hidden by mosquitoes and black flies.

To get beyond the Saskatchewan had strained the system to
the utmost; to go into the high north would necessitate radical
innovations. One necessity was to solve the problem of suste-
nance; as it was, canoes could carry no more food for their crews,
and winterers lived exclusively on fish, smelled like Winnebagos, and sometimes came close to starving. To get to Portage
du Traite, the Frobishers had had to arrange a timed meeting
with canoes bringing them pemmican from the Assiniboins.
And pemmican proved to be the answer, worked out by the
Frobishers and the crabbed genius Peter Pond, Indian pemmican, the greatest of all concentrated foods.

It could be made from any meat and some of that which
Pond bought in the North was made of deer and elk and even
reindeer, but by far the best was buffalo meat. It was dried by
sun and wind or over a slow fire, pulverized, sometimes flavored
with dried fruits and berries, mixed with melted buffalo fat,
and then sealed with more fat in parfleche bags. So packed, it
would last for years and took up a minimum of space. It was a
complete diet in itself, it could be eaten raw or cooked, and it
never palled on a man as the frozen fish always did. The woods
buffalo ranged as high as Great Slave Lake, far beyond Lake
Athabaska, but the North was not a sufficiently large source of
supply as the trade grew larger and a complexly interdependent
system of procurement and transport was worked out. So the
Plains tribes became the principal suppliers, the trade with
them became more important, and for the first time the South
Saskatchewan, which flowed through the plains, became an important highway. Soon the North West Company was itself
manufacturing pemmican in enormous quantities. Production
centered at Pembina on the Red River and was mainly carried
on by halfbreeds, the children of French and English traders
who became the Métis of Canadian history. Supply depots
where it was stored were set up at the posts that held the important crossings of the canoe routes.

Pemmican enabled the Pedlars to enter the farther country

and Pond himself was the first. The stringencies of competition and overhead expense had already suggested such combinations as Henry and the Frobishers formed in 1775. Now in 1778 four Saskatchewan traders pooled their surplus goods and sent Pond north. He went where Thomas Frobisher had been unable to go, to Lake Athabaska. Churchill River, to whose drainage system Lake Isle-à-la-Crosse belongs, flows into Hudson Bay. Pond's route took him from that lake to the one that is now named for him — "very shallow," Mackenzie says, "and navigated with difficulty even by half-laden canoes" and on to Methy River and up that to Methy Lake. From here a thirteen-mile portage led across the height of land; it ended in "a very steep precipice whose ascent and descent appears to be equally impracticable . . . it consists of a succession of eight hills, some of which are almost perpendicular." It was climactic for it led to Arctic drainage: beyond it lay the Clearwater River, which flowed into the big Athabaska River, which emptied into Lake Athabaska. At 59° N., in a different kind of world, Montreal reached its richest subtreasury; it was so full of furs that Pond could take back only half of those he got.

There was a new name now. The combination of 1779, composed of eight partnerships, was temporary but it was called the North West Company. It fell apart; the merchants were not yet convinced that it was stupid to take a quick profit from undercutting one another, nor their winterers convinced that price-slashing, piracy, peonage, subornation, mayhem, and occasional murder would not pay off. But the duress of the facts produced other temporary combinations which became another North West Company in 1784, and three years later a reorganization gave it permanent shape. There would always be opposition, the most brutal originating in the withdrawal of disgruntled partners, but this was the dominant amalgamation. It was Canada's trust, it was trying to become a monopoly in the country of the chartered monopoly, its methods were as ruthless as that country, and it was the answer to a tremendous question, the right instrument for expansion. Or almost the answer. For from the beginning its best minds, and Pond first, saw that another step was necessary for completion, one that would make it bigger and more powerful and enable it to go farther.

Pond organized what was to be known as the Athabaska Department. All the North West Company had an aggressive esprit de corps; management could hardly have got the underpaid lower ranks to spend their lives in labor, hardship, and danger at the extremity of endurance if they had not had pride of craft. But the greatest pride and the lordliest demeanor were those of the Athabaska men. All the other departments were based on Grand Portage. (Later on Kaministiquia, when the company moved its headquarters out of American jurisdiction, back to the earliest head of the route west and built Fort William there.) But not the far northerners. A special depot for them was set up at Rainy Lake and a special caste of pork-eaters carried their goods to it. No one disputed their pre-eminence or challenged their privileges and immunities. They came down to Rainy Lake with the furs as soon as the streams were open, and their renewal by saturnalia had to be brief for the race was to get back before the streams froze again. They left Rainy Lake in August; they and the first film of ice reached Lake Athabaska in October. In January they sent out the Winter Express, the letter and report that began here and circulated, growing as it traveled, among the lower posts and on to headquarters. They were the northern anchor but they were also the northern spearhead and as the years passed they worked on to Slave River, Great Slave Lake, and beyond, and westward to Peace River.

Pond's stay in the North was broken at least twice and it is not known how much of the farther country he himself saw. It does not matter, either, for what counts is his thinking. Genius has chosen few odder habitations than this half-illiterate, quill-bearing brawler. David Thompson was wrong to call him unprincipled but less than severe when he added "of a violent character." If there had been law west of Grand Portage Pond could be said to have murdered a partner and shared in the murder of a competitor, and as it was, the latter violence forced him to withdraw from the country he had pioneered. But in full measure and beyond any of his associates he had the continental consciousness — which must be seen as having two levels, of thought and of the feel for the land's reality that underlies thought. At his post forty miles up Athabaska River from the lake, he directed the rich trade and the meshed activities that supported it. He grew garden stuff, in the shortest of seasons, as an experiment in simplifying the problem of supply. But most of all he was forever talking with Indians who came from the still farther country. His mind was set westward and, if his spelling was eclectic, it was as powerful a mind as any that had ever grappled with the mysteries of the continent. He has to be followed, mostly, in his maps and in the thinking and reports of minds he fructified and the actions of men who took leading from him, but wherever one touches him there is an urgent and even desperate will to find out, to break through, to solve the last problem of Canada. Curiously this intelligence working in the interest of commerce is driven by the desire that is for its own sake only: to know the nature of things.

The last problem was the Pacific Ocean. What routes led to it? How did the Arctic Ocean fit in? Pond learned of a lake farther north than Great Slave Lake, which, God knew, was far enough north of him. The farther one was the huge Great Bear Lake, which extends beyond 67° N. He learned of a river that flowed into the Arctic, and eventually he learned that this was not the Coppermine which Hearne had descended. He learned a lot about the Rockies and a lot that was not true about what lay beyond them. He could not put it together right but he spent years trying to and his thinking has as solid importance as any in the record of the westward thrust. A stiff man whose blood was as cold as his was hot, and who disdained him, was to reap the reward of that thinking, which centered on the great river farther north than his fort at the ridgepole of the hemisphere, and on the possibility that it might lead to the Pacific.[18]

Meanwhile competition with the big rival and the small ones multiplied activity in the regions to the south. Before 1790 a good many posts were built in the sector between Lake Winnipeg and the Saskatchewan. In this country the Hudson's Bay Company could use the advantage that cheap transportation gave it and the additional one provided by the nearness of York Factory, as compared with Grand Portage, which enabled it to open the trading season earlier. That was the sum of its advantages, however; it fought hard but unskillfully and under heavy handicaps. Institutional rigidit/, caste stratification, failure to provide incentives for its servants, unimaginative and even stupid management, slowness to learn from competition, a mild queasiness about violence, disregard for what the customer wanted — all such products of chartered privilege worked against it. Even more basic was the fact that the Pedlars had the inherited French system and the French Canadians who were its foundation, whereas the Company's Orkneymen did not like the country and seemed unable to master its skills. The Northwesters hired away its best men, they had the briskness and ingenuity that it lacked, they had no conscience, and when the Company met them head-on by building posts opposite theirs, they cut prices, intimidated the rival bourgeois, stole their Indians, and ranged deeper into the wilderness to intercept more customers and find new ones. The Company had no choice but to follow and imitate; by the mid-1780's the most stable dividends in British finance had come tumbling down.

There was a cluster of posts at the great crossroads of The Pas and another one at the higher crossroads where the Saskatchewan forked. From there both companies leapfrogged up the river. The northern branch became a highway to the upper country, which had hitherto been reached by way of The Pas, Sturgeon Lake, and Cumberland House. This highway led through Assiniboin country toward the Blackfeet, the suspicious and belligerent customers who would not hunt beaver but did make pemmican and who by now had won ascendancy over the Snakes. The southern branch, which was exploited more slowly, also led to the Blackfeet but by way of their congeners the Fall Indians. The Company's melancholy William Tomison was the first to go very far up the South Saskatchewan but the Northwesters' Peter Pangman leapfrogged him, a man much like Pond, intransigeant, a pioneering explorer, alternately a partner and a leader of tempestuous oppositions.

Farther south was the Assiniboine River. The Northwesters used Vérendrye's route to it, going up Red River from Lake Winnipeg. The Company reached it more directly, west across Lake Winnipeg below, by Lake Winnipegosis and Swan River higher up. Posts spread up the Assiniboine to the big bend, beyond it to the Elbow, in the quadrangle south of Red Deer River, on the Qu'Appelle River. (Crying River: it had a spirit that mourned at night.)

Unbridled and unprincipled competition produced a buyer's

market, but the buyer was controlled by intimidation and debauchery. If the Company was somewhat the slower to use
gunfire, it had as much rum as anyone else. (The Northwesters
averaged two hundred gallons per canoe, most of it "high wine,"
that is rectified spirits, 180-proof.) All traders stole one another's customers as often as possible. Sometimes they took the
Indians' furs from them at the point of a musket, sometimes
they beat up their own or strange Indians on suspicion of trading with someone else. The Indians liked the rise in the price
of furs and the increasing frequency of big drunks, but it was
not bright to humiliate them. So in a country where the white
man had previously had no trouble with Indians there was a
flurry of rebellion in 1780. It was touched off when a trader
solved the problem of a drunken chief by giving him a dose
of laudanum that killed him. It cost the trader a pitched battle,
his goods, and his fort. This was on the North Saskatchewan
and it seemed to Indians so excellent an innovation that they
attacked a couple of posts on the Assiniboine, killed some more
white men, and gave some indication that they might make a
general war. Pedlars and Company men found reasons for a
more cordial amity than they had practiced, abandoned exposed posts, and supported one another.

If any uprising was in the making, the winter of 1781–82 put
a stop to it. War parties from the upper Missouri country
brought smallpox north with them to tribes that had never
had it before, and it spread across the Northwest. The country
came close to being depopulated. "Naught was left to them,"
Alexander Mackenzie wrote, "but to submit in agony and
despair. To aggravate the picture . . . may be added the putrid
carcases which the wolves with a furious voracity dragged
forth from the huts or which were mangled within them by
the dogs. . . . Nor was it uncommon for the father of a family
whom the infection had not reached to call them around . . .
and to incite them to baffle death with all its horrors by their
own poniards. . . . He was himself ready to perform the deed
of mercy with his own hand as the last act of his affection and
instantly to follow them to the common place of rest and refuge
from human evil." Every tribe it struck was decimated; the
Athapascan peoples of the far North were almost wiped out.[19]

(The epidemic changed tribal relations on the upper Mis-

souri. At least half the Mandans died. The survivors abandoned the ill-omened villages at the mouth of Heart River —
the center of the world — and moved sixty-odd miles upstream
to the mouth of Knife River. There they built villages near
those of the Minnetarees, who were also called Gros Ventres
but were not related to the Gros Ventres who were called the
Fall Indians. Their withdrawal left the Missouri open to the
Sioux, who up to now had conducted themselves humbly or
were easily frustrated when they were belligerent, but who were
coming west and embracing the Plains culture in increasing
numbers. The Mandans were never a vigorous people again
and presently they began to be afraid of the Sioux. Within a
few years trading posts for them, their Minnetaree neighbors,
and their visitors from the West had been built on the Souris
River. The first British advance to the Mandan villages, which
was entirely illegal since they were on Spanish soil, appears to
have been made in 1785.)

The trade of the Northwest was at a standstill for a year after
the smallpox. It had already been seriously inconvenienced by
difficulties that had piled up from the Revolution, since the
necessities of defense and the fear of capture by the Americans
on the upper Lakes slowed the flow of goods. The Hudson's Bay
Company suffered a disaster in 1782, when in a last-act oddity
of the war Admiral La Pérouse took a French naval squadron
into Hudson Bay, captured Fort Churchill, and burned York
Factory to the ground. The Northwesters were struggling with
the problem of costs; the overhead, especially transportation
charges, increased multiply and complexly as the distance increased. (As a single item: in the vicinity of the Saskatchewan
the Alaska white birch takes the place of other varieties. The
trees are smaller and have more branches. Bark must be taken
in smaller strips, which meant more labor, and it had more
holes in it. Canoes therefore cost more and did not last so
long.) [20] These were the forces that had impelled the Pedlars
to combine, had produced the North West Company and given
it, in 1787, its final form. But the same forces impelled them
to other actions and plans. The cost of transportation showed
that a connection with the Pacific, and deep-sea freighting, had
become imperative, and the pressure increased with news of
Russian activity on the coast, of Cook's discoveries, and of the

rich trade that the sea otter made possible. As early as 1781 Pond's associate and apt pupil, Alexander Henry, proposed to the influential scientist Joseph Banks, of the Royal Society, an expedition that would reach the Pacific by traveling northwest in the high latitudes.[21] Henry's details were necessarily vague but, using Pond's geography, he was confident that a route from Lake Athabaska could be found. Four years later, in 1785, the North West Company called on the government to give it a monopoly of the trade in the regions which it would discover, in return for an exploration to the Pacific. Henry's letter to Banks had foreshadowed a further step: a demand for a merger of the North West Company and the Hudson's Bay Company or, if the latter would not make terms, then a revision of its charter that would permit others to use the Bay and the rivers that flowed into it for the transport of goods to the Northwest. This idea, revolutionary but as it ultimately proved inevitable, was already in Henry's mind and Pond's and would become Alexander Mackenzie's fixed purpose. But, essential as the Hudson Bay connection was to the full implementation of trade imperialism, the Pacific was even more fundamental. It seemed such a short distance from the markets which the Northwesters now reached by that infinitude of portages from Montreal: to get there had become the paramount necessity.

The geographical problem obsessed Pond at Lake Athabaska. As he went on talking to Indians from beyond, as he studied Cook's text and maps and more imaginative maps by other hands, he changed his ideas. He had first supposed that the still unvisited river which he knew flowed out of Great Slave Lake must empty into the Arctic. He now concluded that, instead, it must flow to the Pacific. It must be, in fact, the big river which Cook had supposed accounted for the fresh water at the eastern end of Cook's Inlet. If it were, then necessarily the Rocky Mountains, that final barrier, must end somewhere south of it. He decided that the Pacific, as Henry put it, "can't be any Great Distance" away. . . . It never had been any great distance away since Champlain, since Verrazano.

Pond's post was in fact only a few degrees south of the latitude of Cook's Inlet but, because he was unable to determine longitude, he thought it was a great deal farther west than it was. At any rate, he prepared to test his theory, to find Cook's

River and go down it to the Pacific. Or rather, since he was fifty, to send down it a man half his age who was his second in command and whom he had made letter-perfect in his ideas. He sketched a complete plan for this young man — he was Alexander Mackenzie. Mackenzie was to go to Great Slave Lake and follow the river which flowed west out of it — Cook's River — to the Pacific. From there he was to go "to Unalaska, and so to Kamskatka, and thence to England through Russia." [22] Mackenzie, that is, was to exactly reverse John Ledyard's dream. But Pond could not stay at Athabaska while he did so. He was held responsible for a clerk's murder of a rival trader and, in a sudden conversion to comparatively peaceful methods which resulted in the final redistribution of North West Company partnerships, feeling against him got so strong that he had to leave the high country.

In 1789 Alexander Mackenzie was twenty-five years old. That he had been a North West Company partner for four years amply proves that he was a first-rate man as well as a well-connected one. He had ample experience of the high country too, for he had spent two winters there, though that was a slight apprenticeship compared to Pond, the ancient of days. He had mastered business principles and the geography of trade. Now he was to work out a fully articulated conception of commercial imperialism. Well in advance of anyone else he determined the commercial axis of the continent and he foresaw the exact shape of the last imperial struggle that was implicit in continentalism. In courage, in the faculty of command, in ability to meet the unforeseen with resources of craft and skill, in the will that cannot be overborne, he has had no superior in the history of American exploration. But he is a hard man to like. The Scots who were nine-tenths of the North West Company's aristocracy average remarkably high in arrogance, caste brutality, and the paralysis of personal emotions that is called reserve. Mackenzie had them in a measure that raised him above the rest. All that softens him is the desire for knowledge — knowledge for its own sake — that got scant respect from the primordial capitalists who were his partners. Clearly he got the desire by infection from Pond, to whom he felt a contemptuous superiority that can still be perceived, a hundred and fifty years later, in the accents of his prose.

In 1789 Mackenzie built a post on the shore of Lake Athabaska, to replace Pond's. Then on June 3, in a single canoe with five *voyageurs* and the squaws of two and an Indian guide, accompanied by a second canoe filled with trade goods and a scattering of other Indians, he set out for the Pacific Ocean. In his mind he was going to go to Cook's Inlet and on to Unalaska: he was beginning a continental crossing. It would have been the first one since Cabeza de Vaca.

One hundred and two days later he was back. Out of Lake Athabaska, the Slave River led to Great Slave Lake, which is larger than Lake Erie. (The name is from the Indians on the lower reach of the river, which was a war road to the Hares still farther north.) Flowing northward out of Slave Lake was one of the biggest rivers on the continent, with a drainage basin of a million square miles. Mackenzie reached it on June 29 and it has his name now, but he took it to be what Pond had supposed it was, Cook's River, flowing to Cook's Inlet from the east. Its first course bore westward as it should, a most hopeful omen. But on July 2 "at nine we perceived a very high mountain ahead, which appeared on our nearer approach to be rather a cluster of mountains stretching as far as our view could reach to the southward, and whose tops were lost in the clouds." They were the Rockies. Rivers came down from them and the big river he was descending never led into them. An enormous range would eventually bear his name, like the river, but the sum of them was what Pond's thinking had been unable to conceive. . . . What all thinking about the western crossing had been unable to conceive. This was the barrier to navigation that traversed the whole extent of the continent from north to south. It was not a slight barrier, a half-day portage for bark canoes, it was something for which no earlier experience could prepare the mind. Mackenzie's was the first attempt that put it to the empirical test.

He drove on past the mouth of Great Bear River, which flows from an even larger lake, past the Hares, past the Quarrelers, into country where the bellicose Eskimos could be expected but never appeared. He traveled with a fury that was imposed on him by the shortness of summer and by the fear which his Indians and *voyageurs* felt increasingly in this most hideous of desolations. Even though he underestimated the distance he was traveling — for it was a swift river and he pushed his men to the utmost — he was disheartened by the long northing that never turned into the mountains at his left or revealed a passage through them. On July 12, with the mountains still there, he reached salt water. It was the Arctic Ocean, though he was not sure he had gone all the way. But he got the point: the river he had traveled was not a route to the Pacific.

But also his route had not intersected the Northwest Passage. He was certain that he had driven the last nail: surely no one could continue to believe in it now. Some did, however, including a number of his partners, and they continued to believe in it after his next expedition and on into the nineteenth century.

He spent three days pushing his canoe among islands and ice floes, chasing whales, and trying to measure tides — tides which he still was not sure were of the open sea. Then he summed up his defeat by naming his river Disappointment and turned back. On the way north he had heard fabulous stories about the far side of the mountains; on the way back he got more solid stuff. The Indians said that beyond the ranges there was a river that ran into the white man's lake — so big a river that compared to it Disappointment "was but a small stream." They must have been talking about the Yukon but when they sketched a map in the sand they rearranged geography and said that the white man had a fort at its mouth. The people there "make canoes larger than ours," which agreed with Cook's report, and "kill a kind of beaver, the skin of which is almost red," which was unmistakably the sea otter. A conclusion was forced on him: "This I took to be Unalaska Fort and consequently the river to the west to be Cook's River, and that the body of water or sea into which this river [Disappointment: the Mackenzie] discharges itself at Whale Island communicates with Norton Sound."

At the farthest north the West continued to generate marvels. The Indians said that the people who lived beyond the mountains were "of gigantic stature" and had wings, though they did not fly with them. They could kill you with the glance of an eye and they ate big birds which were also man-killers. And at the farthest north the way of nature held. His informants were the most miserable Indians he had ever seen, and yet though "to the eye of a European they certainly were objects of disgust . . . there were those among my party who observed some hidden charms in these females which rendered them objects of desire, and means were found, I believe, that very soon dissipated their alarms and subdued their coyness."

He had found and now he phrased the irreducible core: "The sources of those streams which are tributary to both the great rivers are separated by the mountains." It was the end of the pyramidal height of land.

Even so, even though the land westward was upraised in those gigantic peaks, it was not very wide. "By the Indian Account the [Pacific] Sea is but a short way to the westward." And there must be a seaway too; he believed that only the ice would keep one from getting there from the mouth of the Mackenzie River. He would have himself forced a way over those mountains right now but he could not hire anyone to guide him.

Mackenzie's journey to the Arctic had final importance. Unhappily the truth it disclosed was the most frustrating kind, negative proof, and the mystery remained unsolved. And he had unlocked no new profits for the North West Company, to whose summer meeting of partners at Grand Portage he reported his results the following year. Directors believed no more readily then than now that knowledge as such is a capital asset. The imperial vision which Mackenzie was opening to them would have given them a corner on the furs of a hemisphere but he could make no down payment. "My expedition was hardly spoken of but that is what I expected." Thus the partners, though it would be a heroic story at the posts from now on, and a few years later their continuing indifference would drive him from the company into the opposition.

The reasons for his journey still held good and his indomitable spirit had been polarized. The mystery still had to be solved, the facts still had to be dug out, the way to the Pacific still had to be found. Now he would make it exact. "I was not only without the necessary books and instruments [such as star-tables, an accurately calibrated sextant, a chronometer] but also felt myself deficient in the sciences of astronomy and navigation." He determined "to procure the one and acquire the other" and sailed for England.

# Equinoctial Tide

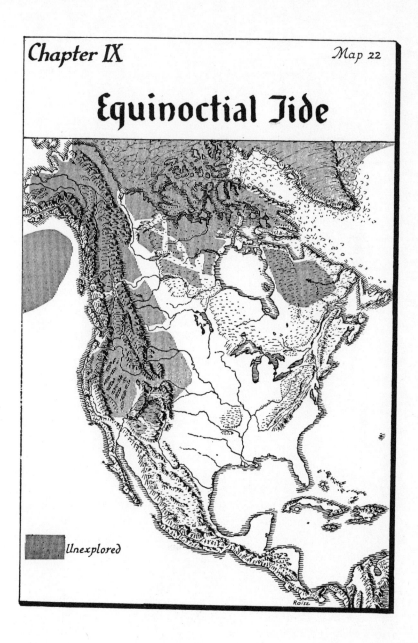

Unexplored

Raisz

# IX

## Equinoctial Tide

T HE PARK RIVER, after meandering eastward across North Dakota, is about forty miles south of the Manitoba boundary when it empties into the Red River. Alexander Henry the younger built a trading post near its mouth when he commanded the Red River Department for the North West Company — he was a nephew of the Henry who has appeared in this narrative and had bought his uncle's shares. He then built a small feeder post farther up the Red at the site of Grand Forks, where the Red Lake River comes in from Minnesota, and another one downstream exactly on the Manitoba line, at a site where already there had been several evanescent posts. The latter was to become famous as Pembina but for a time Henry's headquarters were at Park River.

His customers were Crees from the north, and Assiniboins from west and north, but mostly Chippewas. The Chippewas were at their farthest west here and they were uneasy outside the forest, which ended a little farther east. As a result they were in practically continuous terror of the Sioux. Every movement might be a threat of massacre, even a drift of cloud-shadow across the edge of the plain. Let anyone come in sight above the horizon or along the edge of an oak grove, let a horse stumble in the brush or leave a hoofprint in the mud of the riverbank, let a squaw have a painful dream or a bird dart low over a cook fire — it was enough to start the women screaming and digging foxholes and the braves running in circles and firing muskets at the sky. There were Sioux to be afraid of; they were now coming to the Red, their ancient war road, from

315

the south as well as from the east. But the Chippewas suffered fifty alarms per possible Sioux. A party would leave the vicinity of the post for a hunt, or just on one of its seasonal wanderings, and come stampeding back because someone had heard something. A child playing by the river heard an unfamiliar bird call and everybody got ready to die — but played it safe by trying to get inside the trading post. Anyone could start a panic by merely remarking to someone else that he had a notion the Sioux were coming.[1]

The Red River, the Assiniboine, the Saskatchewan, and their tributaries swarmed with fish. Even the flabby sucker was eaten but the staples were whitefish, catfish, and sturgeon. In the spring sturgeon were so thick that the canoes practically grounded on them and they were big fish, fifty, sixty, up to a hundred and fifty pounds. As soon as the spring floods receded — then as now they inundated hundreds of square miles — you began to seine sturgeon or stretched nets a hundred yards long from bank to bank. Promptly the change from a meat diet gave everyone dysentery.

Spring and fall, such flights of ducks, geese, and swans as no one will ever see again darkened the sky. Pelicans, cormorants, the eagles which the Indians trapped, many other populous species of water and land birds stayed all year. Grouse and quail were as numerous as the fleas you picked up from the grass or the ticks that covered your body with welts. There were even more bears. Grizzlies came hurtling out of the brush on their inexplicable, manic sprint but the regional infestation was black bears, which were everywhere. You could shoot them from horseback, or on the way to the river, or while strolling in the evening. "Bears make prodigious ravages in the brush and willows; the plum trees are torn to pieces and every tree that bears fruit has shared the same fate; the tops of the oaks are also very roughly handled, broken, and torn down to get the acorns . . . their dung lies about in the woods as plentiful as that of the buffalo in the meadows." Henry raised a cub to maturity. The dogs got along with it amiably and it followed the *engagés* across the plain and through the groves. When winter came on it refused to hibernate in the holes Henry dug for it, insisting on scooping out its own den.

A bear carcass tried out many gallons of oil. You boiled the oil to keep it sweet and got a clear fat that kept indefinitely, a good seasoning or soup stock or even a meal in itself. It was a necessity to the Indians' personal life, a foundation cream for paint, a pomatum for the hair, a remedy for all distempers, and a sacred ointment which when rubbed on the body gave them a bear's ferocity and an odor that traveled ahead of them for some distance. Buffalo tallow, which had to be boiled too, was even more valuable. It was indispensable for pemmican but could also be eaten like bread or as a dessert. In the winter cuts and sides of buffalo were frozen and hung in the ice-house — every post had one which was filled as soon as the river and lakes froze. "Piece meat" would not last long in warm weather, so there were always drying stages where it was hung for jerky or pemmican. The squaws brought in many kinds of greens, savories, berries, and fruits. They gathered the "prairie turnip" or "prairie apple" and braided it in long twists that hung from the lodgepoles like onions in your grandmother's cellar. (*Psoralea esculenta* and related species; it was, the white man reported, "windy.") The Sioux and some other neighboring tribes thought dog meat a delicacy but white men fell back on it only in starving times. Best castrate a male as soon as it was killed, to prevent a rank taste.

The great buffalo plains began here; winter or summer there were usually herds at no great distance. A tall oak near Henry's post served him as a watchtower. Climbing it, he could see a dozen small herds on their slow way nowhither or in the rutting season herds too big to be counted, moving like the tide, shaking the earth, bellowing all day and all night. The Indians were always hunting for meat, hides, and the innumerable by-products. The Crees, however, were indifferent leatherworkers and the Assiniboins clumsy, which meant business for the skilled tribes of the upper Missouri, especially the Mandans. The Chippewas did not even use skin tipis but clung to their traditional bark lodges, here where trees were much scarcer than bulls. Buffalo were so easily killed that white men found the sport dull; horses were still uncommon and so they did not get the drama of the chase. One novelty the Indians practiced was to drive them out on the ice when it grew rotten toward spring, and bunch them till they broke through it and drowned; they floated downstream under the ice and at the next opening squaws dragged them out. When the ice started to break up they drowned innumerably and as they decomposed became a delicacy for Indian gourmets, who thought buffalo meat best when it turned green and began to liquefy. At the end of March, "great numbers of dead buffalo from above," and two weeks later, "drowned buffalo still drifting down the river but not in such vast numbers as before." Henry found the stench along the river intolerable and on the way to one of his posts had to skip supper because of it. Coming down the Qu'Appelle past the middle of May, John Macdonnell spent a day counting them. He made it "7360 drowned and mired along the river [thirty miles at most] and in it . . . in one or two places I went on shore and walked from one carcass to the other, where they lay from three to five files deep." . . . Prairie fires were just as hard on them. "Blind buffalo seen every moment wandering about. The poor beasts have all the hair singed off them; even the skin in many places is shrivelled up and terribly burned, and their eyes are swollen and closed fast. It was really pitiful to see them staggering about, sometimes running afoul of a large stone, at other times tumbling down hill and falling into creeks. . . . In one spot we found a whole herd lying dead." But good roast buffalo, if you got there before the wolves.

Above the Qu'Appelle streams might freeze as early as mid-October; below there they were sure to be open two or three weeks longer and sometimes did not freeze till December. Winter closed in with blizzards and abysmal cold spells. (Worst along the Missouri, where for three weeks or more at a time the temperature might not rise to twenty below.) Firewood had been stacked — 120 cords for four fireplaces at Park River — but winter was the time to cut it for next year. All jobs were welcome, meat-making, ice-harvesting, repair and renovation, blacksmithing, whatever could be thought up, for boredom closed in too. The *engagés* went coasting on the sledges that were used for hauling wood, the squaws joining in this odd white man's sport, and the ice games of Canada, Scotland, and the Indians produced strange mixtures of skills. It was a time for hair-raising kinds of hunting and marksmanship contests, for anything that would break the dull, dead-level stretches when the snow and ice deepened and the sun might not show itself for days at a time. "The country affords no tallow for candles nor fish oil for lamps; the light of the fire is what we have to read or work by." There was not much to read; word that someone had a different book would make a man travel a hundred miles, risking blizzards and *mal-de-raquette* (the agonizing strain produced in tendons by unskillful or too violent snowshoeing) to borrow it. There was great satisfaction when a bitch in heat brought the wolf packs around or the post dogs boiled out after a bitch wolf. "Some of my men have amused themselves by watching their motions in the act of copulation; rushing upon them with an axe or club, when the dog, apprehending no danger, would remain quiet and the wolf, unable to run off, could be despatched."

Any Indian ceremony that involved dancing was welcome, and most ceremonies did. If it also involved a medicine man's prestidigitation it had the drawing power of a Broadway opening. The magicians were good. They turned leather belts to snakes which coiled and rattled, then turned them back to leather. They had themselves bound with sinew cords at wrists and ankles and knees, wrapped in a moose or buffalo robe and again bound with rawhide, and inclosed in a small tent from which, after the spirits had howled for a while, they emerged unbound to take a bow. They swallowed broken arrows and coughed up whole ones, cleft each other's skulls with hatchets, sawed the beautiful wench in two, and fired a dozen bullets or thrust a spear through her. She bled gallons and "the spectator really believes that he sees the feathers gradually enter and the arrow, all bloody, come out on the other side, under the shoulder blade."

It was easy to believe things in winter. The windigo walked the frozen snowfields, the spectral cannibal who terrified Indians and made a winterer's spine crawl when a gale out of the north shook the fort at midnight. The Qu'Appelle's spirit wailed. Up at Lake Isle-à-la-Crosse Mr. Grame told John McDonald of Garth that in the light of a full moon he had seen a coach drawn by two white horses drive toward him, pass, and stop where a little river reached the lake; one day next spring they found his canoe at the mouth of that creek but they did not find Mr. Grame, for the coach had taken him away. At Rocky Mountain House even the steel-nerved Alexander

Mackenzie wrote to his cousin Roderic that he could hardly
bear the dreams he had. At Cumberland House Mr. Hallet
bade Mr. King, who was going on a journey, beware of La
Mothe, thus amusing Mr. King a great deal for La Mothe was
a clerk and in the Hudson's Bay Company caste was caste. The
second night he was gone Mr. King's little girl (of course a
halfbreed) told her mother, "There is my father at the foot of
the bed, his neck all red." On the third day a sleigh came back
bearing Mr. King's corpse — La Mothe had shot him in the
neck.

In winter time women, who were usually to be had for a
handful of beads or a civil word, grew coy. John Macdonnell is
scandalized because Garreau, coming in from a long season of
batching it, offered Foutreau a full gallon of high wine — un-
diluted, by God! — for a night with his daughter and then got
talked into adding two quarts more and a half a dozen scalping
knives. (The worst was, before the girl could begin earning
her pay, the whole family got drunk and belligerent on the fee
and Garreau had to be hurried away to save his life.) And
thus Henry, shocked to a sparse statement, "one of my men
gave a mare that cost him G. H. V. P. [Grand Portage] currency
equal to 16£ 13s. 4d., Halifax currency, for one single touch of
a Slave girl."

Even the opposition — Hudson's Bay Company or "the little
companies" — lost some of their villainy in cold weather. At
other times you kept a telescope trained on their post, hired
Indians to report on what was happening there, sent out a man
to follow every small group that set out thence for anywhere,
entered them as "miscreants" or much worse when you mentioned them in the post journal, and charged them (quite
justly) with every kind of fraud and felony the trade had ever
invented. Now you exchanged courtesies and brought the
crews together for a dance, frugally serving them high wine
or the lowest grade of rum you had, but perhaps taking the
bourgeois into your room and getting out your private brandy.
The festival always broke up in fistfights but that heightened it
and everyone felt refreshed.

At any season the arrival of Indians at a post meant a brawl.
Even sober they were the noisiest race. Squaws' casual conversation was in shouts and became fortissimo when there was

work to do. Religious worship was at a roar, with tympani, but was quieter than medical practice and a whisper compared to mourning. ("Great lamentation — must have a keg of liquor to wash the grief from the heart, a fathom of cloth to cover the body, and a quarter of a pound of vermilion to paint it.") And nothing was as noisy as a drinking party: to erase the memory of defeat, to nerve them to make a fight, to get them to come to the post, to welcome them on arrival, to buy their loyalty, to keep it, to induce them to default on the credit they had had from the opposition, to put them in a mood for trade, to inflate the price of goods, to reward them for virtue, to express your everlasting admiration of their brave deeds, or because they drove you crazy begging for it.

They were not lovable when in liquor and they loved neither one another nor anyone else. . . . Cautoquoince bites Terre Grasse's nose off; a relative finds it in the straw and bandages it on again. Wayquetoe shoots an arrow through his wife and with another one wings a gallant who may have been eyeing her. "Indians still drinking. One woman stabbed another with a knife in four places but I supposed none of them dangerous." A four-year-old child has his buttocks shot off while papa is high. A squaw leaves her husband's lodge — a legal divorce — but presently, getting drunk, the old husband and the new one quarrel over the custody of her baby. Each grabbing it by a leg, "they began to pull and haul; on a sudden the father gave a jerk and, the other resisting, the child was torn asunder." Tabeshaw, Henry's most cantankerous Indian, stabs a relative six times, up to the hilt. "Tabeshaw breeding mischief. I had two narrow escapes from being stabbed by him, once in the hall and soon afterward in the shop." "I quarreled with Little Shell and dragged him out of the fort by the hair." The Big Mouth stabs the White Partridge six times but leaves enough life in him so that he starts fighting with his wife, rolls through the fire with her, and bites off her nose. . . . "At night the Indians got drunk and the Cancre [*sic*] had his testicles pulled out by one of his wives, through jealousy. there was but a few fibres that held them, so that his life was almost despaired of." Hugh Faries reporting from Rainy Lake.

Crooked Legs was an old man, "very small and lean," one of whose wives was too young to be prudent. He warned her

about her ways but she clubbed him over the head and ran off
to the Crow's tent, where her husband followed her and
stabbed her three times. "At every motion of the lung the blood
gashed out"; her relatives watched it, "bawling and crying; they
were all blind drunk." They decided to repay the insult by
killing Crooked Legs, who went to his lodge to sing his death
song, but Henry persuaded them to wait till the girl died.
They waited, taking a few drinks to pass the time, and when
she came out of coma made her sit up and have a couple too.
She didn't die and Crooked Legs lovingly expended on her the
afflictions of Indian medicine, "with tears in his eyes, bidding
her to take courage and live." She called him an old dog and
told him he wouldn't like it if she did. That was early October;
the band moved on to hunt. In late November they checked in
at the post again, the young squaw completely recovered. "All
who had any skins to trade held a drinking match, during
which the lady gave her old husband a cruel beating with a
stick and then, throwing him on his back, applied a fire brand
to his privates and rubbed it in until somebody interfered and
took it away. She left him in a shocking condition, with the
parts nearly roasted."

For the Chippewas could be jealous; adultery had to be by
the partner's consent or was disapproved. Gros Bras stabbed
Aupersoi, who wouldn't leave his wife alone; "he never stirred,
although he had a knife in his belt, and died instantly." So
Aupersoi's ten-year-old brother grabbed his gun and killed
Gros Bras, "just as he was reproaching his wife . . . and boasting
of the revenge he had taken." To make a day of it, Little Shell
thereupon stabbed Aupersoi's mother. "Ondainoiache came in,
took the knife, and gave her a second stab. Little Shell, in his
turn taking the knife, gave her a third stab" and finished the
job.

One squaw had her eye on Henry, who did not respond and
had a hard time beating her off. Late one night she got into his
quarters and only liquor would get rid of her. Desperate,
Henry produced his cognac; a glass of it made her pass out, or
pretend to. He revived her and had her carried to the camp,
where she put on her best clothes and hurried back, informing
her husband that she was fed up and intended to stay for good.
Henry again had her carried to the camp; it was breaking up

for departure and she told her husband she was going to stay right here. All right but he would spoil her looks. "He caught up a large fire brand, threw her on her back, and rubbed it in her face with all his might until the fire was extinguished." Now she could have her white man, if she could get him. "Her face was in horrid condition. I was sorry for it; she was really the handsomest woman on the river and not more than 18 years of age. Still, I can say I never had connection with her, as she always told me if I did she would publish it and live with me in spite of everybody."

All this is commonplace. A day's entry in Henry's log: "We found our strayed horses. Indians having asked for liquor and promised to decamp and hunt well all summer, I gave them some. Grand Gueule [the Big Mouth] stabbed Capote Rouge, Le Boeuf stabbed his young wife in the arm, Little Shell almost beat his old mother's brains out with a club, and there was terrible fighting among them. I sowed garden seed."

જ∾

The Convention that had been meeting at Philadelphia so far exceeded its instructions and powers that the document it produced was as revolutionary as the Declaration of Independence. It was called a Constitution of the United States of America. On September 17 1787 Benjamin Franklin's motion that the Convention sign it carried. "Whilst the last members were signing it, Doctr. Franklin looking towards the President's [Washington's] chair, at the back of which a rising sun happened to be painted, observed to a few members near him, that painters had found it difficult to distinguish in their art a rising from a setting sun. I have, said he, often and often in the course of this session, and the vicissitudes of my hopes and fears as to its issue, looked at that behind the President, without being able to tell whether it was rising or setting; but now at length I have the happiness to know that it is a rising and not a setting sun." [2]

Thirteen days later, September 30 1787, a ship of 212 tons burden that had lately been built in Plymouth's North River cleared Boston for the Northwest coast. Her name was *Columbia Rediviva;* the second half was intended as an omen of triumph but everyone has forgotten it. "On her first voyage," Morison says, "the *Columbia* . . . solved the riddle of the China trade. On her second, empire followed in her wake." [3]

On the second voyage she had as fifth mate a Bostonian named John Boit, with a nasal twang that got into his prose style; he was six months short of eighteen when she cruised the Oregon coast in the spring of 1792. His admiration of his captain, Robert Gray, was governable — none of Gray's mates of record thought he was a good mariner but the facts refute them — and he had not liked being ordered to burn the village at Clayoquot Sound, a few miles south of Nootka Sound, as punishment for the Indians' attempt to capture the *Columbia* and massacre its crew. Leaving Clayoquot in the spring she coasted southward, stopping to trade with various tribes who had a lot of furs and an intense abhorrence of sea captains. Almost at the southern boundary of Oregon she turned north again.

Boit's journal entry for April 22 1792 records a latitude of 46° 39′ and says, "Still beating about, in pursuit of anchorage. Sent the boat in shore often, but cou'd find no safe harbour. . . . Experience strong currents setting to the southward. We have frequently seen many appearances of good harbours, but the Currents and squally weather hindered us from a strict examination however Capt. Gray is determin'd to persevere in the pursuit."

Six days later, April 28, Boit says, "this day spoke his Brittanic Majestey's Ships Discovery and Chatham, commanded by Capt. *George Vancover* and Lieutenant W. Broughton, from England on a voyage of discovery. [And to receive from the Spanish commandant at Nootka a small hut and some supposititious acreage, in token of contrition for a useful if uncommitted insult to British honor] . . . we gave them all the information in our power, especially as respected the Straits of Juan de Fuca [the actual one, already named and visited by others before Gray] which they was then in search of."

Captain Vancouver did not visit the *Columbia* himself but sent his Lieutenant Puget and the naturalist Menzies instead. Returning to the *Discovery*, they reported Gray as saying that he had sailed only fifty miles into the Strait of Juan de Fuca and not all the way round the island now called Vancouver, which was what had been reported in England. "He likewise

Dean Channel
Bella Coola R.

Miles
0                    200

52°

Queen Charlotte Sd.

NORTH

VANCOUVER

50°

Nootka Sound
Perez 1773
Clayoquot Sound

PACIFIC OCEAN

Str. of Georgia

Fraser R.

48°

Strait of Juan de Fuca

Puget Sound

Mt Rainier

46°

(Grays Harbor)

(C. Disappointment)
(Pt. Adams)

Columbia R.

44°

(Coos Bay)

42°

Map 23
The Northwest Coast

C. Mendocino
126°

Raisz.

informed them," Vancouver's narrative says under date of April
29, "of his having been off the mouth of a river in the latitude
of 46° 10′, where the outset, or reflex, was so strong as to prevent his entering for nine days. [Nine days mean Gray's summary of his efforts to find a harbor, not to enter this particular
one.] This was, probably, the opening passed by us on the
forenoon of the 27th; and was, apparently, inaccessible, not
from the current but from the breakers that extended across
it."

The breakers that Vancouver had seen on April 27 were
across the mouth of Deception Bay, with Cape Disappointment,
which was Heceta's Cape San Roque, at its northern extremity.
The new names had been bestowed by John Meares, a former
British naval officer who had first come to the Northwest coast
to trade for sea otter in 1786. A first-rate navigator, sometimes
on the gray side of the law, and a vivid but unreliable
chronicler, Meares had set out in 1788 to check Heceta's statement that he had seen the mouth of a river there. The breakers
had convinced him that Heceta was wrong, that "we can now
with safety assert that no such river as that of St. Roc exists, as
laid down in the Spanish charts." As against a Yankee, Vancouver would stand on the judgment of a British officer. He
himself, the day before his ship spoke the *Columbia,* had
identified Deception Bay and Cape Disappointment from
Meares's description. He wrote, "The sea had now changed
from its natural to a river-colored water; the probable consequence of some streams falling into the bay, or into the ocean
from the north of it, through the [coastal] lowland. Not considering this opening worthy of more attention, I continued
our pursuit to the n." The Yankee had been taken in by
appearances.

The *Columbia* continued northward and on May 7 sighted
an entrance that Gray could get into. Boit says, "The Ship
stood in for the weather bar and we soon see from the Mast
head a passage in between the breakers." So they entered the
big indentation of the Washington coast that has been called
Gray's Harbor ever since, anchored, and began to trade with
Indians who "was stout made and very ugly." They were also
charged with the antipathy to sea captains that the tribes had
been learning for seven years. The next day "heard the hooting

of Indians, all hands was immediately under arms," and those who first ventured close got muskets fired over their heads and withdrew. But "at Midnight we heard them again, and soon after as 'twas bright moonlight, we see the Canoes approaching to the ship." Gray fired a warning cannon but "at length a large Canoe with at least 20 Men in her got within ½ pistol shot of the quarter, and with a Nine pounder loaded with langerege and about 10 Musketts, loaded with Buck shot, we dash'd her all to pieces, and no doubt kill'd every soul in her." The Northwest trade had its ruggedness: you killed a large percentage of the customers in order to do business with the rest on terms you considered proper.

When the *Columbia* left Gray's Harbor, she headed south for the river which Gray had been unable to enter the month before, and it was ready to happen now. Boit for May 12 1792:

> N. Latt. 46° 7′ W. Long. 122° 47′. This day saw an appearance of a spacious harbour abrest the Ship, haul'd our wind for itt, observ'd two sand bars making off, with a passage between them to a fine river. Out pinnace and sent her in ahead and followed with the Ship under short sail, carried in from ½ three to 7 f[atho]m, and when over the bar had 10 f[atho]m Water quite fresh. . . . The river at this place [the anchorage] was about 4 miles over. We purchas'd 4 Otter Skins [worth five hundred dollars in China goods at Canton] for a Sheet of Copper, Beaver Skins 2 spikes each, and other land furs 1 Spike each.

And the *Columbia*'s log:

> At four A. M., saw the entrance of our desired port bearing east-south-east, distance six leagues; in steering sails, and hauled our wind in shore. At eight, A. M., being a little to windward of the Harbor, bore away, and run in east-north-east between the breakers, having from five to seven fathoms of water. When we were over the bar, we found this to be a large river of fresh water, up which we steered. . . . At one, P. M., came to with the small bower, in ten fathoms, black and white sand. . . . Vast numbers of natives came alongside; people employed in pumping the salt water out of our water-casks, in order to fill with fresh, while the ship floated in. So ends.

So ends a very great deal: Gray had sailed into the great river that Bruno Heceta had sighted. He named it Columbia's River; the possessive did not last. And with the events so simply narrated much else begins. In the last year of Washington's first term, Mr. Thomas Jefferson being Secretary of State, the Republic had flown its flag in the river of the West. The Western lodestone intensified its tug as the voltage of a current is stepped up by an induction coil, and for the second time the hard Yankee captain had greatly served his country.

In doing so he displayed the attributes of American traders and sea captains which, with the help of the East India Company's restrictions, soon enabled them to take the Northwest trade entirely away from the British who had opened it. He first identified Deception Bay as what it was, the mouth of a river, and then insisted on entering it, succeeding on his second trial. Vancouver saw the same signs and identified "river-colored water" but knew that there could be no more than a congeries of creeks hereabout. He was sure that no one could think there was a river which amounted to anything except a "theoretical geographer" who believed in a mythical "mediterranean sea," a Sea of the West. (He was using language he had learned from his master James Cook, but Cook's skepticism had been exercised more critically.) Therefore there would be no point in investigating and he went about his business as he saw it. Gray would not be deceived or accept disappointment without having a try at it.

So he gave the United States a claim recognized by the polity of nations. Discovery and entrance of a river mouth gave the discovering nation sovereignty over the valley and watershed of the river and over the adjacent coast. The two empires that were pushing westward in the interior toward this same perimeter had met on the Pacific shore. Inland Great Britain was far in the lead — but the Americans had reached Oregon first.

Vancouver went on to the Strait of Juan de Fuca. The English captain William Barkley had discovered and named it in 1787, Meares who knew of Barkley's discovery had seen it the next year and blandly claimed to have discovered and traversed it, and Gray (in 1791) and at least two Spanish mariners before him had entered it. Two Spanish ships were charting some of the related waters when Vancouver sailed up it now. He met them in the Strait of Georgia, which he named but which the Spanish had entered the year before, and here he missed another river. He went on to make the first entrance into Puget

Sound, which he named for one of his lieutenants. He thoroughly explored it, sometimes in the company of the two Spanish mariners. He set up a counter-induction for the westering empires by claiming all his discoveries for Great Britain; many of his names, notably Mount Rainier, proved permanent.

In September, after sailing round his island, Vancouver was back at Nootka, where Gray had meanwhile put in after a profitable trip up the coast. The Spanish commandant, Bodega y Cuadra, told Vancouver that Gray had found a river at Deception Bay after all and gave him a copy of Gray's chart. He sailed south to investigate, though sure that the Yankee could have found only a small river at best. The Columbia confirmed him a little in that the water at the bar which caused the breakers was not as deep as it had been in the spring flood. He sent Lieutenant Broughton in the brig *Chatham* over the bar on October 19 1792 but found no channel he thought it safe to take the *Discovery* into and two days later sailed for California. Broughton found that it could not be called a small river and decided that Gray had not entered it. Gray's chart showed that he had gone thirty-six miles upstream. By triangulation, computation, and divination Broughton scaled this down to fifteen miles and decided that up to here the river had not narrowed enough to be called anything but a sound. A fresh-water sound that opened on the ocean would be unusually interesting geography but it would do to peg down an imperial claim. Broughton anchored in what he took to be one of Gray's anchorages and went on by small boat to a total of a hundred miles above the bar, almost to the mountains and with the river shoaling to three fathoms. That ought to do it, and the names he gave to peaks and other landmarks ought to stick. He claimed for Great Britain the river Gray had named, all its watershed, and the adjacent coast. Vancouver accepted his findings, "having every reason to believe that the subjects of no other civilized nation or state had ever entered this river before." So did British diplomacy down to 1846.

Two expanding empires had now made claim to the Columbia River and the unknown area it drained. For both of them the immediate value was the trade in sea otter furs, the maritime Northwest trade.

The Americans had entered that trade with the *Columbia's*

earlier voyage to this coast.[4] John Ledyard's earliest inheritors were British; the first one to reach the coast for sea otter skins to take to China got there in 1785. The ship cleared from China and in 1786 there were others from Bengal, Bombay, and England. They were by or by leave of the East India Company, a chartered monopoly like the Hudson's Bay Company. The market for furs, Canton, was in its territory. The trouble it could cause independents, the tribute it could extract from them, and the delays and expedients necessary to circumvent it were a force in the loss of the Northwest trade to the United States. But Great Britain would have lost it anyway, for the great American maritime age had begun and ships out of Boston, Salem, Philadelphia, New York, and the Chesapeake ports were outsailing the British in all the seas.

The Yankees had already made much headway in the China trade by 1788, only four years after the *Empress of China* opened it for them, when the *Columbia* reached Nootka for the first time. She was then captained by John Kendrick, Gray having her tender, the sloop *Lady Washington*. They had been sent out by a group of Boston merchants to trade in the pattern that Cook's successor King had sketched out. The ships spent the winter of 1788–89 at Nootka. (The first President of the United States under the Constitution took the oath of office April 30 1789.) The following summer Kendrick, staying behind in the sloop, put the furs they had acquired into the *Columbia* and turned her over to Gray. Gray took her to Canton, exchanged the furs for tea, sailed round the Cape of Good Hope, thus making the first American circumnavigation of the globe, and reached Boston in August 1790.

☙

Toward the end of the *Columbia*'s winter stay the Americans had witnessed the incidents that produced the "Nootka Sound Controversy." Vancouver Island, which was not yet so named, was Spanish and all ships that touched there did so by favor of His Catholic Majesty. (This was quite clear and the East India Company acted in accordance with it, sending its first ventures there under the flag of Portgual, which had treaty privileges.) In 1788 Meares built some earthworks and a shack

on a site which he claimed to have bought from the local chief
for two pistols, and announced that he was setting up in business. In the spring of 1789 the Spanish commandant seized the
shack, searched, seized, and released several British ships, and
ended by again seizing several — one of which belonged to
Meares — and keeping them.

Since the great value of the Northwest trade was now understood, the British government demanded restitution, reparation, and recognition of the right to trade. Spain answered
with a demand that her sovereignty be acknowledged. It was
an explosive situation in a jittery world and would have led to
war — but the beginning of the French Revolution had cut off
the possibility of French help and Spain had no choice but to
accede to the British demands. By the time she could do so,
the growing likelihood that the two nations might have to combine against the destroyers of legitimate dynasties had convinced the British that it was desirable to take no further advantage of the situation so opportunely created.

Vancouver's expedition to the Northwest coast was intended
purely for exploration but his instructions had recognized the
possibility that he might be called on to officiate at the ceremonies of Spanish penitence. He was, and Spain named Cuadra
to perform them in place of the commandant who had committed the original offense. They were conducted with operatic
courtliness and a vast mutual consideration, in so much that
when both representatives found that their instructions could
not be carried out as written, they agreed to refer the questions
to their governments, forgot about them, and got on with the
agreeable business of exploring the coast. They became so
friendly that Vancouver temporarily forgot he was a sea dog
and named the land he had sailed round Cuadra and Vancouver Island.

Thus Spain abandoned her claim to exclusive sovereignty
over Nootka Sound. In effect she relinquished the rights she
had asserted ever since Pope Alexander VI drew his demarca-
tion line in 1493, and the breakup of the Spanish Empire had
begun. The Spanish concession had the utmost importance for
the United States. California remained entirely Spanish but
how far north did California extend? The Republic began to
engross the Northwest trade and, as the wars of the French

Revolution developed, went on to appropriate a steadily increasing part of the world's carrying business. Though it was so weak a political and military power that the crises of the 1790's seemed likely to break it up, it was becoming a great maritime nation. The biggest business it was engaged in was the triangular trade with China. This was based on the Northwest coast, and the Northwest coast was the width of the continent away from the seaboard cities. From now on no American statesman and no continental thinker could forget for a moment that Spanish Louisiana lay between the western boundary of the United States and Captain Gray's river.

Vancouver's instructions from the Admiralty were dated March 8 1791. That was the year when Alexander Mackenzie went to Scotland to get schooling for his final exploration. He did not arrive in time for the Admiralty to share the knowledge he had acquired on his descent of the Mackenzie River. He had, and was always to retain, confused and erroneous ideas about the great mountain ranges that had shut the West away from him as he traveled toward the Arctic. But he did know that those mountains existed and he conjectured that, somehow, they were continuous with the Rockies, about which his North West Company partners had extremely confused ideas. When the Admiralty planned Vancouver's exploration it had only North West Company ideas about the mountains. It pretty thoroughly disregarded them and accepted worse ones.

Some of the Admiralty's intentions must be quoted. Vancouver's first objective was to acquire "accurate information with respect to the nature and extent of any water-communication which may tend, in any considerable degree, to facilitate an intercourse, for the purpose of commerce, between the northwest coast and the country upon the opposite side of the continent which are [*sic*] inhabited or occupied by His Majesty's subjects."

This was a clear recognition of the most advanced North West Company ideas, those of Pond, the elder Alexander Henry, and Mackenzie. The fur trade of the interior was to be given a continental connection with the maritime fur trade, the North West Company with the Northwest coast. The connection was to be by the venerable hope, a "water-communication."

It would be "of great importance," the document went on, if

Vancouver should find "any considerable inlets of the sea, or even large rivers, communicating with the lakes in the interior of the continent." This either erased the mountain barrier or cut a wide and useful corridor through it, and there was no Continental Divide or at most one you could carry a canoe across. That the Admiralty's geographers meant exactly what the words say is patent in the direction to disregard any opening too small for him to sail his ship up. They had the rumor that Gray in the *Lady Washington* had gone into the Strait of Juan de Fuca and sailed round the big island north of it, thus proving that it actually was an island. Vancouver was to determine whether this could in fact be done. And "the discovery of a near communication between any such sea or strait, and any river running into or from the lake of the woods would be particularly useful."

For "lake of the woods" read Lake Winnipeg, and note "into or from." This allows for the possibility that a short portage between Atlantic and Pacific drainage may be required. Nevertheless, in 1791 it still seems to England's best geographers wholly likely that a River of the West may flow directly from the interior to the Pacific. They knew that the Saskatchewan flowed into Lake Winnipeg at its northern end by way of Cedar Lake and the Assiniboine at its southern end by way of Red River. Both these were known to be big eastward-flowing rivers and they must come down from somewhere. Yet a river may flow west to the ocean from Lake Winnipeg, or if not from Lake Winnipeg then from Elsewhere, conceivably in the plains southwest of it between the Assiniboine and the Missouri, which also flows east. For all that had been learned since Vérendrye, and that knowledge stretched to the Rockies and the Arctic, this is the same possibility that was hidden in the unknown when he headed west from Rainy Lake. The Rocky Mountains had not impressed their factual existence on the thinkers in London.

The Admiralty alludes to another route, the one Peter Pond had worked out by postulate and deduction. If Vancouver should find no such continental water route south of Cook's River, "there is the greatest probability that it will be found that the said river rises in some of the lakes already known to the Canadian traders and to the servants of the Hudson's Bay

Company." (Lake Athabaska, Great Slave Lake, Great Bear Lake? Possibly a northwestern outlet of the Lake Winnipeg complex? And on the basis of what the traders and servants had learned by now, just how?) Vancouver is to investigate this "greatest probability" if he is forced to but the geographers hope it won't run to that. "The discovery of any similar communication more to the southward (should any such exist) would be much more advantageous for the purposes of commerce, and should therefore be preferably attended to."

There was no water route leading to the interior for Vancouver to find. There was no gap in the Rockies, no mountain-walled Northwest Passage to Lake Winnipeg or anywhere else. There was, however, something notable which he might have found but didn't. He simply was not a lucky man with rivers.

Following the commandant's seizure of ships at Nootka, the Spanish busied themselves exploring the related waters under the direction of Francisco Eliza. Early in 1792 one of his captains by the name of Narváez followed the Strait of Juan de Fuca to the Strait of Georgia and turned up that. Off its eastern shore, the continental coast, he saw what he took to be unmistakable signs of a river and judged that it must be a big one. Out of supplies, he had to hurry back but in May two more of Eliza's ships, whose captains were named Galiano and Valdés, reached the same mud flats that he had been passing when he saw his signs. "We were already in water which was almost fresh," the report reads, "and we saw floating on it large logs, which indications confirmed us in the idea that the bay which we called Floridablanca [the name Narváez had given it] was the estuary of a considerable river." [5] On June 22 Vancouver met them in these selfsame waters and showed them his sketches for a chart. "They seemed much surprised that we had not found a river" and pointed to the location of their Floridablanca. He was not impressed. He had been there only nine days before and had described the coast as "very low land, apparently a swampy flat" and had seen no such significance as they had in beached "logs of wood and trees innumerable." So he had missed another river, as he had missed the Columbia. One does come down there through three confused mouths, but no one would know anything about it for sixteen years more. Then Simon Fraser would descend it and bequeath it his name. The city named Vancouver includes that swampy mud flat now.

In 1793, after making valuable surveys elsewhere Vancouver was on the coast north of Vancouver Island. On June 3 his small boats made soundings in an inlet and he called "the canal we had thus explored Dean's Canal." It is Dean Channel on modern maps. They were six weeks too early to witness an arrival that would have astonished them. On July 20, in a large, leaky canoe manned by coastal Indians Alexander Mackenzie came down one of the arms of the Bella Coola River and out into salt water in Dean Channel.

ᗡᐁᗡ

The problem of the United States was elementary: to survive. To survive the clash of domestic interests and philosophies, the rivalries of states and the resistances of citizens unused to a federal government, and conspiracies in New England and west of the mountains to take those sections out of the Union. To survive British, French, and Spanish efforts to destroy the nation by breaking it up from within, by making war on it, or by getting it involved in war. To survive the hazards created by the French Revolution, which kept the world at war for a quarter of a century and might have destroyed the Republic in almost any month. The dangers were orchestrated to a climax in 1794. Thereafter there was some confidence that they might be contained but most of them remained threatening through the rest of the decade.

From posts maintained on American soil and protected by the garrisons of illegal forts, British traders dealt with American Indians and crossed Wisconsin and Minnesota to trade illegally with Spanish Indians. At the time of the Nootka incident Washington and his Cabinet had to consider the possibility that Great Britain would request permission to march an army across the American West to take Natchez (illegally held by the Spanish), West Florida, New Orleans, and indeed all Louisiana. Or that it might make the campaign without asking leave. This would have put the British Empire on our western as well as our northern flank. And from the Great Lakes posts, both north and south of the boundary, the governors of Canada and the military commanders continually incited the Indians to make war on the Americans and supplied them with munitions. The minimum hope — preached to the Indians, discussed with other foreign offices, and agitated by the underground — was to keep all the lands north of the Ohio River an Indian Country. That is, to make it a buffer state between the British and American empires, empty of Americans, populated by Indians who would be tributary to British sovereignty and British trade. The full hope was to attach these lands to Canada when the next war should come.

Some thirty thousand emigrants, sprung from God knew where, were crossing the mountains every year. Only a few of them settled north of the Ohio and these few mostly at Marietta and Cincinnati: the everlasting Indian war made settlement elsewhere impossible. The principal tribes engaged were *émigré* Iroquois, the remnant Delawares, several mixed breeds such as the Piankeshaws and the Wyandots, the Kickapoos, the Miamis, and always the Shawnees. They had it pretty much their own way. The Kentucky frontier could not sustain lengthy operations. Under the Articles of Confederation the national military establishment had withered away. The Republic organized only a feeble and bad one. Neither the Americans nor the Indians would make a treaty in good faith; both knew they never would. In 1790 General Josiah Harmar led a kind of militia army against the Shawnees along the Scioto and on to the confederated tribes near the Maumee. He destroyed some abandoned villages and much stored grain, won some skirmishes, lost bigger ones, and failed so badly that the West seemed lost. (In Canada officials interpreted the expedition as a first move to repossess the illegally held posts, prepared for war, and increased the flow of supplies to the warring tribes.) In 1791 Arthur St. Clair, who had been a general in the Revolution and was now Governor of the Northwest Territory, blundered into catastrophe. On the Wabash an Indian confederation under the Miami chief Little Turtle shattered his forces, killed some six hundred of them, and dispersed the rest in panic. These were mostly regulars; the Republic's first army had begun by suffering the most humiliating defeat Americans had ever taken. The Indians, coached by their backers, announced that the United States did not exist west of the Cuyahoga and the Muskingum — Cleveland and Cincinnati.

The British minister refused to discuss the return of the Western posts; the boundaries of the newly created department of Upper (western) Canada were left undefined. Washington and Knox raised a new army, calling Mad Anthony Wayne from retirement to command it.

Kentucky became the fifteenth state in 1792 — Vermont had been admitted a year earlier — but there was no warranty it would stay in the union. Not until Napoleon's more dynamic empire took over Louisiana did Thomas Jefferson write his famous aphorism, "There is on the globe one single spot, the possessor of which is our natural and habitual enemy. It is New Orleans, through which the produce of three-eighths of our territory must pass to market and from its fertility it will ere long yield more than half our whole produce and contain more than half our inhabitants." But he was saying the same thing in other words as early as 1790, and indeed the truth was self-evident. The existence of the United States as an economic entity west of the mountains depended on water transportation, the Mississippi, and on free access to the ocean, the mouth of the Mississippi. The endless procession of emigrants on horseback or in Conestoga wagons, in broadhorns, keelboats, scows, flatboats and more unlikely craft, kept building up a tension that was certain to explode. There is no space here to analyze or even to list the various movements, plots, and conspiracies that the tension created: to seize New Orleans or break the Spanish stranglehold by some other means, to detach the West from the United States and set up a separate nation that could make its own terms or its own war with Spain, to attach it to the Spanish Empire, or to attach it to Canada and the British Empire. But they boiled furiously, in high places and low, agitated by statesmen and adventurers, frontier floaters and the secret agents of three powers. There was a multitude of military schemes, commercial ventures, land speculations, and efforts of the imperial undergrounds to trap the new nation in a war that would destroy it.

The Kentuckians were the first half-horse, half-alligator Americans, nature's premature attempt to create Texans. They viewed their fellow countrymen and the foreigners across the river with equally contemptuous disregard. Of the plots to settle the issue by force some used, and more tried to use, George Rogers Clark. The hero's hour had passed; bankrupt, his just claims unpaid, all his land speculations unsound, he slipped into gaudy imperial fantasies, drunkenness, and sedition. But westward expansion was as implicit in him as courage and he remained a great name on the frontier. He had already suffered betrayal from a lifelong betrayer of everyone, a very small villain on a very large scale, James Wilkinson. Wilkinson was the focus, or front, of many feints and plots to make the West Spanish. He had already converted Spanish faith in his power and connections into a steady source of income, had got valuable commercial privileges at New Orleans as a bait to lure the West into revolution, and was and would remain on the Spanish payroll as a conspirator and spy. He had been in the Revolutionary army — good script-writing attached him to the Quebec expedition with Benedict Arnold and Aaron Burr — and he had usually held some military or political appointment ever since. Now he joined up as Wayne's second in command and of course began to intrigue against Wayne.

While Wayne's army was training, in 1793, the governor of Louisiana (Carondelet) wrote to his chief that the Americans were "advancing with incredible rapidity toward the north and the Mississippi [and they] will beyond doubt force Spain to recognize the Misuri as their boundary within a short time. Perhaps they will even pass that river" unless decisive measures were taken. Spain's need was to prevent the inevitable — to arrest American expansion, if not at the Alleghanies then somewhere east of the Mississippi. Behind this imperial necessity there was, besides the actual American threat, a wholly fantastic fear — of military action against Santa Fe as a step toward the conquest of northern Mexico, where the silver mines were. Those mines had to be defended from St. Louis, San Luis de Ylinoia, as they were being defended from San Francisco Bay. The reports of the Spanish governors, lieutenant-governors, and army officers sandwich Santa Fe between proposals to fortify the east bank of the Mississippi and to keep the Civilized Tribes raiding Tennessee.[6]

Spanish policy was largely conspiratorial, involving a network of espionage agents, agitators, and suborned American officials. This underground cost much more than it was worth, spread across the West, and extended into the American Congress. Policy also kept in a state of exacerbated impotence the negotiations about the boundary between the United States and East and West Florida, employing the Spanish genius for inconclusive diplomacy. It incited and supplied the Southern Indians in a war against the American settlements which though less continuous than that north of the Ohio was sometimes even bloodier. But the most effective weapon was manipulation of the right to navigate the Mississippi: threatening to close the river, closing it, opening it a little for purposes of bribery and subornation, keeping full privileges constantly before Western eyes as a threat and a promise.

To his contemporaries Thomas Jefferson did not seem a bemused ideologue, the role that one of our most durable literary traditions has cast him in. His opponents thought of him as the embodiment of realism and even opportunism, as pragmatism's pope. "Spain is so evidently picking a quarrel with us that we see a war absolutely inevitable with her." Or at least war was inevitable and "we should keep ourselves free to act in this case according to circumstances." The world of the early 1790's was a powder magazine on fire and the flames were spreading. On the outbreak of any war the United States would immediately seize the spot whose possessor was "our natural enemy." Jefferson knew as much and said so. If a war with Spain should break out we would win it, but it would provide a tempting opportunity for France or Great Britain to attack us and was therefore — up to a point — to be avoided. In 1790 as Secretary of State he was willing, in return for the cession of the Floridas and New Orleans, to guarantee Louisiana against British aggression. This was, however, purely because the Nootka incident might enable Great Britain to "completely encircle us with her colonies & fleets," and when the tension abated he repudiated such a guarantee, which was against his instinct and the whole set of his mind.[7] And "were we to give up half our territory rather than engage in a just war to preserve it, we should not keep the other half long." He wanted to acquire the island of New Orleans but would settle — though only temporarily — for freedom of navigation on the Mississippi and the right of deposit at or near its mouth. He confidently counted on the pressure of increasing population in the West to force one or the other solution peacefully (though down to 1793 he would have had only formal objections if the Westerners had gone filibustering) but knew that if that pressure came to a head, the government could not wait upon peace. "The nail will be driven as far as it will go peaceably, and farther the moment that circumstances become favorable."

The West in domestic politics may be oversimplified as: Federalist interests, themselves in part separatist as regarded New England, were antiexpansionist and remained tranquil about Western separation, though not so passive as the Republican West believed. (Jay's despairing proposal to settle the long negotiation by conceding the closure of the Mississippi for a period of years, which really would have amputated the West, met opposition in his own party.) Domestic divisions were intensified, complicated, and confused by the French Revolution. Enthusiasm for the overthrow of despotism and what seemed the extension of our own Revolution by the people who had been our allies in it was dominant in all sections, even New England, till 1793. But the news that Citizen Capet's head had dropped into the sack produced widespread revulsion. The ships that brought it had hardly docked when others arrived with word that the French Republic had declared war on Holland, Great Britain, and Spain.

World war instantly made the chronic crisis acute. At any moment the United States might be forced into war with France, Spain, or Great Britain. With the last two, thus made allies, our relations were already at the breaking point. Postponement of war from day to day while such power as the nation had could be mobilized and some adjustment of the dangers worked out was an absolute necessity. Little more than a month after news of the war arrived American foreign policy was given the base it would retain for more than a century. Washington, backed by the two arch-realists in his Cabinet, Hamilton and Jefferson, issued the Declaration of Neutrality. Before it was published, a minister from France landed at Charleston, Edmond Charles Genet, as odd a figure as ever represented a nation anywhere.

Genet was charged with launching the United States into war with Spain and if possible Great Britain. He and the Committee of Public Security that was governing France (presently superseded by the Committee of Public Safety) apparently expected the United States to become unanimously Girondist if not Jacobin. When it did not immediately do so, Genet undertook to overthrow the government by crying a rescue to the people. His inflammatory foolishness discredited him before he reached Philadelphia and presently it was easy to demand his recall. (In fear of the guillotine, he stayed here and made a rich marriage.) Meanwhile he had made war with Great Britain all but certain by commissioning American vessels as French privateers and basing a campaign against British shipping on American ports. Also he spun an espionage network across the West and commissioned Americans in the French army to raise forces for the conquest of the Floridas, Louisiana, and perhaps Canada. George Rogers Clark thus became a major general of France and in turn commissioned friends of long experience in anti-Spanish intrigue in a paper army of conquest.

The situation worsened as 1794 came on. The war had created a glistening commercial opportunity by making the French West Indies dependent on American trade and of necessity also opening the British islands. The British navy seized American vessels to a total of three hundred and so roused the war spirit in New England, which though it was the center of Federalism was also the center of investment in shipping. With one eye on these seizures and the other on Wayne's now magnificently trained army in the West, the governor of Canada publicly declared that war was inevitable. The Canadian military establishment was increased, the army was readied for the war, and a detachment was sent to build a new fort on American soil squarely in Wayne's probable path, on the Maumee River.

Now domestic turmoil produced what looked like an insurrection. Pennsylvania west of the mountains had mutinously ignored one of Hamilton's revenue measures, a confiscatory tax on whiskey, the one product that could be transported eastward at a profit and indeed the local medium of exchange. By July 1794 federal authority was suspended in four counties. The frontiersmen forcibly adjourned the federal courts, raided the mails, and burned the houses of marshals. In August a half-organized mob marched into Pittsburgh, threatened to burn it, and talked (boastfully, drunkenly, and sometimes in words furnished by Spanish or French agents) of crossing the mountains for direct action.

They confronted the administration, governing a nation unaccustomed to centralized authority, with a challenge which it must meet if it was to survive. But they also gave it a price-less opportunity, the chance it had not so far had to reduce the all-inclusive crisis to manageable size. Washington raised an army of 13,000, only 3000 less than the combined Franco-American force at Yorktown. He named to command it two brilliant Revolutionary generals, Light-Horse Harry Lee and Daniel Morgan, with Hamilton as his personal representative. No such numbers were required to collect an excise. The clue was Pitts-burgh, at the forks of the Ohio River, one of the pivotal sites of continental strategy. The Whiskey Rebellion was soon set-tled in an amiable spirit and without punitive action — and at the Forks of the Ohio there was a large American army to deter Westerners from acting on behalf of France, and to impress conspiratorial Spaniards in New Orleans and belligerent Eng-lishmen in Quebec. Twenty-five hundred of them remained in Pittsburgh, on one route of easy access to Canada, through the winter of 1794–95, while Wayne with a force nearly twice as large was poised on another invasion route farther west. Mean-while in August Wayne climaxed a brilliant campaign by breaking Indian power north of the Ohio when he defeated the confederated tribes at Fallen Timbers. Fleeing Indians sought refuge in the new British fort. Its commander had orders to fight if Wayne should attack and Wayne would have attacked if the fort had been opened to the Indians. It wasn't; Wayne by-passed it, cleaned out the countryside, and began an ulti-mately successful attempt to intimidate and cajole the tribes into submission.

The two armies so close to the border were a blatant, very hard fact. If war should come the United States would take Canada. Whatever else might happen, the possession of Can-ada would put the Americans in a formidably strong trading position at the peace conference. Meanwhile it would give them the fur trade. There could hardly have been a more powerful deterrent to the British war party.

To prevent war was a necessity for the United States, and through 1793 and the early part of the next year it seemed impossible. The situation had all but hopelessly degenerated in April 1794 when Washington made his final, despairing effort. He sent Chief Justice Jay, a veteran if frequently frus-trated diplomat, to England to negotiate the settlement that the American minister had been unable to procure. The negotiation was set before a backdrop of the world war and the Terror. (The 9 Thermidor translates as July 27.) Fallen Timbers was fought on August 20, the new army reached Pittsburgh in November, British distaste for the alliance with Spain grew steadily. The result was Jay's Treaty, which was so offensive to a nation that had three war groups — Western, shipping, the debtor tobacco planters — that Washington feared he could not get it past Congress. But Jay had got from the desperately beset British a good deal more than there was any likelihood he would be able to get. He got the survival of the United States, peace, territorial integrity at last, and the secured and bolstered basis of future power. It was as successful a negotiation as any in our history.

Although the West did not at first understand, it was a victory for the West. In a way similarly obscured at first it was a triumph for Jefferson's ideas too, though he promptly left the Cabinet. As intensely as he wanted anything, the Secretary of State had wanted to get the British off American soil — out of the Western posts which they had continued to hold and from which they not only incited Indian raids but blockaded a large area to American traders. He could have got this early in the negotiation by accepting the British demand in regard to the northern boundary. It was now known certainly that the line fixed in 1783, from the northwest corner of Lake of the Woods west to the Mississippi, was an aberration. Montreal exhausted its political resources to procure a rectification that would make Grand Portage and the canoe route west entirely Canadian, and the British commissioners offered various lines that would join Lake Superior with the Mississippi. Hamilton was willing to sacrifice the canoe route and a portion of Minnesota. Jefferson was not; he held out, Washington supported him, and Jay got the evacuation anyway.[8] But the treaty left a large loophole for the future to rush through. It left the boundary west of Lake of the Woods indeterminate, to be determined by a future commission and survey.

The agreement to evacuate the Western posts thirteen years after the original agreement was the core of the treaty. By this agreement Great Britain, saving Canada without having to fight for it, abandoned forever the dream that the country north of the Ohio could be either neutralized or reacquired. It withdrew armed protection from the Mackinac trade across the

Western border and so opened it to Americans. Finally it threw its unacknowledged mercenaries, the Indians of the Northwest Territory, to the wolves. However rough on the Indians, this cleared the channel down which the main stream of American development would flow.

The treaty had another immediate effect. Thomas Pinckney, dispatched from his post in London to Madrid, got a settlement from Spain which accepted the American definition of the southern boundary, withdrew all but hazardously secret Spanish support from the southern tribes, opened the Mississippi to American commerce, and granted the right of deposit at New Orleans. The Spanish problem, which had risen to a critical peak, now lapsed into mere annoyance.

Besides its political and economic denotation, the phrase "territorial integrity" has a meaning that transcends logic. Few Americans knew anything about the Western marches of the United States. Few could imagine them or envisage any future for them apart from spreadeagle rhetoric. Exceedingly few had ever seen them. But insistence that the political organism called the United States must secure inviolate and hold inviolable the full extent of its sovereignty was, in addition to everything else, an embodiment of the continental consciousness. The nation might know little about the boundary that separated it from the Spanish lands but knew that it was a river which could be crossed, and that beyond it the land continuum stretched to the Pacific.

❧

In the tense year 1793, while the unstable equilibrium of international tensions threatened anew every twenty-four hours to break up in war, Thomas Jefferson arranged his first expedition to explore Louisiana. It has ambiguities which the records do not explain and which do not yield to analysis, for the instructions are in good faith but the preparations are perfunctory. Jefferson had only a small part in the preparations and it is not clear to what extent, if any, he ever expected the expedition to succeed.

His instructions for the conduct of the expedition were written in April,[9] when the protracted and exasperating negotiations with Spain had deteriorated almost to the point of abandonment. This fact is certainly not without bearing on an exploration of Louisiana. But that enterprise had been contemplated by a good many people for a long time — by Jefferson perhaps longer than anyone else — and it was intended to be a peaceful enterprise. Jefferson was acting for an organization that deeply interested him, the American Philosophical Society of Philadelphia. Its membership included many of the best minds in the United States, its influence extended throughout the country, and it was interested in the Western wilderness and in the great additions to geographical knowledge that had been made in the last quarter-century. Now it was raising a fund to support a project proposed to it by a French scientist, André Michaux. A talented botanist, he had been a resident of the United States for nine years and of South Carolina for seven. He had made wilderness journeys in the Blue Ridge and Great Smoky Mountains and the Florida swamps. In 1792, entering another wilderness, he had followed his botanical specialties through northeastern Canada to the vicinity of Hudson Bay. On his return he proposed to the Philosophical Society "to make discoveries in the Western country": to travel overland to the Pacific by way of the Missouri. The route, which harks back to Carver and Rogers and beyond them to Charlevoix, is significant to the utmost. In 1793 the existing knowledge is committed to the Missouri as the best — and the American — way west.

On March 23, as Secretary of State, Jefferson wrote to his commissioners who were negotiating with Spain that the government had learned of a proposed French expedition that was to offer independence to the Spanish colonies on the Mississippi. He observed that France would not be displeased if those on the east bank, in particular West Florida, should be added to the United States.

On April 30, as the Philosophical Society's committeeman, he delivered his instructions for the exploration to Michaux. Earlier in April, his espionage web already spun, the egregious Genet had reached Philadelphia. He promptly enlisted Michaux in the network and, using his well publicized expedition as a front, ordered him to Kentucky to get in touch with Major General George Rogers Clark of the French army and with others, and to help organize an expedition against New Orleans — the expedition of which the government had heard in March. Later, on July 5, Genet told Jefferson, "not as Secy. of state but as Mr. Jeff.," what he was up to. He was careful to say, falsely, that the army which was to be raised would muster outside American territory. "I told him," Jefferson wrote in the *Anas*, "that his enticing officers & souldiers from Kentucky to go against Spain was really putting a halter about their necks, for that they would assuredly be hung, if they commd. hostilities agt. a nation at peace with the U. S. That leaving out that article, I did not care what insurrections should be excited in Louisiana." Not even "as Mr. Jeff." could he say that this might provide the United States with leverage for securing the right of deposit at New Orleans.

By making Michaux a secret agent, Genet aborted one more exploration of Louisiana. Michaux left Philadelphia July 15 1793, his self-esteem so delirious that he was unaware that his secret mission was public knowledge. Everyone knew what he was up to, loafers at crossroads doggeries and the Spanish in New Orleans. His trip through the West had a coloration of Graustark, with innumerable observers reporting every move he made to both the American and the Spanish officials, and innumerable agents of new-budded filibusters and real-estate speculations trying to decide whether he was worth joining. He reached Lexington on September 5 and eventually got to Kaskaskia but got no farther. His great plot was spent and though he was in touch with the French underground in St. Louis he did not even cross the river.

Any attempt on New Orleans would have had to be both American and purely on speculation, for Genet lacked funds to support one. Whether the only one that was seriously intended, Clark's, would have got started in any circumstance is doubtful. The administration took it seriously enough to forbid it, to direct the governor of Kentucky to stop it with militia if it should start, and to order St. Clair to have regular troops ready. One may believe, however, that not many Kentuckians were willing to speculate in futures while paying for their own rations, and the principal interest is the unfortunate Clark. He had now engaged in all but one of the possible variations, the Spanish one. He had had visions of conquering New Orleans for the United States, for Kentucky, and for France, and of detaching from the United States the West he had saved for it in order to make it independent and in order to make it French.

. . . In March 1794 Genet's successor publicly called off the expedition; it had died of anemia long before.

Jefferson's instructions to Michaux while he still intended to be an explorer express the geographical ideas of the best-informed American minds in April 1793. (The *Columbia* would not return with news of Gray's discovery till the end of July.) Michaux is to "find the shortest and most convenient route of communication between the United States and the Pacific ocean, within the temperate latitudes." That route, however, is predetermined and prescribed for him. "The Missouri, so far as it extends, presents itself under circumstances of unquestioned preference. It has, therefore, been declared as a fundamental object of the subscription (not to be dispensed with) that this river shall be considered and explored as a part of the communication sought for." Michaux is to go to the Missouri from Kaskaskia, traveling to the north of St. Louis lest the Spanish stop him. When he reaches it he is to "pursue such of the largest [tributaries] as shall lead by the shortest way and the lowest latitudes to the Pacific ocean."

(At this moment the Spanish in St. Louis were preparing their own exploration of the Missouri. The partners of the North West Company were discussing whether they should not do the same thing, and were only three years short of deciding to. Both the Spanish and, at this time, the Northwesters accepted the same hypothesis about the Missouri as Jefferson did. It has to be called a hypothesis for it was worked out by logical interpretation of what was presumed to be known. Actually, however, it was a guess based on a mental map of the Missouri which, in 1793, was still so fantastic that an entirely blank map would have been more useful.)

And mark an assumption which has not been shaken in two centuries, which in fact has been strengthened by late eighteenth-century theorems of symmetrical geography: "When pursuing these streams, you shall find yourself at the point from whence you may get by the shortest and most convenient route to some principal river of the Pacific ocean, you are to proceed to such river and pursue its course to the ocean." The full assumption is this: the shortest distance between watersheds, and as a corollary therefore the distance most easily to be traveled, is the shortest distance between the headwaters of their principal streams. It is neat and economical; it is persuasive; it can be demonstrated by constructing an ideal model of sand or plaster. But it does not happen to be generally true. Jefferson goes on to note that according to the latest maps, "a river called Oregon interlock[s] with the Missouri for a considerable distance."

He adds that he and his colleagues know the maps cannot be trusted and Michaux must act as the facts he will find out may indicate. But in the thinking of Jefferson and his fellow scientists, two cruxes remain that are unchanged since the discovery of the Missouri. The possibility that the height of land may be wide or may consist of difficult mountain ranges is not imagined, and the Atlantic and Pacific watersheds are assumed actually to interlock.

&⁖&

On the day when Michaux, bound toward his Kentucky operetta, reached the foothills of the Alleghanies, the first transcontinental crossing north of Mexico ended at the Pacific shore. Alexander Mackenzie had found a Northwest Passage, albeit a useless one, by land.

In the summer of 1792 Mackenzie sent men from Fort Chipewyan (at Lake Athabaska) up Peter Pond's other river, the Peace, to prepare timber for a winter post where the Smoke River empties into it. With a small crew for the post and his selected party of pioneers he joined them there in October. Though only some six hundred and fifty miles from the Pacific — since there was a gap in Cook's observations he could only guess at this distance — the site was far north, rising 56°. That was part of the idea: it was still conceivable that he might be going to reach the ocean by Cook's River. He had studied Cook and Meares and probably had their books at his winter post. He had all the knowledge the Northwester had amassed, having himself found out the most useful of it. For what it was worth, he had too all the information they had acquired from Indians. He went on questioning Indians through the fall and winter, knowing that none of what he heard was entirely dependable and that he might have badly misunderstood much of it. That, in short, he knew nothing surely about what he was to face. One trouble was that he kept hearing about two big rivers, not one. . . . He had proved that there was no inland water passage connecting Hudson Bay with the Pacific. But he believed that the River of the West existed, somewhere south of Nootka, perhaps far south.

His adjutant was a North West Company clerk of two years' service, Alexander McKay.[10] He also took two veterans of his Mackenzie River journey, four other *voyageurs*, two Indians for hunters, and a dog. They used a four-fathom — twenty-five-foot — birchbark canoe and their outfit, including goods for presents, weighed three thousand pounds. At seven in the evening May 9 1793 they headed up the Peace River into the unknown and those who stayed at the post "shed tears on the reflection of the dangers which we might encounter." In late afternoon six days later the snowy peaks of the Rockies rose above the horizon. British traders had seen the Rockies at long intervals as far back as Henday. How far the *engagés* of Niverville had ventured from the westernmost French post just before the French and Indian War is not known, but one Northwester had gone all the way to the foothills. Three hundred miles south of Mackenzie's winter post, above the site of Rocky Mountain House where the north branch of the Saskatchewan has become a plains river after issuing from the mountains, Peter Pangman had carved his name on a tree. The tree remained a landmark for years but even Pangman had not entered the Rockies. No white man ever had entered them from the east.

Even in the upper stretch Mackenzie had been traveling, the Peace is a big river, frequently a thousand yards wide or more. Now it narrowed and May 19 is a decisive date. On that day his *voyageurs* took the canoe into Peace River Canyon and the whole complex of skills which had sustained the fur trade for two centuries, and which had brought it from La Chine to the Rockies, collapsed. The *voyageurs* had mastered all the waters up to here but they had never faced a mountain river. The birchbark canoe is not a craft for mountain streams, which in fact are not navigable. Peace River Canyon is not even in the mountains; it is a narrow gulch, a thousand feet deep and twenty miles long, through a detached ridge east of the range, a horror where the river is, in Mackenzie's words, "one white sheet of tumbling water." They knew nothing to do except to force it as they forced the rapids west of Grand Portage, but they had no resources except will and courage. It came to towing the canoe by handlines with the river hurling gigantic masses of swift and deafening water at them from all directions, and to portaging its lading far up clifflike banks that at every moment threatened them with the loss of their outfit and their lives. They did it in six days, in the unintermitted hazard of death. Few except Mackenzie would have kept on. The six days broke the morale of his crew; the *voyageurs*, their tradition rendered impotent, presented two spectacles that were almost inconceivable, terror and rebellion. Even McKay, of the master caste, faltered. Mackenzie held them to the purpose but he himself was in constant fear, not of death but of failure.

From now on they were always in the mountains and most of the going was just as bad. They met Indians, most of whom had never seen white men and were as dangerous as the river. Mackenzie handled them superbly, though taut with anxiety. In deep canyons where toil and terror were at the extremity, they told him about westward-flowing rivers, the sea beyond, the hostile tribes and the white men on the coast, and their own trade routes — told him with artistic disregard of the facts. On June 9 he first saw iron that had come up those trade routes from the ships.

He had talked with an aged Indian during the preceding year and had been impressed by what he said. When he reached the forks of Peace River, he held by the old man's instructions and, against appearances and the judgment of all his party, turned up the one that came in from the south, the Parsnip. It was a crucial decision: the other fork, the Finlay, would have led him to a chaos he could never have threaded. The canoe had been so battered that it had to be patched several times a day, meat was scarce in the canyons, every quarter-hour had its hazard. On June 12, paddling across a little lake, they reached the carrying place their guide had described and "found a beaten path over a low ridge of land eight hundred and seventeen paces in length to another small lake." Here was another climax: the second lake was Pacific water. For the first time north of the tablelands that separate Colorado River drainage from that of the Rio Grande, white men had crossed the Continental Divide.

Next day they were in a stream which the second venturer this way, Simon Fraser, would justly name Bad River. Soon it wrecked the canoe and came within a hair of drowning everyone. "Every yard on the verge of destruction. . . . I was on the outside of the canoe where I remained till everything was got on shore, in a state of great pain from the extreme cold of the water. . . . I [hoped to] obtain an observation of Jupiter and his satellites but I had not a sufficient horizon, from the propinquity of the mountains." His steel will, appeals to the *voyageurs'* pride as Northmen, and a big slug of rum at the nightly campfire kept them from dissolution. Battered in canyons filled with huge rock fragments, carrying canoe and outfit along cliffs, hip-deep in mud, through woods where axemen had to clear a way, they at last reached a big river. Indian guides (they were always getting new ones, who were always deserting) said it flowed west, though for a long distance its course was south or southeast, through canyons that were not majestic but only terrifying. Mackenzie decided that it was one of the rivers he had heard about and thought it might be, at last, the River of the West. . . . It flows down the trough between the Rockies and the Coast Range: the Fraser River. Presently it catapulted them into one of the worst stretches of water in North America, Prince George Canyon.

Now they met some of the Indians who were essential to success, the tribe called by ethnologists the Takulli. Their neighbors called them the Carriers; they cremated their dead and a widow had to carry her husband's ashes on her back for three years before she could remarry. They spoke a Déné language which Mackenzie and his interpreters, familiar with the kindred Chipewyans, could use; one of their women was a Cree in whose tongue most Northwesters were fluent. Hostility so vicious that massacre hung in the balance yielded to Mackenzie's skill; the Carriers would accept presents and be friends. They said that the river did indeed lead to the sea. They added that he could not go down it by canoe or on foot. Better turn around and go back to the long overland trail which they themselves used. And by the way, the coastal Indians were very terrible; they would kill his whole party.

The crisis had come. Mackenzie was convinced that the river he was on "could not empty itself into the ocean to the North of what is called the River of the West." (It must therefore be that river and its mouth might be as far south as San Francisco Bay.) So it must be very long, he could see that it was violent to the utmost, and how could he bring a canoe up it on

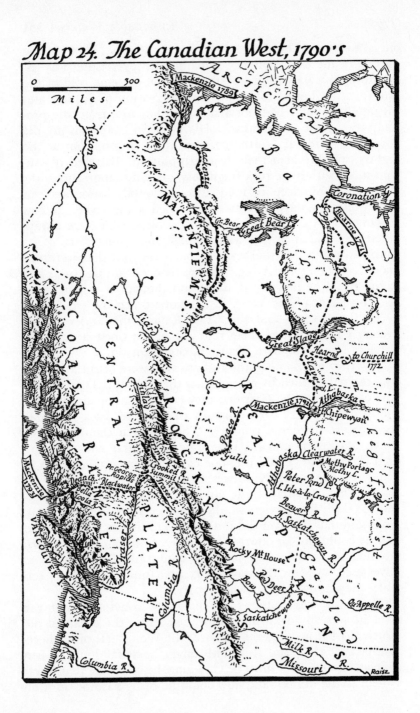

Map 24. The Canadian West, 1790's

his return? If he stuck to it there was no possibility that he could get back to Lake Athabaska this year. His supplies, mostly pemmican, were running out; powder and lead were dangerously low. His men were terrified, had been pleading with him to turn back, had rebelled several times, were on the edge of mutiny, might well slip into the wilderness panic that means death. He himself was hellridden with fear of failure. It seemed to be a full stop. The wilderness had defeated him, and with him the North West Company and Great Britain's effort to reach the Pacific. But there was one possibility still. Perhaps the overland route through the mountains that the Carriers described did exist. Of course they might be lying, and even if they weren't he might take the route and still fail.

Not only river craft had broken down when in the person of Alexander Mackenzie the westering white man reached the Rocky Mountains. Geographical thinking had broken down too, its postulates upset, its extrapolations proved worthless. Neither the French, the British, the Spanish, nor the Americans had taken into account the possibility of a considerable land-traverse. A portage of a couple of miles, or up to half a day, might separate the river routes, ocean to ocean, but the watersheds must interlock and travel would always be by canoe.

Mackenzie would go to the Pacific. He called his men together and told them he intended to retrace the route to where the land-carry started. If they should fail by that route, as the unreliability of guides made more than possible, then he would come back here and start down this big river and go on to its mouth, "whatever the distance might be." If the men would not agree to that hazardous alternative, he would not even try the long portage but would stick to this river now and go down it from this spot.

Throw away the scabbard. "At all events, I declared in the most solemn manner, that I would not abandon my design of reaching the sea, if I made the attempt alone, and that I did not despair of returning in safety to my friends."

The measure of Mackenzie is that he enforced his will on a party of men whose private discipline had been undermined by terror of the unfamiliar. The canoe was now worthless; at the bottom of this ferocious canyon they had the crazy labor of making another one. His about-face alarmed the Carriers who had advised it and for a while they turned hostile again, more deadly than diamondbacks. He got back their belief. Upstream then, amid hostile Indians and the threat of mutiny, to the Blackwater River and up that to the portage. They cached supplies, hid the canoe, and made packs — ninety pounds per man, besides gun and ammunition. On July 4 they started up the first mountainside of the Indian trade route. New guides and pace-setters always appeared in time, the increasingly strange tribes responded to his diplomacy, evidence that he was nearing the ocean and its Indians and its white men multiplied. The forests of the western slope, much rain, much cold and occasional ice, peaks, cliffs, valleys, marshes, lakes — and ebbing strength as the food ran out. Indians clustered round them now, eaters and worshipers of the salmon. They came down to a river "of the color of asses' milk" and a village of antic Indians who were on the whole friendly, the Bella Coola River and the tribe it is named for. On July 18 he succeeded in getting some of them to embark with his party in one of their big canoes. ("I had imagined that the Canadians who accompanied me were the most expert canoemen in the world, but they are very inferior to these people, as they themselves acknowledged, in conducting those vessels.") They had to stop at a larger village, a frantic ant heap of curiosity about them, cluttered with fishing gear and rich in European goods. Mackenzie induced them to return to the river, they made thirty-six miles, lost his little dog (which they would find, half-crazed, on their return), and passed a landmark which he was later able to identify as having been visited by Vancouver's Mr. Johnstone six weeks earlier. Then on July 20, sixty-eight days out from the winter post, "at about eight [A.M.] we got out of the river, which discharges itself by various channels into an arm of the sea. The tide was out and had left a large space covered with seaweed." They went on, passing sea otter, till two in the afternoon, when big swells forced them to land.

It was done. There is no knowing who first dreamed of crossing North America, but it was an old dream in 1534 when Cartier first heard of a big river that would be called the St. Lawrence. Mackenzie had fulfilled the dream. Two hundred and fifty-seven years earlier the amazed slave hunters had led Cabeza de Vaca into Culiacán.

The fear of his men still held, and was intensified by the uneasiness of the Indians who had come with them. Now they met Indians who were aggressive and contemptuous. One of them had a grievance: lately there had been a ship and some white men in these waters and one of them, "Macubah," had fired at him and another one, "Bensins," had struck him with the flat of a sword. Though Mackenzie had no way of knowing it, he was talking about Vancouver and his naturalist Menzies. This was actually the most dangerous moment of the entire journey but Mackenzie insisted on seeing as much of the archipelago as he could, and he would not leave till he had made enough celestial observations to be sure of his figures. His Indians grew more afraid of the coastal tribe and his own men were "panic-struck." On July 22 he had all he wanted, the average of five altitudes and an emersion of Jupiter's third satellite checked by one of its first satellite. "I had now determined my situation." No sail had appeared in Dean Channel and at last he turned back, against a powerfully ebbing tide.

First, however, he mixed some vermilion with melted grease and painted an inscription "in large characters" on the face of a rock. It read: "Alexander Mackenzie, from Canada, by land, the twenty-second of July, one thousand seven hundred and ninety-three."

Mackenzie's book was published in 1801. No one read it more attentively than the new President of the United States, Thomas Jefferson, for it stated Mackenzie's matured plan for transcontinental trade imperialism. The principal reason why he had written and published it, in fact, was to put pressure on the British government to act on his vision and secure this great commerce, the route it must use, and the territory it must cross before the American imperialism could secure them. But an attentive reader of the passage that described this inscription on the rock was George Rogers Clark's younger brother William.

And when William Clark read it both he and its author, and Jefferson as well, believed that the big river which Mackenzie had left in order to make the land-carry, the Fraser River, was the one which, after his return, he learned that Captain Robert Gray had discovered and named the Columbia.

ॐ

The tensions of the 1790's produced in Upper Louisiana, which was now the Spanish Province of Illinois, an effort to strengthen the frontier similar to the one that had been made in the preceding two decades in the Pacific Northwest. It had the same purpose, to protect Santa Fe as the outpost defense of the Mexican silver mines. As the passage (written in 1793) already quoted from Carondelet shows, the officials knew that the Mississippi in itself could not stop the westering Americans. They had "the unmeasured ambition of a new people, who are vigorous, hostile to all subjection, and who have been uniting and multiplying . . . with a remarkable rapidity." They were all frontiersmen: "A carbine and a little cornmeal in a sack is sufficient for an American to range the forests for a month." If they got to the Missouri, or even in force to the Mississippi, nothing could prevent them from going farther. New Mexico was too thinly populated to stop them, and indeed it provided a danger of its own which was abhorrent to Spanish thinking. The New Mexicans would "unite willingly and eagerly with men who, offering them their aid and protection to become independent, to govern themselves, and to impose their own taxes, will flatter them with the spirit of liberty and with a trade free, extensive, and lucrative. In my opinion a general revolution in America threatens Spain. . . ."

The Spanish hold on the Province of Illinois was far from secure. The French who comprised almost all its population thought of themselves as but temporarily subjects of Spain and looked forward confidently to a day when French power would be restored. Though Genet was discredited and his General Clark did not get an invasion started, they alertly watched similar movements in the West and corresponded with all the underground plots, however crackbrained. Some of them turned Jacobin and organized revolutionary clubs. In 1796 a French general who had toured much of Louisiana and the West on what was patently a military reconnaissance was turned back from the Missouri, arrested in New Orleans, and deprived of his notes and sketches. The British military threat subsided for a space but the Canadians forged steadily farther ahead in the trade war.

Handicapped by the incurable lack of funds and by the creeping paralysis that had overtaken Spain's administrative system, the officials at St. Louis went on making inadequate annual presents to many tribes. The Chouteaus and their associates maintained an active trade along the rivers that entered the Missouri from the south. The Osages, the diminished Missouris, the Kansas, the Wichitas, and the populous Pawnees remained Spanish. But the British dominated the trade of the upper Mississippi, they had reached the upper Missouri from Canada, and they had almost monopolized the trade in the region between the Mississippi and the Missouri. They were closing a circle and that fact requires a full statement.

At the beginning of the 1780's the North West Company, the Hudson's Bay Company, and opposition firms traded with the tribes of the upper Missouri at posts on the Assiniboine River. Sometimes they established temporary posts on the Red River and the Souris, and some years before the end of the decade they had made brief visits to the upper Missouri itself, at the villages of the Mandans and the Minnetarees. By this time too competition out of Mackinac was beginning to visit the same region from the Des Moines River, crossing Iowa on the way. Wisconsin and Minnesota as far as the Mississippi were American territory; the rest of Minnesota, Iowa, and Dakota were Spanish.

The culture transformation of the western, migratory Sioux divisions was now complete; they were Plains Indians. They had reached the central stretch of the Missouri River, were warring with the river tribes, and were beginning to raid farther west, in the direction of the Black Hills. They retained their British allegiance and for trade journeyed to the Minnesota River and, presently, to the nearer Des Moines. South of the Sioux were the Iowas. The British traded with them direct and it was through them that British goods reached other Missouri tribes. Those concerned were the Otos, the Omahas, the numerically small but troublesome Poncas, and the schizophrenic Arikaras. They were all displaced victims of the western surge of the Sioux and they all lived near the Missouri.

The Spanish movement up the Missouri was intended to regain the loyalty and trade of these tribes, to clear Spanish territory of the British, to hold the river as a frontier defense of New Mexico — and to discover a route to the Pacific. "All the maps printed both in England and the United States, and in France, are absolutely false, especially in regard to the course of the Misisipi and Misuri Rivers," the governor wrote to the home office. Spanish ignorance of the far country was astonishing. Madrid had always governed the Indies compartmentally, the compartments were not encouraged to communicate with one another, and this was late afternoon getting on toward evening — the empire, sunk in poverty, apathy, cynicism, and graft, was dying. As late as 1787 the mouth of the Big Sioux River was the northern limit of empirical knowledge; few traders had ever passed the Platte. Knowledge of the country north and west of there, except for some distance up the Kansas River, remained what it had been in the Illinois seventy years before, in de Bourgmond's time. Such knowledge as had been acquired since then darkened counsel instead of enlightening it. For whereas the Northwesters were widening the map of North America, the Spanish had contracted it.

The governor describes to the commandant of the Provincias Internas the lands that lie between Louisiana and the Californias. This is 1785.[11] . . . No one has ascended the Missouri beyond the mouth of the Big Sioux River. The Arkansas have told our traders that, at a distance of five hundred miles upstream from their villages, the Missouri comes down a high mountain. The range they are referring to must be the one that begins a little to the east of Santa Fe and runs to Quivira. (No one who had actually been to Quivira — Kansas — had reported mountains there. This conjecture joins the detached Wichita Mountains, which the Louisianans knew in Arkansas, to the Sangre de Cristo of New Mexico, seven hundred miles west of them.) Rivers from the eastern slope of this range empty into the Arkansas and the Missouri, from their western slope into the Rio Grande. (The Front Range of Colorado has been added to the mountain system; exactly this configuration occurs in maps of the third quarter of the sixteenth century, two hundred years earlier.) In this same New Mexican range the Platte, the Niobrara, and the Teton (known only on hearsay) also have their sources. Furthermore the Missouri rises somewhere in the same range and cuts through it. Its source is somewhere north of the Rio Grande headwaters; it flows west, quarters northwest, and as it goes on swinging toward the east it probably makes the big waterfall which the Arikaras have described. West of the mountain range, at its southern end, is the country called Teguayo. (Coronado's Thayguayo of 1540, now generally shown on the Pacific coast near the country of Anian. There is no awareness of anything that has

happened in the Spanish Far West in two and a half centuries.)
But west of Teguayo this range is coastal, and therefore though
it is cut across by the Missouri in one place, the river must rise
in some other part, "because to the west of these mountains the
sea or the Bay of the West almost washes its base." (Ten years
later Carondelet widened this strip west of the mountains,
which resembles what the Bow People had told the Vérendrye
brothers in 1742, to forty leagues, a hundred miles.)

The indications here accepted as certain make it a small
country. The Spanish in New Orleans do not know what those
in New Mexico had learned about the eastern slope of the
Rocky Mountains in southern Colorado. The mountains postulated are those near Santa Fe and there is no conception corresponding to the actual Rocky Mountain system. Such as it is,
this chain of north-south mountains is traversed by the Missouri
about five hundred miles upstream from the Arikara villages.
They were at, approximately, Pierre, South Dakota; five hundred miles farther would be Williston, North Dakota.

That is why the Spanish officials were afraid that the British
would invade New Mexico by boat or by an overland march.
It is also why when they sent men out in the spring to go up
the Missouri to the Pacific, they expected them to be back
before winter closed in. Since they could not have more positively believed in the distances and dangers, some real distances
may be stated. From the Arikara villages near Pierre the air-
line distance to Santa Fe is upward of 700 miles, from Lake
Winnipeg to Santa Fe 1100 miles, from Hudson Bay to Santa Fe
1500 miles. A route feasible for military invasion would be
several hundred miles longer than any of these. When
Spanish St. Louis began to think of the threatened invasion in
terms of actual rivers, it seems first to have conceived the movement as coming down the Missouri to the Platte, ascending the
Platte to a convenient distance from Santa Fe, and then moving on the town direct. Such a movement would have had to
be based on Lake Winnipeg or the Assiniboine River. The invaders would have had to travel from fifteen hundred to nineteen hundred miles, depending on whether they rowed the
pirogues across the plains and desert or over the mountains.

In 1787 a St. Louis fur merchant sent a young man named
Joseph Garreau on a trading venture up the Missouri. How
big a party he took or how far he got is not known. If the

Arikaras were in their villages when he found them, then he was farther upriver than anyone before him had ascended it. He offended the Arikaras by violence and dishonesty and stored up trouble for the future. In 1792 a trader of long experience, Jacques D'Eglise, at last went all the way, closed the gap, and brought the two commercial frontiers into contact. That is, he took his pirogues to the Mandan villages — Stanton, North Dakota, sixty miles above Bismarck — and found them well supplied with British goods from the posts farther north.[12] A Frenchman named Ménard was living with the Mandans and said that he had been for fourteen years. A mysterious, substanceless figure in the literature of the upper river for the next decade, Ménard has no explanation but he was obviously of the type who preferred savagery to civilized life and adopted it. He seems to have got goods from the British posts on the Assiniboine and Souris and to have traded them at the villages; he must have been, irregularly, a contact man for the British companies. David Thompson, who saw him five years after D'Eglise, describes his fair-skinned Mandan woman and says "he was an intelligent man but completely a Frenchman, brave, gay, and boastful; with his gun in one hand and his spear in the other, he stood erect and recounted to the Indians about us all his warlike actions." He liked to count his coups and eventually died of it; now he was boasting about his antiquity and lying about it, for he certainly had not been with the Mandans since 1778.[13]

D'Eglise was back in St. Louis in October 1792. On October 6 some unnamed traders returned from the Kansas Indians, along the Kansas River, bringing with them three Spanish subjects whom their customers had been holding prisoner, a naturalized Frenchman named Vial and two New Mexicans. They had been sent out by the governor of New Mexico to establish communication with St. Louis and by virtue of the Kansas' forbearance in merely capturing them had now done so, fifty-three years after the pioneering expedition of the Mallet brothers in the other direction. (Since then Frenchmen from New Orleans had occasionally reached Santa Fe by more southerly routes; there is evidence that others had tried to get there from the Missouri and perhaps succeeded, but without accomplishing anything.) Vial's march-diary is brief and very vague but he followed, roughly, the route that was to be known as the

Santa Fe Trail. So he may be said to have completed the 1541 journey of Coronado, who when he turned back from Quivira had learned that ahead of him were the Mississippi and the Missouri. Vial had lived in New Mexico for a long time and was an experienced frontiersman, but that the governor had sent so small a party may indicate that distance had shrunk in Santa Fe thinking too since the time of Villasur. Vial confirmed the shrinkage and so confirmed the misconceptions held in St. Louis. He said that if the Kansas had not captured him he could have made the trip handily in twenty-five days. He could not have done it in fifty days, but to the officials in St. Louis here was further evidence that Santa Fe was wide open to British troop movements from Hudson Bay.[14]

Then D'Eglise came back with news that the British were on the upper river. Everything he had to report to the authorities was momentous. The Mandans were rich in furs, he said, and therefore they presented a great commercial opportunity. (He added that they were as white as Europeans; he himself may have been tolerably swarthy.) Furthermore, since they had Mexican saddles, bridles, and other goods, they must trade directly with New Mexico or with tribes that did. The conclusion was as obvious to St. Louis as it had been to the French on the Assiniboine River in Vérendrye's time: here was still another kind of evidence that New Mexico was close to the Missouri. And in this, the first eyewitness account of the upper Missouri that St. Louis had ever had, he made an exceedingly important declaration, one which St. Louis was to hold to and which was to govern its thinking from now on. The lieutenant-governor (who complains that D'Eglise is ignorant and illiterate) reports it: "The Missouri flows always from a western

Map 25  The Santa Fé Trail        Scale 0    100    200 Miles

and northwestern direction, with sufficient water for navigation
by any of the larger boats of these [the locally known] rivers."
By now the locally known rivers were being navigated not only
by large pirogues but by still larger keelboats.

☙

In 1792, the year of D'Eglise's first Missouri journey and the
year before Mackenzie's transcontinental crossing, one of the
most valuable servants the Hudson's Bay Company ever had,
Peter Fidler, set out from its westernmost post, Buckingham
House, which had been built in 1791 on the south branch of
the Saskatchewan River. He crossed Alberta to the Rocky
Mountains somewhere near the head of the Little Bow River,
southwest of Calgary. The first white man who had ever penetrated that region, he traveled extensively, though he cannot
be followed with as much accuracy as would be desirable, and
spent the winter with the Piegans, one of the Blackfoot tribes.
Next year, returning to Cumberland House, he used a different
route. For a long time it was supposed that his journey south
along the foot of the Rockies had taken him below 49°, into
lands that are now American, and this idea still persists as a
legend. He appears, however, to have gone no farther south
than about a hundred miles north of the Montana border.

The reason for this belief was that Fidler, though an accomplished surveyor, had miscalculated some of his latitudes and
distances. Though the Hudson's Bay usually kept its knowledge
securely locked up, it was at this time performing a valuable
public service by turning over some of what it learned to the
London mapmaker Aaron Arrowsmith. Fidler's data were
given to Arrowsmith, as the more important ones of David
Thompson's surveys were presently to be. In 1795 Arrowsmith
published the first edition of his great and influential map of
North America. Fidler's erroneous latitudes appeared on it.
In the event they were to exercise a strong influence on the
thinking of Thomas Jefferson and the actions of his explorers.[15]

☙

The report of D'Eglise, misconceiving or embroidering parts
of what he had seen, called for action and produced it. Mean-

while in 1793 he started up the river again with two pirogues of goods, accompanied by Garreau. Somewhere below the Teton River he ran head-on into a fact more important than any he had discovered last year: either the Sioux or the Arikaras, refusing to let him pass, had closed the river.[16] They had closed it to the Spanish as effectively as the Spanish ever closed the Mississippi to the Americans and the blockade was to last longer. D'Eglise, who covered his backers' investment during the winter trade but made no profit, blamed the disaster on Garreau. He had a "turbulent and libertine spirit" and used his employers' goods "for indecent purposes other than those for which they had been designed." That Garreau was an offensive and unruly man had already been established and he forthwith added to his reputation. D'Eglise's accusation is probably true but Garreau was not the cause of the blockade. As the next few years were to show, this was a contention among the tribes for their neighbors' trade. Neither the Arikaras nor the Sioux wanted their enemies to get guns and powder unless they themselves could control the supply. If there were to be St. Louis goods on the upper river they wanted them in quantity for themselves, and they wanted for themselves the profit to be got from supplying the other river tribes. Linked with this urge was a preference for the better and cheaper British goods that were coming from the Mississippi. And if it was the Sioux who said no, there was also their strong pro-British partisanship.

Tribes not so far upriver had the same idea. As a result of D'Eglise's report the Commercial Company for the Discovery of the Nations of the Upper Missouri had been formed in St. Louis. It is usually called the Missouri Company. An association of St. Louis merchants, it did not include the Chouteaus or their associates and that was one reason for its failure. They were doing well enough on the lower river, they had no taste for distant risks, and they were hostile to competition. The Company was subsidized by the government, though far from enough, and was granted a monopoly of the Missouri trade above the Poncas. The government offered a prize to the first Spaniard who should go all the way and reach the Russian settlements on the Pacific coast. For "Spaniard" read "Spanish subject," for everyone in the Louisiana fur business except Manuel Lisa was French or Canadian English. The discoverer of the Pacific route would travel, naturally, through the notch

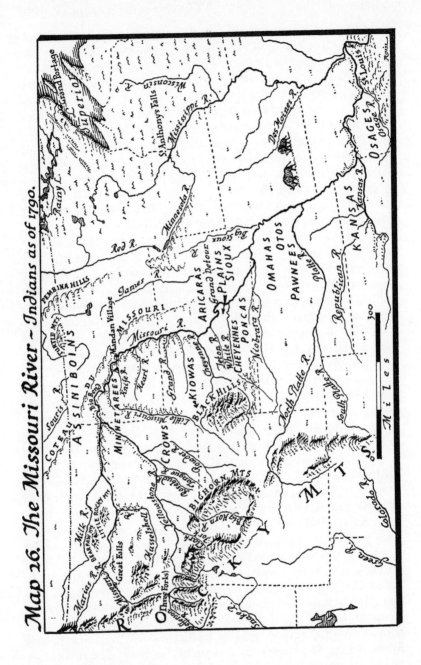

Map 26. The Missouri River – Indians as of 1790.

that the river cut in the Rocky Mountains, and if anvone had consulted a map that showed the latitude of the Aleutians he had made nothing of it.

In 1794 the Missouri Company sent out its first expedition, under Jean Baptiste Truteau.[17] He was the first schoolmaster of St. Louis but had had much experience as a trader on the Des Moines River and elsewhere. In the same year D'Eglise went up the river again and was again stopped by the Arikaras. He returned to St. Louis with the news that his earlier trips had stimulated the British to build a small post between the Mandan and Minnetaree villages, and took with him two French *engagés* who had fled from the British companies.[18] Other traders were on the middle river the same year, Munier at the Poncas where he had first gone in 1789 and one Solomon Petit in the same vicinity.

The attempt was too weak and too penurious. Truteau's party was small, his goods were sparse and of poor quality. He took two pirogues up the unpredictable, ever shifting river, with its innumerable "crossings" of the channel from bank to bank, sandbars and the islands and matted embarras, snags and sawyers and planters, stretches where the boats had to be cordelled by hand, and long periods of *dégradé* when high winds forbade travel. By luck or stealth he got past the Omahas and Poncas, whom the directors had expected to rob him. At the end of September 1794, sixteen weeks out of St. Louis, he was approaching the Grand Detour where the river twice doubles on its course round narrow points of land. He expected to meet his first customers, the Arikaras, just beyond it. But as he reached the lower end a party of Sioux stopped him.

Three lodges of them were Yanktons, at this period comparatively amiable folk, but mainly they were Tetons, who were the quintessential Sioux. They threw the party into just such a panic as Henry's Chippewas were used to, but Truteau did better with them than he ever managed to do with any other Indians. Recognizing Yankton chiefs he had known on the Des Moines, he demanded sanctuary and got it, thus preventing complete pillage and probably massacre as well. When he proclaimed the authority of Spain, they told him that they had no Spanish father. The British alliance held in Louisiana — the Sioux were a resistance movement. They got him to admit that he intended to trade with the Arikaras. (One of his pirogues was intended for them — the other was to supply a post at the Mandans, where he was to clear the country of British and whence he was to reach the Pacific.) They had two comments on that. They intended to be middlemen to the Arikaras — the Grand Detour seems to have been the exact place where the Sioux reached the Missouri and whence they had, recently and timidly, been venturing west. And, they said, there wasn't any Arikara village just ahead. Its inhabitants had emigrated upriver.

This last was true. The Sioux-Arikara wars had taken a decisive turn. . . . Forced west and south by pressure of the westering Prairie Sioux, the Arikaras had long preceded them in acquiring horses and the revolutionary Plains culture. They were supplied with French and British goods by the Sauk-Fox combine and they learned to build the impregnable earth lodges of the upriver tribes.[19] A renaissance followed; they grew prosperous and powerful. When the Sioux first reached the Missouri, perhaps not much before 1760, they trod softly in the presence of the Arikaras, trading for corn and horses. When their native belligerence broke out, the Arikaras usually beat them easily and, when they didn't, holed up comfortably in their Gibraltars. But the Sioux kept coming in greater numbers, better armed, and with more horses, and the smallpox epidemic of 1781 was followed by two others. The Arikaras were reduced from a big tribe to a small one and could not hold their country. In this very year, 1794, they threw in their hand and moved far upriver. They went as far as the Mandans and Min- netarees but after vacillating for a time reversed their drift and finally settled near the Grand River.

Practicing systematic terrorism, the Sioux demanded that Truteau spend the winter with them, promising that their en- tire nation yearned to eat him and would if he should try to go on. But he managed to hold the nightmares of his men in check and eventually got started again, after the Tetons had supplied themselves from his goods and contemptuously tossed him two hundred pounds of beaver in token payment. He cached most of his store, hid his larger pirogue, and went on foot to the Arikara village. It had been abandoned. His men panicked again, divided between urges to stampede back to St. Louis through marauding tribes and to go on and perhaps find safety with the Mandans. The plains winter was coming on. Truteau decided to raise his cache and drop downriver to a place safely south of the Sioux threat and north of the pirates of the middle river. The plan was logical but didn't work.

Building a crude post, he again cached most of his supplies. A winter of terror followed. Omahas arrived from the south under their second chief, Big Rabbit, and then more of them with the head pirate, Blackbird. This was the czar of all the Omahas, swashbuckling, insolent, intelligent, and so terrifying to other tribes that he was able to maintain his own above its station. He would take no opposition from his people or anyone else. His supernaturals had given him a magic that enabled him to destroy any dissenter, after telling him just what was going to happen. It was arsenic, supplied by bargain hunters out of St. Louis, and one story has him giving a feast for sixty of his braves who were all dead by morning. He felt only contempt for white men, knowing his power and seeing how he scared them. As soon as the St. Louis trade reached the Omahas he had established a system of tribute which no one thought or cared to oppose. He first took such goods for himself as his whim ran to, then traded en bloc for the whole tribe, setting his own prices.

French traders had suffered no such humiliation in the old days. The St. Louis French had lost courage as well as enterprise, and the companies on the lower river had skimmed the labor market. Besides, these were Plains Indians, not more ferocious than the forest tribes but fortified in boldness and self- reliance by the horse-revolution. Except under Blackbird the Omahas were never so formidable as the tribes farther west but he gave them a moment in the sun.

He kept Truteau's party in chronic terror. (He could point to what had happened farther south, where the Poncas had completely cleaned out their monopolist Munier.) He let his people precipitate brawls, stopped them, threatened to loose them in worse ones. He supplied them gratis with powder and balls from Truteau's visible supply, or sometimes in a benign mood settled for the levy with a few furs. He harangued Truteau for having sneaked past his village and worked up rages against D'Eglise who had done the same. Bands went away for a while, then came back to bully, beg, extort, steal, and to devour the meat Truteau's hunters had been able to kill. Equally arrogant Poncas began dropping in. Blackbird had gone his boisterous way when one band stopped off but just for amusement he sent them word that Truteau was working a medicine to kill them all. The Poncas painted themselves and worked up a war dance. Truteau managed to appease them.

He got through the winter and in the spring of 1795 went on to the Arikaras. After the Sioux and Blackbird they seemed genteel. D'Eglise had wintered with them and had done an excellent job of disciplining them and regularizing the trade, but also he had got all their furs. Sending his big boat down with his meager take, Truteau decided that he could not reach the Mandans, still less the sea. There was little timber hereabout and none big enough for dugouts. (Lack of enterprise and weight of tradition; it didn't occur to him to trade for horses and go by land.) He accompanied some Arikaras on a journey toward the Black Hills that brought him to the Cheyennes, a promising new market for St. Louis. He sedulously gathered information, much of it untrue, about other Plains tribes, Kiowas, Kiowa-Apaches, Arapahos, and, he believed, Comanches. He got more news of the British among the Mandans. He drew maps and filled his journal with gallic rhetoric.

Whether he returned to St. Louis at the end of the summer of 1795 or spent another winter in the field is not clear. In any event, when he did go back he had financial failure to report and not much else. He did, however, have realistic advice for the Missouri Company. The Company had too little capital and the subsidy was too small. The *engagés* were timorous, unskillful, and unruly. Good discipline was impossible, and the damage to Spanish prestige by the violence of men like Garreau was paralleled by the licentiousness of everyone who got to the notoriously licentious Arikaras. Spanish merchandise was bad: guns all but useless, hatchets and other iron implements soft or brittle, cloth inferior. He had been ordered to "fix a very high price on everything," as anyone who bought Spanish goods had to, and the policy was self-defeating. In short, upriver competition with the British was hopeless unless radical changes could be made. Moreover, the effort to serve two purposes, commerce and exploration, was a mistake.

Truteau recommended bigger investments and larger parties, both as it soon proved impossible. He thought the plan to establish a base among the Mandans sound and a prerequisite

for everything else. He saw the commercial and strategic necessity of maintaining a post among the Omahas, but recommended
a combination of force and tribute-payments that would have
been farcical. He thought that St. Louis could run the British
out of the country if business methods could be modernized and
experienced men obtained.[20]

Of the information and misinformation about the far country and its Indians that Truteau summarized for the Company's use, the most important item was his positive statement
that the largest pirogues could navigate the Missouri all the
way to its source. Since St. Louis knew that the source was five
hundred miles beyond the Mandans, the way to the Pacific
must be fully as simple as it had taken for granted.

৵

The lieutenant-governor and the Missouri Company had already initiated some of the actions that Truteau thought necessary. Alarming news of the British penetration had kept coming in, together with mixed information about the far country.
D'Eglise was back in St. Louis by July 4 1795 and brought with
him the two Canadian Frenchmen who had run away from the
British companies.

They were Northwesters, a novice named Fotman (or Tre-
mont) and a veteran named Jonquard. Officially examined in
the presence of witnesses by the lieutenant-governor, they said
that three companies, established near the junction of the
Souris River and the Assiniboine, were trading with the Mandans and the Minnetarees. They themselves had been attached
to a party under René Jessaume which John Macdonnell had
sent out from Assiniboine House in the fall of 1794.[21] Reaching the Mandans at the end of October, Jessaume turned over
half his stock of goods to the ambiguous Ménard and for the
other half erected a small post midway between a Mandan village and one of the Minnetarees.

. . . Jessaume had taken goods to the Mandans before, probably a number of times. The date of the first British journey to
the Mandans is not known but it was as early as 1785 if not
earlier. In the spring of 1793 Jessaume and several others,
equipped by Cuthbert Grant but acting as free traders, made
the trip; two of them were killed by Sioux on the way back.

In December of that year Macdonnell sent nine men to the villages; the venture made no profit and Jessaume may not have been along. When he went again, in October 1794, he was in the employ of the North West Company.[22] . . .

The party spent the winter in the new post and Fotman was among those whom Jessaume left behind when he returned to the Souris in April 1795. (Instructing them to fly the Union Jack once a week — in Spanish territory.) Fed up with the frontier life and afraid that he might be sent to a worse wilderness, Fotman simply walked away, shot a buffalo, made a bullboat, and set off downriver. D'Eglise was at the Arikara village and so was Truteau, who had sent written notification to Jessaume that if he intended to trade with the Mandans he must buy through St. Louis. Fotman joined D'Eglise; so did Jonquard, who had gone over the hill earlier than he.

They gave the authorities almost identical accounts of the trade and country. It was true that the villages were well supplied with British goods; besides the Northwesters and the Hudson's Bay Company, there was an opposition post on the Souris. (Actually three independents appear to have been there in 1794.) The Arikaras were forwarding goods to the Cheyennes and Jonquard had accompanied a trading party to their country. This much from their own experience; the rest was hearsay.[23]

It was from the Indians among whom they had wintered and from Ménard, and what they said was mostly on hearsay too. It is fundamentally important that neither the Arikaras, who were the principal informants of the Mandans, nor Ménard knew the far country at first hand. It is also important that Fotman and Jonquard had little background experience to understand correctly what was said to them. The result was to confirm the Spanish officials in ideas they had already formed. The country farther on was small, the distances were short, and New Mexico, California, and Mexico itself were open to attack by the British from Ontario and Manitoba. The British could travel all the way by water.

The Spanish now heard about new tribes: the Assiniboins (under two names), the Fall Indians or Atsinas (whom the refugees frequently confused with the Minnetarees, as many travelers and historians would continue to do), the Blackfeet proper, the Sarsis (who lived far to the northeast, neighboring with Crees and Chipewyans). Of these Ménard certainly and the Arikaras probably had seen only the Assiniboins. They located them far from the country they really inhabited. But a more important error originated in their ignorance of intertribal trade. All about them — with the tribes who came to the villages to trade, such as the Cheyennes, and those whom the Arikaras went to visit, conceivably Kiowas and Kiowa-Apaches — they saw horses, mules, bridles, and saddles that were obviously Spanish. (Many of the animals were branded; we might know more if the refugees had named some brands.) Ménard declared that these were ridden straight from New Mexico or California. Well, when the village tribes went west to trade they were back in just a little while, so clearly the mountains and the settlements could be at no great distance. It was just as we thought.

What of that short highway to the Pacific, the Missouri River? Again redoubled hearsay brought confusion. When you traveled upriver some four hundred miles beyond the Mandans, you came to a beautiful river. It issued from some lakes on the slope of a high cliff which the Indians called the Rocky Mountain. (Or as the Spanish alternatively rendered it, the Mountain of Cliffs.) And the Missouri too flowed down this mountain, in a fall more than three hundred feet high.

In Canada the Northwesters had not yet met any Indians who knew the Montana-Wyoming country. (And wouldn't till they got to Shoshones, Kutenais, and Flatheads.) So here, refracted through Ménard's incomprehension is the first mention of the Yellowstone. He had almost doubled its distance from the Mandans, actually 225 miles, and his mountain ranges cannot be rationalized. Since no tribe he knew visited the Yellowstone Park area, his waterfall must be the Great Falls of the Missouri transferred to its biggest tributary. If there were antiquarians in St. Louis they could sagely nod their heads: that high waterfall had been invented long ago. . . . But another noble river had at last come into the white man's awareness.[24]

The Missouri, according to the Indians, "was as navigable on the upper part of the mountain as on the slope." As it stands this means nothing, but the Spanish were thinking of a generalized height of land (with, nevertheless, the Missouri flowing northward between ranges of what were sometimes mountains and sometimes hills) and so it confirmed what D'Eglise had implied and Truteau had positively asserted: the biggest pirogues could go all the way to the source. As for that source, it was "within some higher mountains that were always covered with snow." This too resists interpretation. Maybe the Indians were referring to actual ranges, say the Big Belts near Great Falls, a long way from the Three Forks or their sources. Or maybe, and more probably, this was a statement that farther up the river there were mountains and mountains usually have snow on them.

It made no difference to the Spanish in St. Louis. They were sure to find easy going all the way to Santa Fe, California, and Nootka Sound.

ॐ

The British companies, so the head of the Missouri Company wrote through channels all the way to Madrid, were drawing perilously near. "That of the Hudson Bay, which seemed a dream to us on account of the distance and which today is at our doors; the building of many of these forts on the Osseinbune [Assiniboine] . . . those forts which have not yet been constructed to the west and of which we have only heard are to be built instantly . . . invasion menaces all that extent of our land which separates us from the Pacific Ocean . . . [and] is rapidly moving forward to the frontiers of California." The British must be driven from the Mandan Villages and their flag pulled down. Militia must be raised to garrison a chain of forts that must be built from the Omaha village across the mountain range where the Missouri rose and on to Nootka.[25]

More than the Missouri Company could do must be done, the director said, but it had made a start. It had sent Truteau upriver in 1794, directing him to get rid of the British and go on to the Pacific. And in this present year, 1795, it had continued the campaign by sending out two other parties. The first, led by one Lécuyer, was to winter at the Mandans and then go on to the Pacific the next spring. The second, under James Mackay, was to explore as need be and begin building that chain of forts.

Lécuyer left St. Louis in April, reached the Poncas, and was robbed there. Whether cowardly, treacherous, too much interested in Indian women, or merely unable to command white

men or Indians, he failed altogether. Some of his men joined
Mackay (on whom the journey to the Pacific now devolved)
during the winter, Ponca vainglory increased, and the Missouri
Company had lost 96,000 pesos.

The other expedition of 1795, which started from St. Louis
in August was something else. Mackay was an experienced
Northwester and knew the country he was traveling toward. He
had been on Red River, the Assiniboine, and the Souris, and
from the last of these he had, in 1787, taken an outfit to the
Mandans. He may have been on the upper South Saskatchewan,
for he had met Piegans.[26] Moving from Canada to the United
States, he had gone on to St. Louis, where the Missouri Company promptly hired him. He served it well and with complete
loyalty. He hired an assistant who was straight out of fantasy,
John Evans, twenty-five years old, a Welshman.

The Celtic dreaming that kept the story of Prince Madoc
alive in Wales was intensified in the American diaspora: wherever in the United States there were Welsh, there were true
believers. They heard and believed all the stories and when
the Pawnees proved not to be Welsh some tribe beyond them
must be, very likely the far and terrible Comanches. But by
1780 it was known in the United States, though not in the British Isles, that they were on the upper Missouri, where there had
always been white Indians. (And Chinese Indians and dwarfs
too.) By the end of the decade the American Welsh knew that
they were the Mandans, and as the 1790's came on the ancient
cult had a splendid renaissance.

In London a group of expatriate Welsh intellectuals and
poets fed ravenously on the romanticism that was literature's
new fashion. The antique, the distant, the great heroes of
earlier and purer times, the mysterious wilderness, the splendor
and nobility of primitive men — these were the themes
of the new literature and they were all in the Madoc story.
There was also freedom — the London Welsh rejoiced at its
new birth in France. And the new horizon — they dreamed of
moving to the United States, where freedom had married virtue
and Madoc's people roved the western marches. And yet patriotism too — the Nootka incident roused them to hatred of Spain,
an ancient theme, and Madoc's discovery of America shone with
pre-Columbian splendor. Some of the scholar-poets who were
collecting ancient Welsh texts, and forging them where the

record was deficient or too literal, now created ancient paragraphs about Prince Madoc that regrettably no one had written at the right time. Then in 1791 the story put forth blossoms. A white American pretending to be red, the famous and obstreperous William Bowles, came to London, presented himself as an Indian chief, enchanted the salons, and told what he knew about the Welsh Indians, which was quite a bit. (He said they were the Comanches.) And a learned Welsh clergyman who had spent his life studying Madoc's glory published a book about the hero. It fascinated the Welsh everywhere and galvanized the London expatriates; within a year they had assembled so much additional evidence for the parson that he had to publish another book.

Now the expatriates raised a fund to send someone to America to find the Welsh Indians, and in 1792 the election fell on John Evans, who was then twenty-two and as devout in Methodism as in Madoc. He felt that part of his mission was to reclaim the Madocians to Christianity, for he did not accept the tales of cherished Christian rites and pre-Gutenberg Welsh Bibles. He went not to Canada but to the United States and after a year in the East made his way to St. Louis. (Before he got there informed Welsh antiquaries must have told him that Madoc's people were not the Comanches but the Mandans.) But a young Britisher on such a nonsensical errand was an object of suspicion in San Luis de Ylinoia — this was 1793, the frontier tension with France was at its height, war between Spain and Great Britain was imminent. The officials jailed him. He was delivered from frustration by the efforts of a Welshman who had political connections. Then word of his errand reached an eminent Welshman who was touring the States on behalf of both colonization and the gospel. In the course of his tour he met James Mackay, who was returning to St. Louis from New York, and told him about Evans. Mackay had seen the Mandans and had realistic ideas about the Comanches; a man who knew Indians, he did not believe that any of them were Welsh. But he was willing to help out and when he got to St. Louis he hired Evans for his expedition. The young man was assertive and egotistical but he readily learned the wilderness trade.[27]

With Evans, a crew of thirty men, and four pirogues filled with goods, Mackay started up the Missouri in August 1795. One pirogue was intended for the Arikaras, one for the Sioux,

one for the Mandans (among whom he expected to find Truteau and to spend the winter), and the fourth "to reach the
Rocky Chains with orders to go overland [*sic* but misconstrued]
to the Far West." This was good planning, designed to check
the barrier tribes. On October 14, he reached the mouth of the
Platte — and the Otos. Their hearts were very bad; they intended to do to Mackay just what the Poncas had done to
Lécuyer a year ago. Mackay, however, told them the Poncas
would be in for trouble if they molested him and by a show of
force kept the Otos in line. Still, in order to keep going he had
to break into his small stock of goods and leave some here for
'trade in charge of a clerk.

A day short of the Omaha village the terrible Blackbird came
out to meet him. Word that a man who knew his job was on
the river had traveled ahead, for Blackbird was in a pious pro-
Spanish mood. Again Mackay had to lose time and set up a
trade that had not been contemplated. But he kept a tight rein
on the Ogre of the Missouri, whose orations grew more anti-
British day by day. Winter was closing in. Determining to
spend it here, Mackay unloaded his pirogues and erected a hut
which he called Fort Charles. . . . He had already written his
principals that to control the upriver trade they must begin by
setting up a post for the Otos. He now understood that the
Omahas were crucial, that the site of their village must be held
in force. He told Blackbird that the post he had built would
be permanent, strongly garrisoned, and supplied with artil-
lery — and wrote the Company requesting coehorns.

This post would be the anchor of the projected chain that
was to extend to the Pacific. Like the one among the Otos, it
would dissuade the river tribes from piracy against boats bound
for the upper country. Also it could intercept the British.
Blackbird, his loyal Spanish heart saddened, kept harping on
the size of British presents and the superiority and cheapness of
British goods. He let Mackay know, certainly without scaling
down the statistics, that there were thirty British boats on the
Minnesota River for the midriver tribes and the Sioux. They
would probably cross to the Missouri in the spring, he said,
and a month ago the British on the Des Moines River had sent
a train of twelve pack horses up the Platte for the Pawnees.[28]
He did not explain why as a great lover of the Spanish he had
let the pack train through but did promise to frighten,

negotiate, or bribe all the river tribes and the Sioux into alliance with His Catholic Majesty of Spain. He made clear that this would require great quantities of guns, kettles, blankets, medals, and tobacco. Mackay had no option but to let his goods dribble away to this grandiloquent moocher in hope that some of them would reach the tribes. He wrote the Company that it would certainly perish unless the government subsidized it more adequately. He needed, he said, better interpreters, more men, more arms, and especially more trade goods. But the impoverished province of a bankrupt and dying empire had no more money for the defense of Santa Fe.

Mackay sent messages of defiance to the Poncas. While his post was still under construction he sent a party to tell the Arikaras what he was doing and intended to do. Ice in the river forced them to travel overland and they met a band of Sioux, who scared them back to Mackay. It was idle to try to take the main party farther: too much of his stock had been used up and he now knew that his force was too small. But maybe he could accomplish what was after all the Company's principal objective. Maybe he could get a party through to the Pacific.

By now John Evans had proved himself and Mackay chose him for the forlorn errand. He was to take a small party by land and if, farther up, the overland route should seem the more feasible he was to stick to it. Evans started north from the Omahas — Fort Charles — in February 1796. He met Sioux and had to go back. He tried again in June and this time he made it — not the Pacific but at least the Mandan villages and the British who were trading there.

He was carrying a formal notice from Mackay ordering the British off Spanish soil. He also had a remarkable document, the instructions Mackay had drawn up for his exploration. More must be said about them later; enough now to look at the route prescribed. Mackay thought that before he got to the Mandans the "south fork" of the Missouri might reveal itself as the best way west. This was the Cheyenne, which by now had been described by so many Indians that it must be where they said it was. If it should look promising, Evans was to forget about the Mandans and ascend it toward the mountains and the ocean. If not, then he was to go on to the Mandans and beyond — "where there is a river a few lea[gues] higher that falls into the Missuri from the S West. This river is called

river de Roche Jaune or the river of the Yellow Stone it is said
to afford Good navigation & comes directly from the Stony
Mountains but it is hard to say whether it communicates with
the waters of the west."

Someone, probably an associate of D'Eglise whose report of
the Indians' description Mackay inclosed to Evans, had chris-
tened the big river. This is the first use of the name Roche-
jaune or Yellowstone.[29]

Assuming that Truteau would be with the Mandans, Mackay
directs Evans not to reveal the rest of his mission to him or anyone else. He is to go on to the Pacific and find out what white
nations are on the coast. Most likely their representatives will
prove to be maritime fur traders, Americans as well as Europeans. . . . Evans is the spearhead of the Company for the Discovery of the Nations of the Upper Missouri. Mackay directs
him to tell the sea captains (who have ahead of them that long
voyage round the Horn) that he will gladly take to Mackay to
be forwarded any letters they may have for Europe or the States.
. . . Mackay, who thought that he was from ten to twelve degrees
farther west and about two degrees farther south than he was,[30]
thought of the Rockies as a long ridge, which he understood to
be the Continental Divide. From the ridge to the Pacific was,
he thought, about eight hundred miles. This very realistic im-
provement over the Company's belief that at most it could be
only one hundred miles rests on his knowledge of Canada. But
the map in his mind is impossibly contradictory and the distances he mentions contradict his estimate of the time Evans
would need.

Evans needed a lot of time. It took him two months to reach
the Arikara village from Fort Charles, seven hundred miles by
water.[31] He took six weeks more to talk the Arikaras into letting him go on. That he eventually did so, and took some of
his goods with him, proves his merit. He reached the Mandans
at the end of September, the first man who had broken the
Sioux and Arikara blockade. At once he did what Truteau,
whom he had not met, had been sent by the Missouri Company
to do; he occupied the British trading post and ran up the flag
of Spain.

If there were Northwesters there, he sent them packing.[32] A
few days later a group arrived from the Souris River. Though
he says that he had neither men nor goods enough to oppose

them, he soon "absolutely forced them to leave the Mandane Territory." This must mean that he had won the Mandans to his support. When they left he sent with them copies of Mackay's orders for British traders to stay out of Louisiana. The temporary docility of the British that followed was watchful waiting — they could not tell how big a force and how many goods Mackay might have to support Evans with nor what might be on the way from St. Louis. There was a sudden sunburst of courtesy on the upper Missouri. Cuthbert Grant at the Tremblante post wrote that of course the North West Company would get out, now that a chartered Spanish company had appeared, and asked how to close up the business Jessaume had been running — running at a great loss, he assured Mr. Evans. John Macdonnell, for the Northwesters at Assiniboine House, also would close out and stay away, wanted help in reclaiming some deserters, and was pleased to send books, magazines, and medicines to a fellow countryman so far from home. James Sutherland at Brandon House on the Souris was positive that the Hudson's Bay Company would do nothing improper now, if ever, and might he not send an occasional party to trade for Mandan horses and corn?

So it went through the winter, a winter whose blizzards, cold waves, and Arctic gales must have been hard on a Methodist exhorter deep in wasteland, with three or four men and a few packs of goods, in a crude log hut or at best the Black Cat's odorous earth lodge. Communication with the Souris posts was episodic but urbane. Evans succored some of the British who had been jumped by Pawnees. He permitted the *engagés* to do a little trade in their private capacities and forwarded various bits of British property. (He did nothing about the deserters, who had set up a kind of free port, very sinister in the British view.) Mr. Sutherland warned him against the treacherous Northwesters, wondered if his principal was the Mackay who had once traded on Red River — he was — and delicately inquired whether the Missouri Company intended to explore farther up the river. On his part, Evans learned that the British intended to make the Mandan post a base from which to take the trade to the Rocky Mountains, "a trade that at this Moment is Supposed to be the best on the Continent of America." He tried to buy some of his visitors away from their allegiance but

lacked funds. Meanwhile he impressed the Mandans: they
would have accepted him, St. Louis, and Spain and closed the
road to the British if he had had goods. So at least he thought
and certainly they were to remember him favorably.

But he had only an insignificant amount of goods and could
get no more. So before spring the companies had accurately
appraised the Spanish effort. At the end of February 1797 Macdonnell wrote a supercilious doubt that Evans had authority
from Spain and, in answer to a warning, that "British subjects
are not to be tried by Spanish laws." (The point of Nootka,
after all.) Three weeks later Macdonnell's partisan, Jessaume,
arrived with plenty of goods and, it is obvious, succinct instructions. He was to crush the opposition and he did. Lavish presents bought the Indians' support again. Jessaume tried to buy
a small massacre too but the chiefs, perhaps headed by the
Anglophobe Black Cat, were outraged by their young men's
willingness to violate the code of hospitality. They told Evans
what was afoot and "came to my house to guard me." Thereupon Jessaume tried to shoot him, "but my Interpreter having
perceived his design hindered the execution." But that did it;
he had to get out. A microscopic frontier action with far-reaching implications ended conclusively.

Mackay had already returned to St. Louis from the Omahas,
frustrated by the same lack of goods and men. (He had, however, meanwhile made a pioneering journey to western
Nebraska.) And the Missouri Company had collapsed; a combination with a Montreal firm that might have enabled it to
survive ended when the firm's representative died. Mackay,
under the impression that Evans had started for the Pacific,
reached St. Louis in May 1797, Evans in July. They had done
all that could be done for St. Louis and Spain. The Province
of Illinois, Upper Louisiana, was not going to be linked with
California by way of the Missouri River, and in the end the
experience of Mackay and Evans was to serve not Spain but the
United States.

There was a final bitterness. Evans had to write his eastern
supporters, who had to send the sad tidings on to the poets in
London, that the Mandans were not Welsh, that so far as he
could learn no Indians anywhere were Welsh.

# Chapter X

# A More

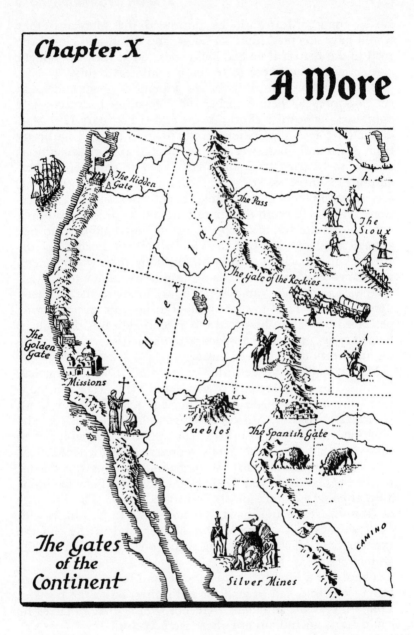

The Hidden Gate

The Pass

The...

The Sioux

The Gate of the Rockies

The Golden Gate

Missions

Unexplored

Pueblos

TAOS

The Spanish Gate

The Gates of the Continent

Silver Mines

CAMINO

Map 27

# perfect Union

The Canadian Gate

The Corridor

Mohawk Gap

The Atlantic Gates

Alleghany Gate

Gap

The Arch of the Continent

The Gate of the Ohio

The Bloody Land of KENTUCKE Cumberld Gap

Mississippi Keelboat

Plantations

The Cotton Gate

REAL

The Master Key

Raisz

# X

# A More Perfect Union

PIERRE CLEMENT LAUSSAT came to New Orleans by command of the First Consul to administer the restored French Empire in North America, the re-arisen New France. There were delays and he smouldered with helpless anger in a still Spanish town. The Spanish gentry and officials began by hating him but soon ignored and despised him. The French inhabitants were of many factions. Some, especially merchants who profited from the smuggling that was universal under the lax administration they were used to, adhered to the Spanish interest. Some looked forward to a Terror to renew the liberty that seemed so antique a notion in the improved world of the Consulate. Some appeared to have been debauched by American ideas. But, impotently waiting for the Province of Louisiana to be transferred to France, he busied himself making plans for its security and development — plans to substitute Napoleon's firm control of maritime commerce for the fatuous Spanish makeshifts under which the Americans had prospered, to nourish French sympathies in the American West, to attach the Indian tribes of the American Southwest to French interest. He forwarded his results, for its information and guidance, to a colonial office where they were tossed into the wastebin of yesterday's debris. When in a moment of curiosity Talleyrand asked what he and his staff might be engaged in, the colonial minister replied, "*Il peut être mort.*"

Almost as soon as Laussat arrived he had to deny a monstrous rumor which the Louisianans and the Americans who were resident among them came increasingly to believe. But it was, he shockingly found out, true. On November 30 1803 he received for the French Republic the enormous domain of Louisiana, transferred by His Catholic Majesty in accordance with a secret treaty made three years before, and this was the first of the duties he had been sent to America to perform. But he was a mere commissioner for the exchange of deeds and he administered Louisiana for just twenty days. On December 20 he added his signature to those of an American governor "with charming private qualities" but "great awkwardness" and a general "full of queer whims and often drunk." The American flag stuck halfway up the pole, then reached the top, and a small crowd of Americans cheered it uncouthly. It signified that a real-estate transaction of incalculable value had been completed, that the French Empire would never return to North America, and that one of the most momentous events in history had occurred.

Upper Louisiana remained. The dream had been that Laussat would direct it toward the prosperity latent in it for the health and strength of the restored Empire. (Who could help dreaming too that some day armies would march from it to raise the tricolor above the Citadel at Quebec where no French banner had flown for forty-four years?) Ended. St. Louis was, on the average, more than two months away and winter had set in. Laussat asked for the name of the American army officer at Kaskaskia whom the awkward governor and the queer general had appointed to receive the transfer of Upper Louisiana.

Thus Amos Stoddard, a Yankee who had been a lawyer but was now a captain in the Corps of the United States Artillerists, became Agent and Commissioner of the French Republic. He accompanied a detachment of that corps which, marching from Kaskaskia, reached Cahokia, across the Mississippi from St. Louis, on February 25 1804. A cold wave came down from the north, the river filled with ice, and it seemed unwise to cross till March 9. They marched to Government House, where the whole population of the town (increased in the last few years by a startling number of Americans) had assembled. Commissioner Stoddard signed the documents for France, everyone made speeches of rejoicing and congratulation, and the flag of Spain was lowered. That of France was raised while

the popguns of the old fort, whose foundation had been laid
when the British threatened St. Louis during the Revolution,
fired a salute. The lieutenant who commanded the American
detachment marched it to the fort, whose garrison was paraded
under arms, and the stars and stripes replaced the tricolor that
had attested·French sovereignty for a few minutes. The Missouri River, all the lands it drained, and all the Indians who
lived in them were American. So was the route to the Pacific
Ocean.[1]

The documents had also been signed by a captain of the First
Infantry, the personal representative of the President of the
United States. He had been on detached duty as Jefferson's
private secretary for two years but he was now in camp a few
miles up the river on the Illinois side. He was Meriwether
Lewis and with George Rogers Clark's younger brother
William he was commanding a United States Army organiza-
tion which the President had ordered, thirteen months before,
to ascend the Missouri and discover the water route to the
Pacific.

ॐ

A dictatorship must be kept in motion. The dictator is a
ruler who must maintain the momentum of events in order to
control them or he will fall. This necessity of absolutisms accounts both for Napoleon's attempt to restore the French Empire in North America and for his abandonment of the attempt.

England was the everlasting enemy. In 1798 General Bona-
parte led to Egypt an army whose eventual purpose was to
shatter the British Empire by reclaiming India for France.
Horatio Nelson cut its lifeline by destroying the French fleet
in the harbor of Abukir, and sea power had frustrated the imperial design. Later, leaving the army to rot or be saved by
diplomacy or by victories in Europe, whose map was now to become the map of France, Napoleon sailed for home with two
new plans matured in his mind. He would build up the French
navy, an enterprise which would require many years, and he
would transfer the imperial contest to the Western Hemisphere.

A month after reaching France he was First Consul. The
energy that no other ruler of men has ever had in equal meas-

ure reached the incandescence it was to maintain until, in Victor Hugo's phrase, God grew bored with him. During the next year while he attacked the victorious Austrians in Italy, crumpling them at Marengo, and launched Moreau against them in Germany to the triumph of Hohenlinden, his mind played with France and with the world as if they were made of building blocks that could be arranged in new structural designs at his desire. He began the internal reorganization of France that was to be his lasting gift to society. Where vortexes of force whirled without direction on the margins of Europe, he touched several so that they would now converge. And as the man of whom Europe had this same summer learned the terror which was to last till Waterloo he approached Charles IV and his queen, Maria Luisa of Parma who was the actual ruler of Spain. In return for creating in Tuscany a Bourbon kingdom for the Duke of Parma, their son-in-law, he demanded possession of Louisiana. The shattering of the British Empire was to occur in North America.

This was the deal ratified by the secret treaty of San Ildefonso in October 1800. Charles IV removed the last stay to the dissolution that had begun in South America and smoothed the way of the liberators and adventurers who within twenty years brought down the whole structure that had been built on the discovery of the Indies. France recovered the vastness which it had given away to keep it from Great Britain and to end a war, and which it had never ceased planning and conspiring to recover. Napoleon repossessed it first of all as an attack on the everlasting enemy. Yet it was also a move in the war he was fighting to bring Europe to an armistice, which in turn was a move toward the unification of Europe. His ruthlessness to Spain was a little the less cynical in that he thought of Spain, seven years too early, as already his. There was, moreover, at least a twofold purpose in regard to the United States, to the Presidency of which Jefferson had just succeeded after an election that had almost torn it apart. The United States, which had fought a bitter and victorious if undeclared naval war with France under John Adams, must be appeased or it would be driven into alliance with Great Britain, and Napoleon put the diplomatic machinery to work at the very moment when he demanded Louisiana from the Spanish king. And possession of

Louisiana was the most powerful force he could direct against the republican and democratic principles which as the post-Thermidor heir of the French Revolution he considered a danger to well-governed societies.

Singularly little ever got written down about the new French Empire in North America. Of the men who have tried to conquer the world, no other had a mind commensurable with Napoleon's, which was infinitely subtle and in much concealed from scrutiny. It played with continents as with dice, with nations as with the markers of the game, and seems to have produced intricate and detailed plans as instantaneously as a reflex. Clearly, his design for the colonial empire in North America which would challenge and eventually overturn the British Empire was as specific as that by which he blueprinted the political and juridical reorganization of France. But what can be said about it is in great part inference — inference from his dispatches and official correspondence and military orders, from the memoirs and conversations at St. Helena, from the memoirs of his ministers, and even less aptly from a handful of state papers.[2]

It may be said that the conception was his own, for it went far beyond what either Talleyrand or the colonials and their conspiracies had urged, and that it derived from a geopolitical concept. He was a conqueror: he saw the shape and contour of empires as issuing from grand strategy. New France had been organized on the axis of the St. Lawrence, whose mouth is one of the strategical keys of North America. The axis of its successor was to be the Mississippi, whose mouth is the master key. It was to have military power not only to maintain itself secure but to defend the Sugar Islands, whose restoration after more than a decade of conquest and revolution was to be completed. Its agricultural wealth was to feed them, and its wealth in furs and raw material was to sustain the French Empire — not only directly under Napoleon's conception of maritime commerce but by competition with whatever of the British Empire could not be conquered.

Strategic security demanded, as well for the Sugar Islands as for the continental empire, control of the Gulf of Mexico and of the highway to it from the south, the Caribbean. Control of the first meant in addition to the mouth of the Mississippi the peninsula of Florida, which another master of geopolitics, President Jefferson, had long been trying to get from Spain for the nation to whose land continuum it belonged. Napoleon had tried to have Florida included in the cession of San Ildefonso but Charles refused. The return to power of Manuel de Godoy, "Prince of the Peace," who was neither afraid of him nor his inferior in guile, kept it out of his hands in the diplomatic game which they played incessantly for the ensuing two and a half years.

Control of the Caribbean required a second stronghold to dominate the island screen. Martinique and Guadeloupe, often fought over but still French, held the Windward Islands at the southern end of the great arc but the central pier of this outpost barrier, instantly seen to be dominant when one looks at a map and even more visibly so in the era of sailing ships, was San Domingo. It was ruled by a Negro genius, Toussaint L'Ouverture, who ostensibly had saved it for Revolutionary France but actually had converted it into a personal and most Napoleonic dictatorship. Horrors and cruelties all but inconceivable, yet inconsiderable compared to those that now lay ahead, had supported the military and diplomatic skill with which Toussaint had manipulated French, Creoles, Spanish, English, Americans, and native blacks till the whole island of Haiti was French in appearance but his in fact. To the American view he had heroic stature in that he had made a Revolution and a Constitution, somewhat less so in that he had freed the slaves, and most of all in that his wars had worked to our safety and had much enriched our merchants. This was the man of whom Wendell Phillips in an orgasmic burst of rhetoric said that the muse of history, dipping her pen in sunlight, would write his name in the clear blue of heaven above those of Brutus, Hampden, and Washington. Napoleon (after perhaps rejecting an idea that he could be used to command a Negro army which would conquer North America) said that by means of him the scepter of the New World might pass into the hands of the blacks. To initiate the first phase of his western empire, he moved to destroy Toussaint and reclaim San Domingo. He prepared the largest amphibious operation the world had yet seen . . . and was to say at St. Helena that it was the greatest act of folly of his life.

Concurrently he prepared an expeditionary force of ten thousand to occupy Louisiana, with the army in San Domingo to make it perhaps twice as large when they had won their victory. The ten thousand (whose start was delayed by the chronic lack of shipping that always hampered Napoleon's maritime plans) were twice as great a force as the sum of the American army and the British garrisons in Canada. No trouble was expected from the Spanish troops in Louisiana; indeed sailing orders were held up till the peaceful cession was assured, and there was talk of hiring them till the army of the Republic could arrive. The purpose of this strength was not only to hold Louisiana for the ten or fifteen years needed to make of the French and tributary fleets a force that could challenge the British navy. It was to insure tractable behavior by the riotously expansive Americans. As for its eventual enlargement from the army in San Domingo, no one knows — but after all New Orleans was the beginning of the road to Quebec.

San Domingo must come first. The army sent there did indeed break Toussaint's armies and capture Toussaint, incidentally destroying what was left of the island's economy. But it was promptly broken in turn by guerrilla warfare, insurrections, and a terrible epidemic of yellow fever. There was nothing to do but to pour into this whirlpool of massacre and pestilence the army that was mustering in Holland for Louisiana and such others as could be raised. The act of folly was completed swiftly. The first army reached the island in January 1802; it was decimated and its commander was dead by November. When, a year later, the commander of its successor, besieged by another Negro general and hemmed in by a British fleet, chose to surrender to the whites, France had lost upward of 50,000 men and Napoleon had sold his western empire to the republicans.

Events must be kept in motion. In March 1802 he had procured the European armistice known as the Treaty of Amiens. The everlasting antagonists regrouped their forces and diplomatic chessmen for a showdown. News of the death of his first commander in San Domingo reached Napoleon in January 1803 and confronted him with the ineluctable fact of failure. In the global war he was a Robert E. Lee who, having in Egypt and the West Indies lost the attacks on the Round Tops and

Culp's Hill, had left no possible action but to break the center at Cemetery Ridge. He abandoned the overseas efforts altogether: the road to empire must run through Germany and across the English Channel. "I renounce Louisiana." But not to Great Britain, the mistress of the seas, whose fleet would seize it as soon as the war he must now make broke out. "To emancipate nations from the commercial tyranny of England, it is necessary to balance her influence by a maritime power that may one day become her rival; that power is the United States." [3]

Godoy had succeeded in delaying the transfer of Louisiana, even after Napoleon had peremptorily demanded it. Meanwhile, the Spanish intendant at New Orleans had seen in the impending transfer a chance to rectify what he considered a mistake in statecraft and a weak surrender to the Americans. Without consulting the governor of Louisiana or the Spanish minister at Washington, he reverted to the policy of the early 1790's and closed the Mississippi to American commerce, which by now had multiplied many times. News that Spain had bound herself to "retrocede" Louisiana had already produced acute tension in the United States. The nation was fully aware that the return of France to the North American continent would, as Jefferson said, "change the face of the world." No American could doubt that the expedition to San Domingo was the beginning of imperial aggression in the Western Hemisphere. And while the one to Louisiana was concentrating at the seaports the American minister at Paris learned what its destination was to be. [4]

The intendant's closure of the Mississippi brought both tensions to crisis. As a first move to resolve it Jefferson sent a minister plenipotentiary to assist the minister's effort, now more than a year old, to buy New Orleans or, failing that, some other site at the mouth of the Mississippi and thus secure an outlet to the Gulf. Talleyrand had ignored the minister, played with him, scorned him, misled him, lied to him. But now, "I renounce Louisiana. . . . Obstinacy in trying to preserve it would be madness." On April 11 1803 the minister, Robert R. Livingston, fatalistically beginning one more routine discussion with Talleyrand on the purchase of New Orleans, was greeted with a decisive question. What, Talleyrand asked him, what would the United States pay for all Louisiana? [5]

❧

Little need be said about the American background of the Louisiana Purchase. This narrative has shown that among the forces which produced the United States, a recurring one had been imperial war. As has been said here and must be emphatically repeated, the United States was itself an empire before it was a nation, though it was first a people who had made a society. Historians have been fond of drawing a distinction: the Peace of Amiens brought the wars of the French Revolution to an end and led on to the Napoleonic wars. It is a distinction for convenience only, for both groups of wars issued, as the American Revolution did, from the collapse of the world order completed by the Seven Years' War, and both were on the way to the stabilization that followed when the final defeat of the unification of Europe at Waterloo permitted the nineteenth century to begin. For a distinction between the groups of wars, for the bridge between them, the Peace of Amiens is perhaps less convenient than the Treaty of San Ildefonso which preceded it. That treaty did most surely, in French and British no less than in American eyes, "change the face of the world." And it brought the greatest military power, under the man who would presently be Emperor and intended to conquer the world, to the western boundary of the Republic that was just twelve years old.

On September 17 1796 George Washington had said, "The period is not far off . . . when belligerent nations, under the impossibility of making acquisitions upon us, will not lightly hazard the giving us provocation." He went on to ask the question which down to this day has lowered like a thunderhead whenever the nation has come in peril, "Why, by interweaving our destiny with that of any part of Europe, entangle our peace and prosperity in the toils of European ambition, rivalship, interest, humor, or caprice?" Always when that cloud has gathered it has been dispelled by the same inexorability that faced Jefferson when Louisiana again became French. It was now that he wrote, and to Robert Livingston as minister, "There is on the globe one single spot, the possessor of which is our natural and habitual enemy." And he went on to say, "The day that France takes possession of New Orleans fixes the sentence which is to restrain her forever within her low water mark. It seals the Union of two nations who in conjunction can maintain exclusive possession of the ocean. From that moment we must marry ourselves to the British fleet and nation." [6]

This realism of the sometime Francophile preceded by thirteen months the British minister's report to his chief that "the most desirable state of things seems to be that France should become mistress of Louisiana, because her influence in the United States would be by that event lost forever, and she could only be dispossessed by a concert between Great Britain and America in a common Cause, which would produce an indissoluble bond of union and amity between the two countries." [7]

Jefferson expected to resolve the problem by peaceful means, as he had been able to do when it was a Spanish problem eight years before.[8] But as President now he no more disregarded the possibility that war with France might become necessary, than as Secretary of State then he had disregarded the possibility that war with Spain might become necessary.[9] Among alternatives, at the extremity he was prepared to abide the presence of France in Louisiana till the force which he knew nothing could withstand should take care of it, "till we shall have planted such a population on the Mississippi as will be able to do their own business, without the necessity of marching men from the shores of the Atlantic." To do their own business: to take New Orleans, open the river, control the French. At almost the exact date of that quotation, the frustrate prefect Laussat was writing from New Orleans, "The Anglo-American [United States] flag eclipses by its number here those of France and Spain. In front of the city and along the quays there are at this moment fifty-five Anglo-American ships to ten French. . . . The Anglo-Americans are the most dreaded rivals in the world in point of commerce. . . . If New Orleans has been peopled and has acquired importance and capital, it is due neither to Spain nor to the Louisianans properly so-called. It is due to three hundred thousand planters who in twenty years have swarmed over the eastern plains of the Mississippi and have cultivated them and have no other outlet than this river and no other port than New Orleans." [10] Either Laussat or Jefferson had only to glance from the Spanish Province of Louisiana to the American Territory of Mississippi just across the river from it. Such a glance, like a single line of type that told the number of emigrants who had crossed the mountains in the last year, was evidence enough that the force would ultimately be irresistible.

But French possession of New Orleans, which meant that the river would continue to be closed, was the extremity. Though Jefferson knew that the extremity would be temporary and calculated that it would prove tolerable, he had no intention of accepting it. He would neither wait for the population that could do its own business nor become the prisoner of events. He intended to be master of events. His sending the plenipotentiary, James Monroe, to support Livingston's efforts to buy the Floridas and New Orleans was a measure of domestic policy. It was to restrain the Westerners from taking matters into their own hands and precipitating a war which he calculated was avoidable, and the eastern Federalists from inciting them to. (It worked out precisely as he had foreseen.) But his letter to Monroe following the appointment was written in full knowledge that the French army in San Domingo had been annihilated and that the crisis between France and Great Britain was sharpening. (And it was written just five days before his secret message to Congress asking funds for an exploring expedition to traverse Louisiana.) He told Monroe that on his success depended "the future destinies of this republic. If we cannot by a purchase of the country insure to ourselves a course of perpetual peace and friendship with all nations, then as war cannot be distant, it behooves us immediately to be preparing for that course, without, however, hastening it, and it may be necessary (on your failure on the continent) to cross the channel." [11] To cross the channel and, as a measure of preparation for a certain and early war, to propose marriage to the British fleet and nation.

It was the French possession of Louisiana that would change the face of the world, but it was the Spanish closure of the Mississippi that precipitated the domestic crisis. Jefferson had been President less than two years, following the first of our political revolutions. The Western interest that had forced

Washington to require of Spain the reopening of the river in 1795 had by now increased many times over. So Eastern Federalism was curiously focused on supporting Western Republicanism, and New England separatism was even more curiously focused on intensifying Western separatism. Jefferson said quite truly that "the fever into which the western mind is thrown by the affair at N. Orleans stimulated by the mercantile, and generally the Federal, interest threatens to overbear our peace." [12] The maritime war which had ended in the last months of Adams's administration might, as a result of that fever, be followed by the war for the mouth of the Mississippi that Washington had averted. Jefferson believed that it could be averted now: he sent Monroe to France to quiet the fever of the Western mind. The rest he confided to the unfolding of events: he saw that in the world turmoil lay the chance of achieving all the results the war could, of achieving them without a war. No statesman ever interpreted events more accurately than Jefferson did at this crisis, or bent them to his ends with more fastidious calculation of the forces they contained. It was Jefferson, not Napoleon, who charted the channel through the whirlpool.

And if the envoys he directed to buy a village secured instead a domain which more than doubled the area of the United States, the surprise, though stunning, was only that this had come about in one step, so easily, and so soon. "I believed the event not very distant but acknolege it came on sooner than I had expected," he wrote to Dr. Priestley. That was afterward but it was not hindsight. He had told Livingston that the British alliance would "make the first cannon which shall be fired in Europe the signal for tearing up any settlement [France] may have made and for holding the two continents of America in sequestration for the common purposes of the United British and American nations." And he had notified the governor of Mississippi Territory on the vital frontier that if France forced us to war, "we should certainly seize and hold [New Orleans and the Floridas] *and much more*." [13] The inescapable fact was that either the alliance with Great Britain or war with France, or anything else that made New Orleans American, must immediately or very soon carry with it the rest of Louisiana. In London Rufus King, the American minister, had freely discussed the military acquisition of all Louisiana, and the correspondence between Livingston and King discusses the same possibility.[14] When nations fight a war the peace settlement will involve more than a free port. There had never been a time when war for the mouth of the Mississippi had not meant war for the western half of the Mississippi Valley.

That was in the event of war. The contingency had been allowed for in circumstances other than war. Madison, the Secretary of State, wrote to the envoys after the great deed was done, "It was not presumed that more could be sought by the United States with a chance of success, or perhaps without being suspected of a greedy ambition, than the island of New Orleans and the Floridas. . . . It might be added that the ample views of the subject carried with him by Mr. Munroe, and the confidence felt that your judicious management would make the most of favorable occurrences, lessened the necessity of multiplying provisions for every turn which your negotiations might possibly take." [15] That is the language of diplomats: at least the possibility had been envisaged. Likewise, proposals to guarantee the western bank of the river to France, to withdraw either the entrepôt or the boundary to Natchez, to accept this or that condition, were items of diplomatic maneuver, pawns authorized to be offered in position play. So far as the administration accepted the very remote possibility that it might be forced to sacrifice any such pawn, it was for a few years only, to hold the position till the advancing West should reach the Mississippi in force.

For neither Jefferson nor Madison nor anyone else who watched the emigration supposed that the frontier would stop at the Mississippi, which the pioneer fringe had in fact already crossed throughout its entire length south of St. Louis. Every chancellery in Europe was on repeated notification from its ministers that, as Carondelet wrote in 1794, the Americans "advancing with an incredible rapidity toward the north and the Mississippi will unquestionably force Spain to recognize the Missouri as their boundary within a short time, and perhaps they will pass over that river. . . . If such men succeed in occupying the shores of the Mississippi or of the Missouri, or to obtain their navigation, there is, beyond doubt, nothing that can prevent them from crossing those rivers and penetrating into our provinces on the other side." That nothing could prevent them was clear to the Spanish governor of Louisiana in 1794. It was no less clear to the President of the United States in 1803. He had, in fact, written to the governor of Virginia in 1801, "However our present interests may restrain us within our own limits, it is impossible not to look forward to distant times, when our rapid multiplication will expand itself beyond those limits & cover the whole northern, if not the southern continent, with a people speaking the same language, governed in similar forms & by similar laws." [16]

King in London had discussed an American Louisiana as a buffer to protect Canada. Livingston had proposed to France, even before the Intendant closed the river, the cession of all Louisiana or, alternatively, all of it north of the Arkansas River. To this proposal he repeatedly came back.[17] And Livingston may be taken as representative: a Republican, an Easterner, in no way a visionary, a member of the committee that had been appointed to draft the Declaration of Independence, a member of Congress, a veteran and very skillful diplomat. He had shared the entire national experience of the United States . . . and he took its expansion across Louisiana as a thing given, only the factor of time being in question. As for that champion of reserved powers, Jefferson, in January 1803 he took up with Gallatin the question of acquiring territory and was sure "that there is no constitutional difficulty." [18] This was two months before Monroe sailed for France, six months before Jefferson had to face the acquisition of the territory in view and hastily sketched an amendment that would sanction it, only to decide that the constitutional question had better not be raised at all. And in that same January he asked Congress for an appropriation "for the purpose of extending the external commerce of the United States" by exploring Louisiana.

❧

Part of the price paid for Louisiana was the assumption by the United States of claims against France by American citizens for damage suffered from depredations in the various naval wars. These later produced the litigation that invariably follows such settlements, and they were not finally extinguished until 1925. In 1939 the Geological Survey, summing up Louisiana, reported that the United States had paid in cash settlement of them the sum of $3,747,268.96. Beyond this "assumption" the price of Louisiana was 60,000,000 francs, $11,250,000. In addition the Survey counts $8,221,320.50 in interest, less discounts of $5021.75. The total is $23,213,567.73.

No one knew what the bonds had bought, for France did not know what had first been transferred to Spain or what had later been repossessed from it. The documents were both ambiguous and contradictory; some specifications they contained had been made futile by events, some meaningless by the progress of discovery. When the minister of finance mentioned this confusion to Napoleon he was told "that if the obscurity did not already exist, it would perhaps be good policy to put it there," [19] and there is sense as well as cynicism in the conqueror's remark. The ambiguities were on the territorial margins, were of the boundaries. Thus though few could believe that France had any claim to land west of the Mississippi drainage, there were some who read the tortuous legalities otherwise. Livingston himself considered that Louisiana reached the Pacific and Dr. Samuel Latham Mitchill, scientist, friend of Jefferson, Congressman from New York, and the author of legislation concerning Louisiana, supposed that it reached the Pacific north of California and, somewhere, south of whatever part of the Northwest coast might be British. Dr. Mitchill, in fact, so reported to the House of Representatives.[20] Again there was a matted tangle of uncertainties about the boundaries of what is now the State of Louisiana. The resolution of disputes about them, honest or factitious, required local uprising, militia occupation, annexation by fiat, the War of 1812, the purchase of Florida, and indeed the Mexican War of 1846. The final definition of Louisiana had to be arbitrary. The nation which Livingston and Monroe represented had an area computed at 869,735 square miles. The three instruments of cession dated April 30 1803 but signed four days later transferred to it, the arbitrary summation says, 909,130 square miles.[21]

It included (subject to the disputes mentioned) parts of Mississippi and Alabama as well as the part of Louisiana that is east of the river. The rest was the actual western extent of the region which on April 9 1682 the Sieur de la Salle had added to the domain of Louis XIV. It was the western half of the Mississippi drainage basin to the as yet undetermined Continental Divide, except as treaties or the acceptance of conventions had trimmed it. . . . Later there would thus be trimmed from it an area instantly and poignantly at issue as soon as the transfer to the United States was made. This is the portion of Missouri River drainage north of 49°, a strip in Alberta and Saskatchewan that amounts to 9715 square miles. . . . Excluded at the south was whatever land was legally Spanish before 1783 — and again this was not determined till the purchase of Florida, when the boundary was established as the Red River to the 100th meridian, north to the Arkansas River, and west along the Arkansas.

*Louisiana:* Minnesota west of the Mississippi. Louisiana west of the Mississippi and the city of New Orleans east of it. Arkansas, Iowa, Nebraska, North and South Dakota. Oklahoma except "the Public Land Strip," that is Oklahoma east of the 100th meridian. Kansas except for the corner west of the 100th meridian and south of the Arkansas River. Colorado north of the Arkansas and east of the Continental Divide. Wyoming and Montana east of the Divide.

When the determinations were made none of the maps they were based on were accurate. In the end, the land itself had to be surveyed to accord with the accepted determinations. An exactly equivalent empiricism governed the American acceptance of the Purchase: all legal, indeed all abstract, questions about it were meaningless and Jefferson was right not to raise the constitutional issue. In the presence of the tremendous fact the laws were silent: the land itself reduced all other meanings to nonentity. And among the meanings thus nonplused was a threefold irony: that Napoleon sold Louisiana before the "retrocession" from Spain was executed, and so it was not his to sell; that he sold it without consulting the Senate of the French Republic and the Legislative Assembly, so that the sale was unconstitutional and therefore invalid; and that the Treaty of San Ildefonso, which alone could give France title to Louisiana, contained an article of absolute reversion to Spain in case there should be any attempt to cede or alienate it, and so whatever warranty or claim he was acting to transfer perished with his act.

The Louisiana Purchase was a resultant. It was the resultant of four systems of imperial energies forced to conform to the unmalleable reality of geographical fact. Because that is what it was, April 30 1803 is one of history's radical dates. "The annexation of Louisiana," Henry Adams wrote, "was an event so portentous as to defy measurement."

In the sector of partial measurements, two remarks of Napoleon's, both quoted by Barbé-Marbois, the minister of finance who conducted the negotiations, are germane. "This accession of territory consolidates the power of the United States forever, and I have given England a maritime rival who sooner or later will humble her pride." [22] The unifier of Europe and the remaker of the world, who had also ended forever the dream of a North American France, was here looking down a long arc of time with great clarity. . . . And yet.

(And yet it is always summer afternoon, they say, on the bank of the Styx to which the Elysian Fields come down. Statesmen there watch without concern the flowing of the asphodel-bordered river of time, tranquil because all doubts are ended and nothing is to be done. So sometimes they fall to talking about *if*'s that have no force. George Canning and John Quincy Adams may thus discuss the Monroe Doctrine in terms of the alliance with Great Britain which Jefferson was willing to envisage, which Napoleon perhaps averted by the cession, and which would have given the nineteenth century, in its third year, an experiment in international discipline based on allied sea power.)

With the threat of France on the Mississippi dissipated, the United States and Great Britain did revert to the antagonism implicit in their respective situations, as the world upheaval went on to define them further. That was the first step in what Napoleon foresaw, and he supplied the pressures that intensified the antagonism to war nine years later. As he foresaw, the maritime rival humbled British power, turned back invasions down the Champlain corridor and up his (and La Salle's) Mississippi route, and as a nation came to understand permanently the natural channels for the application of military force on this continent.

The power added to the United States by the Louisiana Purchase is indeed beyond measurement, and its torque has been exerted on the nations increasingly since 1803. Napoleon's second remark shows that he missed something in it: he missed the core-meaning of what he had done. "Perhaps," he said to Marbois, "perhaps I will also be told in reproach that in two or three centuries the Americans may be found too powerful for Europe, but my forethought cannot encompass such distant fears. Besides, in the future rivalries inside the Union are to be expected. These confederations that are called perpetual last only till one of the confederating parties finds that its interest can be served by breaking them." [23] He was wrong on his oath: Louisiana made the confederation perpetual.

Louisiana welded the implicit significance of the American political experiment to the implicit logic of continental geography. Thereafter they were not to be distinguished from each other. From them came the strength which not only held the confederation secure against exterior force but could not be overcome when the rivalries of its members that Napoleon took for granted produced a revolution intended to break it. When that time came, the land itself forbade.

☙

As Abraham Lincoln understood it must.

Fifty-nine years after the Louisiana Purchase, December 1 1862, President Lincoln sent his second annual message to Congress. He had a dolorous year to look back on when he reported the state of the nation. The invasion of Maryland, it was true, had been stopped at Antietam but the full victory which might have destroyed the Army of Northern Virginia had slipped from McClellan's grasp. There had followed the heartbreaking delays of a general who felt confident that he would be invincible tomorrow and certain that the enemy was invincible today. So McClellan had had to be replaced for the second time. There was the failure that produced the earlier removal to look back on: the glorious promise of the Peninsular Campaign, the timid management of the Seven Days, the magnificent parrying by first Johnston and then Lee, the humiliating withdrawal. There were other humiliations of this year to be remembered: repeated defeats in the Valley, Second Bull Run, smaller actions that seemed everywhere to go against the Union. In the West there had been significant victories in 1862 but they were far away and easily forgotten in the city where every foul bird came abroad and every dirty reptile rose up — where defenders of the Union seemed the most likely to destroy it — where the Secretary of the Treasury preferred the impotence of the government to the continuance in office of the Secretary of State — where part of the Cabinet was hostile to the President and in Congress much of his own party had rebelled against him.

Surely here was a powerful tendency of confederating parties to destroy a confederation. Yet as far back as September 22 Lincoln had read his draft of the Emancipation Proclamation to the Cabinet, telling them they might criticize the phrasing but not the act, which was determined upon, not open to question. Now, on December 1, his message explained to Congress that the Proclamation must be supported, as a war aim, by amendments to the Constitution abolishing slavery. The war, he said . . . "our strife pertains to ourselves, to the passing generations of men — and it can without convulsion be hushed forever with the passing of one generation of men." This was a dark saying to a nation almost at its darkest hour, and in order to make its meaning clear the President had to preface it with another explanation.

He quoted from his inaugural address the moving passage that begins, "Physically speaking we cannot separate. We cannot remove our respective sections from each other nor build an impassable wall between them. A husband and wife may be divorced and go out of the presence and beyond the reach of each other, but the different parts of our country cannot do this." On to the end. When he first addressed that solemn warning to the South there had been no fighting. But now there had been much fighting and God only could sum up how much waste, destruction, agony, and death — and still the inexorable truth of that warning held. So he went on:

> There is no line straight or crooked, suitable for a national boundary upon which to divide. Trace through from east to west, upon the line between the free and slave country, and we shall find that a

little more than one-third of its length are rivers, easy to be crossed and populated or soon to be populated thickly on both sides; while nearly all its remaining length are merely surveyors' lines, over which people may walk back and forth without any consciousness of their presence. No part of this line can be made any more difficult to pass by writing it down on paper as a national boundary. . . . A glance at the map shows that territorially speaking [the vast interior region] is the great body of the Republic. The other parts are but marginal borders to it. . . . And yet this region has no seacoast — touches no ocean anywhere. As part of one nation its people now find, and may forever find, their way to Europe by New York, to South America and Africa by New Orleans, and to Asia by San Francisco. . . . And this is true *wherever* a dividing or boundary line may be fixed. Place it between the now free and slave country, or place it south of Kentucky or north of [the] Ohio, and still the truth remains that none south of it can trade to any port or place north of it, and none north of it can trade to any port or place south of it, except on terms dictated by a government foreign to them. These outlets, east, west, and south, are indispensable to the well-being of the people inhabiting and to inhabit this vast interior region. *Which* of the three may be the best is no proper question. All are better than either and all of right belong to that people and to their successors forever. True to themselves, they will not ask *where* a line of separation shall be, but will vow rather that there shall be no such line. Nor are the marginal regions less interested in these communications to and through them to the great outside world. They too, and each of them, must have access to this Egypt of the West without paying toll at the crossing of any national boundary.

*Our national strife springs not from our permanent part; not from the land we inhabit; not from the national homestead. There is no possible severing of this but would multiply and not mitigate evils among us. In all its adaptations and aptitudes it demands union and abhors separation. In fact it would ere long force reunion, however much of blood and treasure the separation might have cost. Our strife pertains to ourselves, to the passing generations of men. . . .*

Here is the inherence of the physical conditions that have shaped American life, expressed by the man who at the Gettysburg cemetery and in his second inaugural address would express the deepest faith and the highest aspiration of American life. It states what Napoleon's prophecy left out of account, when for sixty million francs he relinquished to the young Republic the West that completes the physical unity.

And there is a striking thing about Thomas Jefferson. He sometimes said that the continental area was too big for the United States to govern. He could tranquilly contemplate the possibility of other republics in the Far West peopled by Americans but independent. "We think we see their happiness in their union [with us] & we wish it. Events may prove it otherwise; and if they see their interest in separation, why should we take side with our Atlantic rather than our Missipi d scendants? It is the elder and the younger son differing." [24] Again, "I confess I look to this duplication of area for the extending a government so free and economical as ours as a great achievement to the mass of happiness which is to ensue. Whether we remain in one confederacy or form into Atlantic and Mississippi confederacies, I believe not very important to the happiness of either part . . . and did I now foresee a separation at some future day, yet I should feel the duty & the desire to promote the western interests as zealously as the eastern." So he could believe and say. Nevertheless it is possible without any distortion whatever to make a chronological sequence of his actions in regard to the American land, from the reports and ordinances he wrote for the government of the Northwest Territory as a member of the Congress of the Confederation, through his ministry to France, his term of Secretary of State, and on to the Louisiana Purchase — and, looking at this sequence, to decide that though he may sometimes have *thought* that the nation could not permanently fill its continental system, he *acted* as if, manifestly, it could have no other destiny.

Or more simply, this: after 1803 the phrase "the United States" in Jefferson's writing, usually plural up to now, begins increasingly to take a singular verb.

ᐸᕓᐳ

History cannot suppose that because the intangibles which alter men's consciousness deposit no documents in the archives they therefore do not affect societies. What constituted the New World a new world, how the new world made Americans out of European stocks, what interactions of men and land established the configuration of American society — these are intangibles which historical thinking has only gingerly considered. Fragmented facts, all commonplaces, await a synthesis. Some are the daily absorption of the humbler sciences immediately at hand and indeed taught to all schoolboys. Others are truisms used by all artists. There is no schoolboy who has not read about "Indian old fields" or does not know that Squanto told his white friends to plant the maize in hills. There is no artist who does not know that the American scythe blade and the American axe handle developed a curve instead of the inherited straight line in response to conditions which the tools must meet. There is no one who does not know that the first-comers must live in, by, and in spite of an abundance of timber of which not even an ancestral memory was left in Europe, and no one who does not know that the syntax of the American tongue differs in structure from the syntaxes of Europe. . . . The poet MacLeish: "She's a tough land under the corn mister: She has changed the bone in the cheek of many races." The poet Benét: "And they ate the white corn-kernels, parched in the sun, And they knew it not but they'd not be English again. . . . And over them was another sort of day, And in their veins was another, a different ghost."

Rivers, mountain ranges, the orientation of glacial lakes, soils, climates, prevailing winds. And, in the late phases of the westering, an acceleration which is the only way time decisively affects the equations. (Time in relation to space; those who thought the space too great to be mastered were miscalculating neither area nor politics but only the rate of acceleration.) These compose an articulation, a pattern, an organic shape. It is not a perfect symmetry nor a perfect unity, but it is incomparably closer to being both than the physical matrix in which any other modern nation developed. The American teleology is geographical.

The Appalachian Highland curves two arms round the state of Maine. Those heavily timbered uplands, incapable of being farmed, were a barrier not worth crossing and therefore an implicit boundary. They remain essentially a wilderness, detoured not crossed, today. The rivers that flow north from them to the St. Lawrence (like those that flow south to it) are short. That is, the valley of the St. Lawrence is narrow. That is further, throughout almost the whole length of the valley agriculture had from the beginning not only a rigid limitation but a visible one. It had another limitation in the northern winter, from which agronomical adaptations could not free it till after the economic pattern had been fixed.

The society of the St. Lawrence Valley, then, must conform to the fisheries, to which the river led eastward — and which the English colonies could reach more easily from Massachusetts than from Maine, where they were nearest to Canada. And it must conform to the fur trade to which the river led west, all the way west, and its tributaries led north, all the way north. By the time settlement reached the first widening of the agricultural horizon, the Ontario peninsula, the society of Canada had conformed to the wealth in furs. There never would be agriculture in the north. The Laurentian Highland made it impossible; the bulldozer of the icecap had pushed the topsoil of Canada into the upper American Middle West. The bulk of Canadian agricultural wealth, apart from the Ontario peninsula, is concentrated west and northwest of Lake Winnipeg. It was hardly developed at all before the middle of the nineteenth century; much of its development has occurred in the twentieth century. By the time its West was settled, Canada had a social organism easily able to assimilate disruptive pressures that could easily have destroyed it a century before.

But the society south of Canada had always been of land, not furs: on an agricultural foundation. Its expansion was at a slower rate and got its power from mass.

After inclosing Maine in parentheses, the Appalachian Highland strikes southwest. The agricultural strip widens toward the south, but movement to occupy it was also movement west. This combined movement is repeated by the wide valleys within the uplands, which are agricultural and trend southwestward.

Mountain ranges impede communication, transportation, and settlement. They therefore concentrate populations, whose institutions integrate before they expand. The Appalachian system produced that effect, thus further slowing the momentum of the agricultural society. Yet the Appalachians are not a difficult barrier. There is one entirely unencumbered way through it, by the Hudson and Mohawk Rivers; a railroad that uses it truthfully calls itself "the water-level route." But a social and military obstacle stretched straight across it, the Iroquois. Similarly, at the southern end of the Appalachians, where again there is no serious topographical barrier to movement, were the Civilized Tribes, who were most civilized in fighting quality. Between these two extremities, straight west of populous colonies lay a number of comparatively easy passages across the mountains. All of them cluster with names from which our legendry has been spun — and all of them are practically continuous with water routes from the littoral. For the rivers that come down from the mountains could hardly have been better designed for access to the interior. Delaware, Susquehanna, Potomac, James, Roanoke, Santee, Savannah — all led to intelligible corridors. From the Susquehanna by way of the Juniata and through moderate passes to the Allegheny, or from the Cheat to the Monongahela — either was a simple and coherent route to a prime focus, the Forks of the Ohio. From the Potomac to the Kanawha the route through the mountains was less simple but equally coherent. Just as coherent, longer and more difficult, but by more profitable stages were the routes that converged on Moccasin Gap and Cumberland Gap and emerged on the Tennessee or the Cumberland River. These truly *crossed* the mountains, yet the mountains were not wide and the streams that cut through them did so by valleys too narrow to be populated. Moreover, all this country could be more easily traversed by horse and eventually vehicular traffic than the Canadian land, very large areas of which prohibited them altogether.

The Appalachian system, then, though a decelerant, implied neither economic nor political discontinuity. The cismontane river system impelled the American society westward or northward to it and westward or southwestward across it. Once across the mountains, the society came into the Mississippi Valley, a geographical unity of tremendous centripetal force. The routes were spun round the Ohio River as if by design. The Great Lakes lay to the north, a water route but also an implicit boundary. The ease of communication between the Lakes and the Ohio has been repeatedly emphasized in this narrative, and so has the ease of communication between the western Lakes and the Mississippi. On a map the rivers look like the veining of a leaf. Miami, Wabash, Illinois, Wisconsin — Kanawha, Kentucky, Cumberland, Tennessee — Ohio, middle Mississippi — and the continental arch through which the Missouri empties into the Mississippi.

To cross the Appalachian system was to come into the Amer-

ican heartland, where nothing could be separated from anything else for very long. Where all cultures and all stocks and
all casts of thought and all habits of emotion mingled. Where
as Lincoln said the dividing lines were either rivers that could
be ferried in a moment or the numbered abstractions of surveyors that could not even be perceived. And where to go west
a mile today was to go twain before sundown tomorrow.

This continuity and integration of the land, it must be repeated, was a centripetal force, a unifying, nation-making force.
It increased as it progressed, so that the centrifugal thrusts, the
separatist actions previous to 1803, could not prevail against
it and those of the next decade were still more futile.

In this force, the element reciprocal with physical design,
though hard to define and all but impossible to isolate, can
never be left out of account. A people that has widened an axehead and changed the angle of its bevel, or that has begun to
shock wheat with one bundle for capstone and covering, has
already developed a different physiology of thought and feeling.
The variations in temperature no less than the abundance of
unowned land, new crops and different growing seasons no less
than a leveling scarcity of labor, differentiations in the usages of
woods, the handicrafts of a wilderness-lapped livelihood, town
government and township surveys, a forest pharmacopoeia, the
skills of clearing and trail, a clergy made rebellious and an
electorate of freehold ownership, the reduction of bog iron or
the presentiment of a wider suffrage — all these are metabolic
processes in the growth of a new consciousness. Lincoln said
that there was no possible severing of the national *homestead;*
he also said that its *adaptations and aptitudes* abhorred separation. They were dynamically connected, functions of each
other. For the whole he had no word except the mystical one,
Union, and no concepts except democracy.

Gathering centripetal energy as it traveled, then, the nation
that was coterminously an empire burst through its western
limit in 1803. Many accessions of power followed, notably an
increase in this same centripetal and nation-making force. Centrifugal, disruptive forces that would oppose it would increase
now too, at first gradually, pellmell later on with the slave-labor
hegemony, but their true nature could not, in 1803, be understood. Much was withholden but it was dramatically clear that

if not in this year then in some other not long distant, the nation would have crossed the Mississippi regardless.

But how far and for how long?

The land remained a continuum. No boundary between nations was ever drawn more justly than the canoe route from Lake Superior to Lake of the Woods. It corresponded exactly to the existing economic systems. Their equipoise sufficed temporarily to make the line of 49° westward from there an acceptable convention to both nations, when it was proved to pass north of the source of the Mississippi. Here, however, there was latent the first possible discontinuity or aberration. It requires a paragraph.

The Red River valley is Hudson Bay drainage and therefore was not geographically a part of Louisiana. (Jefferson, the American commissioners, the British commissioners, almost everyone who thought about those distant regions supposed that 49° had been made the boundary of Louisiana by the Treaty of Utrecht in 1713. Actually it had not been: the treaty provided that the boundary was to be determined by "commissaries," who never determined it.) West of the Red River the great plains begin — and, except climatically, these are as continuous north and south as they are east and west. The southwestern quarter of Manitoba, the southern half of Saskatchewan, perhaps half of Alberta are of the same nature as the Dakotas and Montana and no barrier cut across this enormous area. There was indeed an invisible one, of water flow, climate and especially convertible wealth. Long before the Louisiana Purchase the Canadian economy had determined it: the fur business. Its connection with the Pacific remained an imperial issue — one that had become immediate and urgent even before the Purchase — but the Canadian momentum was furs. Its movement was therefore west and north, not south. By the time the two societies could compete for the Great Plains area both had been integrated past any inner possibility of defying what had been established long before. In the late 1860's and early 1870's the vision that all Western Canada might be American, that the United States might stretch from Lake Winnipeg to Alaska, did indeed excite many Canadian and American minds. Much emotion and some blood was spilled but neither the Red River Rebellion nor any of the lesser plots, dreams, or nightmares ever had a chance. The issue had been settled long before. The acceleration of time had affected the equation and the equipoise held.

The Rio Grande was an implicit boundary between Texas and Mexico. West of the Rio Grande the boundary between the United States and Mexico is even more implicit, the composite of topography, soil, water flow, and climate even more inflexible. The line could have been drawn a few miles north or a few miles south of where it was drawn — but, as the entire history of Spain in the Southwest attested, not many miles.

Northern and southern limits defined a unit that was continentally continuous with the earlier United States. The possible discontinuities, the latent aberrations or refutations of the continental experience, were in the interior West — across the path to the Pacific that the American people had been following since they began to be American, when the Pacific lapped the western foothills of the Cumberland Mountains and could be reached from tidewater in two weeks.

On the map the river system of the eastern half of the Mississippi Valley appears to have been sketched by an artist drawing, freehand, a series of related curves. West of the Mississippi the river system appears to have been blueprinted by a purposive, systematic architectural engineer. That latter appearance is profoundly deceptive. The Red River (of Louisiana) was but imperfectly navigable and was navigable at all only in its lower reaches. The statement is no less true of the Arkansas River. Only the Missouri was a water route — and it was infinitely laborious and infinitely circuitous. Contrary to the traditional experience of the westering Americans, most travel, the principal movement, must be by land. And now the land changed. And the climates.

The Americans would leave the forest behind, then the prairies, then the tall grass, and come into the plains. On a steady gradient the land sloped upward to the unknown Rocky Mountains. The end of the tall grass meant the beginning of aridity and the approach of deserts. The deserts were a much more formidable barrier than the Appalachians. Routes across them must be sounded for like the channels in the shallow lakes of Saskatchewan. The valleys of three rivers threaded the maze of deserts, the Arkansas, the Platte, and the Missouri. Then came the Rockies. They could be crossed in only a few places; only two passes were feasible for wheeled vehicles in the travel season and, so it turned out, only one was actually usable during the critical years of Far Western expansion. The Rockies were a succession of ranges, a zone of mountains, in some places more than two hundred miles wide, everywhere precipitous. Beyond them were additional deserts and then another mountain wall, the Sierra and Cascades, much narrower than the Rockies but even harder to cross.

Deserts and mountains composed just such a barrier as political systems had broken on in Europe. Though the land remained a continuum an ellipsis interrupted its coherence. In the event, the westward thrust reached this barrier with the dynamics of expansion stepped up to the greatest power it ever had, the rush of the 1840's. It hurdled the barrier. The Pacific coast was brought within the American system of energies before the interior West was. Precisely here the acceleration of time had become decisive. Industrial development was the final centripetal force in American expansion: it enabled the single system, the single social and political combination, to absorb the ellipsis and fill out the continuum. There was never a chance that Oregon and California would fall away, as there had once been a strong chance that Kentucky might. The clipper ships, steamships, telegraph lines, railroads, and the subsidiary accelerants of communication and trade developed too fast.

None of this could be foreseen in 1803. In particular no American knew anything about any of the deserts or any of the mountain ranges. But an energy in addition to the uninterrupted westward thrust had been in operation for ten years. A detached portion of the American system existed at and near the known mouth of a river called the Columbia, which was unknown above its mouth. It was as if a whirling sphere had detached an asteroid that traveled in a concentric orbit, and yet the attractive force was in the direction of the asteroid — pulling the sphere toward it. Jefferson truly called the land between "terra incognita." But if a water passage across Louisiana to the Columbia River could be found, then the detached portion of the American system could be brought in circuit . . . or drawn within it. And a feeling of incompleteness, hardly to be diagnosed on the margin of Jefferson's emotion but one of the "adaptations and aptitudes" which the land had wrought as it created the continental consciousness, would be eased.[25]

The determination to send an exploring expedition overland to the Columbia must have been fully matured in Jefferson's mind when he entered on his Presidency in March 1801, for there was no other reason to make Meriwether Lewis his private secretary.[26] Lewis, a family friend, had sought his help when trying to get a place on the abortive expedition of André Michaux and he knew that, as Lewis said, the exploration of Louisiana "had formed a darling project of mine for the last ten years." [27] In March Louisiana remained Spanish (Jefferson did not learn till May that it was becoming French) and he had been dealing with it as a Spanish problem since the beginning of Washington's first term. To what sovereignty the land west of Louisiana belonged, the land drained by the Columbia, was wholly conjectural if not beyond conjecture. Neither the Spanish nor the conjectural sovereignty mattered in the least: he would send out the expedition, wrapping it in the usages of diplomacy. No one who knew the earlier history and no President, least of all our first geopolitician, could doubt that once a water route across Spanish Louisiana had been found the continental issue would be joined. The imperialism is peculiarly, even uniquely, our own kind but the dispatch of the Lewis and Clark expedition was an act of imperial policy. Even while he moved to buy New Orleans the President of the United States was moving to possess Louisiana.

❧

The narrative must turn back forty-seven years, to 1756, and to the Virginia piedmont in from the actual frontier but on the edge of the wilderness. Three years earlier, in 1753, George Washington had been sent to order the French out of the Ohio Valley. The year after that, at the Great Meadows, he had opened the global war in which the mastery of the West was the central issue. And in June 1755, just a year ago, pushing past the Great Meadows and on toward Fort Duquesne at the Forks of the Ohio, General Braddock had suffered a defeat that came close to settling that great issue for France.

That was a year ago. On June 10 1756, in Fredericksville Parish, Louisa County, Virginia, a learned clergyman sits down to write to an uncle of his in England. He reminds the uncle that in an earlier letter he has told how his parish and plantation can be located on the map of Virginia which two of his neighbors, Joshua Fry and Peter Jefferson, published in 1751. Now he has another map to describe, a "map of the middle British colonies in America" drawn by Lewis Evans and published last year. The map, the war, and Braddock's defeat have increased his already intense concern about the Western wilderness. He is thinking continentally on the basis of the information at hand.

His thesis is simple, central, and very old. Whichever nation finds itself "master of [the] Ohio and the [Great] Lakes at the end of [the present war] must in the course of a few years . . . become sole and absolute lord of North America." And it follows that within a few years either the Hudson River or the Potomac will therefore become "the grand emporium of all East Indian commodities."

This idea, the reverend gentleman explains, though it may startle one who lives in England, is not chimerical. It rests on the swiftness and carrying capacity of canoes, and on the affluents of the Mississippi River that lead into the Western wilderness. He specifies. . . . It is not clear whether he himself has read the Baron Lahontan but he reproduces Lahontan's geography, which is taken over in the book that he says he has read, Daniel Coxe's *Carolana*. Also he has read a book that appears to confirm Coxe, "a History of the travels of an Indian towards those regions." His having read it shows that he was indeed interested and alert, for this is Moncacht-Apé — and in the amusing plagiarism by Dumont (or the Abbé Le Mascrier) in *Mémoires sur la Louisiane* published in Paris in 1753, three years before. (The plagiarism preceded the publication of the original invention, Le Page du Pratz's *Histoire de la Louisiane,* 1758.) But Moncacht-Apé is only gratifying support; his reasoning is based on Lahontan's fiction and Coxe's wonderful nonsense. He erroneously makes Coxe say that he has sailed seven hundred miles up the Missouri but otherwise quotes him faithfully.

The headwaters of the Missouri, Coxe had learned from the

Indians, are in a mountain (range) on the western side of which a river flows down to "a large lake called Thayago, which pours its water through a large navigable river into a boundless sea." That sea is the Pacific, and both accounts have described the masted European ships to be seen there. And Coxe's facts and descriptions "are said to have been found by late discoveries, as far as discoveries have been made."

(Coxe had actually said that the Missouri was navigable to the source, "which proceeds from a *ridge of hills* [italics added here] somewhat north of New Mexico, passable by horse, foot or wagon in less than half a day" — and thence the river to the big lake, to the big river, to the South Sea.)

This route, the clergyman says, is certain to make the English plantations, by way of the Hudson or the Potomac, "the general mart of the European World, at least for the rich and costly products of the East, and a mart at which chapmen might be furnished with all those commodities on much easier terms than the tedious and hazardous and expensive navigation to those countries can at present afford." The quest for the Northeast Passage can be abandoned at last. (For this will *be* the Northwest Passage.) There will be no more interminable voyages to the East Indies, no more seamen dying "like rotten sheep" of scurvy. "What an exhaustless fund of wealth would be opened, superior to Potosi and all the other South American mines! What an extent of region! What a — ! But no more!"

No more, certainly, for Sir Humphrey Gilbert, and indeed Columbus, had said it all before.

Not the rector's enthusiasm is the important point, nor even the "aggrandizing and enriching this spot of the globe," but instead "a grand scheme formed here about three years ago." Mark this:

"Some persons were to be sent in search of that river Missouri, if that be the right name for it, in order to discover whether it had any such communication [as Coxe says] with the Pacific Ocean: they were to follow that river if they found it and make exact reports of the country they passed through, the distance they traveled, what sort of navigation those rivers and lakes afforded &c., &c." The outbreak of the war prevented this expedition but it had been organized. (It is still so live a project that this letter is to be kept secret and the writer has directed its bearer to throw it overboard if his ship is attacked by a French privateer.) The head of the expedition was to be "a worthy friend and neighbor of mine," who had already made many discoveries to the westward.

That would be Dr. Thomas Walker. A medical man who was also a piedmont planter, a surveyor, a speculator in lands, he was the first explorer known certainly to have entered Kentucky from Virginia and the discoverer of Cumberland Gap. In 1748 he accompanied a group of land viewers to the Holston River and East Tennessee, but his great exploit began in December 1749 and lasted till the following July. It preceded Christopher Gist's first Kentucky journey by more than a year and took Walker to the Holston again and on to Clinch and Powell's Rivers, through Cumberland Gap, to the headwaters of the Kentucky River and back across New River and the Valley of Virginia. Made on behalf of a land company, this was a notable, influential, and immediately famous journey. And Walker was a notable man and an influential wilderness thinker. He had had a hand in making the map which the clergyman mentions in his letter. . . . He had married the widow of a Meriwether and his oldest daughter married Nicholas Lewis, the uncle and legal guardian of Meriwether Lewis.[28]

The man who wrote that letter in June 1756 was the Reverend James Maury. To increase his income he conducted a small school for the sons of neighboring planters. One of his neighbors and friends was Peter Jefferson, a wealthy planter, the lieutenant of his county, a surveyor and to some degree an explorer of the wilderness. In 1746 Peter Jefferson helped to mark "the Fairfax line," seventy-six miles straight across the Blue Ridge to determine the extent of the great Fairfax grant. In 1749 with his friend and neighbor Joshua Fry he was employed to run the Virginia–North Carolina boundary some ninety miles farther west than the celebrated William Byrd of Westover had run it on the first survey, and this took him deeper into the wilderness. And in 1751 Jefferson and Fry were commissioned to make the map of Virginia to which Maury alludes in his letter and which was by a good deal the most accurate yet drawn.

In August 1757 Peter Jefferson died, attended to the end by Dr. Walker, and the next year his son Thomas, fourteen years old, went to board with the Reverend James Maury and attend his school. As one of the executors of his father's will, Dr. Walker must have had a voice in sending him there.[29] . . . Jefferson's intellectual heritage from the frontier cannot be itemized but it is so well recognized as enormous that nothing need be said about it here. Enough that as early as his tenth year he could have been familiar with proposals to ascend the Missouri as a way of reaching the Pacific — and as a way of possessing Louisiana and so winning the conflict of empires. Jefferson himself expressed his gratitude to the Reverend James Maury for sound instruction in the classical languages. Perhaps the United States should thank him for first planting in Jefferson's mind an idea that matured as the expedition of Lewis and Clark.

❧

The race west was between France and Great Britain when the project Maury describes was formed. Great Britain had succeeded to the role of France and Jefferson was a member of Congress when in 1783 he wrote to George Rogers Clark, saying that funds for an expedition probably could not be raised but asking the most celebrated of Westerners whether he would lead one if they could be.[30] John Ledyard's book had been published by then, as well as more complete accounts of Cook's discoveries. But what Jefferson mentioned to Clark was information that a large sum had been raised in England "for exploring the country from the Missisipi to California." The pretense was that this was "only to promote knoledge," which would be his own pretense twenty years later, but "I am afraid they have thoughts of colonising into that quarter." In 1783, then, the thought of British colonies in Louisiana or on the Pacific disturbed an American statesman who would have had no reason for disturbance if he had not been thinking continentally. The next item in the letter mentions the arrival of the final draft of the peace treaty with Great Britain and the acceptance by Congress of its "cession of the territory West of the Ohio."

It was Great Britain and the prophecy of the sea otter trade in 1786 when Jefferson was Minister to France and Ledyard called on him in Paris.[31] The episode has little weight in Jefferson's writing: a few lines in his sketch of Meriwether Lewis, a few more in his autobiography, scattering allusions in a few letters. Clearly Helen Augur's correction of the record is right and the idea of crossing the continent alone was Ledyard's, not Jefferson's. It would seem certain that the global thinker who was already concerned about the presence of the British in the West had understood the import of the sea otter before Ledyard waited on him. It is clear too that the man who since boyhood had been conversant with what happens on wilderness travel understood how wild the scheme was. "He is a person of ingenuity & information. Unfortunately he has too much imagination." That is just. But "if he escapes safely he will give us new, curious, & useful information." This is penned rather by the author of *Notes on Virginia* than by the geopolitician.

It was, however, Spain and Spain as an immediate neighbor when the Army tried a variation on Ledyard in 1790. Henry Knox had been Secretary of War for only eight months when he ordered the Missouri explored to the source and "all its southern branches" as well. This was forehanded as military thinking — and nakedly expansionist, for the Secretary included such other streams as might be found to empty into the Rio Grande. It was also espionage, G-2 stuff, and the secrecy it was wrapped in is almost impenetrable now. Knox thought that an officer and a noncom would do, if four or five Indians of known loyalty went with them, and he proposed that a second party be sent out after an interval to increase the chance of success. The order went to General Harmar, who ordered a Lieutenant John Armstrong to the post nearest the mouth of the Missouri, Kaskaskia, where he was to begin this impressive errand. No satisfactory Indians could be found but Armstrong went on to St. Louis, in what guise does not appear. (The town was currently afraid of a British attack and very anti-American.) There he copied a mid-century map whose Missouri River region was from Lahontan, and then went up the river, apparently alone, "for some distance above St. Louis." He need not go more than a dozen miles to perceive that his mission was absurd. Returning, he decided that it would be feasible with proper equipment, and a recent student thinks that Harmar and St. Clair were ill-advised in not trying to procure it for him. But it seems likely that the Indians who in the next few years were to stop everyone from Truteau on would have stopped any such attempt just as easily.[32]

The Nootka Sound Incident had occurred and the Americans were in the Northwest trade, though news of Gray's discovery of the Columbia had not yet reached the States, when Jefferson prepared instructions for Michaux in 1793. They are sagacious and ambiguous. Though by this time anyone who knew anything at all about such a venture as Michaux proposed knew that it required companions and a proper outfit, they say nothing about either and all but imply that Michaux is to go alone. (Certainly Michaux, who had gone with fur traders to the Hudson Bay country, intended no such folly.) He is instructed to avoid St. Louis. He is to make comprehensive scientific notes of the kinds that Lewis and Clark were later directed to make, and on behalf of the Philosophical Society Jefferson tells him to pay particular attention to possible mammoths and to determine whether the Peruvian llama ranges as far north as the upper Missouri. Though he has no choice in his route west, which must be by way of the Missouri, he is free to return by any route he may think best. Details are left to his "judgment, zeal, and discretion." Presumably the size of his party and the means of travel were included in this discretion, for in 1793 Jefferson and his fellow scientists knew that for a long distance at best anybody who went up the Missouri must go with traders. In 1813 Jefferson remembered that the expedition was to have consisted of two people only, "to avoid exciting alarm among the Indians." But that is manifestly not true, and Jefferson's whole paragraph here is mistaken.[33]

As for the party, Meriwether Lewis applied to be a member of it. Yet there is something perfunctory and haphazard, and therefore suspicious, about this entire proposal. The funds were insignificant and the preparation nonexistent. How far did Jefferson take it seriously?

And on January 18 1803 why must a message to Congress proposing an unimpeachably proper "literary" — that is scientific — exploration of Louisiana be secret? Why especially since, as the message says, in the last days of its sovereignty Spain could not be much concerned? Even in secret Jefferson explicitly says only that to make the Mississippi River secure as our western boundary, a considerable stretch on the *eastern* bank must be secured from the Indian tribes that now hold title to it. (The legal theory of the United States was that title could not be granted to individual or corporate purchasers till the Indian title — a most complex concept — had been extinguished.) This "planting on the Mississippi itself the means of its own safety" was an entirely proper measure, to which no nation and no domestic faction could object. Disclosure of such an objective could not possibly "embarrass and defeat" the measures needed to attain it, as Jefferson says it would.

The message curtly disposes of one fundamental and compelling reason, leaving its implications unexpressed, and says nothing about another reason not then urgent but equally fundamental. The first relates to Great Britain, the second to Spanish America. Both are inseparable from the sea otter trade and the Northwest coast, the westward thrust of the American people, and the concepts (or emotions) of the continental consciousness. Both must be seen in the light of certain facts.

First, when France took sovereignty over Louisiana, Napoleon's exclusion of British trade from French markets would be extended to it, by fiat if not in enforceable fact. Second, the British trade with Indians which was now almost a monopoly between the upper Mississippi and the Missouri, was a monopoly on the upper Missouri, and was vigorously established on the middle Missouri — this trade, only formally illegal under Spain, would be interdicted. Third, if Louisiana passed to American sovereignty, the British trade would be legalized by the provisions of Jay's Treaty, though subject to regulation and, so far as the administration's economic philosophy would permit, a tariff. Fourth, if the United States should obtain Louisiana from France, whether very soon or later on, a channel for expansion beyond it would be cleared. This last fact was succinctly expressed by the foreign minister of Spain, after the Purchase, in a note of vehement protest to Talleyrand. "The intention" of the retrocession to France, he says, "had been to interpose a strong dyke between the Spanish colonies and the American possessions; now, on the contrary, the doors of Mexico are to stay open to them." [34] For "Mexico" read "New Mexico."

Nothing in politics is so irrecoverable as what Congressional spokesmen of an administration tell their colleagues. (Besides that, the debate on Jefferson's message was in committee of the whole.) Nor do we know the content of his conversations with his Secretaries of State and the Treasury. With the Louisiana Purchase more than any other great episode in our history we lack access to the process of maturation. And yet the crucial part of the secret message is oblique only in that it is brief. The Indians of the Missouri, Jefferson says, carry on an important trade with the British. The British trade is under the handicap of the long winter and the many portages. Contrast this with the "moderate climate" of the Missouri, with "continued navigation to its source," with (unstressed but climactic) possibly "a single portage from the Western Ocean," and with "a choice of channels" from the Mississippi by "the Illinois or Wabash, the [Great] Lakes and Hudson, or Potomac or James rivers, and through the Tennessee and Savannah rivers."

Jefferson is telling Congress that the Lewis and Clark expedition is a preliminary step toward connecting the United States with the Northwest trade and, by utilizing the physical advantages of Louisiana and our continental routes, taking the trade of the interior away from the British. This is January 1803, three months before the Purchase. A President making such a proposal could not expect the trade to be permanently conducted in, or the trade route to lead through, territory belonging to a foreign power. In effect he is notifying Congress that some day Louisiana will be American.

The extreme importance of the point so lightly touched on in the secret message was no secret to Meriwether Lewis, who had been for two years a member of Jefferson's household. It was implicit in the instructions given to him. His application of it may be consulted in his letter to Jefferson from the Mandan villages, in the "Statistical View," and in his "Essay on an Indian Policy" (or "Observations and Reflections on Upper Louisiana" [35]) written after his return. All three envisage taking the trade away from the British and, progressively, each more fully proposes measures to be taken.

A basic purpose of the expedition actually stated in the message proposing it, then, is to prepare the American challenge to Great Britain for the Far Western trade. It seems clear that another basic purpose, though the message touched on it only in the phrase about the single portage at the head of the Missouri, was to get to the mouth of the Columbia overland before the British could.

This narrative has recorded the knowledge of Spanish officials in St. Louis, six years before, that the British intended to move west from the Big Bend of the Missouri. Jefferson and his Cabinet knew of David Thompson's explorations and surveys, and had the liveliest interest in his traverse, in 1798, between the Red River and Lake Superior, the northern boundary of Louisiana.[36] They may not have known that at the summer meeting of its partners in 1800 the North West Company had voted to send an exploring expedition over the mountains, though this seems all but impossible, considering the clearing house of fur trade information at Mackinac. They may not have known that in the fall of that year Thompson and Duncan M'Gillivray made plans to cross the mountains the following year, and that two of Thompson's men, as pioneers, actually did cross, though again ignorance seems implausible. Whether or not either Thompson or M'Gillivray made a crossing in the next year is in dispute, and of the contradictory indications the most trustworthy seem to be that neither did.[37] But M'Gillivray and other partners, though for the time being not successfully, continued to press for the crossing and for the occupation of the Columbia River which it was intended to bring about, and this fact could not long have escaped the administration's knowledge. But if the American diplomatic and consular services were ignorant of that intention or had any remaining doubts about it, if American merchants and traders in contact with the North West Company and competing with other Canadian companies had any doubts or failed to shout their fears at the administration, if any official in the United States failed to read the Canadian signs correctly — nevertheless, everything had been made clear fourteen months before Jefferson's message. It had been made clear by Alexander Mackenzie.

In 1799 Mackenzie seceded from the North West Company he had so notably served and went to England. Eventually he was to return and join the most intense and effective opposition it ever had to meet but in the meantime he published a book and was knighted. The book, *Voyages from Montreal*,[38] was

published toward the end of 1801. In its last chapter he outlines a plan for the organization of the fur trade on a continental and truly imperial scale. He had been maturing the plan
ever since his apprenticeship under Peter Pond, and the short-
sightedness of Simon McTavish in delaying to execute it was
his principal reason for leaving the North West Company. He
calls on the Hudson's Bay Company to merge with the Canadian merchants who had opened the Canadian West, had developed its trade, and in his person had reached the Pacific by an
overland route. (The merger was in fact to be forced on Company and opposition in 1821.) If the Company shall refuse,
then Mackenzie's plan calls for the government to break its
monopoly by opening to the Canadian merchants the cheap
transportation to the West afforded by Hudson Bay and the
Hayes River. This is to the end that a connection may be made
with "the only navigable river in the whole extent of Vancouver's minute survey." (Vancouver's minute survey had been
made, as his instructions from the Admiralty printed in the
book made clear, primarily to find a practical connection between the commerce of the West and the Pacific.) That one
navigable river is the Columbia, whose upper reach Mackenzie
was sure he had himself traveled, though actually he had been
on the Fraser.

Mackenzie believes that access to the Mississippi is also a
right of the British and that, in fact, Canada must be made to
extend south to 45°. But even if this just and rightful extension cannot be made good, the Columbia River must be British.
Then, "by opening this intercourse between the Atlantic and
Pacific Oceans and forming regular establishments through the
interior and at both extremes, as well as along the coasts and
islands, *the entire command of the fur trade of North America
might be obtained,* except that portion of it which the Russians
have in the Pacific. To this may be added the fishing in both
seas and the markets of the four quarters of the globe." But
this continent-spanning unification of the Canadian fur trade is
not all: Mackenzie is thinking with a prophecy of nineteenth-
century British trade imperialism at its purest. His plan calls
on the East India Company to relax its restrictions and permit
the British fur trust which he envisages to trade in Canton without hindrance. This, he says, would drive the Americans from

the maritime Northwest trade — it "would instantly disappear." The maritime Northwest trade was the greatest commercial wealth the Americans had.

The last paragraph in Mackenzie's book reads: "Many political reasons, which it is not necessary here to enumerate, must present themselves to the mind of every man acquainted with the enlarged system and capacities of British commerce in support of the measure which I have very briefly suggested, as promising the most important advantages to the trade of the united kingdoms." [39]

The President of the United States happened to be the most acute geopolitical thinker among its citizens. Even if he had not been, and even if he did not know that the North West Company had already begun the movement which it had refused to make while Mackenzie was a partner, he was after all President of a country which was thus notified that the river which it claimed and the great wealth of its commerce that centered in its vicinity were under threat of alienation.

There was no American-British race for the Columbia River on January 18 1803 when Jefferson sent his secret message to Congress. But there was one as soon as Congress made the appropriation for the Lewis and Clark expedition which the message asked for.

❧

There are striking resemblances between Jefferson's instructions to Lewis and those which James Mackay had drawn up for John Evans when he set out for the Pacific in 1796. This does not demonstrate that Jefferson had seen Mackay's manuscript. Neither does the fact that some of what Jefferson wrote is evidently based on someone's experience of Upper Louisiana. Much knowledge had seeped eastward in the last two or three years, and Jefferson may even then have had the journal of an agent of the Missouri Company which he later abstracted for Lewis. Nevertheless, it is possible that that agent was Mackay himself and that he had Mackay's instructions before him when he wrote his own. He may well have had additional notes or letters by Mackay.[40]

Since Mackay's and Evans's return in 1797 only one St. Louis trader had been able to get past the Sioux, and he only to the Arikaras. Up to that barrier, however, the Indian trade had been flourishing, invigorated by a transfusion of Montreal capital. Lewis and Clark were able to hire a full crew, a "patroon" and either seven or eight *engagés*, of experienced rivermen to take one of their pirogues as far as the Mandan villages. At least four of their own party, and perhaps more, were veterans of the Missouri River trade.[41] Considerably more knowledge of the country had been amassed. (Ideas about the Black Hills, for instance, are now tolerably accurate, whereas previously they had been vague and fantastic.) Much of this information, together with similar data regarding the Indians as well as much misinformation, had crossed the Mississippi. Its bulk was increased by Americans actually in Louisiana, trading on the lower Mississippi, and living along it. How soon after the Louisiana Purchase Jefferson and such people as Mitchill began to make formal requests for such information is not known, but various officials were corresponding with the Chouteaus before the summer of 1803. And in August Lieutenant Zebulon Montgomery Pike, who had been ordered to Kaskaskia early in the year, forwarded information he had been asked to collect for somebody. Mitchill's House Committee on Commerce and Manufactures had studied a mass of new material when it reported in February 1804. Jefferson's "Description of Louisiana," sent to Congress the previous November, a quick summary of what the administration supposed it knew, could not have been written with such assurance as little as two years earlier.

Jefferson assumes as a fact the shortness of the portage between Missouri and Columbia drainage. (This is the durability of Coxe but it is also an assumption of symmetrical geography.) He is strongly interested in the southern affluents of the Missouri and directs Lewis to learn all he can about them. Their headwaters will interlock with the Rio Grande and the Colorado, and Lewis is to determine whether the height of land between is mountainous or a plain. On the connections between these watersheds the Arrowsmith maps Jefferson used — 1795, and the somewhat more erroneous 1796 — offered no help. But the close approach of the Rio Grande to the Missouri had been shown on Delisle maps from 1703 on and was indicated on what was, all things considered, the best map available to him, J. Russell's North America of 1794. To this highly gratifying mingling of the headwaters, those of the Colorado had been added by Bellin in 1743.

To Congress Jefferson said that no dependable map of Louisiana existed. He was clear, however, that the part of it which stretched toward New Mexico was prairie "too rich for the growth of trees." One marvel known to exist was a mountain range 180 miles long and 45 wide, "composed of solid rock salt without any trees or even shrubs on it." This stunner is followed by information, most of it sound, about the Indian tribes on the Missouri. It is sufficiently current to include the smallpox epidemic which had all but annihilated the Omahas a year earlier, and perhaps the least accurate statement in it is the pardonable one that the Pawnees live "not far distant" from Santa Fe.

The rest of what Jefferson says to Congress and asks of his explorers has to do with the needs of a President who has acquired not only an Indian problem, a trade war, and an imperial race to the Pacific but also an area which the United States is going to fill with settlements. It is not geographical. Perhaps therefore ideas currently collected or proposed by Dr. Mitchill may be substituted for Jefferson's. They were fully available to Jefferson and must in great part have corresponded to his own.[42]

Dr. Mitchill says in his own person that "a great chain of mountains" divides Atlantic and Pacific drainage. (This corresponds to Mackay, 1796, but Jefferson was not entirely sure. Mackenzie, whom he had read, did not know how far south his mountains extended and said so; Arrowsmith, 1795, indicated a single chain and in 1796 had a note that there might be five ridges; Russell showed a single chain with a wide break in it convenient to the head of the Missouri.) He adds that the Red River is navigable for a thousand miles — though a later issue mentions the enormous "raft" of driftwood — and that the Arkansas had been "traveled above one thousand miles." (No one had fought the Red in boats for more than 300 miles; its full length is 1018 miles. No one had traveled the Arkansas in a boat as much as 250 miles; no one had gone up its valley more than 500 miles, though some had crossed it much farther west

than that; its full length is 1450 miles.) He bestows on the
Arkansas a tributary (possibly a rumor of the mildly brackish
Salt River) that flows salt water for 600 miles below its source,
and he is sure that there are rich veins of virgin silver and many
mines.

One of Dr. Mitchill's Kentucky correspondents contributes
data gathered by Lieutenant Pike, who has heard that the Missouri has been navigated for 2500 miles but that there is still
"an immense distance to its source." (Length of the Missouri,
2464.4 miles.) And with Pike the Baron Lahontan's interior
lake as the source of the Missouri, and therefore the Western
Sea as well, make their last appearance in American geography.
He hears that this connection is the best way to the Pacific.
Pike has also heard of the salt mountain, which is the size Jefferson gave it in his message to Congress. (He may be Jefferson's source, though a letter from Wilkinson to Jefferson names
Amos Stoddard.) He has heard of a volcano on the upper
Missouri, and in a footnote Dr. Mitchill says he has a piece of
pumice from it that was found floating on the river. (The
volcano — doubtless a burning seam of lignite — was first mentioned by D'Eglise.) Pike says that the mines of Santa Fe are
"nearly on a parallel with the mouth of the Ohio," that he has
met a man who works there and visits his family every year (in
St. Louis or Kaskaskia), and that the trip requires fifteen to
twenty days on foot or ten to twelve on horseback. His information may be of the lead mines at Dubuque; his error
about Santa Fe, though a persistent one, is inexplicable for its
latitude and longitude had been approximately correct on
maps for fifty years.

A senator from Ohio writes to Dr. Mitchill March 8 1804.
He confirms the river of brine and the rich silver mines, adding gold mines and a mountain of pure crystal. He has the
revealing idea that the White River of Arkansas has been navigated for 900 miles. (It is 690 miles long.) In the same issue
of the *Repository* Dr. Mitchill (presumably) sums up his information about the Missouri River. He has the latitude of its
mouth and the distance from Philadelphia exactly, and makes
a wild and quite unjustified guess about its length that comes
close to being right. He describes its water, current, shoals,
quicksands, boils, and sandbars with an exactness that estab-

lishes specific, detailed inquiries sent to St. Louis. He has the Grand Detour right, and the floating pumice. He has the volcano, and though he is partly right about the Platte he is also very seriously wrong, repeating the persistent error of the maps and of deduction about the interlocking headwaters. The source of the South Platte is 750 miles distant from the mouth of the Platte River and not more than 175 miles south of it. But Mitchill says that the Platte "extends southwardly between twelve and fourteen hundred miles, interlocking its headwaters with those of the [Rio Grande], down which stands the city of Santa Fe, where the Spaniards have valuable gold mines." That is the adjacent sources again, an entire failure to imagine what kind of river the Rio Grande was, and a similar failure to imagine the nature of the Colorado and New Mexico Rockies. (The Spanish had been afraid that the British would reach Santa Fe by boat along just that route.) And as for the source of the Missouri it was, after all, "in a chain of mountains at no great distance from California."

Coupling this to Jefferson's single short portage across the Continental Divide reveals the irreducible core not so much of ignorance — for ignorance entirely concealed the details of more than a million square miles — as of inevitable misconception. There were no better minds in the United States than Jefferson and Mitchill and they, with their learned associates, had mastered the existing knowledge.

భాన

Into this unknown area Jefferson sent Lewis and Clark on the expedition that was by far the most fruitful in the history of American exploration. A great deal of its fruitfulness stemmed from the scientific objectives set for it, and mostly phrased in the form of questions, by Jefferson in consultation with Mitchill, Caspar Wistar, Benjamin Rush, and Benjamin Smith Barton, all of the American Philosophical Society. The need for similar knowledge about lower Louisiana was obvious and had been provided for in Mitchill's bill. Jefferson sent his accomplished scientific correspondent William Dunbar into the Ouachita River country (Arkansas) in 1804 and up the Red River in 1806. More knowledge about the upper Mississippi

and the northern border was also urgently needed. Jefferson
had instructed Lewis and Clark to learn what they could about
both and his desire was well known to General James Wilkinson, whom Laussat had reported "full of queer whims and often
drunk." Wilkinson went to Washington in 1804 and was presently appointed Governor of Louisiana Territory. (Roughly
what had been Upper Louisiana; what is now the state of
Louisiana had been organized as the Territory of Orleans.)
When he took office, he acted on Jefferson's desire, though
without consulting Jefferson, Madison the Secretary of State,
or Dearborn the Secretary of War. Late in the summer of
1805 (Lewis and Clark were nearing the Continental Divide)
he sent Zebulon Pike with a sizable party to explore the headwaters of the Mississippi. Thus the new Republic was simul-
taneously making three traverses of its new domain.

Pike was getting experience that would be useful for one of
Wilkinson's designs. The Louisiana Purchase had closed some
opportunities and opened others to the master of treachery. He
was now involved in so much deceit, fraud, betrayal, espionage,
and treason that he could never follow a premeditated course
for very long but must extemporize as occasion might permit —
but something surely could be made of confusion, anger, and
frustration in Louisiana. During his stay in the East he spent
much time with an old friend, the Vice-President of the United
States, Aaron Burr. The orbits of our most tireless small scoundrel and our scoundrel of genius had now reached conjunction.
Burr, who was Jefferson's bitterest enemy, was performing a
political somersault, assisting the New England movement for
secession, and investigating the financial prospects of full-scale
treason. Nothing worked for him, he had begun to fall from
his high estate, and he lost his power when Hamilton procured
his defeat for the governorship of New York. On the morning
of July 11 1804 he ended Hamilton's life and his own career
with a single shot.

The long conferences with Wilkinson were both before and
after that date. What the two conspirators planned will never
be known. Perhaps they planned concretely but little more
than to seize the hour somehow. The world remained at war,
the United States shook with domestic as well as foreign upheaval, the Spanish Empire was dying and Spain was very sick,

there was the new United States west of the Mississippi, beyond it was Santa Fe, and beyond Santa Fe was Mexico. Surely the troubled waters would provide good fishing for two masters of the opportune. They planned something, certainly neither intended honest dealing with the other, and the genius of chicane made an irretrievable mistake in even a little trusting the petty operator. Wilkinson went on to St. Louis as Governor, got his many networks into working order, and began to investigate the possibilities of the Southwest.

One of Pike's objectives concerned the northern boundary of Louisiana in the area he was to explore. The boundary became an even more important consideration for the Missouri River expedition as soon as news of the Purchase reached Washington. (Lewis's letter of June 19 1803, inviting Clark to join the expedition, is the earliest known statement that all Louisiana was being negotiated for.) [43] Apart from a single letter, no written instruction of Jefferson's covering the matter has been found but Lewis had obviously been well schooled in its importance. Among those who did not know just what the United States had bought was the President. He used the slack season, with Congress in recess, to collate what was known and drew up for the Ministers to France and Spain "An Examination into the Boundaries of Louisiana." Dated September 7 1803, it did not discuss the northern boundary; so he took that up in a supplement dated January 15 1804. Jefferson thought (erroneously) that the Treaty of Utrecht, 1713, had specifically extinguished any claim that Great Britain might ever have had to any part of Louisiana south of 49°. There were snarls and defects in other claims, notably in the peace settlement following the Revolution. The negotiation and exploration agreed upon in Jay's Treaty had never been made. David Thompson's surveys had established that the Northwest Angle of Lake of the Woods was not at 49°, as the commissioners had understood in 1783, but north of it, and that the source of the Mississippi was not north of 49°, as had also been understood, but south of it. And Alexander Mackenzie's plan for imperial expansion, to which Jefferson's supplement refers, formally demanded that the boundary be relocated to the south — if not all the way west on 45°, then on such an angle as to make sure that the Columbia River would be British.

There were possibilities of great harm here, but perhaps also possibilities of advantage. Jefferson's memorandum demands that a line be run from the Northwest Angle to the 49th parallel, and that this parallel be adopted as the boundary running west till it reached the Spanish lands. (No map or treaty could tell Jefferson what the true extent of California was. No one knew how much claim to Vancouver Island Spain had retained after settling the Nootka Incident. No one knew what Spanish claim might cloud the American claim to the mouth of the Columbia.) But perhaps — and again no one knew — the 49th parallel might cross the Missouri or one of its tributaries.

How far north the northernmost reach of the Missouri River extended, how much farther north its northern drainage reached — these undetermined matters had absolute importance. Jefferson reverted to the full original conception of Louisiana. He decided that the northern boundary of Louisiana, therefore of the United States, must be so drawn as to loop round and inclose any Missouri drainage which might lie north of 49°. Thereafter it might return south "till it shall again fall into the same parallel or meet the limits of the Spanish province next adjacent." But any Spanish province would be an embarrassment and he ended by practically quoting Napoleon on the value of obscurities: "or, unapprised that Spain has any right as far north as that, & westward of Louisiana, it may be as well to leave the extent of the boundary of 49° indefinite, as was done on the former occasion." Indefinite enough to permit the United States to drive a territorial wedge all the way to the Pacific.

Lewis had certainly heard much deadly serious discussion of this imperial necessity when he left Washington for the West — for the expedition acted on it. It explains the captains' minute questioning of Indians, during the winter at the Mandan villages, about northern affluents of the Missouri. It explains their concentration there on the problem of the White Earth River and their strong interest on their way west in Milk River and the Marias. It explains notations in Lewis's "Summary View of Rivers and Creeks." And it explains why, on the return from the Columbia, Lewis struck north of the route they had followed west and set out to explore the Marias. . . . In December

1804 at the winter fort they had built near the Mandans, they explained to the North West Company's François Antoine Larocque that Louisiana extended all the way north to the Qu'Appelle River, a northern affluent of the Assiniboine. That was faulty geography — they did not know where the height of land was — but it expressed Jefferson's principle. "They make [the boundary] run till it strikes [the Mississippi's] tributary waters," Larocque's journal entry says, "that is, the north branches of the Missouri." [44] And Clark so drew it on his map when they got back to the States.

Unfinished business relating to a national boundary was thus another of the duties with which the Republic charged the party that started up the Missouri River in 1804. They were grave duties. The expedition was directed to discover the water route to the Pacific Ocean — to end the search, more than three centuries old, for the Northwest Passage. To complete the passage to India. It was to take the United States overland to Captain Gray's discovery, the mouth of the Columbia River — the Great River of the West. In doing this it was also to survey an inland route for the Northwest trade and thus give it continuity, celerity, and protection from sea power. It was to strengthen the claim of the United States to the Columbia before Great Britain should be established there — and so to occupy another strategic key to the continent of North America, as a strong point for imperial expansion. It was to take a step preliminary to that expansion by discovering implicit ways of challenging British trade in the American Far West and the Canadian Northwest. As matters subsidiary to the great ones, it was to reconnoiter the trade, to assert American sovereignty over the West, to proclaim American authority over Indian tribes and British traders there, to settle whatever Indian problems it might encounter, and to lay a basis, the hope was, for a solution to the everlasting Indian Problem itself. . . . And it was to fill in a space in the map of the world that had been blank white paper up to now, and to add to the heritage of the Republic and of mankind as much knowledge as might prove possible. Knowledge of the West, its shape and patterns, its Indian tribes, its plants and animals, weathers, seasons, and natural wealth.

It adds up to a heavy national responsibility for this party of forty-five, sent beyond the western frontier of the United States on the energy which had just that moment carried the frontier to the Continental Divide — sent there in fact while the frontier was still the center of the channel of the Mississippi River.[45] And as this spearhead of an expansion that was not to end till 1848 started west, it was outpaced by the same force that had launched it. Clark took the boats into the Missouri River on May 14 1804. In July two traders, backed by a Kaskaskia merchant and equipped with "merchandise and boats," left St. Louis to go up the Missouri and make a trading journey to Santa Fe, presumably by way of the Platte.[46] That much is certain. It may be too that Jacques D'Eglise, the pioneer and veteran of the upper Missouri, also started in 1804 for Santa Fe and its contraband, potentially rich, ultimately revolutionary trade. It may be, even, that the energy making southwestward was in advance of Lewis and Clark, for one authority believes that a trader spent the winter of 1803–4 on the Cedar Island which they did not reach till September 1804 and jumped off from it for Santa Fe in the spring. All this is as may be. But at least two parties of American traders reached Santa Fe from the States in 1805 (if one of these did not get there in 1804) and at least two others started for Santa Fe from St. Louis in 1805. Possibly Wilkinson had a hand in one or more of these ventures, and certainly he made use of them in 1806 when he sent Zebulon Pike and a creditor of one of the traders into the Southwest, to discover the source of the Arkansas River and to be observed by officials of New Mexico, military ones for preference.[47]

Wilkinson was trying to profit from the greatly accelerated energies. Whatever he was planning in the Southwest, on his own and in company with Burr, the alarm of the Spanish authorities would be helpful to him. The strong dyke they had counted on was demolished, the doors to Mexico were open. Following the Louisiana Purchase, the correspondence of the former governor and lieutenant-governor is panicky and on the edge of delusion. In this state of mind they felt sure that Lewis and Clark had been ordered to Santa Fe, and this belief was doubly useful to Wilkinson. Besides furthering his plot, whatever his plot was, it gave him a chance to pick up some money in his capacity as an honest spy for Spain. In February 1804, while he was still in New Orleans, the governor of Florida visited the town and it was through him that Wilkinson fished for a new bribe of $12,000 plus commercial privileges in the West Indies worth much more. Part of his bait was a paper called "Reflections on Louisiana," which he wrote but which passed through the diplomatic channels as written by the governor of Florida. It is a preposterous document of sedition, hardly conceivable outside Hollywood, but was worth hard cash to the despairing Spaniards. One of the empire-saving measures recommended to them in February 1804 by this small-time confidence man was this: "An express ought to be sent to the governor of Santa Fe and another one to the captain-general of Chihuagua [Chihuahua] in order that they may detach a sufficient body of chasseurs to intercept Captain Lewis and his party, who are on the Missouri River, and force them to retire or take them prisoners." [48]

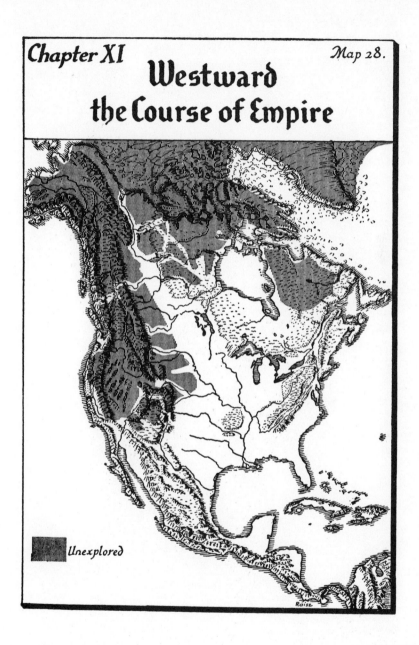

Map 28.

# Westward
# the Course of Empire

Unexplored

Raisz

# XI

# Westward the Course of Empire

W HEN THE expedition entered the Missouri River, May
14 1804, Meriwether Lewis was twenty-nine years old and
William Clark was thirty-three. Both were seasoned army
officers and campaigners, though Lewis had had no combat
experience. The party was an army organization and was always under military discipline, though most formal military
procedures soon disappeared under the pressure of wilderness
conditions. Lewis was a captain of infantry and had served
continuously from the Whiskey Rebellion till detached to act
as Jefferson's secretary. Clark had resigned his lieutenancy in
1796 soon after Wayne's Indian campaign. Lewis had been a
junior officer under his command for a few months "either late
in 1795 or 1796." On Jefferson's authority Lewis promised
him a captaincy when asking him to join the expedition, but
the War Department overruled the President and commissioned him a second lieutenant in the Corps of Artillerists. He
is always "Captain Clark" in the *Journals* and in Lewis's correspondence, however, and so signed official papers, though
without adding his organization as Lewis did. Both were
woodsmen of long experience, both were frontiersmen, and
both were rivermen, though Lewis freely conceded that Clark
had the greater skill and it was he who usually conned the
boats.

Though they exercised equal authority and though both had
in the highest degree the first-rate officer's ability to instantaneously take charge of a situation and control it, the feel of
the *Journals* throughout is that Lewis was the actual commander by the natural set of his personality and that Clark was

the perfect Exec. Lewis had the faculty of command and exercised it by instinct. He was the better educated and was a citizen of the world and a diplomat. His mind was restless, inquiring, speculative, scientific. He was introspective, humorous though rather pompously so, mercurial, moody, and he expressed himself well in elaborate prose. There was not much warmth in him, he was a solitary and a melancholiac, and he was saturated with romantic emotions. He belonged to a type very common in our westering, the complexly introverted personalities who turned to the solitude and beauties and challenges of nature to satisfy a need that human association could not assuage. (The type, which stands stress well, has provided many of our most remarkable wilderness men.) He pondered problems of politics and human fate while patrolling alone with his eye cocked for grizzly bears, new species of plants, and hostile Indians. Though solicitous about the welfare of his men, he lived in the recesses of his own personality and had no deep interest in people. His early death — perhaps by suicide, more likely by murder — cut his career short not long after the triumph of the expedition. It would have been either a distinguished or a tragic career. The highest level of diplomacy would have fulfilled him; otherwise he would have been a frustrate man unless providence had sent him war service or another exploration.

Clark was his complement, a native rather than a seeker of the wilderness, an extrovert, genial and outgiving, untroubled by speculation, uninterested in and probably incapable of abstract thought, a man of many skills, naturally and deeply interested in people. He was a craftsman and especially a wilderness craftsman, a born frontiersman. His terse prose, with a virtuosity of spelling that charms all readers, shows the directness and concreteness of his mind. If Lewis had talent for geography, Clark had genius for it. He had a similar genius for handling Indians — he liked them and he understood the primitive mind. (Though chance entrusted to Lewis the most important Indian negotiation, the first meeting with the Snakes, and he handled it superbly.) And Indians liked him. The expedition gave the Plains and Coast tribes a durable admiration for "the Red Headed Chief"; he was a legend to them from then on. They talked about him at council fires as a friend and their hope of justice and protection; throughout his

life Indians coming to St. Louis sought him out as a friend.
In fact everyone coming to St. Louis sought him out; growing
portly and dignified, he became the town's first national and
international figure. His business ethics were those of his business, the fur trade, but as official and unofficial commissioner
of Western Indian affairs all his life, he notably served the
United States and Indians got from him more understanding
and more help than they ever got from any of his successors.
He had a happy life.

They were remarkable men. And in the end, allowing as
much as need be for unusual leadership, resourcefulness, boldness, and skill, the quality that must be insisted on is intelligence. To read the *Journals* is to realize that no one who went
up the Missouri before them was anywhere near so intelligent,
and that of the westering Canadians only Alexander Mackenzie
and David Thompson had any of their intellectual distinction,
and they in smaller measure.

Jefferson had originally proposed a party of "ten or twelve"
men, which was a bit like Dearborn's belief that Lieutenant
Armstrong and a noncom could explore the Missouri. So few
could not have taken even the keelboat up the river and at St.
Louis Lewis would have been convinced by veterans, even if
by that time he had not himself decided, that a larger party was
necessary. The "Perminent Detachment," which Clark listed
on April 7 1805 when they left the Mandan villages, numbered
twenty-five. Besides these there were the captains themselves,
the hunter and interpreter Drewyer (Drouillard), the interpreter Charbonneau and his wife and infant child, and the
exceedingly useful York, Clark's Negro slave. All the permanent detachment except Charbonneau and his family and
one man enlisted at the Mandan villages were original members of the exploring party at St. Louis; there were, besides,
two privates who were sent back from the villages, one for
desertion, one for insubordination. They traveled on the keelboat, which was navigated by the expert riverman Cruzatte and
his understudy Labiche.[1] In addition a detachment of seven
soldiers commanded by a corporal took a six-oared pirogue,
painted white, as far as the villages, and a party of either eight
or nine rivermen was hired to crew a somewhat larger pirogue,
painted red. Both pirogues had supplementary sails. The
original intention was to send both the boatmen and the bor-

rowed soldiers downriver again in 1804, though in the event the soldiers and some of the boatmen spent the winter at the villages.[2]

Ordway was the most useful of the sergeants; the captains made him next in command and always assigned to him the most important of the jobs they could not themselves supervise. Drewyer was invaluable from the beginning, a magnificent plainsman, riverman, hunter, and scout, a skillful trapper, a master of sign language; he had great courage, versatility and initiative. With him begins an honorable line of French frontiersmen in the western United States; it was to run through the era of the fur trade and into that of the emigration, and to include such men as the Astorians' Dorion (who will be seen presently), Parkman's Chatillon, Stewart's Antoine Clement, Frémont's Basil Lajeunesse and large parts of the rosters of the American Fur Company and the Rocky Mountain Fur Company. Pierre Cruzatte was blind in one eye and could not see very well with the other but he was a superb riverman and hunter and adept at the other frontier skills. The blacksmith Shields and the versatile tinkerer Bratton also stood out from the beginning. Clearly Pryor was a good sergeant and so was Gass, appointed after Floyd's death. Of the rest of the party the captains came to rely most on Shannon, Gibson, Goodrich, Colter, and the two Field brothers.

There were two horses for the hunters when the expedition started. A stray was picked up later but soon died. Add, finally, Lewis's big Newfoundland dog, Scammon.

☙

They learned the river in its easiest stretch, across Missouri, and crossed it at its most beautiful, in May. The boats could be sailed for long stretches and as compared with the middle and upper river there were fewer bends and therefore fewer "crossings." . . . As the current comes into a bend, it strikes the convex side. This turns it — and the channel with it — across the river, where it strikes the concave side and begins to undercut the bank. A shoal and eventually a bar build up on the convex side, which thus straightens and shortens the stretch a little. As the concave side is undermined, it moves downstream a little. Thus every bend is always moving downstream and the river is always shortening itself. One of the innumerable skills of piloting is to identify the crossing, the place where the channel crosses to the far side, with shoal water covering a sandbar just downstream from it.

But the Missouri is seldom easy and can never be trusted and though this stretch looks sluggish it has a powerful current. There was much rain in May and June. Sudden cold alternated with spells of humid, windless heat. Violent squalls struck, the river was a fury of whitecaps that hissed and sometimes roared, the boats were driven toward banks or islands and had to be held off them by frenzied manpower. Where the channel was a "chute" back of an island and where bluffs narrowed the river, the current had to be fought head-on. Staggering knee-deep in mud or waist-deep in water or crashing a way through brush, they dragged the antic craft with tow-ropes. They learned about snags, the mats of driftwood called embarras, sudden surface "boils," crumbling banks and disappearing islands, and the endless sandbars on which the boats grounded, threatening to overturn, and off which they had to be dragged or pried or cursed. They learned too about the snakes, the ticks, and the maddening mosquitoes. But it was a lush life, with the hunters bringing in game beyond their needs. And the captains' notebooks filled with memoranda of the kind required of them: flora, fauna, river courses, topography, astronomical fixes, and data — some of it erroneous — about the country to the west. This last came from their four or more veterans, from the boatmen, from Mackay's map, and perhaps from cruder maps they had found or drawn in St. Louis.

By the middle of June they had met eight parties of traders coming downriver. The most important was the one they met first, May 25, which was headed by Regis Loisel, the inheritor of the upper Missouri trade in succession to the bankrupt Missouri Company. In a series of tangled partnerships he had continued the effort which D'Eglise had begun and Truteau and Mackay continued.[3] He was the proprietor of a post on Cedar Island, just above the Grand Detour, where the trading party for Santa Fe may have spent the preceding winter. (He did not mention them now.) He was returning from the post, after at least his third trip to the upper river. He was the most valuable source of information whom the captains had yet met, but even more important was his clerk Pierre Antoine Tabeau, of whom they had heard in St. Louis and who was currently being bullied and robbed by the Arikaras.

Loisel had had to pay the usual tribute to the tribes below the Sioux — Omahas, Poncas, Otos. St. Louis traders had always acquiesced in this routine piracy, counting it part of the overhead. But the Sioux, who were in increasing numbers on the Missouri and pushing farther up it and farther west, had stopped him cold, as they had stopped D'Eglise and Truteau. They stole his horses and his goods, set ruinous prices for those they condescended to buy, incited the neighboring tribes (meanwhile massacring them) against him, and would not let him go farther, even to their own relatives. "They hope," Tabeau wrote, "themselves to trade the peltries [of the middle-river and upriver tribes] by an intermediary commerce to which they are accustomed, and to sell to us then with profit."[4] Nevertheless, in 1803, by a combination of tribute and stealth Loisel had partly broken their monopoly. He had pushed his associate Heney as far north as the Cheyenne River. (Heney had been his partner and perhaps was still; he represented Canadian capital and perhaps the North West Company; at some time in 1804 he went on up to the Northwesters' post on the Assiniboine River, apparently breaking his association with Loisel.) While Heney traded with the Sioux there, Loisel and Tabeau, who like him was an old Northwester, had taken a sizable party still farther up the Missouri, erected a post, planted a vegetable garden, and traded with the diminished Arikaras. They hoped soon to go all the way to the Mandans, which would have distressed the James River traders by making full connection with the North West Company. After wintering at Cedar Island, Loisel had dispatched Tabeau to the Arikaras and had himself started for St. Louis.

On June 26 the expedition reached "the great river of the Kansas," forty-four days from Wood River, and turned north at last. The narrower bottom remained choked with willows and cottonwoods but now the land beyond it was mostly treeless. On July 21, sixty-nine days out, the Platte hurled its chocolate-colored flood at them through half a dozen channels. They had reached the equator of the Missouri and Clark made it six hundred miles. He was figuring well; it is 611 now. Drewyer knew and Mackay's map showed that some Pawnee villages were not far up the Platte and that the Otos, whose belligerence had been reduced by smallpox and the Sioux, had withdrawn to the same vicinity. They sent Drewyer and Cruzatte to summon them to a council and went on to look for a good site. They found one fifty miles farther on and called it Council Bluff.[5] They guessed that it was twenty-five days from Santa Fe, which would be important if the route to that commerce were to lead up the Platte. They thought it the best place for a Platte River post, which they regarded as inevitable. They were right; in the next twenty years nearly a dozen would be built in the neighborhood and one jump-off of the Oregon Trail would head west from Plattsmouth.

They met their first Indian since a group of Kickapoo hunters encountered soon after the start; he was one of the few remaining Missouris, de Bourgmond's adopted people, who had moved inland and joined the Otos. They sent him and the unexplained Liberté to second the invitation Drewyer and Cruzatte were carrying. The summer buffalo hunt had taken most tribes deep into the plains and not many Otos were about but six minor chiefs and seven others came in. (Well-behaved: many lickings had chastened the once belligerent Otos.) It was a disappointing assembly but they had a job to do and a model for councils to set. They commissioned three of the delegates tribal chiefs in the official view of the United States, filling in the blanks that had been printed for this purpose. They handed out one medal of the "second Grade" and four of the third grade — keepsakes intended to insure the recipient's loyalty and give him distinction among his people. They told the Otos and Missouris that the Great Father lived in Washington now and desired them to stop raiding other tribes and make a general peace. Then they tipped them with presents and sealed the President's message with a bottle of whiskey.

Liberté had not returned and a private named Reed went back to the last campsite to find a knife he had left there. Soon it was apparent that neither was going to reappear. Liberté was an *engagé* but Reed was an enlisted soldier and desertion was a capital offense. So they sent Drewyer and three others to round them up and if Reed "did not give up Peacibly to put him to Death." The assignment would take them to the Oto town again, where they were to invite any chiefs of higher rank who might have returned to pay a visit. They were to come back by way of the Omahas and invite them to come in too and have their deplorable war with the Sioux settled. As far back as June 12 they had met two rafts coming down from the James River Sioux and had hired their proprietor to turn back as an interpreter. He was Pierre Dorion, a casehardened St. Louis trader who had been going to the Mississippi and Des Moines River Sioux since 1781. He has not much pleased modern chroniclers but that he successfully handled Sioux so long and was not afraid of them says a great deal. One reason for hiring him was Jefferson's suggestion that various chiefs be sent to Washington to be impressed and they hoped some Sioux would go.

Above Council Bluffs the bends were shorter and more numerous, there were even more sandbars, the hills beyond the bottom were bigger. Above the mouth of the Little Sioux River they visited the mound which had been built over the grave of the celebrated Blackbird, who would intimidate no more traders. He had died in the smallpox epidemic of 1802 that had reduced the Omahas to nonentity. They passed the site of Mackay's winter post, Fort Charles, halted to wait for the posse, and sent another summons to the Omahas. Labiche came in on the evening of August 17, reporting that they had some Oto chiefs and had taken Reed but that Liberté had escaped after being captured. The next day the others arrived with the prisoner and eight or nine Indians. They tried Reed, sentenced him to run the gauntlet of the whole party four times, and declared him henceforth on sufferance only, no longer a member of the expedition. Sentence was executed in the presence of the shocked Indians, who believed that flogging violated the integrity of the individual. (As the humors of torture did not.) In two days of council the Otos were told that their distresses were their own fault — they had stolen horses from the Omahas and corn from the Pawnee Loups. They certainly knew that stealing always brought reprisals: stop it and join the general prairie truce that had been so hopefully decreed.

Following a distribution of presents, one Oto made an experiment. He returned his imposing certificate and sulked: the Americans were stingy. It didn't work. The Americans were not intimidated, and one of the dissident's friends hastily asked for the return of the paper. "We could not give the Cerft. but rebuked them verry roughly for haveing in object goods and not peace with their neighbours. this language they did not like at first" but they quickly came round. First lesson to the upper river tribes: you can't scare these men. It sank in and began to travel along the prairie telegraph.

That morning the young second sergeant Charles Floyd was stricken with a sudden and severe attack of "Biliose Chorlick." Three weeks before his diary had recorded, "I am verry Sick and Has ben for Sometime but have Recovered my helth again." The earlier attack and the symptoms Clark reports at its recurrence suggest an infected appendix which had now ruptured. The boats started out the next day but toward noon landed again for he was dying. They buried him and "put a red ceeder post" to mark his grave, branded with his name and the date, August 20. Lewis read a funeral service over the grave, they went on, and they named the creek they camped beside that night Floyd's River.[6]

Liberté had succeeded in his desertion and Floyd was dead. Before they built the winter post another private would be discharged for insubordination. After that the expedition was to suffer no further losses.

❧

Four days later, after passing the mouth of the Big Sioux, they paused to investigate an isolated, mesa-like hill some miles from the river. Drewyer, Cruzatte, or some boatmen had told them that the local Indians thought it inhabited by "Deavels" who were "in human form with remarkable large heads and about 18 inches high," bad-tempered and fond of killing Indians. Here were the upper Missouri dwarfs, rumors of whom had reached the French in the Illinois a century earlier and at intervals ever since. On a day so torrid that the dog Scammon suffered a heat stroke, they went to investigate but found no little men.[7]

Patrick Gass was appointed to the vacant sergeancy and as they reached the mouth of the James River, an Indian came swimming out to the boat. A Sioux at last. He said that his village was camped some miles up the James. This meant work for Dorion and they sent him and Sergeant Pryor to call a council, the party itself going on to a likely campsite marked on Evans's map, which they were now using.[8] (It began at Fort Charles.) Late afternoon two days later Dorion rejoined them with his halfbreed son and about seventy-five Yanktons. The son was the Pierre Dorion who would marry the famous

Marie Aioë and in 1810 would go west with the Astorians and meet his death in Oregon.

They met not only the Sioux as such and not only the Sioux as the custodians of the Missouri River. These were also the first Indians they had met who had the Plains culture in full vigor and the first whom they were directed to alienate from the British trade. (But they were the comparatively amiable Yanktons and, as they truthfully reported, the only Sioux hereabout who traded with St. Louis — one Dorion was representing the Chouteaus.) There was a furious questioning and note-taking for two days — personal appearance, dress, arms, dances, range, relatives, alliances, wars, a vocabulary. It would affect the captains' thinking from now on, would be codified in the first report to Jefferson and summarized in the "Statistical View." The council itself was in full ceremony but without melodrama. The Indians staged a dance and the whites showed off their marvels, including a patent airgun and their best performer, York. In their answering orations the chiefs said that the Spanish had been niggardly with presents and that the new Father's representatives did not impress them as lavish, either. We are poor and need everything, we want traders and more presents, we want to stop the next boat that comes up the river. If the last was a threat, it didn't work. They courteously agreed to make peace with everyone and to send a delegation to call on the President. The captains left Dorion to conduct the embassy to Washington, gave him a drawing-account of presents, treated the Indians to some American milk, and went on.

Fair enough: the Yanktons had proved tractable. The Tetons, they reported, were on Bad River, which appeared on Evans's map as the Little Missouri. (Bad River, the Teton, is in South Dakota, the true Little Missouri in North Dakota.) As the boats crawled up an increasingly "sholey" river Clark sketched what he took to be "ancient fortifications" — terraced, barren bluffs in the increasingly arid South Dakota. (There had been cactus for some time and prairie dogs, badgers, coyotes, larger buffalo herds.) He also went to look for the volcano but didn't find it.[9] On September 4 they passed the mouth of the Niobrara, on the 8th. "the house of Troodo where he wintered in 96. Called the Pania House." (Truteau, and Clark was misreading "Ponca House" on Evans's map.) Just beyond White River, which they passed on the 15th, is Pine

Ridge, the true geographical boundary between the middle and the upper Missouri. So they came to the Grand Detour, which they made thirty miles by water and two thousand yards by land across the neck, and on to Cedar Island. Loisel's post (from which, possibly, James Pursley had jumped off for Santa Fe last spring) was deserted but lots of Sioux had recently been camping there. The next day, September 23, three boys swam across to the camp to announce that 140 lodges of their people were not far away. On the 24th they went on to camp just above Bad River, which Clark had renamed the Teton, and that night the first Teton Sioux came in.[10] Military alert: only one-third of the party slept on shore; the rest were on the keelboat, prepared for action. They badly needed Dorion, for the two men who spoke Sioux did not speak it well.

There was forthright action on the 25th. About fifty or sixty Tetons had checked in, headed by Black Buffalo and The Partisan. They had repeatedly roughed up Loisel, and Tabeau considered The Partisan the worst villain on the river. They straightway did their stuff: it had worked with Truteau, Evans, Loisel, and Tabeau — why not with these newcomers? Given their fine red coats, cocked hats, feathers, and tobacco, they announced that this was not enough. They had to have more, they said, and if the Americans wanted to go farther upriver, they had to leave one of the pirogues and its lading with the Sioux. Unimpressed, Lewis invited the two principal chiefs and three others aboard the keelboat, where the routine performance with the patent airgun was repeated and everybody got a slug of whiskey, Clark says a quarter of a glass. It was not enough to get them drunk but they pretended to be and got ugly again. Clark hustled them ashore in a pirogue. One of them refused to get off. Three young men ran down from the barbecue on the bank and took hold of the towrope. The Partisan shouldered Clark and repeated that the presents were cheapjack and that the boats could go no farther.

He was mistaken. Storm-trooper tactics were not going to work for the Sioux had jumped the wrong man. Just this had been foreseen and allowed for, and though the expedition had been charged with a peace mission it would take no bullying. Clark had his sword out, Lewis ordered the three swivels trained on the shore, and Black Buffalo, The Partisan, and a number of other very tough Indians found upward of thirty rifles aimed at them. Of course there were enough Sioux to have annihilated the expedition right there but no Indians would pay the price. They had their bows strung, they had Clark surrounded, and it hung that way for a moment or two. A large stake in American foreign and domestic policy hung suspended too. Clark ordered the pirogue back to the boat, meanwhile answering the chiefs' ferocious threats in kind. "I felt My Self warm," he writes, "& Spoke in verry positive tones." The pirogue came back and twelve infantrymen jumped out of it, their rifles cocked. That was enough. Black Buffalo took the towline from the young men and began to beg: the women and children were so naked that surely the rich captains would spare some clothes. Then he and Buffalo Medicine and two braves waded to the departing pirogue and asked to be taken aboard the boat. They were, the boat moved upstream to an island, and peace broadened down under a heavy guard. The guests may have slept that night but Clark did not try to.

The Sioux had good hearts the next day and tried a different play. The boats again moved upstream a few miles and landed near the main Sioux camp. (Between two and three hundred, Ordway says; if so, then the earlier count by lodge had been wrong, for it would have run to more than six hundred.) Here the Sioux staged their Very Important Person ceremony, first with Lewis, then with Clark. They filled the day and the evening with pageantry. There was a dog feast with pemmican and stewed prairie apple on the side, followed by a scalp dance. The Sioux were as hospitable as might be; "peacable & kind," Private Whitehouse makes them. But their kindred were arriving in quantity and during the day Clark learned that the village held some twenty-five Omaha "Squars and Boys" who had been captured on a big raid a couple of weeks ago. Nearly as many more were with another band and they claimed to have killed seventy-five Omahas, plus a scattering of children. Lewis and Clark both told them that they must report to Dorion, turn over the prisoners to him, and make peace with the Omahas. They had Cruzatte, who spoke Omaha, distribute some trinkets to the prisoners.

One of the Omahas told Cruzatte that the master race were preparing to stop the white dogs where they were. They rejected the first lesson as not making sense; they had met no white dogs along the Missouri who would stand up to them. So the party set itself for trouble through another day of hospitality, oratory, and feasting. Notes on the ethnology of the Plains Indians accumulated in the codices. More Tetons arrived from the Bad River villages. That night there was another dance. Everybody was sleeping on the boat, which was anchored a hundred yards from shore. The Partisan and another chief announced that they wanted to sleep on board again. The steersman of the pirogue that was ferrying them and the captains to her ran afoul of the cable, and Clark's shouted orders brought a couple of hundred Sioux down to the bank. The chiefs helped out by informing them that the Omahas were making an attack. (They were about as afraid of Omahas as of Eskimos.) There was some carefully edited gunfire and the usual howling, and some sixty of them patrolled the bank all night watching for a chance they didn't get. The Americans insisted on keeping the drop on them.

They made another try the next day but it was halfhearted. The American no-appeasement policy had taken the starch out of the Sioux. On September 28, after a fruitless effort to recover the anchor which had been lost in the accident the night before, it was time to get going. A delegation of chiefs, including The Partisan, were in the cabin with Clark, making a stately plea for the expedition to stay on till additional gentle Tetons could arrive and view its wonders. Two hundred braves, armed with muskets, bows, spears, and "cutlashes," were parading the bank to which, lacking an anchor, the boat had been moored. When told to go, the chiefs said they were in no mood to go. The river was closed: "They Sayed we might return with what we had or remain with them, but we could not go up the Missouri any further." Captain Lewis, 1st Inf., had had enough. He ordered the oarsmen to their oars and the rest of the command to battle stations. When a private tried to cast off, some young bucks grabbed the rope. Lewis ordered the chiefs to clear out and again there was a moment. The young men made a dash to get the squaws and children out of the way and, Gass says, "Captain Lewis was near giving orders to cut the rope and to fire on them." Whitehouse adds that Lewis was about to cut the rope with his sword. Clark "took the port fire from the gunner" — prepared to fire a swivel loaded with scrap-iron into the crowd point-blank.

The Sioux quit. They would settle for a little face. Black Buffalo, the head chief, who was still on board would have been the first to depart into the spirit-land. He took the tow-rope from his bully boys and would permit the expedition to get started if paid a carrot of tobacco. Lewis said the hell with him — and Clark made it an insult by throwing the tobacco in his face. The episode was over, the boats got under way, and the Sioux were just beggars again. Parties of them followed along the bank pleading for tobacco and calling the Americans stingy. For two more days various chiefs and delegates came aboard or yelled at them from the bank: stop and see us, we are peaceful and lovable folk, we need a lot of tobacco — the terrors of the Missouri River had been deflated. They all got the same answer: the next stop would be the Arikara villages, the Sioux could have peace or war as they might see fit, and if they had anything more to say, look up Dorion and tell him. Even The Partisan begged to be given a boat ride, to camp with the party at night, to exhibit his white friends to the rest of the Sioux nation. And on October 1 everybody got a drink to celebrate a triumph of foreign policy.

For the Missouri was now open and, so far as the Sioux were concerned, would stay open. On the prairie telegraph tidings that a new era had opened on the upper river traveled ahead of the boats. It reached the Arikaras before the expedition did and sped on to the Mandan villages: Lewis and Clark had made women of the Teton Sioux. The first climax of the expedition had come in those last days of September. It was an extremely important climax, accomplished one of the ob-jectives set for the expedition, and had been won without bloodshed or expense. American authority had been asserted over the Indians of the Missouri.

৫৵০

The effect was evident at the Arikara villages — three of them above the mouth of Grand River — less than two weeks later. The Arikaras treated them with "everry civility" and Clark pronounced them "Durtey, Kind, pore & extravigent." Everybody, in fact, thought them charming Indians. And the grand chief "informed us the road was open & no one dare Shut it, & we might Departe at pleasure."

Clark's adjective "kind" and the chief's remark must have

made strange hearing for Loisel's agent Tabeau, whom they
now met. The gentle, charming Arikaras had bullied him re-
morselessly through the spring and summer, and less than two
months before had climaxed their campaign with a forced levy
on his goods that wiped out Loisel's profits. . . . The captains had
met and questioned various of his employes. One of them,
Joseph Gravelines, they hired temporarily as an Arikara interpreter and would hire again the next spring. With Gravelines
it became fully evident that the upper-river trade from St. Louis
was the British joining hands, for he was from the Assiniboine
River and, like Tabeau, was a veteran Northwester. They did
not meet Loisel's partner or former partner, Hugh Heney, till
six weeks later; when they did he was representing the North
West Company.

So the short stay with the Arikaras was *gemütlich.* Since
everything had been on a war basis with the Sioux there had
been no chance for the gentler pleasures, though after the crisis
one band did send some squaws to solicit loudly from the bank
for two days. The Arikaras had the freest possible sexual customs and the soldiery made the most of them. York here established the amorous pre-eminence that he was to enjoy from
now on, and must have been a very tired Negro when the boats
got under way again. His vaudeville act also reached headline
status and was to retain it. The Indians were forever exam-
ining this novel black man, feasting him, and making assigna-
tions with him. Children flocked round him but ran screaming
when he roared. He "made himself more turribel than we
wished," Clark says and noted an astonishment: the Rees did
not want whiskey.[11]

The expedition had reached the upper river at a time of
many wars. The Sioux had not yet beaten down the Mandans
or their neighbors and affiliates the Minnetarees. In fact they
were frequently getting the worst of it from them still. But
their raids had got them in between the Arikaras, near the mouth
of Grand River, and the Mandans at Knife River, which was
strategically disadvantageous for the Arikaras. The Mandans
fought back and the more bellicose Minnetarees were still contemptuous of Sioux. It was the Rees — the Arikaras — who
suffered. They could not get guns and powder enough from
Tabeau — he did not have enough — or from the Sioux, who
were their trade masters. The Sioux, who needed their corn,

were alternately at peace and at war with them, trading them a few guns at high prices, stealing their horses, and periodically collecting a few scalps just to keep their hand in. The Rees had also lately taken a beating from the Crows, who were in these parts on their annual trade visit to their relatives the Minnetarees. And in July a wife-stealing episode had broken the truce they had for some time maintained with the Mandans and Minnetarees, who had been raiding them ever since. It had dawned on the Rees, and was beginning to dawn on the Mandans and Minnetarees, that this was injudicious, since it helped the common enemy of all three tribes, the Sioux. The peacemaking services of Lewis and Clark were solicited and the Rees sent a chief along with them to make a treaty.

Lewis and Clark met some Cheyennes at the Arikara villages. (Evans had met them there too; Truteau had trekked out toward the Black Hills to find them; possibly Mackay's trip through western Nebraska had had the same purpose.) With them the process of getting information about the Far West began. And in Tabeau and his associates they met the white men best qualified to give them information. Some of it was wrong, some misconceived. In particular almost all the distances were exaggerated and what they said about the headwaters of most streams, about all the long ones, was guesswork. A misconception about the Yellowstone that began here did not get cleared up even when Clark traveled down it. Nevertheless, from this point on the *Journals* are amassing a greater amount of reliable data about the American West than anyone had ever had before.[12]

They had learned about the regular visits of the Cheyennes to the Arikaras and of the Crows to their cousins the Minnetarees. These visits were important in the intertribal trade which reached all the way to the Rockies, and the information about that trade which Tabeau and his men could supply was as important as the geographical data. They named ten tribes that the Arikara trade reached. The names of some were familiar from the Arrowsmith map, which the captains had with them, and included the Arapahos, the Kiowas, and the Pawnees.[13] Tabeau must also have told them something about the westward intertribal trade.

But the year was drawing in. Ever since they entered South Dakota the lavender and purple hazes of autumn had softened the bluffs that bordered the river. (Upper Missouri bluffs, in most places as regular as the teeth of a saw, usually with triangular faces, usually with flat tops, barren except for the dwarf cedars that look black at a distance.) The great flights had been going southward, lighting briefly on the river by the hundreds. The sudden rains were bitter cold, the wind had a honed edge, ice formed in still water at dawn, the clouds were lead-colored in always vaster masses, the gun-metal emptiness of the North made even sunny skies ominous. There had been two short, whirling snowstorms. Flannel shirts had long since been issued and lately the skins of the deer and elk killed for food had been going to the men, though they must have made uncomfortably stiff robes for there was no time for proper tanning. Clark had been knocked out by a sudden attack of rheumatism and several of the men had suffered the same affliction. In mid-September they had not known precisely where they would winter, though they must always have intended to make it near the Mandans. They reached the first Mandan village on October 25, after meeting a good many Mandans and some Minnetarees on hunting parties. On the 29th they held for both tribes the most impressive council they had yet staged. After a fruitless reconnaissance upriver Clark found a satisfactory site downstream, three miles beyond the lower Mandan town. There in a cottonwood grove in the river bottom, at the foot of a high clay cliff with scattered small bluffs leading to bigger ones beyond it, they built their winter post, Fort Mandan. Ice began to run in the river on November 13.[14]

❧

They were at "the Mandan villages," though the Minnetarees proved more important than the Mandans for the future. And they were in touch with the British trade on the Souris and the Assiniboine. That is the significance of the triangular structure of cottonwood logs called Fort Mandan: two rows of four rooms each set at an angle, two storerooms in the angle, a stockade closing the front, a sloping shed roof reared above the rafters to make the outer face eighteen feet high.

There were only two Mandan villages by now, about three miles apart. A mile upstream from the higher one, almost at the mouth of Knife River, was a small village of the people whom the captains called Wattersoons, Ahnahaways, Souliers,

and Shoes Men; the Amahami of ethnologists, they were distinguished from the Minnetarees only by a separate tribal organization. A mile and a half up Knife River were two populous villages of Minnetarees.

The Minnetarees were the ethnologists' Hidatsa, a Siouan-
speaking tribe, close kin of the Crows. They and the Crows, in
fact, had been one people till some not very remote time. Remaining semisedentary when the Crows migrated west, they
practiced agriculture and maintained a remarkably peaceful
association with the Mandans, sharing their middlemen's position in the trade and leagued with them against the Sioux. The
gesture that designated them in the sign language was rendered
Gros Ventres by the French and therefore Big Bellies by the
English. This is one of several reasons why they were frequently
confused by traders, and by historians, with a Canadian tribe
that lived above the south fork of the Saskatchewan River.
These last were an Algonquin-speaking people who were related to the Arapahos but had long since drifted far away from
them and were confederated with the Blackfeet and have appeared in this narrative as the Fall Indians. The Blackfoot
name for them, preserved in the ethnologists' designation,
Atsina, meant "gut people" and this too got translated as Gros
Ventres and Big Bellies.[15] The Atsina had almost as much belligerence and fully as much crabbedness as the Blackfeet; they
were a tough people and, like the Minnetarees, raided far to
the west. The Minnetarees too were tough — they enjoyed
fighting the Sioux — but they shared the Mandans' liking for
white men, as the Atsina assuredly did not. They were of
lighter complexion than most Indians, though not so fair as
the Mandans. Like the Mandans they lived in big, comfortable,
Sioux-proof earth lodges.

The René Jessaume who had squeezed out John Evans was
living with the Mandans and the captains promptly hired him
as an interpreter. He had continued as a free trader affiliated
with the Northwesters on the Assiniboine, and had been David
Thompson's guide and interpreter when he came to the villages in 1797. They also enlisted as a member of the permanent
personnel, to take the place of a private discharged for insubordination, one Baptiste Lepage whom they found here and
who had worked for one of the British companies and claimed
to have been to the Black Hills. The next day "a Mr. Chabonie"

## Map 29. Fur Company Posts

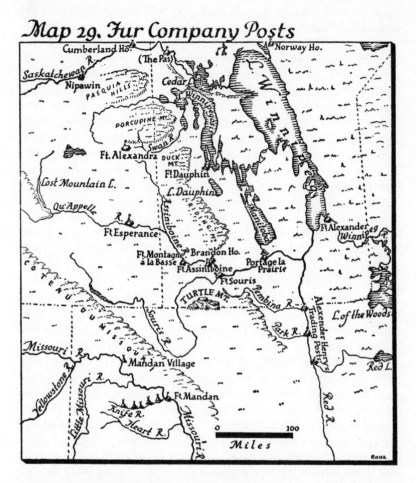

came to Fort Mandan and applied for a job as interpreter to
the Minnetarees, with whom he lived and traded as Jessaume
did with the Mandans. (He had been known to the British on
the Assiniboine as far back as 1793 if not earlier.) He was
Toussaint Charbonneau and he got the job. Presently he was
joined by his wives whom Clark identifies as "two squars of the
Rock mountains, purchased from the Indians." Sergeant Ordway goes further and specifies that one of them — she was
Sacajawea — "belonged to the Snake nation." As yet the fact
that she was a Snake meant nothing to Lewis and Clark.[16]

The hired boatmen built a dugout and some of them went

back to St. Louis in it; others stayed on for the winter to get some beaver and help out at the fort, where help was welcome. It was an active winter — and very cold. "Such frost I never Saw in the States," Ordway wrote. Upper Missouri weather, not much snow but arctic gales blowing up suddenly, followed by the long windless cold waves of the high plains that produced "false suns" — the sundog — and haloes round the moon. The river ice groaned, cottonwood boughs boomed like musketry, the guard had to be changed every hour — on occasion every half-hour. In "blanket cappoes," marveling that the Indians wore only a g-string and robe, the party met all weather but the worst head-on, making long hunting trips, cutting wood, visiting the villages. It took a lot of meat to keep them going but there was plenty at hand. Whitehouse lists a short hunt: "34 deer, 10 elk, 5 buffalo." The fare included porcupine (a delicacy), antelope, and, from the Indians, tallow, jerked meat, pemmican, and the corn, beans, and "persimblans" (squash) lavishly stored in the pits.

The winter must have been unusually snowy in the Rocky Mountains. An obviously unusual amount of water in several affluents of the Missouri, the next year, was responsible for important decisions.

There were always Indians in the fort, gaping, chattering, holding stately debates, recounting their brave deeds, trading corn and tallow, bartering their women, getting their ailments treated, buying metal work and repairs from the two smiths. There was no time for boredom, rumors of Indian raids added spice, and there were plenty of women — the Mandan girls more amiable and pleasing and, it is clear, more adept than the Minnetarees. The soldiery lived high — good food, warm quarters, Cruzatte's fiddle helped out by the bugle and a "tambereen." Cruzatte was a hit with the Indians, so of course was York, and they were much entertained by the dancing of the U.S. Army in the high north, by night before enormous log fires. It would be jigs and squares of the American frontier and the quadrilles of the Louisiana French; the American West's grave tradition of hoedowns with a bandana tied round your arm to show that your partner was to do the leading begins right here.

These were shaped and tempered men by now, not only a force but a company. (The one court-martial at Fort Mandan was the last of the expedition: Private Howard for scaling the gate instead of reporting to the sentry, and thereby inspiring an Indian to do the same. Sentence of fifty lashes remitted by Lewis.) The boys had been made men and soldiers and the soldiers had been made fit instruments for the unknown by the upriver voyage and the encounter with the Sioux. Wintering added the burnish and Lewis summed it up when they got started in the spring: "The party are in excellent health and sperits, zealously attached to the enterprise, and anxious to proceed; not a whisper or murmur of discontent to be heard among them, but all act in unison and with the most perfict harmony." . . . Except, as Clark conscientiously added, "except Venerials Complaints which is verry Common amongst the natives."

Nobody was ever busier than Lewis and Clark themselves. First there was the local peace they were triply instructed to procure. When they reached the villages the Mandans had just hijacked some furs from two of Gravelines's men — apologetically, with the diffident charm that was the Mandan politesse with whites. That had to be attended to and was: traps returned (presumably the furs too) and the Mandans said they had just made an impulsive mistake. They were on the verge of making another one, for the Americans might be bringing a pipeline to rich stores of trade goods. (As Evans had promised but failed to do, they pointed out.) So though the Minnetarees were their honored allies, it would be just as well to keep them from the Americans till they had the agency sewed up — and they told their allies that the captains were going to join the Sioux against them. That was easy: you visited the Minnetarees, smoked with them, and left a standing invitation to drop in for a drink.

More important was the Arikara peace. The Ree chief who had accompanied the boats to the village was presented with the prescribed ceremony and the Mandans explained that nothing was their fault. Their firm policy was to fight only defensive war but of course anyone could be pushed too far. They would greatly love to be able to take off their moccasins at night but that meant that the Rees would have to behave themselves. They sent a Mandan-Minnetaree embassy back with the chief and made another treaty. This pleased Lewis and Clark, who had not learned that the truces of the Plains tribes were even more explosive than those of the forest tribes.

They worked at this diplomatic problem all winter and so did Tabeau at the Ree villages. Unhappily the Rees were satellites of the Sioux, who could use them as guerrillas. At the end of November the Sioux attacked a Mandan-Minnetaree hunting party, killed some of them, and ran off with the horses, and the victims were sure there were Rees with them. There was another flurry of anti-Arikara feeling but it subsided in anxiety about the Americans. For Clark mustered a task-force for pursuit so promptly and with such an evident intention of making it all-out war that the villages were alarmed. Deep winter, they said, was no time for fighting, but the truth was they preferred the more honorable and less costly small raids. At the first of February the Minnetarees started singing to their buffalo skulls again — winter was letting up — but cooled down when directed to. Meanwhile the Sioux, still smarting from the humiliation they had taken and somewhat emboldened by a successful raid on Clark's hunters, notified the Rees and everybody else that, come spring, they were going to stage a universal massacre that would leave only Sioux on the Missouri.

The Arikaras were depressed and believed them. They had been paying too high a tax for occasional Sioux civility. So they fell in with Tabeau's plans and notified Lewis and Clark that they wanted a firm treaty. Their deputation arrived just before the expedition started west in April. After delicious orations, a tripartite alliance was worked out and the captains promised that Washington would back it to the hilt. Heading upriver, they felt confident that they had checked the Sioux. They hadn't, and they had underestimated the Arikaras. Pacifism was not a mood the Rees could feel for very long. This one lasted for about three years, then they were moved to take up the role on the Missouri that the Sioux never quite dared to resume after Lewis and Clark. The Sioux drift was farther west, to the Black Hills and beyond, to bigger buffalo herds and wider fields of activity, and the Arikaras developed into the worst of all river pirates. It was they, plus the independent operations of the Blackfeet farther west, who turned the American fur trade away from Jefferson's water route and the trail Lewis and Clark had blazed, and sent it pioneering up the Platte.[17]

At mid-November a big band of Assiniboins came to the

Minnetarees to trade for corn and the captains learned about
the Assiniboin function in the trade. The British companies,
which bought pemmican from them, had only in part undercut their position as middlemen to the river tribes. For a century their trade with those tribes had been periodically buttressed by brigandage and war, and this was an Assiniboin year
to howl. A Minnetaree raid on the Blackfeet of the South Saskatchewan had got shot up, and on the way home the raiders
salved their discontent by cleaning up some Assiniboins. That
moved the Assiniboins to do something they had not tried for
years: they cut off and killed a small North West Company
party. It was heady stuff and on their way to the Missouri villages they met a small Hudson's Bay Company party bound
to the same place and exacted a big tribute from it.[18] Several
times during the winter Assiniboins appeared at or near the
villages, stole horses and had them stolen back, swaggered char-
acteristically but failed to make a fight of it. Because of them
Lewis and Clark picked up information they might otherwise
have missed. They advised the villages to take such Assiniboin
bullying as they had to till the American trade, about which
they kept up a steady discourse while they worked out a basic
policy for it, could be organized. When some Cheyennes came
up from the Arikara towns, they identified another strand of
the tribal and trade network west of the Missouri.

They heard about the Sioux all winter long and got periodic
rumors that they were going to make war. Even so, they underrated the Sioux. Early in February Clark ranged out from the
fort on a nine-day hunt. His party killed more meat than they
could take back, so when he returned he sent Drewyer with
two horse-drawn sledges and three men to bring it in. To send
so small a party was the most thoughtless act of the whole expedition. Twenty-five miles below the fort they were jumped
by a big band of Sioux, who took three of the horses and the
knives of two men and a tomahawk, though they returned the
tomahawk and one horse in some kind of second thought. The
Sioux dared do no more and Drewyer rightly made his small
detachment hold its fire. Lewis took twenty-four men in pursuit
but caught up with nothing except the smoking embers of an
old hut where Clark had left some of the meat. In Indian terms
it was a glorious victory. The Sioux gained face and the Americans lost it — lost it in the estimation of the village tribes, who

had been careful not to join the pursuit too eagerly, and in those of the visiting Britishers.

It added up. The Teton Sioux were "the vilest miscreants of the savage race and must ever remain the pirates of the Missouri" until coerced into good behavior. So Lewis wrote in the notes sent downriver in April which were digested as the document Jefferson sent to Congress in 1806, "A Statistical View of the Indian Nations Inhabiting the Territory of Louisiana." It describes "the tameness with which the [St. Louis] traders of the Missouri" had submitted to Sioux extortion and plunder. Till the government should firmly regulate the British traders on the Des Moines and the James who supplied them and thereby teach the Sioux respect, "the citizens of the United States can never enjoy but partially the advantages which the Missouri presents." Lewis recommended force: begin by cutting off the trade entirely till the Sioux felt the pinch and realized that they were subject to the will of the United States.

This was an item in the developing Lewis-Clark trade policy. They had previously noted the desirability of a trading post near the mouth of the Platte for the middle-river tribes, where a cluster of them would appear. But the captains were thinking of a government "factory," which if it could not order and regularize the trade could keep traders somewhat in line by underselling them. During the winter they decided that the mouth of the Yellowstone, which they had not yet seen, was another key site for a post. (It was: the American Fur Company would build its pivotal Fort Union there.) They thought that the Mandans, Minnetarees, and "Wattersoons" would locate there if a post were built, for they would be much safer from the Sioux. So perhaps might the Arikaras, who were now only "the farmers and *tenants at will* of that lawless, savage, and rapacious race the Sioux *Teton*, who rob them of their horses, plunder their gardens and fields, and sometimes murder them without opposition." Such protection from the Sioux might also induce the Cheyennes to settle near the Yellowstone. That would bring additional trade to the projected post, probably the Kiowas and a big Crow band they called the Paunches, possibly other distant tribes whose trade they thought might extend as far as New Mexico.

Another thing about the Yellowstone post. It would concentrate all this trade on the Assiniboine flank of the British.

Their information was, too, that it would be on a fine water route to the Saskatchewan and so to the Athabaska country. A long way and a rich trade.

They had achieved an accurate idea of the Falls of the Missouri. They thought that still another post might well be built there. It might attract the trade of various bands and tribes about whom they could learn little. They specified one, the Fall Indians, who they said, making the usual mistake, were a little-known tribe of Minnetarees.

☙

There were further international matters. Even before they reached the villages, the captains had found Hugh McCracken with a party of Mandan hunters down from the Assiniboine, a free trader affiliated with the Northwesters who had accompanied Thompson and Jessaume in 1797 and was even then an oldtimer. Now when he started back to Fort Assiniboine (at the mouth of the Souris) they sent a letter with him to Charles Chaboillez who commanded it. Signed by both of them, it was a discreet document. They were wintering at the villages in the course of a Missouri River exploration, it said, and the inclosed copy of the British Minister's safe-conduct would doubtless assure them courteous treatment from His Majesty's subjects. They themselves would welcome and protect such of those subjects as might visit the Indians of this corner of the United States. The nation they represented did not intend to interfere with the free movement of foreigners throughout its territory. And they would be grateful for any information that would serve their mission.

Before this notification that the United States was traversing its frontier (and nothing said about the Columbia) reached him, Chaboillez had already dispatched François Antoine Larocque and Charles MacKenzie with a small party to the villages. The Montreal partners seem not yet to have notified him that Lewis and Clark were on the way, but did so soon thereafter. Larocque found the small Hudson's Bay Company party among the Minnetarees, prepared to recoup the tribute they had paid the Assiniboins by raising prices beyond the one hundred per cent markup above Assiniboine River figures always assessed against the villages. The Northwesters staged a miniature trade war with their rivals, from cutting prices to stampeding the horses, while learning what they could about the Americans. They had been sent to do some trading but after two weeks Hugh Heney arrived with orders from Chaboillez for Larocque to stay on there for the winter. That would be one result of McCracken's news that the Americans had come. Another was Larocque's request to join them on their western journey. The *Journals* say, "We gave him an answer," and Biddle has added "refused."

Everything was amiable. The Americans bade Larocque keep one of his *engagés* in line — the man had been disparaging them to the Minnetarees. They also directed him to give no medals or flags to his customers, who used to be Spanish Indians but were now American. Possibly they found MacKenzie's British loftiness a little galling (as he thought Lewis a surly Anglophobe) and when Charbonneau grew insubordinate toward the end of the winter they suspected, probably unjustly, that Larocque was responsible. But they freely lent him Charbonneau to interpret for his trade, herded the Northwesters' horses with their own, and kept cordial open house for them at the fort. They were careful, however, to assert the full extent of Jefferson's territorial claim, telling Larocque that the British posts were on American soil. His journal says that they claimed all the way to the Qu'Appelle but must be mistaken. Lewis worked it out in his notes that "the boundary of the United States would pass Red river between the entrance of the Assiniboin[e] and Lake Winnipic, including those rivers almost entirely, and with them the whole of the British trading establishments on the red Lake, Red River and the Assiniboin[e]." [19] They also told Larocque that the trade would be as free to aliens as to Americans, that there would be no exclusive licenses, but that the government might establish posts of its own hereabouts if private enterprise raised prices too high or demoralized the Indians. They were being less than frank. For the trade policy they were working out noted that the British "intend fixing a permanent establishment on the Missouri near the mouth of Knife river in the course of the present summer [1805]." They didn't like the prospect for "if this powerfull and ambitious company are suffered uninterruptedly to prosecute their trade with the nations inhabiting the upper portion of the Missouri, and thus acquire an influence with those people, it is not difficult to conceive the obstructions,

which they might hereafter through the medium of that influence, oppose to the will of our government, or the navigation
of the Missouri." [20]

Here was weighty information for Larocque to take to his
superiors. With uncharacteristic tact the North West Company
had mostly been working through free traders at the villages
while they were still Spanish, only occasionally appearing in
its own person, whereas the Hudson's Bay Company had been
coming every year. If the captains were telling the truth, the
handicap imposed by such carefulness need no longer be assumed. But their trade policy was less important than their
immediate plans. Larocque made a quick trip to Fort Assiniboine in February — on snowshoes and through brutal weather
— to get goods and to report. Chaboillez was not there but the
news that had called him away was. The feud between the
seceding Northwesters called the X Y Company and the parent
firm had ended, following the death of the despotic Simon
McTavish. The two organizations, which had been battling
each other with vindictiveness appalling even for the fur trade,
had worked out a merger. Though Alexander Mackenzie was
excluded from active management, his expansionist ideas would
immediately become policy. The end of the competition released energy for their execution.

So in April Chaboillez notified his Yankee subordinate
Daniel Harmon to prepare for an expedition this coming summer. With a small party and some Mandan guides he was to
explore westward from the villages to the Rocky Mountains
and was to open the trade there. This was what the company
had intended to do for years, what Mackay had reported to St.
Louis, and what Lewis's notes for Jefferson had just predicted.

As it turned out Harmon fell ill and Larocque, who meanwhile had gone back to the villages and told the captains about
the merger, made the trip instead, traveling the valley of the
Yellowstone while they were going up the Missouri.

☙❦

But the most important occupation of this winter was collecting information from the Indians, especially about the country
that must be crossed next summer. Lewis and Clark did an astonishing job; nothing in the history of the exploration of
North America by land can be compared to it. Their approach

to the problem, their attack, resembles that of the great Dablon, but no year of the Jesuit information service ever learned a fiftieth part as much as the Americans did in five months. If it be observed that the acquisition of knowledge was harder in the seventeenth century since the base that could be built on was so much smaller, one turns to the contemporary North West Company. David Thompson's surveys of individual routes, river courses, and boundary lines are indeed superb and his calculations of longitude are consistently more exact than Lewis's. Yet beyond what he himself has seen, his information is sparse, much less copious and more unreliable than what the captains put together.[21] They themselves corrected Fidler. In Alexander Mackenzie's book the thing immediately at hand is reported completely but beyond it there is very little else. And the North West Company had neither made an analytical and critical compendium of the material contained in the letters and reports of its posts nor sought systematically to extend its knowledge or to acquire many of the kinds of knowledge that the captains sought out. The results that Lewis and Clark achieved must be compared, and will stand comparison, with those of the naval expeditions — staffed by corps of specialists and furnished with all the men and equipment that might be needed. Indeed the intelligence working at Fort Mandan has precisely the qualities that made James Cook a great explorer. The era of knowledge of the American Far West begins with the specimens, artifacts, charts, tables, statistics, but most of all the notes consigned to Dearborn and Jefferson from Fort Mandan in April 1805. From Marquette and Jolliet through Mackay and Evans and on to Tableau is preknowledge.[22]

The least reliable information they sent Jefferson was that which they had got from the St. Louis traders, who had got almost all of it as they did, from Indians. That is, the newcomers appraised the Indian testimony better than the veterans had done. Most of this relates to the Black Hills, western affluents of the Missouri below the Knife, the upper reaches of the Platte, and the country south of the Platte stretching toward the Rockies and New Mexico. As a result, though they achieved the first working codification of the Plains tribes ever made, they repeated old errors about them and the country and added new ones.[23] They worked out a sadly inaccurate conception of the Black Hills, much exaggerating their north-south extent and overestimating their distance from the Missouri — and therefore the length of rivers that came down from them — and locating in them the sources of some rivers that do not rise there. They learned an extremely important fact, that the Platte has two principal forks, but they did not get the forks right. As a result they accepted the most serious error of their predecessors. Showing clearly on Clark's map, it represents the persistent failure to imagine the nature of the Rocky Mountains. Since the portion of the Rockies on which the error turns is the gigantic Front Range of Colorado, this in itself is enough to prove that none of their authorities had talked to — or at any rate understood — Indians who knew what they were talking about.

In one of the rectangles of his map, which show one degree of latitude and two of longitude, an area roughly sixty by one hundred and twenty miles, Clark shows headwater branches of the South Platte, the Arkansas, and the Rio Grande and they have no mountains separating them. In the next rectangle west he shows other headwater branches of these three rivers and by his own wildest misconception brings the Yellowstone River south till it is only some ten miles from the Platte and twenty from the Rio Grande. And in the rectangle next west of that one the Rio Grande, the Yellowstone, and the Colorado approach one another, though here a mountain range does come in between. (But no mountains separate the Colorado from the Yellowstone.) [24] This is a sum of old geography, part residual Delisle, part Le Page du Pratz, part St. Louis rumor, part deduction, part extrapolation. Some of the Arkansas and more of the Red River (of Louisiana, which barely enters Clark's map) would be improved on by Pike. But not much improved. Nobody so far had had accurate ideas about this country; nobody would have very accurate ones till American trappers began to explore it twenty years later.

At the villages they had no trouble working out the relationships of the Red, Assiniboine, and Saskatchewan Rivers and Lake Winnipeg. Their main effort was to determine what kind of country lay ahead of them and what routes they should take in it. They questioned all comers but of course mostly the Mandans and Minnetarees. Through many evenings before the big log fires at Fort Mandan and the smaller ones in earth lodges the Indians drew charts with a stick on the ground or with a piece of charcoal on a hide, heaping sand to make mountains, sketching canyons on the air with vivid gestures. It is certain that the more warlike and more wide-ranging Minnetarees told them most.[25] They learned that the Minnetarees made frequent western forays, often all the way to the Rocky Mountains and sometimes, they said, even across the Divide. They went primarily to steal horses but also to trade with tribes on the way and to pick up some scalps. What Lewis and Clark learned about the country beyond had been gleaned on these expeditions.

Now they acquired a fundamental idea, one which had not been determinable even from Alexander Mackenzie's book though James Mackay had got an inkling of it: that the Rocky Mountains were multiple, that there were several ranges. (Arrowsmith, with Thompson and Fidler as well as Mackenzie to draw on, made them a single ridge.) They counted on three ranges through which they would find the Missouri had cut its way, then a fourth range. The fourth would be the Continental Divide and just at its western base they would find a big northward-flowing river. And now the Great Falls of the Missouri came out of rumor — or out of legend. If the stories which St. Louis traders had heard referred to the Great Falls at all, which is not certain, they now lost their fabulous quality — the river no longer slides down an immense cliff. (Even so, the captains exaggerated their height, conceivably because the Indians insisted so on the noise of the waterfall, and heard about a nonexistent passageway back of the water's arc.) And another fundamental fact emerged: the Missouri divided into three forks. The northernmost of these would lead them to the range that carried the Continental Divide. And here — how cheering is divine philosophy, or how strong is the hold of a fixed idea — here they would reach Jefferson's, and Coxe's half-day portage: "The Indians assert that they can pass in half a day from the foot of this mountain on its East side to a large river which washes its Western base." This river, they concluded, must be the south fork of the Columbia. No one before them had supposed that the Columbia had a south fork. But rivers flow into other rivers, and they flow into rivers that flow into the sea.

At last, one hundred and thirty-one years after Jolliet and Marquette, the white man was getting correct information about the farthest reaches of the Missouri River, and about the route to its source that many expeditions had been ordered to find. And the white man was still, in part, misconstruing it. For Lewis and Clark were given two sets of facts. They blended most of them together and made one set. The rest they thought unimportant and therefore disregarded until the significance of what they had been told forced itself on them in the field.

In September 1805 they crossed from the Bitterroot Valley to the Clearwater River, starting that traverse by going up Lolo (Traveller's Rest) Creek, which is a few miles south of Missoula, Montana. The trail they took west there was the middle one of three trails which the Nez Percé Indians used to reach the plains where they hunted buffalo. During the winter the Minnetarees had described two different ways of getting to this trail from the Great Falls. But Lewis and Clark appear to have blended most of the facts they were told and to have made a single route of them.

There were two northward-flowing rivers west of the Continental Divide and the Minnetarees told them about both. One was the Bitterroot, which flows north to Clark's Fork; the other is the Lemhi, which flows north into the Salmon, and the Salmon itself for a space thereafter flows north. That Lewis and Clark concentrated on the second one appears to have been due to a combination of reasons. The knowledge of the country which the Minnetarees possessed stopped short with these rivers. The valleys through which they flow are similar and therefore cursory descriptions of one could easily be worked into a mental image of the other. When a coherent image has been formed there is a tendency to reject data that do not accord with it, especially when they are vague, and the details translated from an Indian language or crudely sketched on an Indian map could easily be rendered vaguely. Anyone who had decided to go on from the Great Falls to the Three Forks would tend to assimilate to that route whatever information could plausibly be assimilated to it, and would tend not to explore the significance of unassimilable information that dealt with the other route.

That would appear to be what happened. The first route, which the Minnetarees described, the one which the captains disregarded, led from the Great Falls up Medicine (Sun) River or up Dearborn River to Cadotte's Pass or to Lewis and Clark's Pass, and from there to the Blackfoot River, down the Blackfoot through Hell Gate, and on to the mouth of Bitterroot River. The other led on up the Missouri from the Great Falls to the Three Forks and up the Jefferson River. From the Jefferson there was a trail to Lemhi Pass and down to Lemhi River and the Salmon. From the Jefferson also there was a trail to Gibbon's Pass and on to the Bitterroot River, but that the Minnetarees told them about this does not appear. (They learned about it from the Snakes, after the Salmon proved to be not a feasible route.) At any rate, they concentrated on the route to Lemhi Pass and they attached to it some data that belonged to the first route. In the outcome they were occasionally puzzled and occasionally altogether deceived. When, after the long circuitous detour that ended with frustration at the Salmon, they came down to the Bitterroot River, they could see the significance of what they had disregarded.[26]

The Yellowstone River came out of the fog. They learned most of its tributaries — all the major ones — and got names for three that have been permanent, Powder River, the Tongue, and the Big Horn. They got — and this is astonishing — one of the key facts of the geography when they declared that where the Yellowstone issues from its mountains "it is said to be no more than 20 miles distant from the most southernly of the three forks of the Missouri." (The Indians probably said "one sun" or "one sleep," which is about the distance of the Gallatin River from Livingston, Montana; they had all but heard of Bozeman Pass.) Up to that point too they were right in believing it navigable (though with as much difficulty as the Missouri) by small dugouts. And before they saw the Yellowstone they worked out its importance and decided that a trading post must be built at its mouth. But they erred seriously when they extrapolated. The Yellowstone "takes it's rise in the Rocky Mountains, with the waters of a river on which the Spaniards reside; but whether this stream be the *N. river* or the waters of the Gulph of California [the Rio Grande or the Colorado] our information does not enable us to determine." They were stirring in with the Indians' accounts both Delisle maps and St. Louis speculations, and the mixture was responsible for Clark's bringing the Yellowstone on his map to about the deserts of southern Utah.[27]

The Minnetarees raided westward for, among other things, horses and the captains learned that they would need horses. To transport their outfit across the fifteen-mile portage to Columbia waters "and to continue our march by land down the river untill it becomes [navigable] or to the Pacific Ocean." Both the Flatheads and the Snakes had horses in quantity, they learned, and right there the knowledge of the Minnetarees ran out. So Charbonneau's young squaw Sacajawea, who had given birth to a son on February 11 (Lewis easing her labor with some powdered rattlesnake rattle), proved to have a previously unsuspected importance. For she was a Snake — a Shoshone — who had been captured by the Minnetarees on a raid to the Three Forks five years before. She spoke Shoshone and so could interpret for them, and she must know some, perhaps much, of the country they would have to traverse.[28]

In 1804 no white man knew much about the Snakes. (Arrowsmith's map, which the expedition was using, showed them only a few miles west of the site of Fort Mandan.) Today no one knows much about their movements during the eighteenth century. They belonged to the large linguistic family that included such Plains tribes as the Comanches, Utes, and Kiowas, the sedentary Hopis of the Southwest, and such farther tribes as the Pimas and Papagos; it is supposed to be the stock from which the Mayas and the Aztecs developed. The North West Company's all but total ignorance of them is surprising, for their westernmost customers warred and traded with them. They ranged widely in the far plains and in the mountains, but less widely than their Ute and Comanche relatives. They appear to have stayed behind when those tribes began their great southern migration in the seventeenth century. From that migration, however, they got horses at a very early date and so became for a time a powerful people. This may have been as early as 1700, perhaps earlier, and for some decades there followed a historically amazing spectacle, for as horsemen they terrorized the tribes who would eventually be the terrors of the Far West, the Blackfeet.[29] They raided as far north as the Blackfoot country, which was then the upper reaches of the South Saskatchewan, and massacred all comers. But when the Blackfeet got guns they terrorized the Snakes in turn, raiding them primarily for horses as the Minnetarees did. By 1804 there were no Snakes in Canada or in Montana; it is not certain that any were then living east of the Continental Divide.[30] Those whom Lewis and Clark heard about lived most of the year in Idaho but had to cross the mountains to get to the buffalo herds after the summer salmon run was over. There the Minnetarees appear to have habitually expected them in the vicinity of the Three Forks, as the Blackfeet lurked round Hell Gate to jump Flatheads and Nez Percés, who also had big horse herds and came to the plains for buffalo. The Minnetarees sometimes crossed the Divide to get at them, to the Lemhi and perhaps the Salmon, as farther north the Blackfeet were by now raiding the Kutenais on the far side of the mountains.

And with that river, the Lemhi or the Salmon, and with the undifferentiated Bitterroot River where the Flatheads lived, the Minnetarees' knowledge of the country ended. Because they said it was a large river (as none of the three is) the captains made it the previously unguessed south fork of the Columbia, probably a product of theory but possibly a deduction from the sum of things heard but not set down. Clark drew it as such on his map, taking the Cascade Mountains and the Columbia eastward to them from Vancouver. The rest of the Columbia proper he took from Alexander Mackenzie's map, adding some of Arrowsmith's wholly supposititious tributaries. He called the upper Columbia Tacoutche, from Mackenzie's alternate name, Tacoutche Tesse. . . . The Minnetarees said that the country west of the northward-flowing river was open plains with occasional mesas; they could not have been more wrong.

Long questioning got the Missouri beyond Fort Mandan established even more accurately than the Yellowstone. They carried the northernmost fork too far to the south. They got the Musselshell (which they named, presumably by translation) sixty miles nearer than it was. They shortened the distance to the Three Forks two hundred miles and that to the Salmon River (or the Bitterroot) two hundred and fifty.[31] And they laid up a future anxiety for themselves by misinterpreting what they heard about a northern tributary. This may have resulted from their strong desire to discover one that would providentially enable the United States to reach the Assiniboine and Saskatchewan trade. Repeated memoranda and entries in the *Journals* point out the desirability of such access by an American river. It would connect the trade of both rivers and that of the Athabaska country too — the whole treasury of the Canadian fur trade — with Jefferson's portage-free and supposedly ice-free water route, the Missouri plus all those splendid rivers of the East. And this would be connecting them, as well, with his transcontinental water route to the Columbia and the Northwest trade. It would give the American trade an enormous advantage over the North West Company — and take it square across the path of Canadian expansion.[32]

At any rate, they got an idea that a very important river emptied into the Missouri from the north (and necessarily from north of 49°) just three miles downstream from the mouth of the Yellowstone. It "interlocked" with waters of the South Saskatchewan, not far "from the establishment of the N. West Company" on that river and was "navigable nearly to it's source." This river was the principal though not the sole reason for their early assertion to Larocque that the Assiniboine posts were on American soil. And, they added in their notes, "should the portage between the Saskashawin and *White earth* river prove not to be very distant or difficult, it is easy to conceive the superior advantages which the Missouri offers as a rout to the Athabasca country, compared with that commonly traveled by the traders of Canada." The phrase "White Earth" is a translation of an Indian designation — and it made for trouble. There was a creek precisely where they located their providential river; it held a sizable flood of spring runoff when they reached it. But they had attributed to this river data that (apparently) were being given to them separately about two much more important northern affluents of the Missouri, which they were themselves to name the Milk and the Marias, and about the Souris, which is a tributary of the Assiniboine.

Nevertheless, when they started out again, they had incomparably fuller, clearer, and more reliable information about the country ahead of them as far as the Divide than any white men before them had ever had about the West, than any North American exploration had had. They traveled with complete assurance and were justified in it. Except for the serious doubt that arose when they reached the Marias, which was the result of their White Earth creation, and for the failure to differentiate the two routes to Columbia waters, they could hardly have traveled with more confidence if they had had a twentieth-century map.

They retained two curious areas of ignorance. They had heard little about the Blackfeet and so were unaware of their implacable hostility. The Snakes and the Nez Percés would mend that ignorance. And they had traveled and were to travel by streams whose bottoms were lush with vegetation, and though they ranged across the adjacent plains they took most of them to be richly fertile, except some in the Missouri Breaks and some along the Jefferson. The Idaho and transmontane Montana valleys they saw were lush indeed, and though the Snake and the upper Columbia flow through deserts and near-deserts they do so by narrow valleys or deep canyons. So the captains got hardly a hint of the West's deserts and much underrated its aridity.

❧

While the expedition exchanged visits with the jovial, sometimes fair-skinned Mandans, back in the States a gorgeous fantasy was renewed. An English immigrant who had hoped to find tidings of the Welsh Indians had at last found some in Kentucky. A Frankfort newspaper published his story about them in December — one of the longest and surely the most circumstantial in all that glorious congeries of lies. Though basically the same tale the Reverend Morgan Jones had told sixty-five years before, it was much more splendid. It related the adventures of a still living Kentuckian who, beginning in 1764, had gone up the Missouri River — possibly farther, the Englishman said, than Captains Lewis and Clark would be able to go. Deep in those distant marches, among whose fauna was a swift-running animal nine or ten feet tall without tusks or horns, the Kentuckian's party had found the Welsh Indians. And in great numbers, for their cities spread along the Missouri for fifty miles and they could muster fifty thousand fighting men, which made them the most populous Indian nation. Their skins were pure white for they had kept their blood pure, refusing to intermarry with other tribes. Their language was proper Welsh but unhappily they had declined to the neolithic culture of their neighbors, being without metal or domesticated animals, even dogs or horses. The narrator was, the Englishman assured his readers, a sturdy honest man and of the strictest veracity.

Some exchanges reprinted this marvel and it spread to England and Wales where it was rapturously embraced. It came to the attention of one of Dr. Mitchill's correspondents, who sent him something like a confirmation. Presently the *Medical*

*Repository* published it: an account, already "more than twenty-five years old," by some Canadian traders of an expedition they had made long before, in fact at about the time the Kentuckian met the Welsh Indians. This last party, the account ran, had been sent to explore the upper Missouri and had succeeded in reaching a place about 600 or 700 leagues above its mouth. Six hundred would be spang at the Mandans, 700 well beyond the Yellowstone, and the marvel they found was not ten-foot animals but plants and trees "of astonishing height." Here, however, they met a large nation of Indians of "a complexion as white as that of Europeans" and speaking a language unlike any that Indians were known to speak. These Indians had white beards. They were hostile too; long before the Sioux or the Arikaras tried, they closed the Missouri to aliens. Dr. Mitchill's correspondent felt sure that the Kentuckian's Welshmen were the same Indians. . . . A mistake pardonable in so early an ethnologist. Two myths had temporarily made contact but they were entirely distinct. These were the Bearded Indians, a more humble people than the Madocians, though almost as widely dispersed over North America and fully as hard to kill. The document does not enable us to determine if the Canadians had met the subgroup of them known as the Lost Tribes of Israel.

It was hoped that Lewis and Clark would settle the Welsh Indian problem but they did not. That they found no Welsh Indians (and no bearded ones) on the Missouri established nothing. Their fellow officer Amos Stoddard put the case in *Historical Sketches of Louisiana,* which he published in 1812. He could not doubt that Madoc's descendants roved the Western plains and his candidates were apparently the Comanches though he gave them a Ute name. Lewis and Clark had missed them by taking the wrong fork at the Three Forks. It was clear to Major Stoddard that the Kentuckian had taken one of the other forks. That one was probably a more expeditious route to the Pacific anyway and there, he felt sure, the country of the Madocians was located. . . . The logic that satisfied Amos Stoddard in 1812 has satisfied believers ever since, down to Madoc's latest biography, published in 1951.

Besides fantasy there was rumor. On November 5 1804 Stoddard's commander, Major James Bruff of the Corps of Artillerists, who was military commandant of Upper Louisiana, wrote a long letter to his chief, General Wilkinson. In the course of it he reported: "No news since the 4. August from Capts. Lewis & Clark — they were then at the mouth of the river Platt where two of their boatmen deserted: and it is reported by several Canadians who happened there at the time that the others were much dissatisfied & complained of too regid a discipline. I am not, however, disposed to give *full* credit to their story, as they report other unfavorable circumstances that cannot be true — Such as a difference between the Captains &c."

This is without explanation. The *Journals* do not record any meeting with white men between June 14, when the expedition was below the mouth of the Kansas, and August 2, when "Mr. Fairfong" arrived at Council Bluff. No other Frenchman is mentioned as coming with him or being in the vicinity. They had reached the Platte on July 21 and August 4 was the day when Liberté should have returned and when Reed deserted. Bruff's account is just distorted enough to suggest that his Canadians had met not the expedition but Liberté.[33]

೮⌁೦

On April 7 1805 the keelboat started back to St. Louis, in

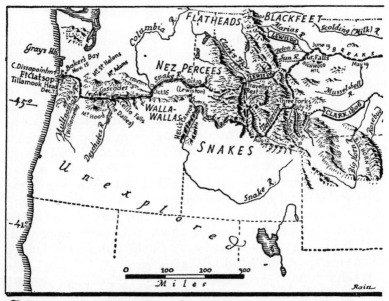

*Map 30. The Route of Lewis and Clark*

charge of Corporal Warfington. One of his squad had been enlisted for the exploration, leaving him five soldiers. He took the two (Reed and Newman) who had been sentenced to discharge and also two of the wintering Frenchmen. At the Arikara village he was to pick up Tabeau, four of his *engagés*, and his furs. An Arikara chief was to come aboard there too, beginning the long journey to see the Father in Washington, to be impressed (the idea was) by American strength and wealth, and so to preach peaceful co-operation when he returned. Tabeau's Joseph Gravelines would go with him as cicerone and interpreter and would be the keelboat's pilot.[34] Warfington's heaviest responsibility, however, was the letters, reports, maps, and notes the captains were sending to Dearborn and Jefferson. They were accompanied by five boxes of specimens — Indian artifacts, furs, dried plants, prepared skeletons, horns, Heney's root for snake bite — and living prairie dogs, magpies, and a prairie hen.

The expedition would not get back to the States this year, Lewis told Jefferson, but they expected "to reach the Pacific Ocean and return as far as the head of the Missouri, or perhaps to this place, before winter."

And "Entertaining as I do the most confident hope of suc-

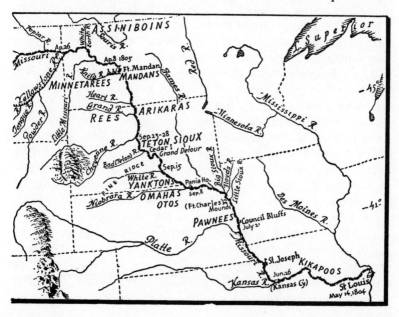

ceeding in a voyage which has formed a darling project of mine for the last ten years, I could but esteem this moment of my departure as among the most happy of my life." Thus Meriwether Lewis, who had the wilderness in his heart, on April 8 1805 when they headed into the unknown.

Besides Lewis and Clark, the exploring party consisted of three sergeants, twenty-three privates, York, Drewyer, and Charbonneau, Sacajawea, and her two-month-old son Baptiste or Pomp. They had the two pirogues which grew steadily less handy as the river grew steadily more difficult, and to take the place of the keelboat had built six cottonwood dugouts. These were unquestionably the best craft for the swift and shallow Missouri, choked with snags and sandbars and forever catapulting logs at them. They were rugged but their narrow beam (much narrower than that of canoes of equal length) made them exceedingly unstable and their lines prevented fast handling. No yachtsman would have called them yare; the Nootkas would have laughed at them. Lewis did not feel easy about them except when they were being towed. That became most of the time as they went farther upstream and practically all the time above the Three Forks. And towing meant staggering through bush and marsh and mire, and climbing or careening along the gullied edges of bluffs when the channel was inshore. That and jumping out to shove the heavy craft off the myriad bars were cold work in April. The water never got warmer for they advanced toward the mountains as the season advanced.

Though there were mosquitoes at once not a tree had budded. The great flights of wildfowl were arriving, however. One who travels those reaches of the Missouri by boat today can still see remnants of those flights big enough to give him some sense of what the unspoiled time was like — black cormorants trailing their legs as they fly, large white and stately pelicans, herons, teal, pintails, greenheads, "duckinmallards," geese — everywhere the flash of woodpeckers through matted willows, kingfishers, owls, hawks, gulls, sometimes an eagle. But though many beaver dive from the bank as the boat approaches, there is nothing else to suggest the incredible hosts of game the expedition found in 1805. They had only to walk inland from the bank to kill anything they might want in quantity, Lewis says. It was mere sport for him or Clark to kill meat for thirty-one in the course of a stroll — buffalo, elk, deer, antelope in fabulous numbers, and soon, amazed, they met the bighorn and presently the grizzlies, to which Lewis felt superior till he saw the second one.

Indeed from here to the Three Forks is, despite the intensifying labor of the passage, a fabulous period. Up to the Breaks the river forever repeats the same pattern: banks only two or three feet higher than the water and with sliced vertical sides, then a flood plain thick with willows and cottonwood, bluffs beyond the flood plain on one side and on the other usually near and sometimes at the bank. The bluffs are innumerably repeated — pyramidal, truncated, domed, writhen with erosion, some steep sides gullied like the ruffles of a child's dress; white, gray, yellow, red, rust, cobalt, or veined horizontally with gray and black; bare or a little grassed or drab with greasewood or black where the dwarf cedar can grow. The banks crumble with a splash or a groan or a roar — Clark marveled that they never got a towing crew or its boat — and the olive-drab water boils and a small or a big wave swells toward the other shore. Tree roots snapped by the collapse stick straight out from the new raw bank and the trees float off to strand at the next bend, or moor themselves as sawyers, or build new courses on old snags. Sandbars are wedges, the point downstream, and it is toward the point that the driftwood gathers. Some bars submerge, some dissolve, some grow to islands and the islands narrow and elongate.

Over river, shore, and bluffs is enormousness, the endless sky and the clouds that join and climb. Clouds and a quarter or a half of the sky turn black with inconceivable suddenness. A wind strikes out of them, the water begins to hiss, whitecaps or big waves leap at the boat, and navigation is over till the wind drops half an hour from now, tomorrow noon, day after tomorrow. The wind is as cold as it is violent; snow comes with it, or hail, or a battering and strangling rain. Against the trailing black gauzes of the storm, cottonwood trunks are a pale pure silver and their buds the pale green of the sky at daybreak. . . . Round such a bend, against such cottonwoods, up the dun-colored water, two pirogues, six dugouts, thirty-one minute figures in drenched and very cold buckskins while the sky flaws with rain.

Few boats ever travel the river through the Breaks now —

not, at least, beyond the hundred-mile reservoir of Fort Peck. An airplane flight, with the pelican-galleons white below one instead of above, shows why they have been called breaks instead of badlands: occasional wide intervals of meadows or even timber interrupt the frantic chaos of the buttes. From above or below, however, they are chaos, tumbled, precipitous, bizarrely modeled, bright-colored, drab, monochrome, or splashed with white or mottled with brilliant reds, and transformed with every change of light in a vastness where light is never long the same. When the expedition reached the Breaks their strangeness assailed these poised men. In the *Journals* almost for the first time the word "beautiful" means something other than "fine country for settlement."

Wool uniforms and cowhide boots were going or already gone; from now on it would be buckskins and moccasins. Three "Frenchmen," presumably some of those who had wintered at Ford Mandan, traveled with them for a while to trap these entirely untrapped waters but were left behind on April 13. On April 14, a week out, beyond the mouth of the Little Missouri, coming into the last arc of the Big Bend that turns the river full west, they named a small watercourse Charbonneau's Creek, since he had once camped on it with the Minnetarees. And Lewis wrote, "This was the highest point to which any whiteman had ever ascended except two Frenchmen [one of whom was the Lepage they had enlisted at the villages] who having lost their way had straggled a few miles further." They went on into the wholly new.

The river was beginning its spring rise but fluctuated and there was never enough water where it was needed, which meant the towropes. It was flawed spring weather, with snow flurries bursting into days of summer heat and sometimes ice forming on the oar blades as they rowed. They reached the greatly desired White Earth River on April 21 and it disappointed them — too small. But it widened and had flood water upstream and Lewis wrote that it must extend at least to 50° and must be navigable to its source and must be "no great distance from the Saskashawan."

They *must* have such a river but the tug of the Yellowstone was greater still and Lewis hurried overland to it ahead of Clark, who brought the boats to the great confluence on April 26. Now first seen by white men, it was a noble river. The country it flowed through seemed magnificent and both captains summarized in their notebooks all the lore about it they had stored up and written down before. (Including the assurance that its sources and those of its affluents the Tongue and Big Horn, and of the Platte and a southern branch of the Columbia lay close together to the southwest.) They noted that its curiously colored water was without sediment — today it carries far more silt than the Missouri — and debated the best site for the post that must be built here. It must be a stone or brick post for there was little timber.

On May 3, six days beyond the Yellowstone, the urgency to get behind the British posts found promising substance at last. They reached what seemed to be just such a river leading north as their winter fantasy had sketched. It was the inconsiderable stream now called Poplar River but the heavy winter snows were melting and its flood plain, like that of the White Earth, was handsomely full of runoff. They needed it, and Lewis did not doubt that its source was "not far from the main body of the Suskashawan river and that it is probably navigable 150 miles; perhaps not far distant from that river." By big boats too, and "should this be the case it would afford a very favorable communication to the Athebaskay country, from which the British N. W. Company derive so large a portion of their valuable furs." [35]

On May 8 a bigger river came quietly flooding in from the north and, a hard blow holding them up, they reconnoitered it. It "must water a large extent of country" — from a pinnacle Clark thought he could see fifty or sixty miles "level and beautiful," with big herds of buffalo. It forked and one fork made straight north. So "perhaps this river also might furnish a practicable and advantageous communication with the Saskashiwan river." Now they had three, which ought to be enough. This was a hundred miles short of where they had expected to find the one they had called the Scolding River, or as the Minnetarees put it "the river which scoalds at all others." Nevertheless it must be that river — they must have estimated the mileage wrong last winter — for the Indians had described only two northern tributaries large enough to satisfy their requirements. Disregarding the White Earth as perhaps too small or too short, they had now passed both, the Poplar and the Scolding. They were satisfied but they rechristened this one, giving it a name for its "peculiar whiteness," like a cup of tea with a tablespoon of milk added to it. They made it Milk River, and again flood stage had deceived them.

Long habituated to drinking Missouri water, they found the alkali of increasingly bitter creeks strongly purgative when they tried it. They saw some Indian sign but no Indians and were content, for hereabout they would probably have been Assiniboins and therefore troublemakers. By now rattlesnakes were an old story and the fury of grizzlies was just as familiar and had taught everyone respect. And Sacajawea had long since proved her value. She was always busy, digging edible roots they were familiar with and finding strange ones, tailoring buckskins, making moccasins, explaining novelties, cool and swift in emergencies. . . . She is remarkable, this girl who may have been no more than seventeen in 1805, and of whom no word is directly reported since she could speak only Shoshone and Minnetaree. She is better known than any other Indian woman and she has unusual resourcefulness, staunchness, and loyalty. a warmth that can be felt through the rugged prose of men writing with their minds on a stern job, and a gaiety that is childlike and yet not the childishness we have associated with primitives. All the diarists liked her, Lewis gave her a detached admiration, Clark delighted in her, and she has charmed the imagination of readers ever since Nicholas Biddle wrote his book. In so much that she has become a popular heroine and something of a myth, a Shoshone Deirdre created out of desire. And from Bismarck to the sea many antiquaries and most trail markers have believed that Lewis, Clark, and their command were privileged to assist in the Sacajawea Expedition, which is not quite true.

On May 19 they saw peaks to the north, the isolated range called the Little Rockies which eager travelers always and mistakenly believe to be the beginning of the Rockies. That day a wounded beaver which the dog Scammon was retrieving bit his leg so badly that Lewis feared he would bleed to death. The next day they reached the Musselshell and their information was vindicated for they had expected to find it just where it was. In the Breaks loose soil made sandstorms of all gusts of wind and an epidemic of sore eyes ensued. Experience was adding up; Lewis amended ideas about climates and soils that had been too generous and Clark refined Indian geography to

closer approximations. On May 25 Clark thought he saw more
mountains and the next day climbed a hill and made sure.
Like the Musselshell, they were where the data said they should
be; they must be the Indians' first range. Perhaps those to the
north were the ones now called the Bear Paws and those to the
southwest the Highwoods, but the angle of sight joined them
to other ranges and they were so far away that only snowy tips
could be seen. Lewis wrote, "I felt a secret pleasure in finding
myself so near the head of the heretofore conceived boundless
Missouri." But a premonition of the strain to come struck
through his exultation.

It was nearly always the towrope now and the wind forever
full of sand. At the end of the day fatigue felled a man like a
club but dinner of buffalo ribs or bighorn chops or boiled
beaver-tail was restorative. Driftwood fires bowed and flared
to the high wind and Cruzatte would fiddle for dancing till the
flames sank and only the glow of embers was reflected from the
water. On May 29 a clear bold river came in from the south.
Lewis named it Big Horn but Clark pointed out that the map
already had a Big Horn emptying into the Yellowstone. He
called it Judith for Judy Hancock, who was only thirteen now
but whom he was destined to marry in 1808. There was much
Indian sign, an encampment two weeks old and big — 126
lodges, Lewis made it — and another, older camp near by.
Sacajawea identified Atsina moccasins, the Fall Indians, probably raiding her own people. The Breaks were magnificent
scenery but "the men are compelled to be in the water even to
their arm pits" a full quarter of the time and, when they
weren't, had to take the rope slitheringly across mire barefooted
and drag "the heavy burthen of a canoe . . . walking acasionally
for several hundred yards over the sharp fragments of rock"
that had fallen from the cliffs above. They came out to plains
country again and now the mountains were not so distant and
wild roses had bloomed in the bottom and cactus on the banks.
On June 2 no one had been able to wear moccasins while towing
for the last several days and "many of them have their feet so
mangled and bruised . . . that they can scarcely walk or stand."
On that day the current was more gentle, a grizzly chased both
Drewyer and Charbonneau before Drewyer shot it through the
brain — and they reached "a very considerable river."

So considerable a river that the Missouri seemed to have

forked. To encounter a big river was both dramatic and stunning. For the Missouri was not known to fork below the Three Forks. Above the Scolding River, which they assumed they had passed a month ago, no important tributary entered from the north. Above the Musselshell, two weeks behind them, no tributary of even moderate size entered from the south. Months of inquiry and of analysis had made this clear, and maps sketched in sand by war chiefs and on paper by William Clark had plotted the Missouri all the way to the mountains, the Snake Indians, their horses for the land traverse, and the easy pass leading to the southern branch of the Columbia. None of these maps showed such a crossroads as this. And once more it was a big river only because of the big runoff — actually it is small.

So they must determine which fork led to the mountains, which was the main stream of the Missouri, and they must decide correctly. For if they decided on the wrong one and, going up it, eventually had to come back to the other — then that error "might defeat the expedition altogether." With so much time lost there could be no reaching the Pacific, perhaps no getting across the mountains, before winter. Their supplies being steadily depleted, they would have to go back to St. Louis in failure, the continent uncrossed, no route discovered, Jefferson frustrated, and the foreign and domestic policy of the United States perverted.

Almost the entire party, including the expert rivermen whose top authority was Cruzatte, decided during the first afternoon that the north fork was the Missouri. Dissenting from the first, Lewis and Clark were convinced by the end of the next day that it was not, though convinced too that both must be explored. All but conclusive arguments against both could be drawn from the Indians' information and from the analysis the captains had subjected it to. In every respect except that the Missouri had not been known to trend north (and even this might have been sanctioned by data acquired after Clark made his map) the north fork *looked* like the Missouri. Its water had the same brown-mud color that the Missouri had all the way from the Mississippi to this confluence. Its current appeared to have the same velocity they had recently been facing. Its bed was the exhaustingly familiar muddy bottom of the Missouri. One would say that it was the Missouri continuing. To the contrary, the south fork was "perfectly transparent." It had a swifter current. It flowed over a bed of "flat smooth stones." It was not continuous with the Missouri but an abrupt change and novelty. Yet it was the bigger river, it bore southwest which was the anticipated direction, and the very novelties that changed its character argued that it was the Missouri. For rocky beds and clear water were characteristic of rivers that came down from mountains not far away, and the Minnetarees had made it sure that the Missouri entered mountains at a not much greater distance than the expedition had now traveled. They had insisted too that it traveled through mountains for a long distance. The turbidity and the mud bottom of the north fork argued that it did not penetrate the mountains, or that it made too slight a penetration to bring it as near the parting of the waters as the Missouri was known to come, or that on emerging from the mountains it traveled across plains for a much greater distance than the Missouri was known to up to this point. And for this conclusion there was support in the size of the south fork; it could get so much water only from deep penetration in mountains, not from country so dry as they knew that which lay between it and the Yellowstone River to be. (The barrenness, the succession of tiny creeks, and the big dry runoff channels all indicated the same thing.) Again, the Minnetarees had said that the Great Falls were on a bearing a little to the south of sunset from Fort Mandan. Determination of the latitude showed that this confluence was only a few minutes north of Fort Mandan. The strong southwestward bearing of the south fork squared with these data, the strong northerly bearing of the other fork would not square. But against the south fork was the weighty fact that, so their Arrowsmith map showed, Peter Fidler was believed to have ranged down the eastern foot of the Rocky Mountains to 45° and he had not encountered a stream anywhere near so large. The latitude they had taken was 47° 24'. If Fidler had reached 45° then this fork must enter the mountains even farther south than that — which was farther south than they had calculated the head of the Missouri and the all-important pass to be.

If the north fork was the Missouri, why hadn't the Indians mentioned the south fork? If the south fork was the Missouri, then the north fork must be the Scolding River — and how could the Indians, or they themselves, have so seriously erred about its distance from Fort Mandan?

Thus Lewis and Clark on the first afternoon and the next

day. They went to the top of the highest hill — not very high, only the ground swell of the plains. Snowcapped mountain ranges rimmed the horizon in a gigantic arc southwest, west, and northwest. Above and beyond them was the faintly sketched, spectral substance of farther ranges shining in the diamond air. The rest was "one vast plain," with buffalo herds, "their sheppards the wolves," elk herds, "solatary antelope," and the yellow and pink flowers of opuntia. But they could not make out the courses of the two rivers. . . . They talked it out exhaustively: "Thus have our cogitating faculties been busily employed all day." As often as they cared to re-examine their information and the conclusions they had drawn from it, they were compelled to decide, on the sum of the evidence, that the north fork was a plains river and the south fork a mountain river. They would therefore eliminate the north fork. It must rise east of the Rockies and south of the Saskatchewan. It could not penetrate the first range of the Rockies. Most of its sources must be north rather than west, toward the lower and middle reaches of the Saskatchewan. (That British river, the key to much of their thinking.) The north fork *was* the Scolding River. The south fork was the Missouri.[36]

The commanders of a momentous exploration thus answered the question that must be answered right if the expedition was not to fail. At this second critical turning point, with immediate failure hanging in the balance, the answer is co-operative and joint. It is Lewis who writes down the reasoning in the *Journals,* but this analysis is as clearly Clark's as his, for it was Clark who had formulated the winter's data, had checked and rechecked and tabulated it, and must now first check it again for unperceived possibilities of error in calculation or in reasoning. And this joint effort is a remarkable act of the mind. Considering all that went into it and all that depended on it, it must be conceded a distinguished place in the history of thought. It is the basic method of science; they searched the information for what it was and what it contained, and what it must therefore be adjudged to mean and imply.

The methods of agreement and of differences had been applied and a conclusion reached. Now for verification. . . . They had sent squads up each fork in a dugout and another one overland; no significant observations had resulted. Now they would themselves go up the forks till they could be sure. They started on June 4, Clark up the south fork, Lewis up the north one. Clark was sure on the second day, after fifty-five miles of travel, forty-five on a true course. Throughout that time the south fork continued to be a mountain river. Its bearing was always much to the west of south and therefore towards the mountains, which were the Big Belts. It nowhere made so southerly a bearing as it must if Fidler had correctly stated his farthest south. It must be the river that led into the Rockies and on toward Columbia waters: it must be the Missouri. Lewis was sure on the third day. The north fork "had its direction too far north" — it did not make enough westering — to take them a significant distance toward Columbia waters. It could not be the Missouri.

. . . Lewis was right. Yet the river would have led them to Marias Pass, which crosses the Continental Divide. Except for the great portal of the West, South Pass through which the American emigration moved to the Pacific, it is the only pass by which wagons could cross the Divide — and the short summer of the north closed it to emigrants. Lewis and Clark could not have taken their boats into Marias Pass, and beyond it to the west they must have traveled a sore distance before taking to a river route again. Yet, theoretically, a water-crossing to the Pacific, that fantasy which had so powerful a hold on the imagination of men, could have been found by that approach. Theoretically only, but truly. From Marias Pass they would have come down to the Flathead River and that would have led them to Clark's Fork, which eventually was as navigable as the Clearwater down which they took their dugouts to the Snake. Clark's Fork would have taken them to Pend Oreille Lake and beyond that its extension, which is sometimes called Pend Oreille River, would have taken them to the Columbia at almost exactly 49°, more than three hundred miles above the mouth of the Snake.

Though not the Missouri, it was in Lewis's judgment a noble river, and he named it the Marias, to commemorate one of his many romantic attachments. He was confident that the land it watered was as fertile as he had seen it was picturesque. It must be rich in furs too. And once more he sounded the compelling — and revealing — theme. With so northerly a bearing it must rise north of 49° and so was "destined to become in my opinion an object of contention between the great powers of

America and Great Britin with rispect to the adjustment of the Northwestwardly boundary of the former." And it "most probably furnishes a safe and direct communication to that productive country of valuable furs exclusively enjoyed at present by the subjects of his Britanic Majesty."

This was the fourth river that Lewis and Clark had decided would enable the United States to get behind the North West Company, flank it, and tap the richest beaver trade. They had been looking for one ever since they started talking to the Minnetarees last November. Had they been explicitly instructed to look for one? Why?

Back at camp they put it all together again on June 9 and were, once more, sure. The south fork was the Missouri River and was therefore the route to the Rockies. It must reach them north of 45°. Mr. Fidler must have been wrong. During the winter, on Clark's map, they had already moved his farthest south up to 46°, relying on the sum of what they had learned. Now they decided that he could not have gone farther south than 47°. (Actually he did not reach 49°.) They committed the expedition: they would go up the south folk. The men unanimously believed them wrong. They determined to leave the larger pirogue here, secured it, and made caches for their heaviest tools and for sizable stores of concentrated foods, lead, and powder.

If they were right, then the Great Falls could not be very distant; in fact Clark could not have been far short of them when he turned back. On June 11, leaving Clark to bring up the boats, Lewis started overland to find out, taking Drewyer, the Field brothers, and Goodrich with him. Suspense was strong on the 12th but the plains were majestic, the distant mountains beautiful, and they killed two grizzlies at the first fire which had never happened before, and that night they dined on mountain trout, for Goodrich was a mighty fisherman. Well before noon the next day Lewis heard the sound of a waterfall and, far away, saw spray like a column of smoke rise above the plain, drift, disappear, and rise again. (The river flows in a sunken valley.) He pressed on in the direction of that fluctuant mist and soon heard "a roaring too tremendious to be mistaken for any cause short of the great falls of the Missouri. here I arrived about 12 OClock having traveled by estimate 15. miles."

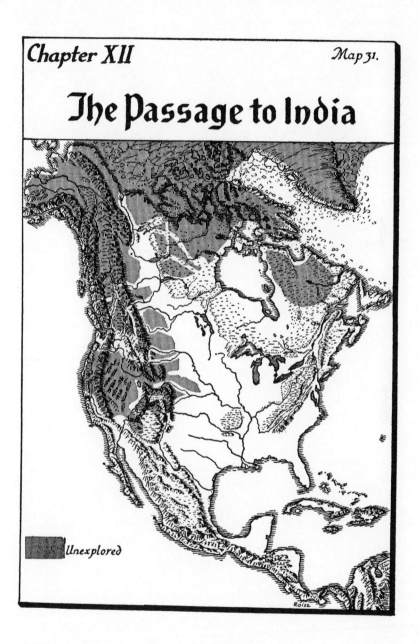

Chapter XII

Map 31.

# The Passage to India

Unexplored

# XII

## The Passage to India

THE SHELVES over which the Missouri pours in five falls and a series of rapids are the true beginning of the Rocky Mountains. No one will ever again see the beauty or feel the majesty that overwhelmed Lewis. Nowhere in the entire length of the river has industrial civilization, if the noun be permitted, more hideously defaced the scene than a power development has done here.

Lewis reached the falls June 13, Clark took the boats into the creek where the portage round them must begin on June 16. (Sacajawea was so sick that they thought she might die. They bled her, gave her appalling purges, and in the end cured her with the water of a mineral spring that stank so bad it must be therapeutic. The Montana health department now posts it as unsafe to drink.) The portage, surveyed by Clark, was 18¼ miles, to an island which they named White Bear three miles above the mouth of a stream which at the villages they had named Medicine River (Sun River). The remaining pirogue had to be left below the falls. The dugouts and all the outfit except what they buried in caches for the return journey had to be pulled eighteen miles on crude truck-frames. They found just one cottonwood big enough to make wheels; the wheels, tongues, hounds, and body repeatedly collapsed. It took more than a month but they paused long enough on July 4 to drink the last of their liquor.

Lewis remarked that a spirit of fatality possessed the place. He had evidence. The labor of the portage exceeded anything that had preceded it; stopping for a breather, men would drop to the ground and instantly fall asleep. They double-soled their moccasins with parfleche but no buffalo hide would turn the universal cactus. The plains heat that crushes like a great weight made them work naked but hailstorms struck out of it, felling them and bloodying their scalps. Cloudbursts scattered the outfit along the trail, turned the trail into a gumbo glue, and filled the ravines with water; one of them came close to drowning Clark, Sacajawea, Charbonneau, and the baby. The plain had more buffalo than ants and more grizzlies than mosquitoes. Both species had the charging-madness, so that sidestepping big game was as routine as the daily tax of wounds from the axes whose helves broke and the disposition of other paraphernalia to go to pieces right here. (But they would soon look back on this vast game preserve with the longing of hungry men.) The beauty that surrounded them was ominous. The ranked ranges of mountains to the west, wearing midsummer stoles of snow, had to be crossed and they knew they had no forecast of what crossing them might involve. Suddenly time had accelerated and was mortally precious. As early as June 30 Lewis faced the double-edged fact that, almost three months out from the villages, they were still short of the Rockies and had no way of computing the travel to come. He decided "that we shall not reach Fort Mandan again this season, if we even return from the Ocean to the Snake Indians." Realism was considerably diluted still with desire.

They could not send a detachment back to the States from here, as Lewis had told Jefferson he would. Their command had been just large enough for the labor of this portage; they could not approach similar jobs with a smaller one. Except for the Snakes and Flatheads and the underestimated Blackfeet they did not even know the names of the farther tribes, and they needed all the military strength they had. They thought too that to send back a squad to home, family, white girls, and comfort might weaken the will of those who would be left in immensity facing another year, of which nothing could be surely predicted except that it would be harder. . . . So the arrival at St. Louis of Warfington, Gravelines, and the keelboat toward the end of May was the last word the United States got from its explorers till they themselves reached St. Louis in September of the next year. Before they got there they learned

that the failure of the expected dispatches to arrive had convinced everyone that the wilderness had wiped them out; few
besides Jefferson still hoped to see them again.

By holding to the Missouri they were disregarding one of the
routes which the Minnetarees had described. That route led
west from the Great Falls to the Blackfoot River and on to
Bitterroot Valley. Its first stretch had two optional variants,
close together. One of these optional trails led up Sun River,
just below White Bear Island where they put their boats into
the water again, the other up a river whose mouth they passed
on the third day and which they named Dearborn's River for
the Secretary of War. And this first stretch would cross the
Continental Divide by either one of two passes. One, which
Lewis traveled from the west a year later, is called Lewis and
Clark's Pass now; the other, a little farther south and nearer
Dearborn River, is called Cadotte's Pass. Either would have
taken them to the Blackfoot River, to its junction with Hellgate
River, through Hell Gate, and to the mouth of the Bitterroot
River.[1]

The Bitterroot is the stream which they named Flathead
River when they reached it after a long detour and a weary
time but renamed Clark's River. It was the other northward-
flowing river which the Indians had described and the captains
had merged it with the Lemhi-Salmon. Traveller's Rest, where
they left the Bitterroot to head west over the mountains, is
only about six miles up from its mouth, which they would have
reached a few miles after emerging from Hell Gate. . . . They
could not have taken this route in any event without gambling
that an advance party could get horses from the Flatheads, a
gamble made prohibitive by their information that the Snakes
were nearer and would be reached sooner. But they had not
in fact truly understood that there was a route here. And the
whole bent of their minds, the whole drive of their purpose,
was to press on to the Three Forks and by the easily navigable
waters of the northern fork reach the portage that in a day's
land travel would take them to Columbia waters.

From now on this thinking would be slowly undermined,
though they took a long time to become conscious of what was
happening and a longer one to accept the implications.

With the six old dugouts and two new ones they started up

the Missouri on July 15. The next day, in the mountains at last, they were traveling a clear and very swift river through the beautiful Missouri Canyon. It was everybody ashore except those needed to row the dugouts, with repeated use of the braided elkskin towropes and sometimes sheer cliffs to the edge of furious water. On the 19th, after another canyon had taken them beneath more towering peaks, they reached the spectacular formation that Lewis named the Gate of the Rocky Mountains, a vertically walled gash which truly is a portal. Going through it — as an exit, not an entrance — they came into the wide, desolate plain through which the river twists from the Three Forks. . . . This is near Helena, Montana, and here too another, shorter land route would have taken them to Traveller's Rest on the Bitterroot. It was twenty-seven miles from water to water, though the second water would have been even less navigable than the Jefferson is where they left it. They would have been nearest the intervening range about ten miles upstream from the Gate of the Mountains, say at the stream they named Pryor's Creek. Several passes, of which Mullan's would have provided the easiest travel, lead across the Continental Divide and presently they would have reached the Little Blackfoot and gone down it to the Blackfoot and through Hell Gate to the Bitterroot.

The meeting of plains and mountains always makes a big country and this is an even bigger one than that at the Great Falls. At no time since they left St. Louis had the commanders displayed anything that could be described as anxiety but now tension settled in. Approaching the Three Forks, they were approaching the test: this was the place where the Snakes and their horses would be found, if found at all, and whence the trail to the Columbia must be blazed. It was nearly always poling the boats now or towing them against a powerful current, midges and mosquitoes so thick that all faces were bloated and there could be little sleep at night till the dawn chill drove them off. Cactus lacerated the feet of the towers, the hunters who ranged for game, the captains on their daylong search for Indian sign, and even the whimpering Scammon. Clark's feet were specially bad but, judging himself the better scout, he refused to let Lewis relieve him. Well in advance of the boats, he reached the Forks early in the morning of July 25. The born geographer decided almost at first glance that the Indians were right, that the northern fork was the route the expedition must take. Leaving there a couple of men who were more tired than he and sorer of foot, he pressed twenty miles up the north fork and then twelve miles crosscountry to climb a small peak and spy out the land. Everything he could see supported his first decision. He crossed to the middle fork for further verification, came back across the plain without finding any recent Indian sign, and fell ill — an exhausted man. Lewis brought the boats up July 27, they hurried scouting parties out, and that night they camped on the spot where Sacajawea's band of Snakes had been in camp when the Minnetarees jumped them five years before. The Snakes scattered and the horses got most of them away but she was run down on a shoal in the middle of the north fork while trying to cross it.

. . . Without prescience three marked men went about their jobs at the Three Forks, John Potts, John Colter, and George Drewyer. In August 1806 as the returning party came together again at the mouth of the Yellowstone after traveling thither in several divisions, they met two Americans named Dixon and Handcock. The first of the host that was to come, they were trappers following the trace of Lewis and Clark and had crossed this amazing length of it alone. The expedition was less than six weeks' travel from St. Louis, which it had left in May 1804. The two trappers would like a man to go on with them and would especially like one from this company of first-trained masters of the Western wilderness. John Colter would go back if he could get his discharge, and on the promise of the others that no more would ask, the commanders gave it to him. They outfitted him as well as they could and he turned again into the wilderness of mountains that now had a trail. That fall and winter he (who had had tuition in the trapper's trade from George Drewyer) trapped with his two companions, perhaps along the Yellowstone he had just traveled down with Clark. In the spring of 1807 this twice-tempered mountain man set out down the Missouri in a dugout, homeward bound, quite alone, remembering deeds and adventures that would have enriched literature if there had been anyone to write them down.

Colter reached the equator of the Missouri, the mouth of the

Platte and there met a big party coming upriver, bound for the rich beaver country which the return of Lewis and Clark had publicized. It was commanded by the veteran St. Louis trader Manuel Lisa, the shrewdest mind in the business and by far the most successful bourgeois the American trade was to have during its first period. Lisa had hired George Drewyer as his guide and principal dependence in the new country. He had also hired the John Potts and Peter Wiser who had been with Lewis and Clark, and now he got another member of the great company. For again Colter turned back to the wilderness, and this time he marched into American legendry. Where the Big Horn River empties into the Yellowstone, a site that had been duly recommended by William Clark, Lisa built the first post of the Western fur trade. Thence he sent Colter to find the Blackfeet and bring them in to trade. Without companions Colter made a journey so remarkable that it makes one's breath catch. His route has been much disputed and will never be confidently established, but he went up the Big Horn and found the Stinking-Water River (more gently the Shoshone in these days) and the neighboring bitumen springs and thermal phenomena that were called "Colter's Hell" thereafter — and were laughed at by greenhorns back home as a damned amusing lie. He crossed the Divide, the first white man to do so after the expedition, and traveled across part of the more spectacular hell that is now Yellowstone National Park. Before he met the Blackfeet he met their implacable enemies the Crows and traveled with them. So when the enemies met he had to fight with the Crows, killing some Blackfeet.

Some writers have held that this encounter sealed or perhaps created the murderous hostility to American trappers that the Blackfeet maintained for many years. Probably it was more complex than that, and perhaps a great part of it was the natural belligerence of the Blackfeet who never kept the peace with anyone for very long, not even with themselves, but there is no doubt that the anti-American policy existed. And in the next year and a half Colter helped a good many Blackfeet find the sunset trail. So in 1809, probably not a dozen miles from the Three Forks where he is camped this night of July 25 1805, they jumped him and his partner Potts. They killed Potts but they stripped Colter naked and told him to start running, giving him a sporting start before loosing their fleetest runners to catch and kill him. He ran barefoot across the cactus that had torn the parfleche-protected feet of his companions four years before, ran till blood gushed from his nostrils as from a dying buffalo's, ran six miles all told. All his pursuers but one fell behind, he killed that one, and he ran on till he reached the northernmost of the Three Forks. He dived into it and came up under a matted raft of snags and driftwood. Frenzied Blackfeet howled and whooped round him all day long, sometimes leaping to the very logs between which he was watching them. That night he swam downstream some distance, then climbed out of the river and struck out overland. He traveled all that night and seven days more, naked, barefoot, and unarmed, till he reached Lisa's post on the Big Horn and some amazed companion, glimpsing a strange beast in the plain, could run out and give him a shoulder. Tradition's creative art soon transformed his matted driftwood into a beaver house and had his dive carrying him up the chute from its water-gate to stretch out on the shelf beneath the mud dome while the frustrate Blackfeet ran over the roof, but his own facts will do.[2]

And George Drewyer, who slept sound the night Lewis and Clark camped where Sacajawea's people had camped, was sleeping on the site of his own grave. He went back to St. Louis with Lisa in 1808 and came up the Missouri again in 1809 with a larger company, dispatched by a firm in which Lisa was one of several partners. One of them was traveling with this party now, Andrew Henry, who would cross the Divide first after Colter, give his name to one branch of Lewis's River, and eventually be General William Ashley's partner in the venture that shifted the mountain trade to the Platte River route. They wintered at Lisa's Big Horn post and in the spring crossed over to build a new post at the Three Forks, where Lewis and Clark had said there were "emence" quantities of beaver. (And where they had recommended that a post be built.) The Blackfeet hovered and struck, killed some of them, feinted, lay in ambush, and struck again. Drewyer was out with a small hunting party getting meat for the fort when they rode at him out of the brush. He fought back so tremendously that this drama too passed into legend, but they killed him and took his scalp.

☙❧

At the Three Forks, an "essential point in the geography of this western part of the Continent," the captains decided that the Missouri ended — or rather began. They named the all-important northern fork the Jefferson for the author of their enterprise, and the others Madison and Gallatin for his Secretaries of State and the Treasury. They started up the Jefferson on July 30 under grievous strain — having seen no Indians and no fresh Indian sign. It took them into a canyon through the Tobacco Root Mountains and on to a wide plain; the river was narrower, more twisting, and much swifter than the Missouri had been; it called for prolonged, desperate towing. On August 3 Clark found the trail of a solitary Indian who had responded to the signal fires they had lighted by climbing a butte to get a look at them — and they realized that instead of attracting the Snakes they might frighten them into flight. That day Reuben Fields killed a puma near a small stream which they therefore called Panther Creek. Passing this creek, which is now called Pipestone, they passed one more route that would have taken them to the Bitterroot. Above its head the gently graded Pipestone Pass leads across the Continental Divide and down to the plain that has the city of Butte in it. That plain opens out northward as Deer Lodge Prairie, a great buffalo country for the Indians and long a wonderland for white sportsmen. Presently they would have struck a small stream that grows in size and gets a number of names. Beginning as the Silverbow, it becomes Deer Lodge River and flows north to meet the Blackfoot River (or Big Blackfoot) down which the other routes that have been specified all led. From this junction on the enlarged stream is Hellgate River but it has become the Missoula when the Bitterroot flows into it.

Two things were happening as they labored the boats up the Jefferson: to their sharpening anxiety the Snake Indians were eluding them, and against a will that was charged with three centuries of desire, the water route of commerce to the Columbia River was being proved an unfruitful dream. They knew now what mountain rivers were: too swift, shallow, and narrow for anything but the towrope, full of boulders, choked with gravel islands, overhung with brush. Furs and Indian goods and China tea would have to be pulled up and down these streams in unwieldy dugouts at a pace that even roaming

hunters on the shores exceeded far. ". . . they therefore proceeded but slowly and with great pain as the men had become
very languid from working in the [chilling] water and many of
their feet swolen and so painfull that they could scarcely walk."
And they "were so much fatiegued today that they wished much
that navigation was at an end and that they might go by land."
This could never be a Rainy River route with *voyageurs* carrying birchbark canoes across short portages.

And still the will denied the fact. Where the Jefferson forks
for the second time, in a single entry Lewis decides that "it
would be vain to attempt the navigation of either [fork] any
further," and also remarks that the world cannot "furnish an
example of a river running to the extent which the Missouri
and Jefferson rivers do . . . [yet] so navigable as they are."
Properly speaking, the Jefferson was not navigable at all; only
inherited faith could say that the Missouri was navigable above
the Great Falls. But Lewis adds that if the Columbia does as
well by them, "a communication across the continent by water
will be practicable and safe."

Sacajawea had been identifying familiar landmarks, as the
failure of the Snakes to appear grew intolerable. (The "guiding" that romances attribute to her is only recognizing.) Before they reached the second forks she told them they were nearing the appointed place: her people would be on the larger
fork or across the range from it on westward-flowing waters.
Clark's torn feet had ulcerated. He had to leave it to Lewis,
who determined to cross the mountains to those waters. "In
short it is my resolution to find them or some others who have
horses if it should cause me a trip of one month." Thus the
force of a wilderness commander, fifteen months out from his
base, deeper in the continent than anyone had ever gone before.
This was August 8 and he started with the indispensable
Drewyer, Shields, and McNeal. They easily found the trail
that led toward Lemhi Pass across the Beaverhead Range, and
caught agonizing glimpses of Indians who slipped away like
startled lizards. On August 12 a spring bubbling up on a
grassy slope became a rivulet and was flowing west. It was
Columbia water. The four had crossed the Continental Divide,
the first Americans who had ever done so, the first white men
who had done so between New Mexico and Alberta. "I had

accomplished one of those great objects on which my mind has been unalterably fixed for many years."

And he had crossed from Louisiana, from the United States, to . . . to the unowned.

There followed a tense, disaster-charged time while the terrified Snakes, who knew nothing of white men and took these invaders to be Blackfeet,[3] spied on them and quaked. Then a tenser time when at last Lewis caught up with a band of them and with a skill beyond praise (and with Drewyer's sign talk) contrived to lull the terror-saturated suspicion of the primitive mind just enough and no more. Tension grew while he persuaded them to go back with him to find Clark, the rich wealth of manufactured goods (it amounted to about half a basketful by now), the marvel of a man who had a black skin, and a woman of their own blood. The suspense ended when a dramatic contrivance not permitted dramatists revealed that Lewis had met the very band from which Sacajawea had been raped and that its chief was her brother. "She instantly jumped up and ran and embraced him, throwing over him her blanket and weeping profusely."

They got the horses, though keeping them and keeping the Snakes to their promised help in freighting the outfit through the pass were anxious matters, considering the capricious changeability of the Indian temperament. They also got a mass of information about the country west and south; some of what they extracted from it served them, more gave them false ideas. They had crossed the Divide by Lemhi Pass; from, roughly, Dillon, Montana, to Idaho just north of its wide southern half. A salmon which was further evidence that these were Columbia waters was stranded in a creek that enters the small Lemhi River, which presently flows into Salmon River. The Snakes knew from the Nez Percés, of whose existence the captains now heard for the first time, that the Salmon River eventually reached the sea, where there were white men. This news confirmed the data of the Minnetarees and the geographical reasoning of the captains: the Salmon must be the south branch of the Columbia which they had posited. Clark named it Lewis's River and eventually they knew that it emptied into the Snake River, which inherited the name. But though in the end they smelted some of the impurities out of these data they never did get a true idea of the Snake River's extent and course.

Their principal informant was an old man who had originally belonged to a division of the Shoshone group known as
the Bannocks.[4] They lived twenty sleeps away, he said, beyond what the captains had no way of realizing was a desert of
jagged lava. (That is, they lived on the Snake River on its long
western course across southern Idaho; for orientation, one may
think of Twin Falls.) When he wondrously said that this was
"not far from" the Spanish settlements, he further solidified
their sufficiently theory-fast idea that the Columbia headwaters
which they knew they were on "interlocked" with those of the
great Colorado River and those of Delisle's synthetic Rio de
los Apostolos. This drastic compression of the Southwest led
them to interpret the route he advised them to take to the
Pacific — he was telling them to head southwest, to strike the
Snake River at (say) Boise, and to keep going down it from
there — as a trail "to the Vermillion Sea or gulph of California." By the end of another year Clark's geographical sense
had got much of this country correctly oriented, but he never
did pull the Southwest out of this compression.[5]

The old man described rivers identifiable, though not by
Lewis and Clark, as the Snake after it turns north toward the
Columbia and the Owyhee, and two others incapable of identification and no doubt imaginary. He and the chief said, quite
truly, that the expedition could not travel on the Salmon River
or down its valley. This was shocking, even disastrous news,
violently upsetting their expectations. (They had not differentiated the Bitterroot River in what the Minnetarees told
them, and the Minnetarees had no knowledge of the country
west of a northward-flowing river.) Asked how this new tribe,
the Nez Percés, crossed the mountains, the old man said, again
with an accuracy which the expedition would sorely verify, that
they traveled by a bad road — bad because, among several
reasons, you nearly starved on it. Everybody's advice was to
stay here with these merely underfed people for the winter
and try the tough mountains and the lava plain next spring.

Impossible: to do so would mean failure. Lewis decided "instantly" that if the Salmon River indeed proved to be an impossible route he would commit the expedition to the Nez Percé
road. If Indians could travel it, his men could.

This band of Snakes were about to make their desperate annual venture to the buffalo plains, and for greater security against the Blackfeet were going to travel with a band of Flatheads, about whose tribe Lewis now got further information. He stayed on with the Snakes, trading for horses and completing preparations, while Clark went off to determine whether the dismaying news about the Salmon River was true. In four days he found that it was all too true. No boats could be taken down that wild water roaring round immense rocks, and horses could not possibly travel along the steep sides of the canyons through which it flowed. Even today the full extent of those canyons is traveled by boat but rarely and then for sport only, in craft specially designed for the tour de force and used exclusively for the downstream voyage.[6] An upstream voyage is so nearly impossible that the regional dude wranglers advertise the Salmon as "The River of No Return."

An exactly similar total situation had cowed all of Alexander Mackenzie's party except him. Its seriousness could not be overstated but Clark had no thought of abandoning the exploration, and Lewis had none when he heard about it. Both had heard, separately, of a trail east of the Divide that would take them northward from the place where they had hidden the dugouts, toward the Nez Percé road. (It led through Big Hole Basin, across the Divide by Gibbon's Pass, and down to the valley of the Bitterroot River, which they now began faintly to separate from the Salmon.) And on his reconnaissance Clark's guide had told him of a route north to the southern and secondary Nez Percé trail and, more vaguely, of another route that led to the main one, off Bitterroot Valley. He summarized what he had seen and heard and sent his results to Lewis, along with the guide whom Lewis might question further. Also he sent his recommendations. One was sufficiently drastic: to send picked men scrambling on foot down the sides of the Salmon River canyons and to send the rest on horses across the less precipitous mountains back from those canyons, the divisions to get in touch whenever the river might permit. The second, loosely phrased but soon adopted, was to hire the old man who had been guiding him (perhaps the same old man who had instructed them in geography) and who henceforth was addressed as "Toby," to guide them to the Nez Percé trail. If on that trail food proved to be as scarce as Toby said, they could eat some of the horses they had bought.

Clark also made a third recommendation which, though it is crossed out in the *Journals,* is revealing. They might divide the party in thirds. Two of the divisions would force the mountains somehow, leapfrogging each other. The other one was to go back to the Great Falls and get supplies from the caches — and from there it was to come west up Medicine (Sun) River. Clark was beginning to see his way through the information from the Minnetarees that they had misunderstood or disregarded.

They had promised the Snakes to bring the trade to them — all the wealth in wool, metal, and firearms that would enable them to stand up to the Blackfeet. (A formal assurance by the President's agents that the United States would carry the trade beyond Louisiana to foreign soil. Or to the soil whose bound· aries Jefferson thought it wise to keep obscure.) They had got their outfit packed through Lemhi Pass. They had bought twenty-nine horses, not so many as they needed but all the Snakes would sell on the eve of the big hunt. They were good horses but Indian-owned and therefore refractory, and likely to have running sores on their backs. On August 20 they started out down the Lemhi River, a land expedition now, guided by Toby and one of his sons. In the high country late August is the ebb of autumn. Birches, willows, cottonwoods would be golden and the aspens would fly their bright silver against the lowering evergreens. Toby took them on a trail that led down the Salmon for a space, then away from it where it curved westward and up defiles and over shoulders of the Bitterroot Mountains. The high, steep trail failed to resemble anything the captains had seen in the Alleghanies or the Blue Ridge. "Several horses fell, Some turned over," others slid down glazed rock faces. Rain in the canyons was snow higher up. They marched through sleet storms on slippery rock and sliding gravel and on September 3 "passed over emence hils and Some of the worst roads that ever horses passed." The next day took them over the last ridge of the Bitterroots, down to a creek that eventually led to Ross's Hole and to a band of Flatheads, who were friendly and very rich in horses.

The Flatheads always were friendly to white men and white

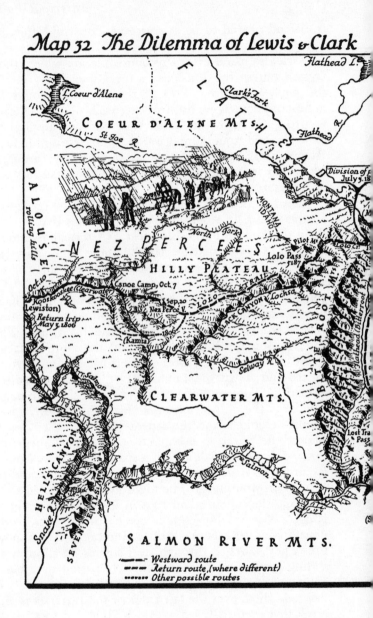

Map 32  The Dilemma of Lewis & Clark

Flathead L.

FLAT

Clark's Fork

L. Coeur d'Alene

COEUR D'ALENE MTS.H

Flathead R.

St Joe R.

PALOUSE rolling hills

Division of
July 5, 18

North Fork

NEZ PERCEES

Pilot Mt.

Lolo Cr.

MONTANA
IDAHO

HILLY PLATEAU

Lolo Pass
5187

LOLO TRAIL

Oct 10

Kooskooskie (Clearwater)
(Lewiston)

Canoe Camp, Oct 7

Sep.20

Nez Percé V.

BLACK CANYON

Lochsa

Return trip
May 3, 1806

(Kamia)

1806

Selway R.

BITTERROOT MTS.

Flathead (Bitterroot)

CLEARWATER MTS.

Salmon R.

Lost Tra
Pass

HELL'S CANYON

Salmon R.

Snake R.

SEVEN DEVILS MTS.

SALMON RIVER MTS.

Westward route
Return route, (where different)
Other possible routes

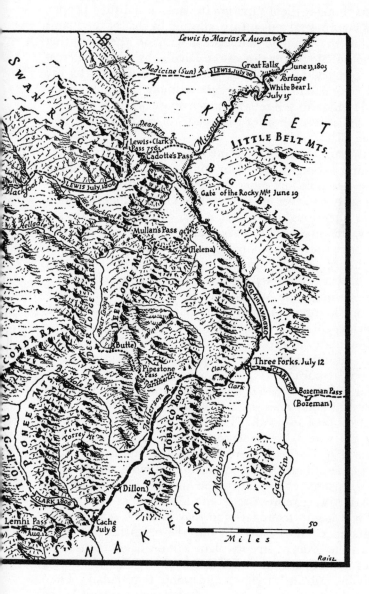

SWAN RANGE

BLACKFEET

Lewis to Marias R. Aug.12. '06

Medicine (Sun) R.    LEWIS July '06    Great Falls    June 13, 1805
Portage
White Bear I.
July 15

Dearborn R.

Lewis-Clark's
Pass 7561
Cadotte's Pass

LITTLE BELT MTS.

Missouri R.

Blackfoot R.    LEWIS July 1806

Little Blackfoot R.

Gate of the Rocky Mts. June 19

BIG    BELT    MTS.

Hellgate

Clark's R

Mullan's Pass

(Helena)

DEER LODGE MTS.

Bear Gooce R.

(Butte)

CONDA RA.

PIONEER MTS.

Big Hole R.

Torrey Mt.

BIG-HOLE

CLARK 1806

Lemhi Pass
Aug. 12

Deer Lodge R.

Pipestone
Pass
Panther R.

Jefferson R.

TOBACCO ROOT RA.

RUBY RA.

Dillon

R.

Cache
July 8

SNAKES

SMITH'S R.

Three Forks. July 12
Clark
Clark    CLARK '06    Bozeman Pass
(Bozeman)

Madison R.

Gallatin R.

0    50
Miles

Raisz

men always thought them superior Indians. A Salishan people, they had a vigorous mixed culture, affected by that of the Plains tribes whom they fought and traded with on their hunting trips to the buffalo-plains east of the Divide, by the Nez Percés and their congeners west of the Bitterroots, and even by the river and coastal tribes who had the salmon economy. They were good-humored, self-reliant, much more warlike than the Snakes but miserably short of guns and iron and therefore harried by the Blackfeet.

The Salishan tongue, it immediately developed, was "a gurgling kind of language." Toby could speak it but it did not in the least resemble any other Indian language that any member of the party had ever heard. So an old notion flickered for a moment in these toughly realistic minds and Lewis conducted an inquiry, making a vocabulary and listening most carefully to this enunciation which Ordway says was like "an Impedement in their Speech or brogue on their tongue." For that moment the expedition was on the rainbow edge of fantasy: this language might be Welsh and these people, "light complected more So than Common for Indians," might be descendants of Prince Madoc. The notion, which would have impressed thousands of devout Celts in the States who knew that the Rocky Mountains were a likely place to find Madocians, flickered out at once.[7]

The Flatheads readily sold some horses from their big herds and exchanged good ones for a few that were lame or balky, taking a little boot in goods. The party now had forty horses and three colts, to be considered not only pack animals but provisions. The Flatheads were glad to help in any way they could and especially they had sound information. It mortised perfectly with some of the things the Snakes had said and, the captains now saw, it reinterpreted the Minnetarees' data in the way they had begun to understand they must be read.

On September 7 they came out into the beautiful Bitterroot Valley with the orderly, steeply planed peaks of the Bitterroot Mountains making a north-south line at their left. (Still west of the Divide, still on foreign soil beyond Louisiana.) They had named the creek they were traveling down Flathead River but would eventually rechristen it Clark's River, Clark adding an annotation, strange in one who had seen so many new rivers, that he was the first white man who ever saw its waters. At the southern end of the big valley it was joined by a fork that came in from the southeast and they kept the name for the enlarged stream. The Flathead was a splendid mountain river, wide, swift, clear, singing. It flowed north, down that magnificent valley. It was the river that has repeatedly been called in this text, as it is on modern maps, the Bitterroot. The Minne-tarees had told them about it.

They traveled beside it down the valley till on September 9 they reached a creek which came tumbling from the western mountains and up whose valley, the guide Toby said, they would reach the Nez Percé trail. Deciding to stay here for a day, they called the site Traveller's Rest. And here they codified the new and revised information. Toby and the Flatheads did not know where the Bitterroot, which they were calling the Flathead and would presently call Clark's River, led in the end but so far as they knew it, it flowed north. (More or less: on the average northwest and frequently west.) It was navigable here by such boats as those they had left at the Jefferson and clearly it must be part of the Columbia watershed. The ab-sence of salmon, however, convinced Lewis and Clark that somewhere in it there must be a high fall. Toby said that only a few miles farther down the valley it joined a stream almost as large that came in from the east. This stream — it was Hell-gate River — crossed a big plain by which the Missouri could be easily reached. The trail that led up the Hellgate, Toby (or someone else) said, reached the Missouri some thirty miles up-stream from the Gate of the Mountains. (He was describing the second route mentioned previously; this trail would lead up the Little Blackfoot and across the Divide by Mullan's Pass.) It was easy going and the travel time from the Bitterroot to the Missouri was five days. . . . At Traveller's Rest they were fifty-three days out from the Gate of the Mountains.

Clark went on down the Bitterroot till he reached the river that came in from the east. They tentatively called it Valley Plain River, then settled for the East Fork of Clark's River. (It is the Missoula, but has been the Hellgate from the mouth of the Blackfoot down to this confluence.) After further thought and analysis they were sure that the Minnetarees had told them about this river — had condensed it as a northward-flowing

river along the base of the Rocky Mountains "at no great distance from the sources of Medicine river." (Sun River, just
above whose mouth, at White Bear Island, the portage round
the Great Falls had ended. They were fifty-eight days from
Sun River.) They decided that it must flow north to the latitude of Sun River — which they made only a third of a degree
higher than they had now reached — and then turn west and
fall into the Columbia. As Clark's Fork it does flow west but
as the Pend Oreille, beyond Lake Pend Oreille, it reaches the
Columbia at 49°. . . . If they had yet differentiated the route
up Sun River and down the Blackfoot from the one through
Mullan's Pass and down the Little Blackfoot, the journals do
not say so. Nor is there any suggestion that they will explore
the route on their return.

Some Flatheads dropped in for a chat. One of them, who
agreed to go along as a guide but didn't, reported that some of
his relatives, probably meaning another Flathead band, lived
at the western foot of the Bitterroot Mountains, on a river that
must be the Columbia. Last fall some of them had gone down
to the sea, where an old man who lived alone there had given
them just such handkerchiefs as the captains had among their
goods. It was easy navigation all the way, he said, and the trip
would require five days. Whether he meant five days from
Traveller's Rest to the sea or five days from the Walla Walla
plains to the sea, he was composing an eloquent traveler's tale.

In June of the following year, 1806, when they were diligently collecting food for the return passage along the trail
they now faced, Lewis wrote, "not any of us have yet forgotten
our suffering in those mountains September last, and I think it
probable we never shall." Though Toby and his son contrived
to get them off it at times, the Nez Percé trail, which was a road
from their country by way of Hell Gate to the buffalo at Deer
Lodge Prairie, was well marked and ought to have been unmistakable while free of snow. The trouble was that whereas
white men made a road along the floor of a valley, Indians held
to the ridges. They traveled the ridges now, sometimes at an
elevation of 7500 feet. They took the trail up Lolo Creek
(their Traveller's Rest Creek), across the first Bitterroot ridge
by Lolo Pass, on through a maze of mountains, and eventually
down to plateau country and the upper creeks of the Clear-

water River. They left Traveller's Rest on September 11 and reached plateau country on the 19th, so that the worst of it was only nine days. The worst, however, was pretty bad.

They had been on short rations ever since leaving the Jefferson but now there was practically no game. And they were in the mountains, "the most terrible mountains I ever beheld," Sergeant Gass wrote. That meant not only heaving lungs and pounding hearts, and not only slopes and cliffs and heavy timber choked with underbrush — it also meant piercing cold after months in the furnace-climate of the plains. They chased the pack-wary, accursed horses through brush that was saturated with rain. Clouds settled down to treetop height and you could see neither the vast slopes below nor the trail ahead. Rain that no coat would turn became hail, then sleet, then snow — eight inches of snow one day on top of four inches of old crust, ten inches one night. Straining ahead for deer or anything else that could be eaten, Clark was "wet and as cold in every part as I ever was in my life, indeed I was at one time fearfull my feet would freeze in the thin Mockirsons which I wore." (That day he quit early to build big fires so that the men could get warm fast when they came up.) For the first time Lewis broke out the iron-ration, "portable soup," a military experiment in concentrated or desiccated stock that was at least edible. They named one stream Colt-Killed Creek from that day's supper and another one Hungry Creek. With "inexpressable joy" they came down to flatter, more open country and warmer weather. It remained broken, however, with more cliffs, more hills and even mountains, tangled timber — and no game.

For the first time the journals record low morale. "The want of provisions together with the dificul[t]y of passing those emence mountains dampened the sperits of the party which induced us to resort to Some plan of reviving ther sperits." That could only be food. Clark took six hunters and hurried ahead to find game. Coming across a stray horse, he killed it and hung "the greater part" from a tree for those behind. Thus it was that on September 20 he met a village of the admirable Nez Percés. There was hardly a moment of strain or hesitation. Voluble in sign talk, the advance party was soon in the chief's tipi dining enormously on dried berries, dried salmon, and cakes made from the camas root that was a Nez Percé staple. Clark supposed that simple overeating was what made him sick that night; he soon found out otherwise. He bought "a horse load" of this strange but welcome food and sent it back to the party, which had dined that day on coyote and had, quite suddenly, begun to weaken. He hurried back too, taking more camas and dried salmon and managing to kill three deer on the way. The deer helped but not enough, and this new diet proved disastrous. As the whole party staggered down to the Nez Percé village and on to the Clearwater River, Clark's journal became a hospital morning-report. "Cap^t. Lewis scercely able to ride on a jentle horse which was furnished by the Chief. Several men So unwell that they were compelled to lie on the Side of the road for Some time others obliged to be put on horses. . . . Cap^t. Lewis verry sick. . . . Several men taken sick on the way down. . . . Several taken sick at work. . . . Drewyer sick. . . . 8 or 9 sick. . . . 3 parts of Party sick. . . . nearly all the men sick."

The trouble was gripes and dysentery: "Their bowels, a heaviness at the Stomach & Lax." No doubt there were bacteria in the dried fish. Far from well himself, Clark bombarded them with explosives, "Rush's pills," the mixture of ten grains of calomel and ten grains of jalap which the most famous American physician of the time relied on to cow the disordered system. But the ailment lingered on and he brought up heavier guns. "I gave Some Salts & Tartar *emetic*. . . . I gave Pukes Salts &c to several. . . . I administered *Salts* Pils Galip Tarter emetic &c." They mended as they got used to the diet and were helped by occasional meat, adding horse-beef to whatever the hunters could find. But they had to go on eating roots and dried fish, which the men complained acted like Glauber's salts. There was an additional embarrassment: "Cap^t Lewis and my selfe eate a supper of roots boiled, which filled us so full of wind that we were scercely able to Breathe all night." Eventually Sacajawea taught them how to prevent the flatulence with a salad of herbs.

The Nez Percés navigated the Clearwater, which the captains called by the Flathead name Kooskooskee, with rafts and dugouts. They would make dugouts, then, though they must use pine, a poor wood for the purpose.[8] They went down to where

the north fork of the Clearwater came in and started felling
trees, gladdened by the thought of taking to boats again after
the mountains. They kept getting geographical information
from the Nez Percés.

The Nez Percés were the charming tribe, a highly intelligent people who had an advanced culture. Like the Flatheads,
they had absorbed much Plains and Coastal culture, and had
paid in kind. Not only the camas and kouse (the arrowhead)
which they dug and dried passed along the trade routes, but
their basketry, their woven bags, and especially their bows.
The captains had seen Nez Percé bows at the Mandan villages
— short, with a double curve, usually of horn and always
backed with sinew. The most powerful bow used by Indians
in the area of the United States, it brought premium prices.
It was a bow for horsemen — and horses were the Nez Percés'
principal celebrity. In the broad Idaho valleys, a fine stock
country, they raised great herds of them. Almost the only Indians who practiced selective breeding, they had developed a
tough, strong, fast stock. When first the trappers and then the
emigrants followed Lewis and Clark west, the Nez Percé horse
made its breeders affluent for it outsold the scrubby "ponies" of
other tribes. One strain they isolated became famous, lapsed
from public knowledge, and has again been brought back —
the strong and graceful Appaloosa with its characteristic mark-
ings, usually a spotted rump. The first description of the Appaloosa is in Lewis's journal.

Neither Toby, his son, nor Sacajawea could speak Nez Percé.
The explorers could talk with them only by signs, through
Drewyer for best effect though Lewis and Clark were adept by
now. In spite of this handicap, they got additional information
about the routes they had missed. They got more when they
returned to the Nez Percés the following spring, found the Lolo
trail blocked by snow, and made a longer stay here than they
made anywhere else except at the two winter posts. Between
visits they had picked up further data and had mulled the
whole problem over during the winter.

The point it indicated was clear and precise: the route they
had taken west was certainly not the shortest and probably not
the best one.

∽

They reached that conclusion the following winter, the winter of 1805–6, which they spent at the log structure they called Fort Clatsop, a few miles down the coast from the mouth of the Columbia River and a few miles inland from the Pacific.

Here the chronological order of this narrative must be broken.

The conclusion that they had missed a cardinal part of the direct route which they had been sent to find was extremely important. It was responsible for their dividing the party on the journey east in 1806 and exploring the area which they had detoured, the area where the missing section of the direct route would be found, if indeed there were one. The decision will be clear only if it is seen in relation to the events just narrated. Its consequences too belong here. Moreover, actions and lines of force outside the westward journey but in circuit with it have been too long ignored in the narrative. If the relationships are to be clear they must now be brought within the system of forces and actions, and their consequences with them. It is oddly necessary to bring the expedition back from the Pacific now, before it gets there.

We will come back to them here, in this first week of October 1805, at the forks of the Clearwater River. (Where in 1836 the Reverend Henry Spalding was to establish the Nez Percé mission.) They called the site Canoe Camp, and we will take them on from there at the proper time. But we must now leap over the miles and months between, to the dank winter at Fort Clatsop.

৵

Only twelve days of that winter were free of rain, only six were sunny. Everything, especially their clothes and bedding, mildewed and rotted. Their monotonous food was elk meat, which the damp also spoiled. Always drenched themselves as soon as they left the fort, they had no liquor to give them an occasional lift. Tobacco got so low that no more could be issued to the men, since the remnant had to be kept for currency on the way east. Some of the local Indians were amusing in a wan way but all were a decadent stock and a good many were treacherous. They were the fringe-tribes of the Northwest culture that was much more vigorous to the northward, and they had been debased by the maritime trade. Dull-witted, thievish, lying, rotten with gonorrhoea and the pox, they were a depressing contrast to the Mandans and Minnetarees of last winter. But if all this spelled monotony for the men, there was more than enough work to keep the captains busy. Lewis systematized, expanded, and entered in the journals a large mass of notes they had made on the flora, fauna, and Indians between Fort Mandan and Fort Clatsop. Clark, similarly ordering their route sketches and notes on courses and geography, drew a series of maps covering the same area, from Fort Mandan to the mouth of the Columbia.

The decision about the best route west is another triumph of analysis. It is announced in both Lewis's and Clark's journals, Clark's entry being much the more detailed, for February 14 1806. Several items must be specified, beginning with the final residue of geographical error. This is concentrated in the area which they had not seen and about which everyone's thinking had been erroneous — the country south and southwest of the route they had traveled, between the Continental Divide and the California coast. In their thinking this area remains compressed, with shorter distances than it actually has and an orientation which they have worked out by deduction from imperfect data. One river that traverses it is the Snake, which on their western journey they had entered at the mouth of the Clearwater. They call the Snake the Southeast Fork of the Columbia. They have decided that it cannot be used as a route through the Rockies to the Columbia, for the Indians say that above the Clearwater it is not navigable. Nevertheless its sources are close to those of the Madison and the Jefferson.[9]

They have learned by now that there is another large affluent of the Columbia which they did not know existed when they passed its mouth on their way west. (The mouth was hidden by a long island and masked by a series of marshes.) They call it the Multnomah; it is the Willamette of today. The Multnomah, they have decided, rises in the mountains south of the sources of the Jefferson River and near the Spanish settlements (New Mexico, not California). That is, its eastern fork does, for they have decided that it must fork, and the source of this fork is also contiguous to that of the Yellowstone River.

This fork flows through barren plains and cannot be navigated. The western fork of the Multnomah "probably" heads with the Rio Grande or the Colorado. (A cluster of serious errors. Some of this is data that relate to the long western course of the Snake River through southern Idaho, some the unquestioned assumptions of symmetrical geography, some their own false premises acquired from the Snakes at Lemhi.) But though the Multnomah is impracticable as a way of reaching the Columbia from the Missouri, they suggest that perhaps it may "afford a practicable land communication" with New Mexico by way of its western fork.

So much for a route to the mouth of the Columbia south of the one they had taken in 1805. Since leaving the Clearwater, they have learned that Clark's River (Clark's Fork plus its extension, the Pend Oreille) reaches the Columbia far to the north of where they thought it must when they speculated about it at Traveller's Rest. They have also confirmed their deduction, based on the lack of salmon, that it is broken by an enormous waterfall, and this obstacle cannot be passed either by water or by land. (This must be Metaline Falls.) Clark's River, therefore, must be ruled out too as a route from the Missouri to the Columbia; it is unnavigable and besides it is too long and would be too circuitous.

But close criticism of their last year's route has forced on them the conclusion that it will not do, either. The Missouri above the Great Falls will not do: its crookedness and the weary labor of towing boats forbid. And the route is much too long. From the Great Falls to the forks of the Jefferson where they abandoned the boats is, they make it, 521 miles of water travel. From there to Traveller's Rest is 220 miles of land travel. This is "further by abt 600 miles" than the direct land route, which they now understand, from the Great Falls to Traveller's Rest where the Nez Percé trail begins its traverse of the Bitterroot Mountains. In addition, if they can believe the Indians, their land passage from the Jefferson to Traveller's Rest is "a much worse rout," being more mountainous.[10]

This decision of February 1806 signalizes westering man's final abandonment of the hope that there may be, the mystical certainty that there must be, a direct water communication between the Atlantic and the Pacific. A combination of water

and land travel must be accepted instead. "We now discover
that we have found the most practicable and navigable passage across the continent of North America." It is the route
they traveled west, except for that laborious and time-consum-
ing stretch from the Great Falls to the Nez Percé trail where it
comes down to the Bitterroot Valley. The "best and most
practicable rout" will simply not try to cross any part of the
Rocky Mountain system by boat. It will abandon boats and
take to horses at the "entrance of Dearborn's river," will head
overland almost due west to Traveller's Rest, and will cross the
mountains and go down to the Clearwater as the expedition
did, by the Nez Percé trail.

This momentous conclusion must be tested. That was fully
determined at Fort Clatsop and was one of several reasons why
the party must be divided on the return journey. The country
of the upper Yellowstone River must be looked into, for there
are those possibly convenient approaches to the Jefferson,
Madison, and Gallatin Rivers. Clark will make that exploration, determining also what the indications are of those even
more convenient approaches to the Spanish settlements, the Rio
Grande, the Platte, and the Colorado. Lewis will explore the
smaller and easier area which the direct land route from the
Great Falls is to traverse. After doing so he will look into something more important and very promising. The Marias River
was the last of several approaches to the great Canadian
fur country. Lewis was sure that it must come down from north
of 49° and this most valuable probability must also be put to
the empirical test.

◌∿◌

Tying up these ends was the last necessary business at Fort
Clatsop. Their winter stay in the rain had accomplished every
objective except one: they had met no American ship (or any
other) that was in the maritime trade. One was actually trading in the vicinity soon after they built the winter post but the
Indians never mentioned it. They had, however, got the names
of several American ship-captains and various memoranda
about their peculiarities and methods. They had seen an impressive variety of trade results, good and bad. Picking up a
little of the Clatsop language and of the "Chinook jargon," the

trade-Esperanto of the Northwest, they had got direct Indian testimony about it. They judged it unwise to leave a detach· ment here to find a trader and take reports and messages to the States by sea: the stay might be so long that the expedition would get home by land first.[11] It would have been exceedingly desirable to speak a ship, not only to formally close the trade circle but to get goods to pay their way home. For by now the conquerors of the wilderness were paupers. Of the baubles and small hardware that were pocket-change they had left, Lewis recorded, no more than could be wrapped in two handkerchiefs. The rest of their stock consisted of seven trade blankets, one coat and hat of the gaudy army uniforms much desired by war chiefs, "and a few old cloaths trimmed with ribbon" to make them more salable. Some skins and furs of their own tanning could be added to the drawing-account. And this was all they could spend for service, food, horses, and diplomacy till they could reach the Nez Percés, where they had left their own horses and cached a handful of goods. There was another small cache at the head of navigation on the Jefferson River, but the largest one — not very big at that — was at the Three Forks.

They left various rosters and announcements in the hands of various Indians to be shown to traders when they should appear. They carved and branded trees and affixed notices to them, recounting their achievement. (No one said so but in the polity of nations this was ritual to buttress the claim which the United States had to the Columbia drainage through Robert Gray's discovery. The ritual announced that they had traversed the country and had occupied it.) Then in some of the dugouts they had made at the Clearwater and two of the much better seafaring canoes of the Columbia River tribes, they abandoned Fort Clatsop on March 23 1806.

They found the Columbia beginning to flood. They located the Multnomah — the Willamette — 142 miles upstream and Clark took a party some distance up it. What he saw and what the resident Indians said and sketched for him convinced him that it was an important river. It must drain the area they had supposed it did and its California source must be as far south as 37° if not farther. (It is north of 43°.) That was the most important Columbia business. The rest was fighting the furious river, harder work than going down it had been, though God

knows that had been hard enough. The tribes who lived
along the river were even more objectionable than they had
been last year — of a curdled and spiteful disposition, repulsive with lice and trachoma and venereal scabs, whining and
jeering and arrogant toward white men who had no goods,
profiteering on what few necessities they condescended to sell,
ambitious to be pirates and murderers but only daring to be
thieves.

At the Cascades, the first-reached of the furious rapids, one
of the big canoes was swept away. Also the local residents tried
to take back from Shields a dog he had bought for food — he
drew his knife and they ran. They stole an axe, which the
owner wrested away from them, and they most injudiciously
kidnapped Scammon. Lewis went after them in such a fury that
the mere sight of him terrified them; they dropped his friend
and ran. "Our men seem well disposed to kill a few of them,"
he wrote and he doubled the guard and told the chiefs that he
would let the men do just that unless manners improved at
once. Beyond The Dalles, however, they were in a decent
climate again; apparently it had not rained here for at least
a week, which seemed miraculous. A big trading fair was in
progress, with bands from upriver and downriver and the inland country exchanging what they had. There were plenty of
horses and, fed up with fighting the Columbia, the captains
wanted to buy some but at the extortionate prices demanded
they could afford only four.

At the lower end of the Narrows of the Columbia they
thought that they might not be able to get the boats through.
So they again tried to convert into horses their remaining shreds
of the white man's miracle, manufactured goods. The Indians
— they were Skilloots, a sizable Chinookian tribe — derided
them and kited prices higher. (It is here that Clark says of
one squaw, "I took her to be a sulky bitch.") Then they decided that there was no chance whatever of getting up the
Narrows: they would have to go by land. That was April 19
1806, and the decision was one more blow to the inland water
route, for eastbound commerce would always have to travel
up the Columbia.

High prices, surliness, contempt, and incessant theft kept
Lewis in a rage. He personally beat up one Indian whom he

caught stealing — the only personal violence the expedition had yet inflicted — and told everyone that from now on he would have thieves shot and their villages burned. But they managed to get enough horses and they found a visiting Nez Percé who said he could find a road, then another who had some horses and would lend them. They were land travelers again when on April 27 they met a roaming band of Walla Wallas (kindred of the Nez Percés) and their chief. The captains, who called him Yellept, had met him last fall, given him a medal, and made a friend. Now he had his people bring them firewood (they had been paying odd bits of metal for odd bits of driftwood), plenty of dogs for food, and "a very eligant white horse." He would have liked a kettle but they had only their own mess kettles left, one for every eight men, and the grateful Clark gave him his sword instead and a handsome quantity of powder and balls. They traveled with these pleasant people to the mouth of the Walla Walla, dancing for them by night to Cruzatte's fiddle.

The Walla Walla empties into the Columbia from the east some distance downstream from the Snake by which they had entered the great river last October. In 1818 the North West Company would build a post at its mouth. In 1836 Marcus Whitman would establish his mission for the Cayuse Indians some twenty miles above the mouth, and would begin his ministry not only to the Cayuses but to the emigrant trains of his countrymen that had been all but shattered by the crossing of the Blue Mountains. Both Yellept and the two Nez Percés told them that there was a trail from the mouth of the Walla Walla to the Clearwater, south of the Snake River. So after practicing some hospitable deceptions to keep these fabulous white men with him a little longer, Yellept had them ferried across the Columbia, and on April 30 they took the trail. On May 3 they reached the Clearwater and the best Indians, the Nez Percés.

They stayed with them till June 24, though meanwhile they had started out to cross the Bitterroot Mountains and had been turned back again by snow. When they made their final departure everyone could speak the language a little and a lot of Nez Percé girls were pregnant. Not so many as the tribal legends count today, however, for there probably were not that many girls.

A desperate idea developed during that otherwise pleasant interlude must be mentioned. Time was now riding Lewis with spur and quirt and he must get his command back to the States this year. Unable to stand the delay longer, he determined to force the trail and determined too that, if forcing wouldn't work, he would take to a route which both Snake and Nez Percé information, so he thought, showed to be free of snow. Free of snow, the information ran, but very bad. It was bad indeed. His idea was to follow the main Snake River to its hypothetical "interlocking" with the Madison or the Gallatin. Almost at once he would have been in the deepest gorge in the United States, Hell's Canyon, whose upper portion was to turn Captain Bonneville back later on. Then there were the long sunken canyons that take the Snake through the lava deserts and the barrenest of sagebrush deserts. In those canyons the river is wholly unnavigable, filled with rapids and falls, including the Caldron Linn that was to wreck the Astorians. On the eastern edge of Idaho and on into Wyoming there were other canyons, besides a maze of mountains that would have to be crossed ridge by ridge.

And note this. The Nez Percés, like the Walla Wallas, lived in a country that lay outside the boundary of the United States. They were probably Spanish Indians, though no one could be wholly sure, and there was a possibility that they might soon become British Indians. That possibility would turn on the event and could be prevented — but assuredly they were not American Indians unless Robert Gray and Lewis and Clark had made them so. And Lewis and Clark told both tribes that the American trade would soon be with them, coming over the Continental Divide, the boundary of the United States.

The salmon had not yet come. They supplemented the game that was so hard to find with camas and kouse, with horse-beef, and with dog, which they had learned to consider a delicacy but of which the Nez Percés fastidiously disapproved. The horses and saddles they had left here last fall had been well taken care of, and the Indians turned over to them the contents of a cache that flood water had exposed. Now they got accurate information about the Blackfeet, the "Pahkees," whom Lewis was sure to meet. He undertook to make the Blackfoot–Nez Percé truce and so would be on a peace mission as well as an exploration. Also they got the last details about the route he

was to travel, the disregarded overland trail between the Blackfoot River and the Great Falls.

They were sorry to leave these fine Indians. And the Nez
Percés were sorry to have them go. They were staunch friends
of the Americans from then on. . . . Till one culminating American treachery, after many others. It was no worse than any
routine dishonor by which the United States government
broke its engagements with Indians but it proved one too many
for the great man who is called Chief Joseph in the histories.
He led a part of his tribe on a 1200-mile retreat from the
United States army, repeatedly outfighting and outmarching it.
The telegraph and the Montana winter stopped him just short
of freedom. He came out of his frozen trench, this man of wisdom, compassion, and war, to meet General Miles, a figure of
less dignity, and raised his hand and made his oath: from where
the sun then stood he would fight the white man no more. So
the white man made his own oath and promptly broke another
agreement with the Nez Percés.

With sixty-five horses, ample food, and two Nez Percé guides
they started for Traveller's Rest June 24. Five days of going that
was much easier than they had expected brought them down to
the hot springs of Lolo Pass, where whites and Indians soothed
their muscles. And to, at last, a country where there was game.
They reached Traveller's Rest on June 30. Here they broke up
into two divisions, which were to become four. They were all
to rendezvous at the mouth of the Yellowstone River.

Clark's party started up Bitterroot Valley July 3. He wanted
to find the road to the Jefferson River which they had heard
about while toiling along the Salmon last year. Shields found
it and on July 6 it took them across the Continental Divide
by what is now called Gibbon's Pass. (Not far in miles from
the Lost Trail Pass by which they had crossed from the Salmon
to Bitterroot Valley.) It led on down to Big Hole Valley,
where Sacajawea could really be a guide. Holding to her landmarks, on July 8 they reached the place where they had sunk
the dugouts in a pond last August 23. There was a furious digging to raise the cache — after more than six months of absti-
nence they could chew and smoke tobacco again. Clark embarked most of the party in the boats for the Three Forks,
leaving Ordway to take the horses there by land. Downstream

by boat was a different matter from the dreadful upstream tug. They made ninety-seven miles the first day and reached the Three Forks on the third.

Here Clark divided his party. Taking eight men, besides Charbonneau, Sacajawea, and her eighteen-month-old child, he struck overland with the herd of forty-nine horses, heading for the Yellowstone. Ordway took the rest of the command down the Missouri in the boats toward the first rendezvous, the Great Falls. Three days brought Ordway to the Gate of the Mountains and three more to the head of the portage, White Bear Island, which he reached on July 19.

Lewis, traveling the direct route from Traveller's Rest, had of course reached the portage first. He had left there all his party except Drewyer and the two Fields brothers, putting Sergeant Gass in command. He had also left horses, and with them to pull last year's truck-frames the job was easy. They raised the caches, transported the whole outfit across the eighteen-mile portage, and got the white pirogue into the water. They took eight days, as against more than a month last year. Lewis was to meet them at the mouth of the Marias, which he was now exploring, and on July 27 Gass started thither with the horses and Ordway headed the boats downstream. The next day, well above the mouth of the Marias, they saw Lewis and his three men who had just reached the Missouri after a hell-for-leather ride, and fired a swivel to welcome them.

Meanwhile Clark's little party had had an interlude of pure enjoyment, the most comfortable traveling of the whole expedition. On July 15 he reached the Yellowstone and saw the majestic mountains through which it flows north from Yellowstone Park. He could find no timber big enough for dugouts and the river was too swift, he decided, for the bullboats he had thought of making. So, looking for better timber, he struck off down the beautiful Yellowstone Valley by pack train. Signal smokes showed that he was being watched and he thought the watchers must be Crows. He found some suitable trees and made two dugouts. They were 28 feet long, 16 or 18 inches deep, 16 to 24 inches wide — in such craft, axe-hewn and hollowed by fire, the great Yellowstone was first navigated by white men. The Crows, honored by all Plains tribes as the most expert horse-thieves, crawled up one night and got half the horses, taking them away over hard and gravelly ground so that their trail could not be followed. The first American horses the Crows ever got.

Clark ordered Sergeant Pryor, with three men, to take the rest of the *remuda* to the rendezvous at the mouth of the Yellowstone and on to the Mandan villages. (They were money; they could be traded to the village tribes for corn and skin-clothing.) In addition, tying up the expedition's loose ends as he traveled, he gave Pryor a letter to the Northwester who had most impressed the captains, Hugh Heney. Pryor was to take it to the posts on the Assiniboine River, as soon as he could get there from the Mandans. It offered Heney, who had traded with the Teton Sioux, the job of inducing several Sioux chiefs to accompany Lewis and Clark to Washington, to be impressed and pacified. Clark offered the going wage — and, with a look ahead, the job of agent for the Sioux at the post which Lewis had advised the government to establish on the Cheyenne River.

Heney was given, half explicitly, an additional mission: to counteract among his Sioux clients the influence of the British traders from the Minnesota River. And the letter had two further purposes. Clark told Heney that the expedition had accomplished its objective by descending the Columbia to its mouth. He added that there was now in progress an exploration of a northern affluent of the Missouri which it had discovered and named the Marias. By this means the North West Company, and Montreal, would learn at the earliest possible moment that the United States had reached the Pacific, had nailed down its claim to the Columbia country, and had investigated its northern boundary.

This was first-rate policy but it did not get started. On Pryor's second night out, the accomplished Crows managed to steal all the horses he was driving. That left no currency to buy goods to give to Heney for presents to the Sioux, and the letter was never delivered. . . . How well these men had mastered the West is demonstrated by Pryor's action. Unhorsed deep in the wilderness, with his command and commander outdistancing him farther every day, he was entirely assured. He shot buffalo and made two Mandan bullboats. Two, not one. One would have comfortably transported the four men and their small outfit, but if it had capsized all the rifles would have been lost. Pryor embarked his detachment in these novel craft and floated off down the Yellowstone, soon mastering their use. They floated up to where Clark was waiting for them on August 8. Pryor reported that bullboats were better for such a river than dugouts.

Clark, however, had taken his party down in his two new dugouts, lashing them together for greater stability, a trick he had seen used on the middle Missouri. There were no difficulties and the party lived high on buffalo meat. He ventured to name a southern affluent Clark's Fork. (Not to be confused with the much bigger Clark's Fork of the Columbia, the extension of the Blackfoot, Hellgate, and Bitterroot.) It is a beautiful stream that comes down from the Absaroka Mountains, passing through a spectacular gorge that is soon to be ruined by a dam. He carved his name on a rock formation which he named Pompey's Pillar for Sacajawea's child, who traveled in his dugout and had become "my boy Pomp." He passed and identified the mouth of the Big Horn on July 6, and repeated the Minnetarees' idea that it was very long and headed with the Platte. (Manuel Lisa was to build his post here a year later and send Colter, who was now with Clark, to determine whether the Minnetarees were right.) He passed the Rosebud on July 28, the Tongue on the 29th, and Powder River, which he seems not to have identified, on the 30th. On August 3, fatly fed, comfortable, an easy master of the job, he brought the expedition's biggest loop full circle, reaching the Missouri. They had come to this confluence on the upstream journey April 26 1805.

He summed up — or rather Lewis presently wrote a summary from Clark's notes. It could be called The Error of the Southwest. The Yellowstone, said the summary, rose on the border of New Mexico, whence a short, good road led to the Spanish settlements. (These unseen settlements were always Santa Fe.) Its source was adjacent to those of the Willamette, the Snake, the Platte, the Bitterroot (!), and the Big Horn. Pirogues would be the best craft for the Yellowstone but even "batteaux," big boats, could go all the way to the mountains. (There must have been a lot of water in the river that year too.) The Big Horn and Clark's Fork must be navigable for a considerable distance. (Entirely wrong.) Besides the Big Horn, the mouth of Clark's Fork would be a good site for a trading post. The transmontane Indians would probably visit such a post, for it would be fairly safe from the Blackfeet. This last suggestion seems, temporarily, to undercut the promise to the Snakes, the Nez Percés, and various Columbia River tribes that the American trade would seek them out. But only temporarily.

The journal says that the Yellowstone has "a considerable fall" somewhere in the mountains. This statement has been crossed out and an explicit "No" has been written after it. Both the statement and the denial are beyond explanation. Clark had talked to no Indians while he was coming down the river. There had been no mention of the tremendous Yellowstone Falls in any of the notes made at the villages or among the Snakes.

At the rendezvous mosquitoes were so numerous and buffalo momentarily so scarce that Clark dropped down the Missouri for some distance, leaving a note on a post for Pryor and Lewis. On August 11 the two trappers, Dixon and Handcock, the first followers, rowed their dugout round the bend and stopped at Clark's camp. Here was the first news of anything downriver. The peace which the captains had negotiated at the villages had lasted only a little longer than the echoes of the orations: the Mandans and the Minnetarees were fighting the Arikaras. Also the Assiniboins were on the prod: they had killed a Northwester on the Souris, had vaingloriously forbidden his company to go to the villages (which was just talk), and had sent out word that they were going to get Charles MacKenzie.

The next day, August 12, Lewis's party came down the Missouri and, yelling, turned in toward the camp. Lewis was stretched out in the bottom of a dugout, temporarily crippled by a gunshot wound in the thigh.

ᛟ

He had had quite a time. His reconnaissance of the Marias River was the most daring venture of the entire expedition. And on that venture Meriwether Lewis is at his best.

With the Nez Percé guides his party started out from Traveller's Rest when all the others did, July 3, marching north down the Bitterroot, and reached the Missoula. They traveled up it — it becomes Hellgate River now — for a space and camped. The next day the Nez Percés, who had been uncomfortable, refused to go through Hell Gate, afraid that there might be Blackfeet. Lewis had meat killed for them, smoked a last pipe with them, gave them what small presents he could, and said goodbye to these good friends. It was July 4. The Nez Percés were sure that the Blackfeet would wipe out his party.

They had told him that he couldn't miss the road, and he didn't. Already certain that a number of routes from the east converged in Bitterroot Valley, he convinced himself on the first day, if he was not already convinced, that no route to the Columbia by way of Clark's River could be so good as the one that led west from Traveller's Rest. He took every military precaution against Blackfeet and went on, turning up Blackfoot River when the Hellgate forked. On July 6 he saw the trail which, he knew by now, led to Dearborn River. This was, he also knew, the one which the Minnetarees had talked about and it was probably the best route between the Missouri and Traveller's Rest. But he wanted buffalo as soon as possible, for meat and for hides, and he could be sure of them on Sun River. So he took the other trail, which he had heard about from the Snakes, the Flatheads, and the Nez Percés. On July 7 it took him across the Continental Divide — into the United States — by way of the easy saddle that has ever since, and inappropriately, been called Lewis and Clark's Pass. Like Clark reaching Big Hole Basin, he had crossed from Columbia to Missouri drainage. On July 8 he reached Sun River and, after a pause to make bullboats, on July 12 the Missouri. The next day they camped opposite White Bear Island, at the head of the portage round the Great Falls. Nothing had ever been so beautiful as the plains or so musical as the once-loathed bellowing of the bulls.

Nine days from Traveller's Rest to the portage, no more than six if he had not stopped to hunt and make the boats, possibly only five if he had pushed himself. (Five days was what the Nez Percés had said.) The thing was proved. No doubt could ever be raised again: the transcontinental water route must be broken by a land passage from the Great Falls of the Missouri to the head of navigation on the Clearwater River.

There remained the question of the Marias River and its access to the Saskatchewan and Athabaska furs.

Some of the horses had been stolen, so Lewis could not take with him so large a party as he had intended to. Leaving everyone else to make the portage, he took the best men he had, Drewyer and the Fields brothers. The four of them, with six horses, started out on July 16 — and hoped to meet no Blackfeet, whose reputation they now knew about. Two days later, July 18, they reached Marias River and started north, keeping it in sight.

Lewis was in the biggest country in the United States. The main chain of the Rockies was at his left, the Lewis Range that has been named for him. In this immensity, where the Great Plains roll a series of vast ground swells toward the mountains and break up in a spray of buttes, desire told him that some of the purple-shadowed defiles he saw to the northwest must be flowing Saskatchewan water. (None was.) The Marias, which at its mouth had so promising a northern bearing, disheartened him by turning west toward the mountains, much to the south of where it ought to. On July 21 he reached a creek which entered it from the north and which, as he could see, came down from the mountains. With sorrowful realism he decided that this creek, which is called Cut Bank, would take him "to the most no[r]thern point to which the waters of Marias river extend which I now fear will not be as far north as I wished and expected." Note: *fear* and *wished*.

The next day, July 22, he was sure. He must give up the hope but he would take what comfort remained. "I believe that the waters of the Suskashawan approach the borders of this [Marias] river very nearly. [Perhaps, that is, a portage would do the intensely necessary job.] I now have lost all hope of the waters of this river ever extending to N. Latitude 50° [into North West Company country] though I still hope and think it more than probable that both *white earth* river and milk river extend as far north as lat$^d$. 50°."

One river or another river, but somehow.

Now in the farthest corner of the United States, desire to get the exploration ended surged up in Meriwether Lewis, bringing anxiety with it. Could he get his command back this year? He could barely hold himself here for the weather to clear for the necessary observations, though he knew he was south of 49°. When it did clear, his chronometer had stopped — and he headed for the mouth of the Marias. He had told the rest of his party to wait for him there, after finishing the portage and getting the boats into the Missouri. (But if Captain Lewis did not get there by September 1, they were to forget about his whitening bones and go on to the mouth of the Yellowstone and rejoin Clark.) On the way there this party of four ran into, at last, the Blackfeet.[12]

Eight of the most belligerent Plains Indians. (Lewis at first suspected that others were near by.) There was nothing for the small group of whites to do but to put up a bold front and they did so, breaking out a flag and advancing toward the Indians. The Blackfeet, certainly much more scared, willingly talked sign with Lewis and Drewyer. They said — as a precaution — that a big band of their people were not far off, and insisted on camping with their new white friends. That night there was big talk. Lewis told the Blackfeet about the Americans and they told him about the posts on the Saskatchewan, six days from here they said, and the fine goods and wonderful drinking-liquor they got there. They said that there was a white man, presumably a trader, with the big band; if there was his name has never been learned. Lewis issued a general invitation for a council at the mouth of the Marias, to talk peace and trade. Confident that they would not attack four armed men but sure that they would steal the horses if a chance offered, he sternly cautioned his men. He took the first watch himself, then went to sleep beside Drewyer. Drewyer slept that night in a tipi of the people who would lift his scalp at the Three Forks in 1810.

Joseph Fields was a first-rate man, a master of the wilderness skills, long trained and hardened. But, though this was no time to lay his rifle down, he did just that — on the dawn watch. Instantly one of the Blackfeet seized it, another of them seized that of his sleeping brother Reuben, others grabbed Lewis's and Drewyer's, and everybody was running. Joseph Fields ran after the now armed thief, followed by the instantly awakened

Reuben. They caught up with him at sixty paces and it was Reuben Fields who seized the gun by the barrel and sank a knife into him, killing him. Recovering their rifles, the two men rushed back to the tipi, whence Lewis and Drewyer had rushed out in a chase of their own. Drewyer wrested his rifle away from a very lucky Blackfoot and set out after Lewis, joined now by the Fields brothers. Lewis, his rifle in his hands again, yelled at them to shoot any Indian who might try for the horses, and kept running. He made a snap shot and brought down one of them. The return fire fanned his hair but he too had killed an Indian.

That ended it, two Blackfeet dead and six running hellbent across the plain. The party reassembled and, in Indian values, had won a victory as great as Waterloo. They took the pick of the Indian horses and booted the rest off into the sagebrush. They helped themselves to some meat, took one of the two Indian muskets and some other souvenirs, and then burned the saddles, the bows and arrows, and all the miscellany. If any brave had had a medicine bundle with him, a blight would waste his family from now on.

The skirmish took place at dawn July 27. The Blackfeet had said that they had a big force half a day's journey away. The combined detachments of Ordway and Gass were, or soon would be, coming down the Missouri, unaware that there were Blackfeet about. Lewis ordered his men to mount and took them toward the mouth of the Marias as fast as they could push the horses, picking a route by instinct. They rode till nightfall, then stopped for a couple of hours and killed a buffalo for supper. Mounting again, they took a slower pace across a plain populous with buffalo herds and fitfully lighted by a moon that kept going behind storm clouds. When they bivouacked at 2 A.M., they had ridden just a hundred miles.

Waking at daybreak, they could hardly stand but the horses had not been broken by the long drive. Supposing that the big, but probably imaginary, band of Blackfeet were pursuing them, Lewis told his companions that if it came to a fight they would tie the four bridles together and shoot it out. They pressed on toward the Missouri and presently heard a shot. Perhaps the Blackfleet and Ordway had met. Nothing more while they rode another eight miles, then several shots, irregu-

larly spaced, echoing across the empty plain. They were rifle
shots, which meant white men, and no musketry fire answered
them. It was Ordway's party. His hunters were killing meat
along the banks while the others enjoyed the boating on the
Missouri. Hurrying down to the dugouts, Lewis and his men
threw their few belongings in them and gave the horses "a final
discharge" — he was through with irregular cavalry operations.
They slid on down to the mouth of the Marias, where almost at
once Gass's detachment caught up with them.

The old hope glowed again when on August 4 they passed the
mouth of Milk River, bankfull. Lewis had no doubt that it
"extends itself to a considerable distance North." August 7
brought them to the rendezvous, the mouth of the Yellowstone, but Clark had gone on farther. On August 11, Lewis
went hunting with the invaluable Cruzatte, because the meat
had spoiled and they had spotted a herd of elk. They followed
them into a willow thicket and there the half-blind Cruzatte
fired at some movement or at what he took to be an elk, and
hit Lewis in the left thigh. Lewis supposed that he had been
ambushed by Blackfeet or Assiniboins, managed to reach the
boats and order a party to land and rescue Cruzatte, but could
accompany them for only a little way before his leg gave out.
He ordered the others on and crept or crawled back to the boats
for his pistols and the airgun, a complete Indian-fighter though
disabled. . . . Gass dressed the wound and there was no permanent damage but it did not heal for over a month.

❧

The two individual explorations of Lewis and Clark completed the geographical mission. Clark's results added up to a
negative confirmation. Though there were plausible crossings
to the Bitterroot Valley from the Yellowstone River, to turn
the transcontinental route up the Yellowstone would only increase its length. Lewis's reconnaissance gave positive, empirical confirmation to what the two of them had worked out
before he began it. It established that the water route must
have a long land-traverse across the mountains, and that the
best plan was to leave the Missouri at or near the Great Falls

and to make straight west from there for the Nez Percé buffalo road.

Clark summed it up.[13] The route finally determined on shortens the one they first traveled, in 1805, by 579 miles. By this route the distance from the Mississippi River up the Missouri to the Pacific Ocean is 3555 miles. The proper route, however, will leave the Missouri at the Great Falls and head straight for Bitterroot Valley, picking up the first one again at Traveller's Rest. The land-traverse is to go by way of Dearborn River, through Lewis and Clark's Pass to Blackfoot River, and down the Blackfoot to the Bitterroot. By this route the land passage from the Great Falls to the Clearwater is 340 miles, "200 miles of which is good road [to Lolo Pass], 140 miles over a tremendious Mountain Steep and broken, 60 miles of which is covered Several feet deep with snow [in June]. . . ."

Clark says that this route will enable the fur trade to be carried on by way of the Missouri and the Columbia "much cheaper than by any rout by which it can be conveyed from the East indias." There must be a post at the mouth of the Yellowstone, for the furs of that region, and another one below the Great Falls, probably at the mouth of the Marias. The Snakes, the Flatheads, and other transmontane tribes will visit those posts and will sell the indispensable horses cheaply. The carriage of western furs (and commodities, the China trade) across the mountains is entirely practicable from June 20 to the end of September. And "you may leave those establishments on the Missouri [at the Yellowstone or the Marias] 15 or 20 of June and arive on the Kooskooskee [Clearwater] river between the 1st. & 5th. of July. from that time you have untill the middle of September to decend the [Columbia] River and return to the mountains in time to pass them before the Snow becomes too Deep to cross them."

Thus William Clark, a triumphant explorer and here entirely accurate in his geography, but economically naïve and fantastically optimistic about the time factor. And convinced that western man has herein found at last the foredestined route of his commercial relations with the Orient.

Lewis was tentatively more cautious but, in the end, far more daring in his inferences, implications, and recommendations. The day after the expedition reached St. Louis, he wrote a letter to Jefferson, sending ahead to have the post, which had already left, held for it on the Illinois side.[14] . . . He summarizes the route whose details Clark has stated. Then he outlines the trade that is to come. Sloops can navigate the Columbia from the mouth to the Cascades. From there to the Narrows large "batteaux," presumably like his own keelboat, must be used. (Thereafter, presumably, by small boat.) "This passage across the continent" offers immense advantages to the interior fur trade but not such immense ones, Lewis fears, to the China and Indies trade. Much of that trade will always be more profitably conducted by way of the Cape of Good Hope, though "many articles not bulky brittle nor . . . perishable" can be transported more quickly and more cheaply by the Columbia and the Missouri.

Here Lewis makes the first of his points. And that point is that the Missouri and Rocky Mountain country "is richer in beaver and Otter than any country on earth." All the furs taken in this great expanse can be transported along the inland route to the mouth of the Columbia. So can those of the Minnesota River, the Red, and the Assiniboine — the last a British river, the trade of the others dominated by British companies. All these furs can reach the mouth of the Columbia by August 1 every year and, shipped thence, they will reach Canton "earlier than the furs which are annually shiped from Montreal arrive in England." (This would give the American trade an advantage in time equal to the duration of a voyage from London to Canton.) Lewis does not say so but the North West Company would become impregnable if it secured that advantage: if Great Britain were to possess the Columbia, as Mackenzie had proclaimed it must, or even if the Company were to dominate it as it did the Red River. What he does say is that if it is permitted to (and those words should be italicized) the Company can use this American trade route to get its Athabaska, Saskatchewan, and Winnipeg furs to salt water and cheap ocean transport.

And now this. In its infancy, Lewis says, and again the words should be italicized, this trade beyond the Rockies will be carried on from posts east of the Divide. (Here his ideas leap far ahead of those Clark had expressed.) This fact will slow things up while it holds true. For the parties that take the furs to the Pacific will not be able to get back to the Great Falls of the Missouri, taking with them the products of the Indies, till about October 1. By that late in the year there is the danger, which Jefferson had not foreseen, that the Missouri will probably freeze over before boats can reach St. Louis. "But," he says, "this dificulty will vanish when establishments are made on the Columbia and a sufficient number of men employed at them to convey the East India commodities to the upper establishment [on the Columbia River] and there exchanging them with the men of the Missouri for their furs in the beginning of July. by these means the furs not only of the Missouri but those of the Columbia may be shiped to Canton by the season before mentioned and the comodit[i]es of the East Indies arrive at St. Louis by the last of September in each year. . . . "

So far as the fur trade of the United States and Canada is concerned this is fact of granitic solidity, fact which was soon to be verified by both nations. The trade with the Indies was only a prismatic vision, the rainbow substances of three centuries of hope. But consider:

The representative of the President of the United States foresees and anticipates an American commercial establishment west of the Continental Divide. That establishment is to be so strongly developed that it will be able, by using the American trade route, to undercut the British trust. All its operations are to be conducted in territory of which not an acre was American soil when the President's representative reported to him posthaste. . . . Unless Robert Gray and Lewis and Clark had made it American soil. Was the Columbia River country, in Lewis's mind or anyone else's, to remain unpossessed for very long?

The plan which Lewis expounded was an open, explicit form of trade imperialism. Was it more than that? Bits of evidence that have been cited in this narrative add up to a tolerably impressive sum. It would be possible to read the sum as meaning that even before Louisiana was American Jefferson had considered the possibility of Oregon's becoming American soil, by attraction if not by overt action. And had acted.

৽

While Jefferson's explorers were on their way west, the North West Company, the British trade with which the American trade whose advance party they were was to compete, resumed its own effort to reach the Pacific.

At the end of June 1805 Lewis and Clark were still in camp opposite White Bear Island and their fatigue-drunk men were portaging the outfit round the Great Falls. Before they finished, François Antoine Larocque started west from the Mandan villages, where he had been their guest last winter. The North West Company had appointed him in place of Daniel Harmon, who would have done a better job, to blaze a trail to the Rocky Mountains and open the trade with the tribes there. He and a very small party, with a small stock of goods, traveled with some Crows, who were returning from their annual trade visit to the Minnetarees.[15] It was a horseback trip but it was intended to explore the Yellowstone River. Larocque's journal could have been of little help to his company; he was incurious, learned little, wrote down practically nothing. He reached the Big Horn River at its lower canyon but did not record a single item of the kind that Clark would have gathered by the pageful. But he had skirted the Black Hills along their northern edge and from there on may be said to have opened a white man's trail across Wyoming, though he certainly did not blaze it. He picked up some furs, gave various Indians his sales talk, and in early September reached the Yellowstone River — ten months before Clark got there. (On the way he garnered one fact that Clark missed: there *were* big cataracts on this river, well, somewhere on it.) He wandered along it, left it, returned to it, and on September 30 reached its mouth. (Lewis and Clark had left here, bound for the Great Falls, on April 27.) He went on to the Souris, where Charles Chaboillez probably dressed him down for having accomplished exactly nothing.

David Thompson was something else: he was a great man. After the North West Company had absorbed the opposition that had most destructively interfered with it, it sent Thompson on an errand much more serious than Larocque's: to establish the trade west of the mountains. The Company was resuming a plan that had been allowed to lapse and this time it was in earnest. The essence of the plan had been advocated by a series of partners from Peter Pond and the older Alexander Henry on. They may be characterized as the more daringly speculative partners, who believed that rich potential profits justified the risk of far western expansion (as opposed to those who believed in conservatively safeguarding the system as it was) and those who felt most concern about the increasing cost of getting goods from Montreal by canoe. The deep-sea freight that lowered Hudson's Bay Company costs and the similarly cheap transport that a Pacific outlet would afford were always in their minds. The developing plan was the reason for Mackenzie's transcontinental crossing, but his failure to find a feasible route had cost it support.

The reluctance of Montreal partners, especially Simon McTavish, was the principal reason for Mackenzie's withdrawal from the North West Company in 1799. Meanwhile Vancouver's expedition, though it had not found the convenient route the Admiralty had hoped it would, had powerfully illuminated the possibilities of the plan. By now Mackenzie had worked it out with economic and managerial genius and, going to England, he wrote his book and petitioned the colonial office to amend the monopoly charters of the Hudson's Bay Company and the East India Company so that the plan could be effected. He was back in 1802, reorganized the opposition, and directed its truculent tactics. Coming back into the North West Company with the merger of 1804 that followed McTavish's death, he worked furiously to bring about what he had so long regarded as inevitable. And now Duncan M'Gillivray became the Company's Montreal agent, that is its operating head, so far as it could be said to have one.

M'Gillivray was McTavish's nephew but he was Mackenzie's convert and disciple. It was he who with David Thompson had made the abortive attempt of 1801 and 1802 to cross the mountains and open the trade.[16] When he moved up to managerial power in Montreal he undertook to make the Pacific enterprise Company policy. The colonial office, declining to work for the amendment of the Hudson's Bay Company charter, had suggested to Mackenzie the almost inconceivable expedient of buying control of the Hudson's Bay Company and thus effecting the great combination by private finance. His diligent efforts had of course failed. Now in 1805 he and Mackenzie succeeded in getting the North West Company to authorize a step fully as revolutionary, an offer to lease from the Hudson's Bay Company the right of transit across its charter-protected routes. The first negotiations were made in 1805, the year in which Lewis and Clark reached the Pacific. The negotiations were protracted and ultimately they failed. The final stage of competition between the two great companies was beginning, the stage that would produce organized warfare and eventually the very merger that Mackenzie had envisioned. He measurably advanced that stage when, following the refusal of the Hudson's Bay Company to grant the easement, he made another effort to get control of its stock and brought Lord Selkirk into it.

The same meeting that voted to lease the right of transit voted also to cross the mountains and make a firm establishment west of them. The establishment was to embrace the Columbia River, which was known (above its mouth) only in Mackenzie's experience and his geographical speculations. Two separate but equally important moves followed, from Mackenzie's old base on Peace River and from the Saskatchewan.

An experienced clerk in Mackenzie's old department was Simon Fraser, Vermont-born but one of the stolid, tenacious, fearless, almost inarticulate Scots who were the type-model of the North West Company winterer. In 1805 — Lewis and Clark were crossing to the Pacific and Larocque was sauntering toward the Big Horn River — in 1805 Fraser went up Peace River, crossed the mountains, and reached Fraser River. (Not yet so named, thought to be the Columbia, known only in Mackenzie's report.) He was following Mackenzie's trail, twelve years later. He built a post which he called Fort McLeod. The Canadians thus had a fur trade post west of the Continental Divide before Lewis and Clark built Fort Clatsop on the Pacific shore, two years before Manuel Lisa built his post at the mouth of the Big Horn.

Going back to Peace River to replenish his supplies, Fraser left Fort McLeod in charge of his clerk, James McDougall. This young man promptly set off down the Fraser in a birchbark canoe and reached the mouth of the Nechako River. When he turned up the Nechako he left Mackenzie's trail and entered country that was not only previously unvisited by white men but unheard of. How far he went in this land of big and little lakes and great peaks is not known: to Stuart and Fraser Lakes, perhaps beyond them. From that region he returned to Fort

McLeod. So while Lewis and Clark spent their dank winter at Fort Clatsop in Oregon, a party of Northwesters were wintering in fierce northern weather west of the Divide in British Columbia.

In 1806 Fraser crossed the mountains again, joined McDougall, and set out to verify this pioneering. Meriwether Lewis had surveyed the land-traverse of the American transcontinental route and had killed his Blackfoot when Fraser reached the lake country that McDougall had discovered. During the

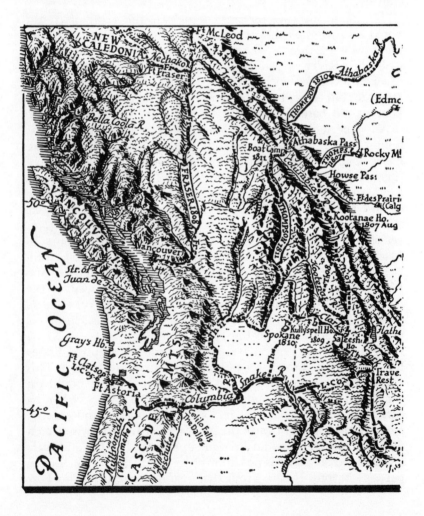

exploration which he now proceeded to make Fraser named this country New Caledonia for the homeland. The name was to last for a long time and to shimmer forever in the annals of the fur trade. It was not only a beautiful country, it was a virgin field. The Indians there had never so much as heard a rumor of the white man or of the white man's manufactured goods that made life easier. So in that summer of 1806 Fraser built and manned two posts in New Caledonia. The North West Company thus had three posts west of the mountains by the time

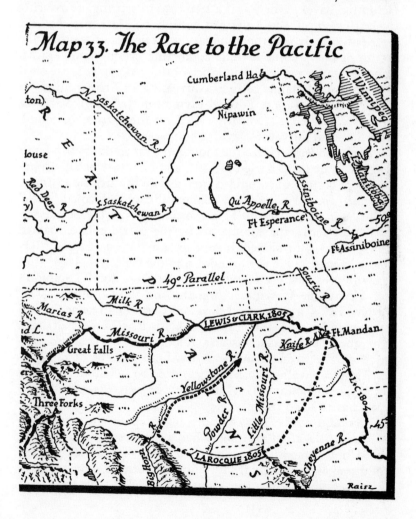

Map 33. The Race to the Pacific

Thomas Jefferson received a letter from his representative urging the utmost possible haste in establishing the trade in that foreign region, the Columbia River country. (Going to the utmost extreme possible for a Republican, Lewis urged the United States government to "aid even [if only] on a limited scale the enterprise of her Citizens.")

New Caledonia, however, was in the high north, above 55°, and it afforded no route to the Pacific. And opening the trade there, though profitable, was not a step toward securing the Columbia River. Nor was it a step toward settling, or assuring the settlement of, the American boundary. Mackenzie had declared absolutely that the boundary *must* be drawn south of the mouth of the Columbia. The conflicting or ambiguous provisions of treaties made in ignorance of the geography sanctioned arguments and claims fully as convincing to British minds as their contraries were to Edward Livingston, Jefferson, or Lewis. So the North West Company called on the master of boundaries, its great surveyor David Thompson.

Thompson was to cross the mountains far to the south, by or near the route he had tried in 1801, and he was to occupy country where occupation would count heavily toward the ultimate outcome. In 1806 a halfbreed named Finlay and called "Jaco" cut what was intended to be a horse-trail from the South Saskatchewan across the mountains by the corridor later named Howse Pass. (But named, inappropriately, for a Hudson's Bay Company man.) He went on down to the Columbia itself (not recognized as such) and built a canoe there for Thompson's use. Finlay, the forerunner, was thus on the Columbia — far north of 49° — a year after Lewis and Clark first reached it and a few months after they came back up it. (Lewis had been right about the Saskatchewan's importance but had concentrated on the wrong branch. A northern affluent of the Missouri would have had to be long indeed to neighbor with the North Saskatchewan. Not 50° but 52° was the parallel he would have had to reach.)

Thompson took Finlay's badly blazed path the next year, 1807. His Blackfoot customers had been determined that he should not make the crossing — that he should not get the profits they were getting from trading with transmontane tribes nor supply those tribes with guns and powder. But the Lewis

and Clark expedition had served him well by killing two Blackfeet the preceding summer. For the tribe raged south to avenge
those murders — on anyone — and so left the road open. He
reached the upper Columbia, north of 51°, on June 30.
(Manuel Lisa with George Drewyer as his adjutant was taking
his keelboat up the Missouri, bound for the Yellowstone and
the Big Horn.)[17] He determined his longitude and realized
from Mackenzie's figures that he was not on the Columbia —
not, that is, on Mackenzie's river. The advance of science thus
actually abetted geographical misconception: both Thompson
and Mackenzie were too exact. Mackenzie had been on the
Fraser, which he took to be the Columbia, at a longitude no
great distance west of Thompson's position. But it was too
great a distance to be within the range of error possible to two
such experts.

Thompson therefore understood that this, the true Columbia, was a different river and bestowed on it the name of a
resident tribe whom the Blackfeet massacred and traded with.
He spelled it Kootanae and sometimes Kootanie; American
ethnologists and geographers like to make it Kutenai; in Canada it is Kootenay. After making canoes from inferior Western
bark, Thompson moved *up* the Columbia. Just short of
the lake in British Columbia that is now called Windermere
he built a post which he named Kootanae House. This was
early August 1807. Lisa was building his Big Horn post at the
same time but that was on Missouri waters, whereas Thompson
was on the Columbia. There would be no American post on
Pacific drainage till 1810 when Andrew Henry, driven from
the Three Forks by Blackfeet, crossed the Divide into Idaho
and erected one on the branch of the Snake River that has ever
since been known as Henry's Fork.

The British were far ahead with the trade but they had accepted a time handicap in strategy and exploration, one which
Thompson did little to overcome.[18] In the spring of 1808 he
crossed into northern Idaho, the first Englishman to go south of
49° west of the Divide. A year later he went deeper into the
contested territory, to Pend Oreille Lake, whose location Lewis
and Clark had fairly well established, and across it to the
stream they had named Clark's River, now Clark's Fork. He
may have been directed to this, the country of the Flatheads,

by Patrick Gass's journal, the first account of the Lewis and Clark expedition, which had been published in 1807. (If he was not familiar with it now, the clear presumption is that he was soon afterward, and three years later his notes cite information he could have got nowhere else.)[19] Upstream from the mouth he built a post which he called Kullyspell House. In November of the same year, 1809, following Clark's Fork farther upstream, he crossed into Montana and built a second post for the Flathead trade, calling it Saleesh (Salish) House. He wintered there — by now there was a sizable number of North West Company *engages* south of 49° — and in March 1810 he traveled up Clark's Fork well beyond the mouth of the Flathead River. Here he was less than three days' ride from Traveller's Rest, which the American captains had returned to four years before. (Later on he got within six miles of it.) Later in that year he set out from Pend Oreille Lake by its outlet, the extension of Clark's Fork that is alternatively called Pend Oreille River. From indirect evidence and from what the Indians told them the Americans had decided that this river could not be navigable and Thompson now undertook to find out. It wasn't navigable all the way, and he stopped half a day short of its junction with the Columbia. And here the North West Company missed out again and badly. To have reached the Columbia at this confluence, in eastern Washington practically on the international border, would have been immensely important, whether Thompson identified the big river with his own Kootanae or recognized it as the Columbia. He went back to Kullyspell House.

All this time he was burning with an obviously temperate fire to reach the mouth of the Columbia — the mouth, that is, of Mackenzie's Tacoutche Tesse. To reach the river mouth which the canyons had prevented Mackenzie from reaching by canoe, which the Yankee sea captain Robert Gray claimed to have entered, which Captain Vancouver said Gray had not entered, which Lieutenant Broughton most certainly had entered. An inexplicable leisureliness banked Thompson's fire. Now unquestionably his instructions and first duty were to develop the trade with the Kutenais, the Flatheads, and the neighboring tribes, and it had proved busy and profitable. The route back to the Saskatchewan which he and his goods had to travel was uncommon hard. The Blackfeet were determined to stop this trade with their customers and enemies, outguessing and defying them was exhausting and full of danger, he was forced to make the terrible and magnificent winter exploration in the north during which he discovered Athabaska Pass. All that is true — and still there is no explaining why David Thompson did not devote three or four weeks to going down the Columbia.

Farther north a less intelligent, less imaginative man was determined to solve the problem both had been set. Simon Fraser, developing the New Caledonia market and trying to get for the North West Company profits that the maritime traders had been taking at Nootka, was also learning about the country beyond. By 1807 he had made many local explorations and had considered and rejected several theoretically possible routes to the Pacific. (He had no horses in the high north; land travel was excluded.) In the summer of 1807 he made a wide, laborious, only negatively informing circuit to the source of the Parsnip, Mackenzie's Bad River, the Nechako, and roundabout to Stuart and Fraser Lakes. It confirmed his earlier belief that there was no way out but on. The Pacific must be reached by way of the Columbia.

That is, by Mackenzie's Columbia, the Tacoutche Tesse. It had defeated Mackenzie, whom Fraser enviously tended to disparage, but it must not be permitted to defeat the North West Company. The Company backed Fraser's decision promptly and to the full, which suggests that Thompson would have got backing if he had asked for it.

So in 1808 Simon Fraser took a party of twenty-four down the Tacoutche Tesse in thirty-five days. There is no space here to describe that savage journey. Lewis and Clark had faced no such waters — no one traveling a North American river ever had. The voyage, in his own canoes, on foot, in Indian canoes, is incredible, it happened, and in Fraser's stolid prose it seems nothing much. The point is that, having taken his party through, across, around, and along those gorges and cataracts, he came on July 2 1808 to salt water. To, in fact, Vancouver's Strait of Georgia.

It was staggeringly, heart-pulverizingly wrong. Fraser took the latitude; it proved to be "49° nearly." Whereas, he knew by Vancouver's charts, "that of the Columbia is 46° 20'." He wrote in his notebook, "This river therefore is not the Columbia." Simon Fraser had discovered a great new river, which now bears his name. The North West Company had again been turned back from its Pacific outlet. And Great Britain was losing a race.

That was 1808. Even so, knowing now that trade would never move along the Fraser River, that its mouth was too far north to count for sovereignty, that Lewis and Clark had found what everyone else had missed — knowing all this, Thompson went on doing a profitable business on the upper watershed for two more years. It was in July 1811 that, having in the meanwhile worked out the geography, he launched a canoe on the Columbia, to go all the way. He did go all the way to the mouth, making a faster passage than Lewis and Clark. He thus ended the voyage which his great predecessor Mackenzie had believed he was beginning eighteen years before. But when Thompson floated down the widening estuary that had so battered Lewis and Clark in 1805, there was a trading post on the Columbia's left bank and it was flying an American flag. Fort Astoria.

Mr. John Jacob Astor's Pacific Fur Company had got there first by just four months. No more a wilderness man than any Montreal fur merchant, Mr. Astor was a better business man than the best of them. Adept at the international alliances of finance, he could also think in terms of continental and world trade quite as cogently as Alexander Mackenzie. Jefferson, a sophisticate in politics, was an innocent in commercial methods. He told Lewis that Astor was "a most excellent man" and he had pledged Astor "every reasonable patronage and facility in the power of the executive." [20] For this excellent man might be the agency that would secure to the United States "exclusive possession of the Indian commerce."

Astor believed as devoutly as Jefferson in exclusive possession. His plan, Kenneth Porter says, was "even more exclusive than that of the President. It was his purpose to concentrate the Western fur trade in the hands of only such American citizens as had been born in Waldorf, Germany, in 1763 and had arrived in the United States from London in the spring of 1784." But the plan was not therefore any the less brilliant or formidable. The man who dealt in all the world markets for fur, and whose financial alliances bound him as intricately to Canton and St. Petersburg as to Montreal and St. Louis — this creator of the first American trust was preparing an encirclement of the greatest magnitude and the greatest simplicity. It was as large and as bold as Lewis's proposals, which in fact it duplicated under a single ownership, and as simple as the lessons to be learned from the expedition Lewis had commanded. Astor intended to encircle and undersell not only the North West Company (to which he had first proposed a joint arrangement) but all Montreal by utilizing the direct transport of the Lewis and Clark trail, by maintaining trading posts at the indicated sites along that trail, and by securing the key to everything, the mouth of the Columbia River.

To that end the American Fur Company was organized, the Pacific Fur Company was added to it, the Astorian enterprise was launched, the continent was crossed, and Astoria was built. Astoria failed. War, incompetence, treachery, and a complex sum of causes that can be generalized with the adjective "premature" — these not only ruined it but delivered it over for a time to the custody of its foreign rivals. But though Astoria failed as a step in Astor's creation of a fur trust, it did not fail Jefferson or the United States. For it was built in time: by just four months. Since it was built in time it placed the Columbia River country, all the vast area that was to be called Oregon, in the domain of international relations. And it kept them there, a counterpoise in the equilibrium of two empires. That equilibrium need only be preserved; if it were not upset, then the westward advance of the American people was quite sure to make Oregon a part of the United States. Astoria was the final addition to the counterpoise against which Great Britain, at the height of her imperial expansion, could make but insufficient headway. If Jefferson be excluded from the enumeration as a personal agency, it is fourfold: the ship *Columbia*, the Northwest trade she created, the expedition of Lewis and Clark, and Astoria.

And Astoria followed from the expedition of Lewis and Clark as the flight of an arrow follows the release of the bowstring.

⌘

The attainment of their Pacific boundary by the American people, the completion of the continental unit, is later on and the work of thousands of men and women. But it is right for a moment to look at the firstcomers, as one might look at figures on a frieze.

It would be one of the days in early September 1806 when, near the mouth of the Platte River, the dugouts bound for St. Louis turn shoreward, having sighted a boat bound upstream, which also heads for the bank. The far-wanderers leap ashore to shake the hands of their own kind and ask how fares it with the United States. If there is whiskey, and most boats they spoke had whiskey, they have earned their drink. Give them names.

Charles Floyd has been dead these two years and is buried on a bluff upriver; we passed his grave a few days back and found that Indians had opened it. John Colter is not here, having turned back to the Yellowstone. Charbonneau and Sacajawea and the little boy are also absent from this noon parade; they are at the Minnetaree village on Knife River, where Clark will presently address Charbonneau, asking the family to move to St. Louis so that he may have "my boy Pomp" educated. So there are Meriwether Lewis, Capt. 1st. Inf., commanding, and William Clark, Lt. U.S. Corps of Artillerists. The hunter George Drewyer. Three sergeants: John Ordway, Nathaniel Pryor, Patrick Gass. Enlisted men: William Bratton, John Collins, Pierre Cruzatte, Joseph Fields, Reuben Fields, Robert Frazier, George Gibson, Silas Goodrich, Hugh Hall, Thomas Howard, Francis Labiche, Baptiste Lepage, Hugh McNeal, John Potts, George Shannon, John Shields, John Thompson, William Werner, Joseph Whitehouse, Alexander Willard, Richard Windsor, Peter Wiser. And with the dancer by many campfires, the slave York, all are accounted for. Most of them were to have no further place in events except the events of their own villages and farmsteads.

These in bold relief. There will be other figures in low relief, as many as the sculptor may care to use. Francisco Vásquez de Coronado, who first heard of the river whose olive-drab water slurs round the sterns of the anchored boats. Jolliet and Marquette who first saw it — "I have seen nothing more frightful." Many who traveled desirously toward the Western Sea:

Cartier, Hudson, John Smith, Champlain, Brulé, Nicolet, Duluth, La Salle, the defeated Vérendrye (many defeated men,
drowned or scalped or dead in bed so long ago no one remem-
bers them), the thin-nosed Jonathan Carver, John Evans who
left Wales to find Madoc's lost colony. There are as many as
the sculptor may desire: they are ghosts, nothing of them lived
on but their desire or dream. None saw the Western Sea, or the
place where the Great River of the West reached the South
Sea, or whither the Missouri led, or the mountains whose stones
shone night and day. The figures carved in bold relief have
seen.

Talking to men who but lately had kissed their wives good-
night and slept under storm-tight roofs, they must have had a
look in their eyes and a way of standing. Their shirts and
breeches of buckskin or elkskin had many patches sewed on
with sinew, were worn thin between patches, were black from
many campfires and greasy from many meals. They were
threadbare and filthy, they smelled bad, and any Mandan had a
lighter skin. They gulped rather than ate the tripes of buffalo.
They had forgotten the use of chairs. Words and phrases,
mostly obscene, of Nez Percé, Clatsop, Mandan, Chinook came
naturally to their tongues when they asked what word from
Kentucky and was John Bull still fighting the frogs.

And still, men who by guts and skill had mastered the farthest wilderness, they must have had a way of standing and a
look in their eyes. While they scanned the faces of white men,
their glance took in the movement of river and willows, of
background and distance. While they talked as men talk nearing home and meeting someone newly come from there, their
minds watched a scroll of forever-changing images. What they
had done, what they had seen, heard, felt, feared — the places,
the sounds, the colors, the cold, the darkness, the emptiness,
the bleakness, the beauty. Till they died this stream of memory
would set them apart, if imperceptibly to anyone but themselves, from everyone else. For they had crossed the continent
and come back, the first of all.

How the earth shook when the herd passed and how the bulls
bellowed all night in the rutting season, bellowed so loud a
man could not sleep. The mosquitoes would not let a man
sleep either, and gnats got through the net he wore over his

cap and crawled under his eyelids; the dog Scammon was bitten so sore he raised his swollen muzzle and howled. Grizzlies came charging out of the brush and men leaped into the river or crouched under an overhanging bluff, a grizzly would run half a mile with a dozen bullets in his vitals and then claw a three-foot hole in the ground to die in. Above the Mandans going west we had canvas tents but the captains slept in a skin tipi: Drewyer slept in it too, and Charbonneau and the squaw. Sometimes the baby squalled at night but the squaw pinched its nostrils shut; a child must learn that to make a sound might mean death, telling the Blackfeet where you were. In April the child's throat swelled and Captain Clark feared he would die.

There were no tents coming home. Along the Columbia and the Clearwater a man learned to roll out of the pool of rain that had dripped from him and go back to sleep. Beyond the mountains are more mountains, ridge after ridge farther than you can see; the snow is deep on them, very beautiful, but the high country is cold and game is hard to come by. All the rivers, all the running water. The Nodaway, the Kansas, the Platte, the Teton; beyond the Milk were the Musselshell and Judith and Marias, beyond the Salmon were the Kooskooskee, Lewis River, the Multnomah; coming down the Yellowstone you pass the Big Horn, Powder River, the Tongue. But there is no river so great as the big river, the Columbia; the falls are much whiter and angrier than the Great Falls of the Missouri but we let the boats down some of them by handline and, while those who could not swim watched from the rocks, we by God ran the others. There was always white water and the boats to be thrust from the rock edge at the last second. But the buffalo wallows stink with urine and a dry gulch can pour a wall of water at you from a cloudburst so far away you did not know there had been thunder.

Below the Sioux we all had boils; at Fort Clatsop we were always coughing. A man's bowels writhe from the alkali of poison creeks, from rotted salmon, from the blowflies in the arrowhead flour. Boiled dog is fair eating, the loin chops of bighorn are very fine, but we lived high in the buffalo country. The night is full of wolves' howling. Mountain night is so cold ice forms on your beard; you hear bits of stone fall from the cliffs. But the sound of all our nights is running water, soft and strong as the Missouri, a rush and humming as the Columbia, but most beautiful as the hurrying of mountain creeks. There were dreads that grew on you of danger at hand or to come but no danger a man would not face and no moment of fear or action we did not come through knowing we had done well. It was cold work towing the boats up the Jefferson; below Celilo there was always rain. Our feet swelled from the cactus at the long portage and we were like to rot at Fort Clatsop. But there was strangeness, beauty, wonder, and afterward a great content. We were the first.

They will always be the first. The frieze will show them coming home, the West at their backs. The others came afterward. A dugout leaves no wake in the water but the boat they spoke above the Platte was following in theirs.

જ⊷જ

But that is afterward and their most important achievement remains to be told.

Henry Adams said that the annexation of Louisiana was too portentous an event to be measured. The Lewis and Clark expedition produced results too various to be enumerated and too disparate to be appraised here. Their importance went rippling out through the history of the West and of the United States. But the most important achievement can be stated simply: they got there.

So, the epicycles having been traced, we may now return to the westward crossing. And to the expedition where we left it, Canoe Camp at the forks of the Clearwater River, in Idaho, October 1805.

They made five dugouts. They dug caches for their packsaddles, powder and lead, a few small articles of trade. They branded Lewis's name and rank on the thirty-eight horses left in the *remuda* and turned them over to the Nez Percé chief who had promised to herd them with his own. This was their favorite, whose name they translated Twisted Hair. He would go on with them for a time, to point out the way and introduce them to acquaintances. So would a younger chief, Tetoharsky. These went for a small fee and in the dugouts, but various other Nez Percés rode down the banks for some days, loping in to gossip at halts or to announce startling but altogether trivial tidings. . . . On the third day out Toby, the genial Snake who had served them so well, quietly turned homeward, and his son with him. Since the two were with the shore party, the captains did not learn about their desertion till evening. Grieved because Toby had not made a ceremonious farewell nor, surprisingly, asked for his wages, they asked Twisted Hair to send one of the outriders after him so that he could at least be paid. Why bother? the chief asked: his own people would hijack two solitary wayfarers who had anything worth taking. But at least Toby and his son helped themselves to a couple of U.S. horses when they reached the herd, and rode home.

The last stretch began Monday, October 7 1805. On October 10 the Clearwater, after boiling through two long rapids, brought them into a larger and nobler river, the Snake. It is flowing east of north here — Lewiston, Idaho, at the foot of dramatic hills — after emerging from a series of fearful canyons, and the Salmon has joined it some fifty miles upstream. His sense of the Salmon's flow led Clark to say that the Snake River "is the one we were on with the *Snake* or *So-so-nee* nation." He duly entered in his notes the by now familiar datum that it had a southern fork. That fork is the main stream and Clark never did get it right.

Between the Mandan villages and Lemhi Pass, the expedition had met no Indians. Since Lemhi they had not lacked company and from now on they would always be with Indians. Too many Indians and too troublesome. The best of them were here and around the mouth of the Snake, Salishan or Shahaptan subtribes, relatives of the Flatheads or the Nez Percés. Below the mouth of the Snake they got steadily worse. They were all fishermen who took the salmon in many ingenious ways. They dried it and stored it for winter and the trade; and, like all fisheaters, they smelled high. The expedition would live principally on fish now but blessed the multitude of tribal dogs, for dog-meat stew had become holiday fare. Contriving to shoot some teal, Clark noted that they furnished the only good dinner he had had in three weeks. Since the last colt.

The barren canyons into which the Snake took them are much less rugged than those in the Seven Devils Mountains upstream, but they are full of furious water swirling round fragmented cliffs. The Indians advised them to portage the worst stretches but the season was "far advanced and time precious with us," and they had to run the rapids. One of them wrecked Drewyer's dugout and there was a strained moment while the swimmers, including Twisted Hair, pulled the nonswimmers out and then dived for the outfit. A wide plain succeeded the canyons, with tips of peaks on the southern horizon, the Blue Mountains. But the Nez Percé guides and some casuals of the river warned them of worse rapids ahead. They earned their pay by piloting the boats through them safely, though Pryor's filled with water. That was October 16 and below the rapid the Snake flowed smooth and quiet for six miles. Then it rippled over a bar and entered a much larger river, one which was smooth and quiet at this place too but from here on would merit either adjective but fleetingly. A big river, here crossing a desert plain that was cut off east and west by low hills. A river that had been flowing in men's imagining for at least a century and a half as the Great River of the West, and in their desire since 1792 as the Columbia.

Lewis and Clark had reached the river they had been ordered to find. At Monticello Thomas Jefferson could now carry the line inked across blank and hypothetical paper to these massive, beautiful waters. St. Louis to the Mandan villages, to the Bitterroot, to the Snake, to the Columbia. At the known mouth an inked line led eastward to where the *Columbia* had anchored and on to the mountains that Mr. Broughton's longboat had almost reached. The small remaining gap would now be filled and Lewis and Clark had taken white men's boats into the upper Columbia for the first time. With the boats they took an impalpable sovereignty as well. The river would not be formally American for a long time to come. But from this moment — no matter the North West Company, His Majesty's Navy, Sir George Simpson and the Hudson's Bay Company — a persistent sense of earliest possession kept it from being anything other than American. . . . The journal has no mention of emotion or drama at this great arrival but the local Indians celebrated for the United States. Some two hundred of them danced and chanted at a big campfire. The captains smoked with them, expressed Mr. Jefferson's goodwill, and bought seven dogs.

A day later it was forty dogs, for they were wary of the proffered fish, having seen hundreds of rotting salmon on the banks. On October 18 the boats "proceeded on down the great Columbia river." They met the Walla Wallas and their chief Yellept, who would be so neighborly on the return trip, but thereafter no Indians they could like and few who were not troublemakers. As the Columbia turned west into its first canyons, the banks were littered with their drying-stages for fish, their weirs and nets, their ugly bark hovels. . . . The country here is bleak; the river flows in a narrow valley when it is not actually a canyon. The walls are mostly black or dark brown with lava, the Washington side more plateau-like and achingly barren, the Oregon side more likely to break up in hills. But the river was beautiful beyond expressing. Its water was perfectly transparent, faintly green, and it spoke in a deep bass, the voice of incalculable power. Where it narrowed and churned itself to white eddies careening at impossible angles, a shrillness overcame the other voice, but still there was a deep undertone, almost of drums.

The second day out, October 19, when the valley widened, they saw far to the west the perfect cone of Mount Hood but, reading their Arrowsmith or a Vancouver map, took it to be St. Helens. There were a lot of rapids now — and not rapids in the small Clearwater or even the Missouri but huge and violent beyond anything they had imagined. Each one must be examined, with the authority of Cruzatte, Labiche, and Drewyer to be consulted — and these were first-rate men for though they too had never dreamed of such mighty water they always found the right way through. Pausing for such an inspection on October 21, they made their first contact with the Pacific. The Indians who lived here had two scarlet blankets and a blue one, and one was wearing a sailor's peajacket.

Most Indians, they observed, lived on the northern bank and they attributed this to fear of the Bannocks, about whom they kept getting word — or so they thought. The Nez Percés understood the river languages so far and translated to the captains by signs. But now there was a perfectly intelligible word, "Timm." It meant some big falls a few miles down which everyone had talked about. They heard the falls naming themselves Timm when they reached them on October 22, fifteen days out from the Clearwater. They were Celilo Falls. The

valley contracts, the river is two hundred feet deep, and this immensity of water flings itself insanely among projections of
black mountains in swirls and sucks and whirlpools hypnoti-
cally violent.

The worst of this could be portaged. Through the rest of it
the boats had to be lined. One got torn away and swept through
that combing rage. The local residents were waiting below for
just such a happy chance and collected salvage. The residents
were horribly infested with fleas, in so much that all their by-
paths and campsites crawled with them. The men who were
lining the boats had to cross some of these places and were
forced to strip naked so that they could deal with fleas. They
were talkative Indians, but sullen. Twisted Hair got word that
they intended a massacre and that if they changed their minds
the next tribe downstream would take over the job. They
didn't try — the routine security measures dissuaded them —
but they did some expert stealing.

Coming into quiet water below Celilo Falls but knowing that
as bad or worse lay immediately ahead, they got new reminders
that journey's end was near. These thieves had big handsome
canoes with carved figureheads, "built for riding the waves."
They had reached the Northwest watermen, though these were
the least skillful of them and had the smallest cedarwood canoes.
Lewis bought one for a dugout and a hatchet; the seller said he
had got it from a white man downriver. And in this quiet
water were "great numbers of *Sea Otters*," at play or peacefully floating. They saw at last the placid, winning animal
whose rich fur had been the starter's signal for the race that
the captains were now winning.

Mount Hood was visible to the waist or below, but from
Celilo it looked rather pyramidal than a cone — when visible.
For clouds shut it out or mist closed round it every hour or so.
And the sunset had thickened with autumn haze.

They walked down from Celilo Falls to inspect the next
hazard, later called The Dalles. There were twelve miles of
hazards, the first a fall followed by a mile-long rapid and then
another fall where, in Clark's words, "the whole of the Current of this great river must . . . pass thro' this narrow chanel
of 45 yards wide." It was the Short Narrows. There was no
way of portaging: the banks were cliffs. Clark said the nar-

rows must be run, Cruzatte said they could be. And, "notwithstanding the horrid appearance of this agitated gut swelling, boiling & whorling in every direction," they were run. Clark acknowledged that it looked worse from the water than it had from the land. The lower Dalles, the Long Narrows, could be in part portaged but they ended by packing the lading round and running the boats through. This was October 25 and now Twisted Hair and his companions would go home. They knew no further languages and could help no more. So after "a parting smoke with our two faithful friends," they left.

The Dalles Indians stole everything that wasn't nailed down and furiously resented being caught at it. The Columbia widened but seemed just as swift. Cloudy weather was a portent but so were the Indians, who had a musket, a sword, and some brass kettles. Then there was an oddity, one who not only wore a peajacket and a sailor hat but had queued his hair. "The country begins to be thinly timbered" — they were nearing the Cascade Mountains, the Columbia Cascades, and the rains. There was vehement sign talk round the nightly fire, and however cussed the tribes might be they kept Cruzatte playing his fiddle and York dancing. Cedar houses were getting larger and beginning to be made of planks. So on October 30 they reached the Cascades and promptly it began to rain.

There are no Cascades now; the water impounded by Bonneville Dam covers them. But this is the gorge which the Columbia cuts through the Cascade Mountains and before Bonneville there were the Upper and Lower Cascades. "Great danger," the notebook comments. They portaged the upper one on November 1, rolling the boats on logs from rock to rock, and the lower one on the morning of November 2. That day they traveled twenty-nine miles. They thought the river two and a half miles wide where they camped that night, the mountains were east of them, and there was a tide. President Jefferson could now ink in the rest of that line, for coming west they had reached Broughton's farthest east.

The Columbia now became what Huck Finn called a powerful big river, though it was powerfully different from Huck's river. And from here on the Indians were an infestation. They were always swarming about, short, squat, crawling with fleas and lice. They were always stealing, too. They had to be scared away. But, like their rich relatives who had impressed Mackenzie, they were sailors. With casual skill they took their big canoes through waters which the white men simply dared not attempt.

So the conquerors of the cactus plains, the masters of mountains and canyons, became deepwater navigators. The river widened several miles more and was a choppy sea that became a tempestuous sea with every wind. Fogs settled in at night and would not lift till midmorning or noon. The rains were here — "Rained all the after part of last night. . . . A cold wet morning. . . . A cool wet raney morning." On November 7 intensity breaks through. In the notebook which was always on his knee for changes of course, Clark suddenly writes "Ocian in view! O! the joy." In the journal this becomes "Great joy in camp. we are in *view* of the Ocian." They weren't and wouldn't be for quite a while. It was just that the river had widened again, to about a dozen miles, had become a river reaching out for the sea — and they saw big waves rolling in straight from the west. Saw them from dugouts which had been made to travel mountain rivers. They progressed with a corkscrew motion and the subduers of the Rocky Mountains retched over the sides. Sacajawea too, a Shoshone girl, was seasick this long way from the Three Forks, from the Minnetaree villages, traveling the big water with the acceptance of red womanhood but probably with wonder too. "The water is too salt to drink," Clark wrote But that was no problem for they were in the Northwest winter: "We use rain water."

Whenever they landed, the big canoes came in. Their rowers had venereal scabs and diseased eyes. But they had all the trade goods too — woollen clothing, blankets, hats, muskets, pistols, tin powder flasks, axes, wire, copper. Some of them had spurts of English words. They had to be tolerated for they might bring news of ships or white settlements. For some time the sign talk had indicated that there was a white man's town at the mouth of the river. It could not be explained but if it existed it must be inquired into. A new British trading post or, happily, one set up by the Boston men? Much turned on that — but either way it would mean white men's food, replenishment of the trade-currency, and new clothes and blankets to replace the leather ones which the rain was rotting.

There was no town, of course. There was an American ship, the *Lydia* out of Boston, Captain Hill. She came over the bar and past some of their waterlogged campsites two or three weeks after they moved down to the Netul River and erected Fort Clatsop. The customers told Captain Hill about the white men who had been so wet and showed him some of Mr. Jefferson's medals. They carefully did not tell him that these white men were still in the vicinity. The *Lydia* went on to Nootka and traded out of there to the northward, and came down the coast again in the spring. Now the Indians, who had said nothing about her at Fort Clatsop, brought Captain Hill one of the summaries and formal notifications which the explorers had left with them for just that purpose.[21]

On November 8 they were trying to cross the indentation of the Washington shore that is called Gray's Bay when a gale whipped up waves too high to be faced. They camped on a narrow shelf — the bank beyond was vertical. The gale kept blowing and there was intermittent rain. When the tide rose it swamped the boats and projected driftwood logs at them, the great spruce and fir and cedar of the region, bigger trees than any of them had ever seen before, "nearly 200 feet long and 4 to 7 feet through." The whole camp was under water and the whole party was too, and hardly able to save the boats. The gale increased, blowing the whole sea at them. In the same weather they had to stay there the next day and the next night. On the 10th they started out again but another gale drove them to shore. Again the shore was cliffs but it was broken at the beach by a small "drain," where runoff water came down. The opening made a somewhat better shelter for the boats and at tidemark there was a driftwood raft of those big logs. On the logs they made a kind of camp and built a fire to try to dry their clothes and bedding. But it rained all day and "nothing to eate but Pounded fish."

Gales and breakers forced them to stay there till November 15. High tide floated their campsite till it seemed likely to drift off with the ebb and forced water between the logs in geyser-like gushes. The rain became torrential, the gales a hurricane. But nothing could keep the Indians away. They pushed their canoes through those waves, traded salmon for fishhooks, lied about the white men farther down, stole loose objects, and turned back into the waves. "Their canoe is small, maney times they were out of sight." Life varied now by stones which the rain had loosened, bombarding them from the cliffs, the captains went on making notes about Indian languages and artifacts while the rain fell and the wind blew.

They sent men out into the rain to hunt and went themselves. But though with much effort the bank could be climbed, the forest of the Northwest coast was choked with "intolerable thickets." On November 14 the wind moderated enough so that Colter and two others could get a canoe past the headland which they had so far been unable to round. The next day he walked back to report that before another roaring sea stopped them they had found a gravel beach where a safe camp, if not a dry one, could be made. Lewis could stand inaction no longer. He took a party out to see that beach — and to look for those white men.

On November 15 Clark's journal entry, which must have been written the next day, included, "The rainey weather continued without a longer intermition than 2 hours at a time from the 5ᵗʰ. in the mornᵍ. untill the 16ᵗʰ. is *eleven* days rain, and the most disagreeable time I have experienced confined on a tempiest coast wet, where I can neither get out to hunt, return to a better situation, or proceed on." But that day, the 15th, the unbelievable happened. The sun shone all morning long, so that they could get some bedding a little dry. Another gale roared up out of the southwest but it moderated. At three in the afternoon the wind dropped altogether and the river calmed. At once Clark ordered the boats loaded and set out. He found the beach — beautiful sand, a small stream of fresh water, and a lot of abandoned Indian huts full of fleas. Presently Shannon came in, sent by Lewis who had taken his party farther on. He reported that the Indians had contrived to slide his and Willard's rifles from beneath their necks while they slept, but that Lewis had come up in time to scare them into returning them.

Here on tolerably high ground Clark made a camp, for "this I could plainly see would be the extent of our journey by water, as the waves were too high at any stage to proceed any farther down." The campsite was on the Washington shore, in the indentation called Baker's Bay. They could see Cape

Adams, Cape Disappointment, the whole width of the river mouth. Between those headlands and beyond them, the open sea.

Here the expedition stayed till November 25. Lewis and his squad had already gone down to the seashore. Clark went down too, taking with him everyone who wanted to dip a finger in the Pacific. York and nine more desired to go all the way, "all others being well contented with what part of the Ocean & curiosities which could be seen from the vicinity of our Camp." Filthy weather battered them again and game was very scarce, but the Indians brought their girls. A sailor had tattooed "J. Bowmon" on the arm of one, and that they had made the acquaintance of many other sailors was evident from their sores. But though there would be use for the mercury in the medicine chest soon, these were men of eager blood. "We divided some ribin between the men of our party to bestow on their favorite Lasses, this plan to save the knives & more valueable articles."

There was so little game here that they determined to cross to the southern shore and, after more gales, finally did so. The hunters found only elk but plenty of them. Elk is a coarse meat and rather tasteless in the winter leanness but it would be better than the fish they had been dolefully living on. And the skins would replace the clothing, blankets, tarpaulins, and buffalo robes that the rain had ruined. Lewis went looking for a suitable campsite, finding the land south of the Columbia all salt marsh and hummocks. Eventually he located a site that was a little higher than the swamp and had good timber. It was some miles up one of the listless tidal streams that wandered down to the sea, Netul River on early maps, Lewis and Clark River now. It was during this interval that Clark, remembering Mackenzie and realizing the import of his own crossing, carved on a big pine the legend that says it all:

"William Clark December 3rd 1805. By land from the U. States in 1804 & 1805."

But they are better seen at the camp on the Washington shore, whence they looked out under a lead-colored sky to the lead-colored sea. There Clark, who from day to day seldom captioned the journal entries, suddenly gave one day's entry a heading charged with pride and remembrance: "Cape Disappointment at the Enterance of the Columbia River into the

Great *South Sea* or Pacific Ocean." That is what it was. The waves pounding over the bar which had deceived George Vancouver but which Robert Gray had crossed were coming in from China. Night closed in out of the rain clouds, daybreak showed above the mountains, and the company who had crossed from the U. States saw the waves coming in from China. Here was the great South Sea, the Pacific Ocean, and they had brought the United States to its shore.

From China and India but also from Cathay: on the far shore were not only the Canton merchants who bought the sea otter but Prester John and the Grand Khan who ruled kingdoms of marvel. There had been but there would be no longer; for if this camp was a beginning it was also a final end. They had filled out the map and when the map is made there is no room or use for dreams. The darkness into which the sentries peered till dawn was only night, not the mystery through which for three centuries the mind of western man had groped toward the horizon lands, the islands of the sea, the Golden Chersonese, Anian, Quivira. Yet if the dream faded from men's minds forever, this sleeping company who had made a trail across America had fulfilled it. It had had much beauty: they had brought it to completion. If they were the firstcomers to this shore they were also the lastcomers, and they had been led here by all who had sought the fact in the dream.

On December 7 1805 the Lewis and Clark expedition moved to the Netul River and went up it to the site Lewis had selected. A detachment went down to Tillamook Head on the Pacific beach, to make salt. The others began to cut timber for the winter post, Fort Clatsop.

౼

*Most Christian, most exalted, most excellent and powerful Princes, King and Queen of the Spains and of the islands of the sea: Your Highnesses determined to send me to the countries of India, so that I might see what they were like, the lands and the peoples, and might seek out and know the nature of everything that is there. And you ordered me not to travel to the East, not to journey to the Indies by the land route that everyone had taken before me, but instead to take a route to the*

*West, which so far as anyone knows no man had ever attempted. Therefore you granted me great favors and bestowed noble rank on me, so that I would thenceforth be called Don. You named me High Admiral of the Ocean Sea, and Viceroy and Governor of all the islands and continents which I might discover and conquer, or which anyone who came after me might discover and conquer in the Ocean Sea. So on Saturday the twelfth of May 1492 I set out from Granada, and I traveled to the seaport of Palos. There I fitted out three vessels and got crews for them and supplied them well with provisions. And on the third of August that same year, a Friday, I left Palos and stood out to sea, half an hour before sunrise. . . .*

Comparative Dates

Notes

Bibliography, Annotation, and Minutiae

Index

# Comparative Dates

| | |
|---|---|
| 1603 | Martin Pring along New England coast |
| 1605 | Weymouth along New England coast |
| 1607 | Jamestown founded |
| 1608 | Quebec founded |
| 1610 | Santa Fe founded |
| 1613 | Champlain up the Ottawa River |
| 1615 | Champlain to Georgian Bay |
| 1620 | Landing at Plymouth |
| 1623 | Brulé to the Sault |
| 1630 | Great Migration (to Massachusetts Bay) begins |
| 1634 | Nicolet to Green Bay |
| 1640 | Supposed date of de Fonte's voyage |
| 1654–56 | Groseilliers among the Far Nations |
| 1659–60 | Radisson and Groseilliers in the Lake Superior country |
| 1668–70 | Radisson and Groseilliers at Hudson Bay |
| 1671 | St. Lusson at the Sault |
| | Batts and Fallam across the Appalachians |
| 1673 | Jolliet and Marquette down the Mississippi |
| 1679 | La Salle builds Fort Crèvecoeur |
| 1682 | La Salle to the mouth of the Mississippi |
| 1684 | La Salle to the Texas coast |
| 1688 | de Noyon to Lake of the Woods |
| *1689–97* | *War of the League of Augsburg (King William's War)* |
| 1689 | Pensacola founded |
| 1690–92 | Kelsey in the Saskatchewan River country |
| 1699 | Cahokia and Biloxi founded |
| *1702–13* | *War of the Spanish Succession (Queen Anne's War)* |
| 1712–17 | de Bourgmond along the Missouri |
| 1718 | New Orleans founded |
| 1720 | Villasur expedition |
| 1734 | Vérendrye to Lake Winnipeg |
| 1738 | Vérendrye to the Mandan villages |
| 1742–43 | Vérendrye's sons to the Black Hills |
| *1743–48* | *War of the Austrian Succession (King George's War)* |
| 1750 | Walker through Cumberland Gap to Kentucky |
| 1754–55 | Henday in the Blackfoot country |
| *1755–63* | *Seven Years' War (French and Indian War)* |

| | |
|---|---|
| 1764 | St. Louis founded |
| 1766–67 | Carver in Minnesota |
| 1771 | Hearne down the Coppermine River to the Arctic |
| 1772–73 | Cocking to the Saskatchewan |
| 1773 | Perez to Nootka Sound |
| 1774 | Joseph Frobisher to Portage du Traite |
| | Hearne builds Cumberland House |
| *1775–83* | *American Revolution* |
| 1776 | Escalante in the Great Basin |
| 1784 | North West Company organized |
| 1789 | Mackenzie down the Mackenzie River to the Arctic |
| 1792 | Gray to the Columbia River |
| 1793 | Mackenzie to the Pacific |

# Notes

(See Statement on Bibliography and Annotation, page 632)

## CHAPTER I: THE CHILDREN OF THE SUN

DOCUMENTARY CABEZA DE VACA: Ramusio, *Navigationi e Viaggi;* Hallenbeck (2); Buckingham Smith in F. W. Hodge, ed., (1) *Spanish Explorers in the Southern United States* (N.Y., 1907). FRAY MARCOS: Hallenbeck (1); George P. Hammond and Agapito Rey, *Narratives of the Coronado Expedition, 1540–1542* (Albuquerque, 1940). DE SOTO: Gentleman of Elvas in Hodge (1); Luís Hernando de Biedma in B. F. French, ed., *Historical Collections of Louisiana,* Part II (Philadelphia, 1850); Rodrigo Ranjel in Oviedo, *Historia Natural y General de las Indias;* Garcilaso de la Vega, *The Florida of the Inca,* trans. John Grier Varner and Jeannette Johnson Varner (Austin, 1951). CORONADO: Hammond and Rey, above; Castañeda, *Relación de la Jornada de Cíbola,* text and translation in George Parker Winship, "The Coronado Expedition," Bureau of Ethnology, *Fourteenth Annual Report,* Part I (Washington, 1896); Hammond and Rey. Additional: Hakluyt, *Virginia Richly Valued* and *Principall Navigations;* Purchas, *Purchas His Pilgrimes;* Raleigh, *Discovery of the Empyre of Guiana;* John Smith, *Generall Historie of Virginia.* Columbus quotation from Torre, ed., *Raccolta Completa degli Scritti di Cristoforo Colombo,* my translation and I have condensed it; the same for quotation at end of Chap. XII.

SECONDARY H. H. Bancroft: *Arizona and New Mexico; California; Mexico; North Mexican States and Texas; Northwest Coast.* Herbert E. Bolton, (1) *Coronado: Knight of Pueblos and Plains* (N.Y. and Albuquerque, n.d. [1950]). John B. Brebner, *The Explorers of North America* (N.Y., 1933); cited throughout this book

as Brebner. Cleve Hallenbeck: (1) *Alvar Nuñez Cabeza de Vaca*
(Glendale, 1940); (2) *The Journey of Fray Marcos de Niza* (Dallas,
1949). Samuel E. Morison, *Admiral of the Ocean Sea* (Boston,
1942). Prescott: *The Conquest of Mexico* and *The Conquest of
Peru.* Carl O. Sauer: (1) *The Road to Cíbola* (Berkeley, 1932);
(2) "The Discovery of New Mexico Reconsidered," *New Mexico
Historical Review* (hereafter cited as NMHR), XII (1937). Henry
W. Wagner: (1) *California Voyages* (San Francisco, 1925); (2)
*Spanish Voyages to the Northwest Coast* (San Francisco, 1929); (3)
*The Cartography of the Northwest Coast* (Berkeley, 1937). Justin
Winsor, ed., (1) *Narrative and Critical History*, II. Lawrence C.
Wroth, *The Early Cartography of the Pacific* (N.Y., 1944).

1. Purchas, translating Ramusio's account of Guzmán.

2. Most of my quotations from Cabeza de Vaca in this chapter
are from Buckingham Smith's translation in Hodge (1) which
Hodge slightly touched up, but one is from Smith's first edition
and one from Purchas.

3. Not, it must be made clear, the sign language of the Plains
Indians, the exceedingly effective means of communication that will
be important in this narrative later on.

4. Here and in my statements about Cabeza de Vaca's route I
follow Hallenbeck (1). From the northern border of Sonora on,
Hallenbeck follows Sauer. Elsewhere I make a general statement
about routes.

5. These are the figures arrived at by the United States De Soto
Expedition Commission in its *Final Report.* In the main my account of the expedition follows the Gentleman of Elvas, supplementing him with Biedma and Ranjel and with a few details from
Garcilaso.

6. The arquebus had the additional superiority that its ball
could not be seen and therefore could not be dodged, as arrows
frequently could be. It is hard to generalize about this arm, not
only because individual weapons varied enormously but because
gunpowder varied infinitely more. It may be reasonably thought
of, however, as about three feet long, with a weight of ten pounds
and an effective range of at most a hundred yards. Throughout
the 16th century (in the latter half of which the musket came into
use) the most effective hand-weapon in the world remained the
English longbow. This, however, could be used only by experts,
who must have years of training and constant practice, whereas
anyone could soon learn to use a musket.

7. "In the southern part of what is now Pontotoc County or
the northern part of Chickasaw County." U.S. De Soto Expdn.

Commission, *Final Report*, p. 225. The Commission determines the site of the crossing as "near the present Sunflower Landing." Many maps do not show it; it is in Coahoma County, above Island No. 66. Note, however, that the river has been much displaced and the terrain much changed since De Soto.

8. "Probably near either Camden [Ouachita County, Arkansas] or Calion [Union County]," *ibid.*, p. 257. The Commission doubts that De Soto encountered the Quapaws, though most other au-'iorities believe it did, and does not sanction my conjecture that t may have seen the Wichitas.

9. Whether or not Fray Marcos went all the way to a hill overlooking Háwikuh and how much of what he said was true have been debated ever since Coronado. Bolton (1) declines to say, expressing a hope that more evidence will some day turn up. In such cases throughout this narrative I have regarded the best test as the analysis of travel time in relation to distance. Hallenbeck (2) makes such an analysis and I regard it as entirely conclusive. See also Sauer.

10. This is Bolton's count (1), p. 68. He agrees with Hammond and Rey.

11. F. W. Hodge, "The Six Cities of Cíbola," NMHR, I (1926), may be said to have established that there were only six all told when Coronado got there, and all recent students have followed him. Nevertheless Coronado and all his companions who wrote accounts made it seven.

12. From here on the route rests on Bolton, who buttresses his own studies with the archeological discoveries of Waldo R. Wedel. Why he calls the here shallow, sluggish, sometimes stagnant Arkansas "a noble stream" baffles me.

13. All the documents that anyone except a specialist will need to read are translated in Hammond and Rey.

14. One of my quotations from Castañeda ("The land is the shape of a ball . . .") is from the translation of Hammond and Rey. This is more literal and closer to the original than any other and my narrative follows it in contexts which it alone makes clear. All my other quotations from Castañeda, however, are from Winship, who has a much finer literary quality. At the cruxes I have checked both against my own rendering of the original.

CHAPTER II: THE SPECTRUM OF KNOWLEDGE

DOCUMENTARY   In addition to most citations for Chap. I (but especially Hakluyt, Oviedo, and Purchas), Charlevoix and Hennepin as cited later. Georges Marie Butel-Dumont ("M. Dumont"),

*Mémoires Historiques sur la Louisiane* (Paris, 1753), ghostwritten by Jean Baptiste Le Mascrier ("M. Le M."). Francisco López de Gómara, *Historia General de las Indias.* Antonio de Herrera (a French translation of) *Historia General.* Louis-Armand Lahontan. I have mostly used Thwaites's reprint of the first English edition, *New Voyages to North-America.* Bartolomé de Las Casas, *Breve Relación de la Destrucción de las Indias;* also the scarehead translation by "J.P.," *The Tears of the Indians.* Pierre Margry, ed., *Découvertes et Établissements des Français . . . de l'Amérique Septentrionale,* cited hereafter as Margry. Le Page du Pratz, *Histoire de la Louisiane* (Paris, 1758). R. G. Thwaites, ed., *The Jesuit Relations and Allied Documents,* Vols. 44, 50, 53, 68; cited hereafter as JR.

SECONDARY Parkman. Prescott; Wagner (1) and (2); Winsor (1), Wroth. Bancroft: *California; Central America; Northwest Coast.* William H. Babcock, *Legendary Islands of the Atlantic* (N.Y., 1922). Adolph F. Bandelier, *The Gilded Man* (N.Y., 1893). Benjamin F. Bowen, *America Discovered by the Welsh* (Philadelphia, 1876). Paul L. Cooper, "Recent Investigations . . . in South Dakota," *American Antiquity,* April 1949. Clark B. Firestone, *The Coasts of Illusion* (N.Y., 1924). John Filson, *The Discovery, Settlement and Present State of Kentucke.* John Fiske, *The Discovery of America.* Louise Phelps Kellogg, (1) *The French Régime in Wisconsin and the Northwest* (Madison, 1925). Paul S. Martin, George I. Quimby, and Donald Collier, *Indians before Columbus* (Chicago, n.d. [1947?]). Thomas Stephens, *Madoc* (London, 1893). Waldo R. Wedel: *Culture Sequences in the Great Plains* (Washington, 1940); *Prehistory and the Missouri Valley Development Program . . . in 1946* (Washington, 1947); *Prehistory [etc.] . . . in 1947* (Washington, 1948); "Some Provisional Correlations," *American Antiquity,* April 1949.

1. This is the translation of Hammond and Rey, pp. 263–64, and is almost literal. But Winship's reading, which is entirely true to the text, is more revealing. The operative part, p. 529, reads: "The sources were not visited because, according to what they [the Wichitas] said, it comes from a very distant country in the mountains of the South Sea, from the part that sheds its waters onto the plains. It flows across all the level country and breaks through the mountains of the North sea, and comes out where the people with Don Fernando De Soto navigated it." (Spanish text, Winship, p. 457.) The supporting analysis is too long for a note: it must suffice to say that Castañeda's references to Cabeza de Vaca

and the passages in which he orients Coronado's expedition with the larger geography produce exactly this idea.

2. I am not writing the history of American geography but I must insist on the importance of this concept, whose consequences have by no means been sufficiently appreciated. See Mercator's map, one of the most famous ever drawn, the relevant part of which is reproduced as Pl. 15 in Paullin's *Atlas of the Historical Geography of the United States*. Here, in 1569, the east-west range is one of the dominant features of the continent. Paullin's Pl. 24 shows the range persisting east of the Mississippi in a Delisle map of 1718 — by which time the French had enough empirical knowledge to obliterate it — and Pl. 23 shows it on a Delisle map of 1750. But the importance and persistence of the idea, and the misconceptions caused by its variations can be appreciated only by studying a chronological sequence of 16th-, 17th-, and 18th-century maps.

Fragments of this east-west range linger on after the idea of a continuous range became impossible. (There are enough maps in Paullin to show this.) Coronelli's "North America" in his Atlas of 1692 and his sheet map of 1688 show mountains of both banks of the Mississippi from the mouth of the Illinois River southward past the mouth of the Missouri, which is located with fair accuracy. But the most revealing expression of the idea I have seen is in a group of maps best represented by a North America of Giacomo de Rossi's, Rome, 1677. Its prime meridian is Ferro and its longitude is expressed exclusively as east. (Subtract Ferro East from 378° to get Greenwich West.) The error in the longitude of then known places varies but is sometimes as low as 15°, which is excellent for the time.

On this map a mountain range trends southwest across Virginia till it reaches the 295th meridian well north of the 40th parallel. Here it turns straight south for a little and then, still north of 40°, heads due west as the northern boundary of Florida. Its east-west extent is from 295° to 270°, that is from 83° W. to 113° W. Here it turns sharply southward from above 40° N. to 30°. Below 30° N. it again heads west to about 263°, where it turns straight north till it crosses 40°.

This is a single uninterrupted mountain chain. Florida and the Mississippi are both south and east of it. But it makes a deep embayment, southward through more than ten degrees of latitude and after crossing 13° of longitude northward again — beginning at de Rossi's 295°, which may be corrected to about 95° W. *true* longitude. The area contained in this embayment, west, north, and east of the mountain range, is part of de Rossi's New Mexico, which also extends south and southwest of the range, to Mexico and the Gulf of California. In the embayment Santa Fe, Zuñi, and Cíbola are north of the mountains. His error at Santa Fe, which had been founded sixty-eight years before his map, is about 18° of longitude and about 2° of latitude. . . . Needless to say, by 1677 the Spanish had enough empirical knowledge to prove this entire conception of New Mexico and its mountains absurd.

3. To avoid confusion I always refer to it in my text as the Western Sea. It was frequently called the Western Ocean, however, and for a long time the names were interchangeable in the thinking of cartographers and explorers. But the term Western Ocean was also used to designate the Pacific, or more usually some undefined portion of it, without reference to the Western Sea.

4. Kellogg (1), p. 11. Verrazano's discoveries were much depreciated and doubted until his own account of them was discovered in 1909. This is printed in the Fifteenth Annual *Proceedings* (1910) of the American Scenic and Historic Society of New York.

5. Aubrey Diller, "Maps of the Missouri River before Lewis and Clark," in *Studies and Essays in the History of Science and Learning Offered in Homage to George Sarton,* an article so important, as will become apparent later on, that it may be permitted this idealism.

6. The story of Moncacht-Apé is in Vol. III of Le Page du Pratz. It is translated in Andrew McFarland Davis, "The Journey of Moncacht-Apé," American Antiquarian Society *Proceedings,* N.S., II (1883), which is apparently the only analysis of any length ever made, except for an article which Davis is discussing, Quatrefages in *Revue d'Anthropologie,* IV (1881). Davis is on the whole skeptical but believes that the story may represent a genuine journey. That belief is not tenable and I find incomprehensible Nellis M. Crouse's acceptance of the story as substantially true. I therefore add a brief analysis.

When du Pratz left Louisiana, 1734, the French had not gone up the Missouri above the mouth of the Platte. Communication with New France was usually by the Wisconsin River or the Illinois River. The story of Moncacht-Apé is written from the understanding of a well-informed inhabitant of Louisiana at exactly this time — one, furthermore, who had not seen New France or gone up the Mississippi much beyond Natchez but who had talked with people who knew both and had read all the books. His book is redolent of Hennepin, Lahontan, and Charlevoix.

Moncacht-Apé's eastern journey is from those books and is entirely free of descriptive detail in the area from the upper Ohio to Lake Ontario which was then unknown to writers. As for his journey to the Pacific, the points made in my text show that it is a fabrication but I may enlarge on them. No natural feature whatever can be recognized in the story, not even what I arbitrarily state was the mouth of the Kansas River. The very few natural features that the story mentions are all from maps not later than 1740, which were in this area completely conjectural and erroneous. Moncacht-Apé sees mountains throughout the last stage of his thirty-day journey from the mouth of the Kansas; there are none there to be seen, though some appear on the maps du Pratz knew. Thirty days on foot might have taken him to the South Dakota line. Nine days later, perhaps at Pierre, he leaves the river, which thereafter is on his left. (Davis speculates that he may have been traveling up the Platte, not the Missouri. If so we may suppose him to have left it and turned north somewhere between Grand Island and North Platte, though he would not then have had the river on his left — and a glance at any modern map will show that the rest of his journey has become preposterous.) After five days, perhaps in North Dakota now, he strikes the Beautiful River, which flows west to the Pacific. Time and distances may be juggled to any extent one will, to allow for a journey which du Pratz says took eighteen months though neither he nor Moncacht-Apé states any travel times except those I give in my account. But no matter what adjustments may be made the story goes to pieces when the hero reaches the Beautiful River, for north of the Missouri and east of the Continental Divide no river flows west — and Moncacht-Apé crosses no height of land.

Belle Rivière was a designation of the Ohio. There is no record of such a name for a western river, unless this is a reference to the Belle Fourche, the northern branch of the Cheyenne, which flows east and which in 1758 the French had neither seen nor named. The Sioux called the Cheyenne below the forks Good River, as opposed to Bad River which was the Teton, but the Sioux did not reach it till more than seventy-five years after Moncacht-Apé's journey. He says that the Beautiful River flows northwest. No river of the United States or Canada that falls into the Pacific has such a course; some flow south of west, none north of west. After leaving Omaha, or perhaps Vermillion, the heart of the plains, he does not mention mountains; but there is no way of reaching any river that flows into the Pacific without crossing many ranges. Here too du Pratz is faithful to the maps, none of which show mountains where the Rockies are, though some have conjectural ranges on the coast. (He obviously had not heard of the Vérendryes.) Finally, the Beautiful River parallels the course of the Missouri, though the two rivers flow in opposite directions. This is precisely what certain mapmakers whose geography du Pratz appears to be championing were doing with the River of the West. On some maps it appears as a short middle portion of a westward-flowing river whose source and mouth are not shown; in this form it is not far to the northwest of the postulated sources of the Mississippi, which at this time lie farther west than those of the Missouri. On others it appears as a river which flows west out of the farthest lake of the chain that Lahontan's Long River drains to the east. On still others it is the Long River moved north, reversed, and made to flow west.

The only Indian tribe known as the Otters originally lived north of Lake Huron. They were massacred and disappeared during the Iroquois wars of the 1650's; a few survivors settled on Lake Michigan and Green Bay. The stunted, very white men with beards and strange garments are fable. The state of anthropology was such in 1883 that Davis wondered if they might not be some little-known Chinese or Japanese tribe, though why they would be so strikingly white is not clear, nor why they should be sailing obviously European ships. So far as is known, neither Japanese nor Chinese sailed to the Northwest coast till long after 1700. There is no dyewood, yellow or any other color, on the Pacific coast. But brazilwood, which gives a purple dye after exactly the treatment Moncacht-Apé describes, was still an object of search in Louisiana when du Pratz lived there.

7. The authoritative study of the legend is Thomas Stephens's *Madoc*. It was submitted in a competition at the Llangollen Eisteddfod (bardic and scholarly festival) of 1858, of which the assigned subject was "the discovery of America in the 12th century by Prince Madoc ab Owain Gwynedd." It was declared ineligible because it found the Madoc story false, and it was not published till 1893. As late as 1858, then, the historical scholarship of Wales accepted the myth.

8. In an effort merely to make out the morphology of the myth, not to compile a complete bibliography, I have found many American appearances not cited by Stephens, which means that most of them were published later than 1858. In August 1947 the Associated Press carried a story saying that the Kutenai Indians of British Columbia had been identified as descendants of the Madoc colony. In 1950 the Lookout Publishing Company of Chattanooga published Zella Armstrong's *Who Discovered America? The Amazing Story of Madoc*, which faithfully repeats all the familiar details as historically true.

9. No date has ever been given for the Armageddon of the Welsh Indians in Kentucky. Learned or pious readers will perceive that this folk migration parallels, though in reverse, two earlier population movements in the Americas. Following the fall of the Tower

of Babel (2247 B.C.) a prophet named Jared led his people through Armenia and across Europe to Spain, whence they sailed to the Gulf of Honduras. They gave central America and adjacent Mexico (then called Moron) a splendid civilization, extended it to Yucatan, and moved on to The Land Northward, which was the United States. They built great cities at New Orleans, St. Louis, and Cincinnati, and spread through the river valleys and built the mounds. Their culture lasted for 1600 years but at last civil war broke out. In the year 600 B.C. the entire nation, in two hostile camps, fought a final battle and slew one another to the last man, save only two who fled south leaving the land empty.

This final battle was fought on a hill near the village of Palmyra, New York.

That was the Jaredite Armageddon. One year before it, a prophet in Jerusalem, foreseeing the wars that were to come, led his people to the shore of the Indian Ocean and from there sailed to the coast of Chile. They too were a vigorous people and they spread out in several directions, taking their advanced culture with them. Unhappily one branch of them had evil hearts: these became the American Indians. The good people founded the civilizations of South America, including the Inca, and going on to Mexico recognized the ruins of their predecessors there and brought the Aztec Empire in. Their migration continued. Reaching the United States, they followed the trails blazed by the *émigrés* from Babel, building great cities and advancing up the Ohio Valley toward New York. (Everywhere else too, from the Gulf to Hudson Bay, from the Atlantic to the Pacific.) The Indians warred on them and like the Celts they died, their cities smoking and their land laid waste. A long time passed. Finally in A.D. 384 an entire nation again gathered on that hill near Palmyra — its name is Cumorah. In that year and on that hill the Indians massacred them.

All but a few. One named Mormon lived to write their story. (The historian of the Babel folk had been named Ether, possibly a comment on his prose style.) A few others fled down the Ohio, up the Mississippi, and into the Missouri. Like the Madocians eight hundred years after them, they started up the Missouri and the mist hid them.

The historical, archeological, and ethnological proof of all this is given in *Report of the Committee on American Archeology*, Reorganized Church of Latter-Day Saints (Lamoni, Iowa, 1910).

10. Nowadays such axeheads were dropped by one of Paul Knutson's explorers fleeing from the Sioux toward Deacon Arnold's round stone tower at Newport. (It is not clear why they traveled west from Minnesota to reach Narragansett Bay.) They never got

there; they paused at the Great Bend of the Missouri and became the Norse Indians. Eight hundred years before them the Mormon Indians had reached the same place and settled there; they were exhibiting striking biological traits when the Norse got there to amalgamate with them.

11. I count thirteen in the literature and it is certain that many additional tribes were rumored to be Welsh.

12. Some school texts of American history published later than 1900 carried the statement that the Mandans were descended from Madoc's colony. It is worth noting that in 1865 the Smithsonian Institution still regarded the Madoc story as an open question, and two years earlier the American Antiquarian Society (to which a benefactor had presented a relic of the Welsh Indians) had been told that the future must decide. In 1862 the American Ethnological Society felt that considerable evidence supported the story. John G. Palfrey was of the same opinion in 1858 and made no change when a revised edition was published in 1876. Benjamin F. De Costa, probably the most respected authority of his time on pre-Columbian voyages to America, accepted the Madoc story and retained his belief in his last book, 1901. Even Justin Winsor, as skeptical a historian as ever lived, preferred to dodge the issue in the great work he edited. See Winsor (1), I, 109–11.

13. Today the Yellowstone enters the Missouri as an opaque, dark brown stream that maintains its identity in the clearer water it has reached for five or six miles. All this silt comes from its southern tributaries and of these the Big Horn is the worst. The first white man who described the Yellowstone was François Larocque and the second William Clark. Their journals (see Chap. XII) show that all these streams were as clear as plains rivers ever are. Larocque especially insists on the limpidity of the Big Horn.

14. This statement, which sets an earlier date (1738) than standard texts give, will be justified later on.

CHAPTER III: THE IRON MEN

DOCUMENTARY As before: Hakluyt (trans. of Cartier); Margry; JR, Vols. 18, 38, 40, 41, 45, 46, 47, 48, 50, 54, 55, 56, 58, 59. H. P. Biggar, ed.: (1) *The Voyages of Jacques Cartier* (Ottawa, 1924); (2) *Samuel de Champlain's Works* (Toronto, 1922–26). Emma H. Blair, ed., *The Indian Tribes of the Upper Mississippi Valley* (Cleveland, 1911). Pierre F. X. Charlevoix, (1) *Histoire et Description Générale de la Nouvelle France* (Paris, 1744). *Documents Relative to the Colonial History of the State of New York*, Vols. 5, 9;

hereafter cited as NYCD. Louise Phelps Kellogg, ed., (2) *Early Narratives of the Northwest* (N.Y., 1917). *Minutes of the Hudson's Bay Company* (Toronto, 1942). *Voyages of Peter Esprit Radisson,* reprint of Prince Society edition (N.Y., 1943). John Gilmary Shea, ed.: (1) *Discovery and Exploration of the Mississippi Valley,* 2d ed. (Albany, 1903); (2) *Early Voyages Up and Down the Mississippi* (Albany, 1861).

SECONDARY Parkman. As before: Kellogg (1); Winsor (1). C. W. Alvord and L. Bidgood, *First Explorations of the Trans-Allegheny Region* (Cleveland, 1912). Morris Bishop, *Champlain* (N.Y., 1948). George Bryce, *The Remarkable History of the Hudson's Bay Company,* 3d ed. (N.Y., 1903). Jean Delanglez, (1) *Life and Voyages of Louis Jolliet* (Chicago, 1848). George T. Hunt, *The Wars of the Iroquois* (Madison, 1940). Harold A. Innis, (1) *The Fur Trade in Canada* (New Haven, 1930). Henri Lorin, *Le Comte de Frontenac* (Paris, 1895). Grace Lee Nute, (1) *Caesars of the Wilderness* (N.Y., 1943). Francis Borgia Steck, *The Jolliet-Marquette Expedition, 1673* (Washington, 1927). Baron Marc de Villiers, (1) *La Découverte du Missouri* (Paris, 1925). Justin Winsor, (2) *Cartier to Frontenac* (Boston, 1894).

1. Batts and Fullam. But of course in the valley of New River they found initials carved on trees. Nameless men have always preceded the firstcomers.

2. Oscar Lewis, *The Effects of White Contact upon Blackfoot Culture* (N.Y., 1942).

3. In this summary I follow Hunt in the work cited above, a valuable book which has brought a long-needed realism to the study of the Iroquois. Its emphasis on the commercial character of the Iroquois wars is a wholesome corrective but, in my judgment, becomes overemphasis in the end. I cannot follow Mr. Hunt in his characterization of the Iroquois. He makes them a kind of 20th-century board of trade and slights elements of Indian psychology, Indian and Christian religion, Indian and white irrationality that had enormous effects on imperial grand strategy. Moreover, he has not so drastically revised Parkman as some of his remarks assert. Most of his analysis of the fur trade is explicit in Parkman, though not concentrated in a single place; much of the rest is implicit. Finally he adopts a cliché of the last generation of American historians when he condescends to Parkman's understanding of Indians; this generation has learned better.

4. One does not question Dr. Grace Lee Nute's decision that, contrary to what had been previously believed, Radisson wrote his

account in French, not English, and that it was translated on commission from the Hudson's Bay Company. This discovery, however, makes more mysterious the odd and charming language of the book. The translator, whoever he was, could not cope with English idiom and syntax and was repeatedly baffled by English vocabulary.

5. Nute (1), p. 23.

6. *Voyages of Peter Esprit Radisson,* 1943 reprint of Prince Society edition, pp. 167–68.

7. Of recent ones, Dr. Nute allows the possibility that the unnamed companion may have seen the Mississippi, whereas Father Delanglez rejects it and denies that this is a reference to the Mississippi.

8. Father Steck, a Franciscan, on denials by some historians, principally Jesuits, that the Order designed to organize the interior as a North American Paraguay: "It cannot be denied . . . that the plan was to erect in New France so-called reductions, such as existed in Paraguay." Steck, *Jolliet-Marquette Expdn.,* p. 52.

9. Nute (1), p. 61.

10. Parkman, *The Old Régime,* Chap. VI. (All my citations from Parkman are from his last revisions and I use the Centenary Edition.)

11. Steck, *op. cit.;* Delanglez (1). In 1927 Father Steck proved that the narrative was not written by Marquette and suggested that Dablon was the author. Father Delanglez, a Jesuit, later established Dablon's authorship and in a series of articles in *Mid-America,* which are summarized in his book, furiously attacked Steck's hypothesis that Dablon used Jolliet's journal. In "The 'Real Author' of the *Récit,*" *The Americas,* V (April 1948), Steck answered Delanglez with even greater heat and, in my judgment, conclusively as regards Jolliet's journal. See also his "What Became of Jolliet's Journal?" *The Americas,* V (Oct. 1948), and "Father Marquette's Place in American History," *ibid.,* VI (April 1949). Although the narrative very frequently fails to be specific, it is nevertheless specific so often about places, locations, and dates that I cannot believe Dablon got only oral information from Jolliet, as Delanglez maintains he did apart from the letters to Frontenac and Laval. Furthermore the point of view of the narrative is that of a layman, except for pious passages that Steck holds to be Dablon's interpolations. I follow Steck.

12. Steck's correction; the narrative says June 15.

13. The mouth of the Iowa River has also been suggested and there is no way of determining which, if either, is right. One of the two Illinois villages here was called Moing-wena and is so

named on Jolliet's map. Presently the Des Moines is called the the Moingona on maps; in fact the present name is a corruption of that one.

14. Steck prints the letter to Frontenac, pp. 171–72, and the "relation" to Dablon, pp. 173–80.

15. Their coiffures and the clothes of their women resembled those of the Hurons, which suggests an Iroquoian tribe. Shea thinks they were Tuscaroras; in this place Cherokees would be more likely.

16. Delanglez, who adjusts Jolliet's latitudes, is unwilling to commit himself to anything beyond two hundred miles downstream from Memphis; Steck, who accepts the latitudes, would bring them below Vicksburg.

17. Three leading authorities have held that the route thus reported by the Illinois to Jolliet is essentially correct, that the Indians were describing a route up the Platte and the South Platte and down the Colorado. That finding is not tenable. It is true that the headwaters of the South Platte come within three hundred miles of Santa Fe and within five or six hundred miles of Mexican silver mines, are separated from the headwaters of the Grand by only a single mountain range, and are less than a thousand miles from the Gulf of California toward which the Grand, which was not the Colorado River till the U.S. government defied geography by transferring the name to it, leads down from the mountains. But Jolliet, together with everyone afterward who held this durable illusion, was thinking of a water route, a route up which boats, even birchbark canoes, could be taken. The Indians who told him about this route did not know the geography involved and had never met any Indians who knew it — there were none. Scholars who accept it ignore the Rocky Mountains.

The statements in the narrative itself make the idea preposterous. The river five or six days' journey up the Missouri may be the Osage, though it flows east, not west as the narrative has it. If it is, then what the Illinois said merely shows that they had never been to the Osage, though they fought and traded with tribes that had been there. Of course nowhere in the Missouri drainage basin is there a river that flows to the Pacific: what is the Continental Divide? Nowhere in that basin or beyond it is a lake from which a river flows to the Pacific or the Gulf of California.

18. NYCD, IX, 72–73.

19. This is the correction by Delanglez (1), pp. 15–16, of the date usually given as June 14.

CHAPTER IV: THE HEARTLAND

DOCUMENTARY As before: Blair, Charlevoix (1), French, Kellogg (2), Lahontan, Margry, NYCD, Radisson, Shea (1) and (2), Hennepin: (1) *Description de la Louisiane;* (2) *A New Discovery of a Vast Country,* R. G. Thwaites, ed. (Chicago, 1903).

SECONDARY Parkman. As before: Alvord and Bidgood, Brebner, Bryce, Hunt, Innis (1), Kellogg (1), Lorin, Nute (1), Winsor (1) and (2). Theodore C. Blegen, *The Land Lies Open* (Minneapolis, 1949). Jean Delanglez, (2) *Some La Salle Journeys* (Chicago, 1938). Robert E. Pinkerton, *Hudson's Bay Company* (N.Y., 1931). Baron Marc de Villiers, (2) *L'Expédition de Cavelier de la Salle dans le Golfe du Mexique* (Paris, 1931).

1. Parkman, *La Salle and the Discovery of the Great West,* speculated about this possibility, which might explain aspects of La Salle's last expedition which he found inexplicable. He might have rejected it, however, if he had been acquainted with the Peñalosa material that has since been brought to light. De Villiers (2) has no doubt that La Salle was insane and has found a psychiatrist who says that he was always *"méfiant, orgueilleux, autoritaire et égocentriste,"* that he was a paranoid who would have ended in a madhouse if he had lived in France but found useful fulfillment in the wilderness, and that in this last stage he became *"aliéné, c'est-à-dire étranger à lui-même"* (p. 178). Retrospective diagnoses by psychiatrists are sometimes open to dissent. De Villiers' own understanding of La Salle's character does not inspire confidence. He is capable of saying, for instance, that all his life La Salle "lacked the qualities that make a leader loved." That would explain Tonty's devotion.

2. In my account of La Salle's effort to colonize Louisiana I have mainly followed de Villiers (2) but from the assumptions of Nute (1). The authorities do not agree about the Peñalosa episode and have failed to make it clear; in fact, a very great deal about La Salle's last venture remains to be cleared up. It badly needs the learning, indefatigability, and ingenuity of Dr. Nute. The documents in Margry must be used with extreme care. Delanglez (2) is invaluable, though marred by the author's conviction that La Salle must be reburied outside consecrated ground. See also Parkman, *La Salle,* Chaps. XXIII–XXIX; Dunn, *Spanish and French Rivalry in the Gulf Region* (Austin, 1917); E. T. Miller, "The Connection of Peñalosa with the La Salle Expedition," Texas Historical Assn. *Quarterly,* V (1901); C. W. Hackett, "New Light on Don Diego de Peñalosa," *Mississippi Valley Historical Review*

(hereafter cited as MVHR), VI (Dec. 1919); Dr. Nute's introduction to Marion E. Cross's edition of Hennepin.

3. Annual fur fairs continued to be held at Three Rivers and Tadoussac, for the tribes north and east of them.

4. Several students have assumed that this salt was from Great Salt Lake and concluded that the Sioux had been there. The assumption is unlikely to an extreme, the conclusion quite untenable. It is not known what Indians lived in the Great Basin at this time. The area does not come into history for another century; when it does, there is but little trade out of it, and that is with New Mexico and western Colorado by an infinitely laborious route. Salt from there would have had to reach the Sioux by so many exchanges as to make it most probable that none did. There is neither evidence nor tribal tradition that the Sioux had ever ventured as far west as the Red River. It is, however, entirely possible that they did — but it is all but certain that they had gone no farther. As for crossing the plains beyond the Red, on foot and through such tribes as the Pawnees, the Kiowas, and the Comanches, it is inconceivable — there is no suggestion that they had yet seen even the Mandans. There were a number of salt springs in the Red River country and this seems the probable origin of the salt which Duluth's men saw. The Sioux could have got it directly or by trade.

5. *"Leur langue n'aucune ressemblance avec celle des autres Sauvages. Elle tient des prononciations chinoises."* Antoine Raudot in 1710: Margry, VI, 15.

6. This hard fact is what recent essays in the economic interpretation of the Iroquois Confederacy fail to account for. Professor Hunt, for instance treats the Iroquois as if they were Economic Man implementing a business policy with the detachment of Standard Oil organizing a new territory, and he ridicules Parkman's idea that they had a natural liking for war. But they did. The murderous raiding that we perhaps inaccurately call war, though that is what the Indians called it, was metabolic and interstitial in the Indian way of life. Killing was sport, it obtained religious grace, and it was a way to social and political distinction. It is idle to represent murder and the collection of scalps as a trade device for cornering a market or increasing the markup on goods. Economic Man would not torture and kill the customer nor would economic diplomacy sanction the murder of an ally. The Iroquois did not torture or kill the customer to get his trade but because religious belief and magical antisepsis required it, and because they enjoyed torture and killing. What led the Iroquois to repeatedly make war on the tribes they were trying to bring into a peaceful alliance was not the shortcomings of their political institu-

tions nor their mistakes in economic theory but the fact that there are stronger motives than the economic one. In a neolithic society which exalts killing the elders may understand the desirability of curbing them temporarily in the interest of prosperity but there are always the young men.

7. They were linguistically related. As Professor Hunt points out, the term "Illinois" in the literature is very loose; the French used it as inclusively as they did "Ottawa." The same is true of "Miami."

8. Lahontan was on the expedition. His detailed report of the oratory is in *New Voyages*, pp. 66 ff.

9. Kellogg (1), p. 230.

## CHAPTER V: CONVERGING FRONTIERS

DOCUMENTARY   As before: Charlevoix (1); JR, 64, 65, 66, 67; Kellogg (2); Shea (1) and (2); Margry. Herbert E. Bolton, ed., (2) *Spanish Exploration in the Southwest* (N.Y., 1930). Pierre F. X. Charlevoix, (2) *Historical Journal* (N.Y., 1851). Arthur G. Doughty and Chester Martin, eds., *The Kelsey Papers* (Ottawa, 1929). Alfred Barnaby Thomas, ed., (1) *After Coronado* (Norman, 1935). J. B. Tyrrell, ed., (1) *David Thompson's Narrative* (Toronto, 1916).

SECONDARY   Parkman. As before: Brebner; Bryce; Hunt; Innis (1); Kellogg (1); Nute (1); de Villiers (1); Winsor (1) and (2); Lawrence J. Burpee, (1) *The Search for the Western Sea*, rev. ed. (Toronto, 1935); Charles Gayarré: (1) *Louisiana: Its History as a French Colony* (N.Y., 1852); (2) *History of Louisiana* (N.Y., 1854). George E. Hyde, *Red Cloud's Folk* (Norman, 1937). Arthur S. Morton, (1) *A History of the Canadian West to 1870–71* (London, 1939). Rupert Norval Richardson, *The Comanche Barrier to South Plains Settlement* (Glendale, 1933). Alfred Barnaby Thomas, (2) *The Plains Indians and New Mexico, 1751–1778* (Norman, 1940). Justin Winsor, (3) *The Mississippi Basin* (Boston, 1895).

1. It may be well to point out that at this period none of the Sioux had acquired the Plains culture. Though the French sometimes called them "the nation of the buffalo," the buffalo they hunted were in southern Wisconsin and Minnesota.

2. Here the scanty documents are vague: it is not certain that de Noyon reached Lake of the Woods.

3. Charles Napier Bell made a minute study of Kelsey's account in the *Transactions* of the Historical and Scientific Society of

Manitoba for May 1928. Doughty and Martin, the editors of *The Kelsey Papers* (1929), question his results. I follow them.

4. Except for the "Nahathaways," who were woods Crees, and the "Stone Indians," Assiniboins, the tribes he mentions cannot be confidently identified. By another branch of the Crees I mean the portion of the nation who by this time had moved out of the forest permanently and were learning the Plains culture. For them see David G. Mandelbaum's *The Plains Cree* (N.Y., 1940), one of the most valuable of all anthropological studies of the Plains tribes. (Curiously, however, Mandelbaum appears not to have known of Kelsey's journey.) Kelsey calls a distant tribe he eagerly wanted to meet but did not the "mountain poets" — "poet" is his rendering of a Cree suffix. Arthur S. Morton conjectures that they may have been the Fall Indians, the Atsina. Brebner, footnote p. 295, suggests that the tribe which I believe were the Plains Crees were Assiniboins or Mandans. But Kelsey's "Stone Indians," whom he so designates, must surely have been Assiniboins and the Mandans seem to me entirely improbable. They are not known to have traveled to the Saskatchewan and forty years later, in Vérendrye's time, they are still a mysterious people far to the south of Lake Winnipeg about whom the Crees and Assiniboins are relating fabulous yarns.

5. This discounts the finding of a recent book which identifies as a grizzly one of the fabulous animals in David Ingram's bestiary. Ingram is the man mentioned, but not named, on my p. 61 as having influenced the geographical ideas of Sir Humphrey Gilbert. I regard the story of his wandering as fictitious. Even if it were true, Ingram's route cannot be located on the map, and it would have been an odd course indeed that took him to New Brunswick through grizzly country. And I cannot identify his beast as a grizzly.

6. For the acquisition of horses, see Francis Haines, "The Northward Spread of Horses among the Plains Indians," *American Anthropologist,* Vol. 40 (1938), supplemented by his *The Appaloosa Horse* (Lewiston, n.d. [1951?]). The literature of the fur trade contains many vivid descriptions of the change which the horse wrought in Plains culture. See for instance the account of the Blackfeet in Thompson's *Narrative,* ed. Tyrrell.

7. I cannot say how early the French heard allusions to mountains which we must recognize as having been a part of the Rocky Mountain system. (There is nothing in the literature that I feel sure about, however, till long after 1700.) This does not matter, since they misconceived what they heard. It must be remembered that few French explorers of the West had seen the Laurentians

or the New England mountains; none had seen the Alleghanies. Another trouble is that the word *"montagne"* connoted hills as readily as mountains. Thus portions of early 18th-century maps that rest on empirical knowledge may show as mountains such low relief as the bluffs along the Mississippi in Wisconsin.

Typically, a map of North America of this period will show mountains in the vicinity of New Mexico. These are not actual, they are at random, vestiges of the east-west range and its embayment discussed in note 2 to Chap. II, above. In the terra incognita between the Mississippi (and the Missouri when its mouth and a little more begin to be shown) and the Pacific, no north-south range is shown: the concept did not exist. The concept of a height of land was established — geography, especially symmetrical geography, required it — but it is always confused and in most texts it is a contradiction.

If there were an interior water route to the Pacific, the Northwest Passage in any of its forms, it could cross no mountain range; well past this period, in fact, it was usually assumed to be a sea-level route. The earliest conceptions of the River of the West had it flowing from a height of land whose eastern slope was drained by the Mississippi, and this divide was believed to be very near the western end of Lake Superior. Later conceptions, when they did not have it flowing west parallel to eastward-flowing rivers, had its sources so near to known country that no room for a mountain range was left.

In addition, the recurring idea that the interior contained the Western Sea, with either a river or a strait as its western outlet, negated the possibility of a mountain barrier. I have seen no map of the Delisle school that shows a north-south range in the interior West except such fragmentary ones as are certainly vestiges of the east-west range and its embayment. That is also what the "high mountains" between the Gnacsitares and the Mozeemleks are on Lahontan's map, as his text shows.

8. Henri Folmer makes it the mouth of the Niobrara, "Etienne Veniard de Bourgmond in the Missouri Country," *Missouri Historical Review*, Vol. 36 (1941). He is following de Villiers (1), Chap. III. Nasatir makes it the mouth of the Big Sioux, which is downstream from the Niobrara. De Villiers' identification of the *"Rivière Fumeuse"* as the Niobrara may be questioned, and in any event de Bourgmond places the mythical *"Mahas blancs"* below the Arikara villages, which demonstrates that he did not reach the latter, and his description of both the Missouri and the country it flows through above the Platte are self-evidently hearsay. His own report has been lost but de Villiers prints *"Routte qu'il faut*

*tenir monter la Rivière du Missoury,"* which is a summary of a daily log, and selections from a *"Description de la Louisiane"* which is based on his data. A full report of his later expedition, probably written by the engineer Renaudière, is in Margry, VI.

9. Some accepted assertions about the early appearances of the French on the Missouri River rest on tolerably thin evidence. Bolton is certainly right in saying that in "the western country traders from Canada roamed far and wide at an early date" but it would be helpful to know at how early a date and how far and wide, and one must remember that before 1713 no one can easily be imagined to have reached the Missouri from anywhere north of the Chicago River. Bolton does not support his further statements that "a Canadian is known to have reached the Rio Grande overland" (whence and at what point?) and that "by 1694 Canadian traders were among the Missouri and Osage tribes." I have seen no evidence that they were in the country of those tribes so early. A much repeated statement that twenty Canadians left the Tamaroa village (near the mouth of the Illinois River) in 1703 to find New Mexico appears to rest on a single sentence in a letter of Iberville's which gives no further details and does not say how far they got. On a single, similarly vague sentence from a summary of a letter of Bienville's rests the doubtful assertion that there were 110 Canadians on the Mississippi and the Missouri in 1704. One Laurain who is reported in 1705 to have met on the Missouri tribes who lived on the border of New Mexico is an entry in a journal and at second hand; it is self-contradictory and no tribes that lived on the New Mexican border came to the Missouri at that time. Nicolas La Salle's statement in 1706 that the Missouri has been ascended for 750 or 1000 miles is quite impossible; note too that the description of the Missouri country which accompanies it is wholly imaginary.

10. Bolton (2), pp. 201–2. All that is known about the journey, which was unauthorized, is the testimony of one Indian who was the sole survivor. His deposition and interrogation were five years after the events but the country and the tribes that he reports are recognizable and Oñate heard confirming testimony at Quivira.

11. New Mexico and Chihuahua ranches were the original sources from which the Plains Indians got horses, stealing or trading for them or picking up strays. Occasionally they took them from the herds of wild horses that had originated in the same way. The idea that strays from the expeditions of Coronado and De Soto bred up the wild herds or were bred by Indians who captured them is untenable. See note 6 above for Haines, *loc. cit.*

12. It is difficult to differentiate the tribes whom the Spanish

called Apaches at this time and sometimes impossible to identify them. Those which were currently harassing the settlements at this period were, mainly, Faraones to the southeast and Navahos to the west.

13. Thomas (1), p. 13, *passim;* F. W. Hodge, "French Intrusion toward New Mexico in 1695," NMHR, IV (1929), 72–76.

14. A guess could make the originators of the yarn the Iowas, between whom and the Colorado Apaches there were at least seven tribes. They were currently being forced westward across the state named for them. Conceivably the traders had met them on the Des Moines River, the Moingona, whose lower reach was more familiar to the French at this time than that of the Missouri.

15. For a summary of these yarns see Thomas (1), p. 14. They must be scaled down to the possible possession by some Navahos of goods that had come down the native trade routes. One must reject entirely the jewels and cannon listed among them. Where would jewels and cannon have come from?

16. Thomas locates El Cuartelejo in Otero or Kiowa County, Colorado. See also Henry W. Hough, "Picking up the Trails of Ulibarri and Villasur," in *The Brand Book,* Oct. 1948.

17. Thomas (1), pp. 66 ff. In this book Thomas appears to accept the account as true, in (2) he appears to be skeptical. It comes down to one Frenchman whom the Spanish did not see and one gun which they did see. The French deserter who was with Ulibarri claimed that he recognized the gun as having belonged to a relative of his, and Ulibarri supposed that that was why the Indians changed their story. But this identification made the story even more nonsensical; if the deserter was right, then the gun must have belonged to La Salle's party and tribes had been passing it about for twenty years.

18. Father Gilbert Garraghan, *Chapters in Frontier History,* pp. 62 ff., believes that the Sieur Derbanne may have reached the Platte and gone beyond it in 1706 or 1707. He quotes an unpublished letter written eighteen years later in which Derbanne claims to have gone 400 leagues, a thousand miles, up the Missouri, which would take him above the White River. Such a distance at such a date must be a sizable exaggeration. Conceivably Lemaire's reference of 1714 is to him: "*On a remonté le Missouri plus de 400 lieues sans recontre aucune habitations espagnole, et ce n'est qu'a quelque 500 lieues qu'on commence a en avoir des nouvelles par des Sauvages, qui font la guerre avec eux.*" (Margry, VI, 185.) Five hundred leagues up the river the only Indians who made war on the Spanish that would be heard about were the Comanches, and they could be heard about only a few leagues distance up it. Actu-

ally, this is evidence of the vagueness of the available information.

19. Documents and editorial introduction and notes, Thomas (1), which accepts all these stories except that of the invading six thousand. Thomas (2), published five years later, reduces most of them to the status of rumors.

20. Thomas (1), p. 287, locates the site of the massacre on "the south side of the North Platte near the town of North Platte." See also his "Massacre of the Villasur Expedition," *Nebraska History*, VII (1924), 68 ff. Most students agree with him. De Villiers makes it, untenably, the junction of the Platte and its Loup Fork, much farther east, "Le Massacre de l'Expédition Espagnole du Missouri," *Journal de la Société des Américanistes de Paris*, N.S., XIII. Father Garraghan follows him.

21. Boisbriant's report, quoted in de Villiers (1), p. 71. Bienville's later report is in Margry, VI, 386.

22. It is harder to determine distances in the plains than anywhere else. De Bourgmond says that he marched on a compass course. De Villiers decides that the Comanche village was in Rice County, Kansas, just east of the Big Bend of the Arkansas. The massacre of Villasur's party had taken place more than 200 miles to the northwest.

23. Margry, VI, 455–92. See also Henri Folmer, "The Mallet Expedition of 1739," *The Colorado Magazine*, Vol. 16 (1939) pp. 161–73. The route I give in the text is the one determined by Folmer.

CHAPTER VI: THE WORLD TURNED UPSIDE DOWN

DOCUMENTARY   As before: Charlevoix (1) and (2); JR, 67, 68, 69; Kellogg (2); Margry, VI; Tyrrell (1). Lawrence J. Burpee, ed., (2) *Journals and Letters of Pierre Gaultier de Varennes de la Vérendrye* (Toronto, 1927); ed., (3) "York Factory to the Blackfeet Country" (Henday's journal), Royal Soc. of Can. *Proceedings*, 3d Ser., I (1907). Elliott Coues, ed., (1) *New Light on the Early History of the Greater Northwest* (N.Y., 1897). Arthur Dobbs, *An Account of the Countries Adjoining to Hudson's Bay* (London, 1744). *Report on Canadian Archives, 1886* (St. Pierre's journal). *Wisconsin Historical Collections*, XVI, XVII.

SECONDARY   Parkman. As before: Brebner; Burpee (1); Gayarré (1); Innis (1); Kellogg (1); Morton (1); Winsor (1) and (3). Lawrence H. Gipson, *The British Empire before the Revolution* (Caldwell and N.Y., 1936–49). Charles A. Hanna, *The Wilderness Trail* (N.Y., 1911). F. O. Libby, "Some Vérendrye Enigmas,"

MVHR, III (1916). Grace Lee Nute, (2) *The Voyageur's Highway* (St. Paul, 1947); (3) *Rainy River Country* (St. Paul, 1950). L. A. Prud'homme, (1) "Pierre Gaultier de Varennes . . . Découvreur du Nord-Ouest," Royal Soc. Can. *Proceedings,* 2d Ser., XL (1905); (2) "Les Successeurs de la Vérendrye," *Proceedings,* 1906. Doane Robinson (1) and Charles E. De Land, "The Vérendrye Explorations and Discoveries," *South Dakota Historical Collections,* VII (1914); Robinson, (2) "The Vérendrye Plate," Mississippi Valley Historical Association (hereafter cited as MVHA) *Proceedings,* VII (1913–14).

1. Burpee (2), p. 64, n. 2, speaks of a memoir 1706 in which Guillaume Delisle says he has established the existence of the Sea of the West and that he had laid it down on a manuscript map. I cannot find this memoir and though it seems to be well known to geographers those I have consulted have not been able to direct me to it. Delisle's thinking dominated geography at this period. I have seen no map by him which shows a Sea of the West but the American Geographical Society has a photostat of a page of an unidentified atlas which it dates "c. 1700." The legend reads (in French) "Conjectures on the existence of a sea in the western part of Canada, and Mississippi [meaning Louisiana]. By G. Delisle . . ." Like Delisle's "North America" of 1703, this map has been strongly influenced by Lahontan's geography and so can be no earlier than 1703, when Lahontan's book was published. It shows the Mississippi with, for the period, fair accuracy; the Missouri, which has a wholly conjectural course, is longer than it is on most contemporary maps. There is a River of the West which flows out of an unnamed lake, possibly Lahontan's, to the Sea of the West. A northern bay of this Sea extends so far that it runs off the map at the top. From the southwestern shore of the Sea a strait leads to the Pacific at 45° N. This is a perfect visualization of the purest form of the idea. I may add that 45° is north of the latitude below which the Pacific coast was conceded to be Spanish.

2. Aguilar was with the Vizcaíno expedition that was sent from Mexico, 1602–3, to explore the western coast. A storm blew his ship northward, nobody knew how far but presumably beyond Cape Blanco at 43°. He seems to have thought that the big entrance which he claimed to have seen was in the neighborhood of 41°, an awkward place if his latitude was right, for it was too far north for the Golden Gate and too far south for the Columbia or even Rogue River. See Bolton (2), pp. 47–48, 97, 101–2. Aguilar understood that he had seen a river, not the Strait of Anian, toward

which Vizcaíno at one time thought he was being borne by storms and currents.

3. See p. 190 above.

4. "The Lake of the Assiniboins" most often meant Lake Winnipeg by a few years after Charlevoix. Sometimes it meant a lake beyond Winnipeg, either Manitoba or Winnipegosis. Sometimes it meant all three of them as one, sometimes a rumored one somewhere else. Sometimes there is no telling what it meant.

5. Charlevoix's recommendations are not in his *Historical Journal* but in letters to the Count of Toulouse and the Count of Morville. See Margry, VI, 531–35. They are also among the documents supplementing J. Edmond Roy's "Essai sur Charlevoix," Royal Soc. of Can. *Proceedings,* 3d Ser., I (1907).

Note that, as one who had been on the ground and had worked hard to clear up the confusion, Charlevoix knew that the existing maps were grossly unreliable. "In order to identify the Europeans whom the savages have seen on the coast of the Western Sea," he tells the Count of Toulouse, "we should have to have Spanish maps." The powers, however, kept their colonial maps as secret as their diplomatic ciphers, and Charlevoix adds that a map of the Spanish Indies which he has heard is in the Escorial must be unique. . . . One of the rumors he accepted has it that two priests had reached Lake of the Woods from these Europeans on the coast.

6. The Assiniboins currently had the valuable position of middlemen who purveyed English and French goods to farther tribes. In addition to this there was the very old intertribal trade with its fixed routes and habits. They ranged far enough west to exchange goods, in some way hard to be sure about but probably by means of intervening tribes, with tribes of the Pacific slope. They told Vérendrye and later white men that they themselves crossed the Divide and went down to the Pacific, to trade and make war. So far as I can see there is no evidence beyond this statement that they did. I have consulted all the studies of ethnologists without finding that they have turned up any evidence beyond the statement itself, which they usually accept. Surely it is open to the most serious question. By what route did the Assiniboins cross the Rockies, what tribes did they go through on the way, and what tribes on the western slope did they know? As will become apparent in the rest of this book, they did not during the rest of the 18th century know any route west that was south of 49°. They were not notable for courage or initiative in war and any route through Montana would certainly have exposed them to the Snakes who

were then at the height of their power, probably to the very belligerent Kiowas and Kiowa-Apaches, and possibly to the Comanches
— all of whom had horses, as the Assiniboins did not, and were
more warlike than they. The Canadian routes over the mountains
which they could have used led up the North or the South Saskatchewan to such passes as Howse and White Man's — they did
not range far enough north, it seems, to take the Peace River
route. Here they must encounter not only the Snakes but the
Blackfeet and the Gros Ventres, with whom it is clear they traded
but who, it seems equally clear, would and could keep them from
the direct trade farther west. I cannot find that they displayed any
first-hand knowledge of the Canadian Rockies or the country
beyond them.

7. The crucial passage in Vérendrye's text is Burpee (2), pp. 244–
49, though his earlier conceptions must be considered with it. The
text must be studied in connection with the map that represents
what "the most experienced" Cree and Assiniboin chiefs had told
him; it is opposite p. 116. The confusion is great but not so great
as what I take to be misreadings by Burpee have made it out to
be. I have no choice but to submit my own interpretation, and
the reader is notified that it differs not only from Burpee but
from Brebner and Morton.

The map does not square with the compass and this fact increases the difficulties. The eastern portion, though distances are
telescoped (70 miles from Kaministiquia to Rainy Lake as against
150 in fact), represents the geographical relationships pretty well.
Rainy Lake and Lake of the Woods, though wrong in relative
size and in their north-south bearings, are in the right relationship
and the source of the Mississippi is in the right relationship to
them. Vérendrye believes that Red River flows out of Red Lake,
which is actually the source of its affluent Red Lake River, and
knows nothing about its upper reaches. His lower Red River is
satisfactory; so is the mouth of the Assiniboine, which is not
named, and the Assiniboine has a fork, which may represent the
Souris. The Winnipeg River is shown correctly. The long axis
of Lake Winnipeg is east-west, whereas it should be north-south.
Though Vérendrye by now knew in days' travel the true length
of the Nelson River, it is much shortened on his map, its direction is wrong, and Hudson Bay is on an incorrect bearing. Lake
Winnipeg has a second outlet to Hudson Bay, presumably the
Hayes River. The mysterious Rivière du Brochet (Pike River,
but there were pike in all these streams) flows into the Nelson
and the crucial Rivière Blanche, which is east of it and drains
Lac des Glaizes, flows into Lake Winnipeg from the south. (It

might be possible to argue that in this portion of the map south should be translated southwest.)

A river which Vérendrye does not mention in his text, called "Grande R. de la Nation des Couhatelle" (a tribe which also is not mentioned in the text but presumably the Mandans, in still another and here detached avatar) flows east and then south from the height of land in a tolerable approximation of the actual Missouri. Clearly it was not the River of the Mandans or the River of the West or both as one, and he had heard nothing which gave the Missouri (whose actual entrance into the Mississippi he knew about) such bearings. It could be a river from maps familiar to him, possibly a variant of Lahontan's Long River, possibly a guess about the upper Des Moines as the Moingona.

The height of land, remember, is the fundamental divide. Vérendrye's text says that it is a range of mountains which runs (west — "as far as the unknown country") from north of Lake Superior. I cannot in the least explain this. He knew that there were no such mountains: the Lake Nipigon and Lake of the Woods country was entirely familiar to him. And in fact the map shows the mountains not only southwest of Lake Superior but west and south of Lake Winnipeg. At their western end is Lac de La Hauteur, the Lake of the Divide. It has, so he says and so he shows on the map, three outlets. One empties into Lake Winnipeg and this must be, as Burpee says, the Saskatchewan. (Nevertheless, Vérendrye adds elements of Red River to it.) This he names Rivière Blanche. Another outlet flows west-southwest, to *"Mer Inconnue,"* the unknown sea, which can be understood *only* as the Western Sea, not the Pacific. This is the River of the Mandans and on the map it is called *"R. du Couchant."* Of the third outlet, which has no name, Vérendrye says only that it flows south, and so the map shows it. But here, I believe, Burpee misreads the text, which runs: *". . . et troisième au Sud, a l'embouchure de la rivière Blanche, il y a un rapide le plus de demi-lieue toûjours grand eau . . ."* Surely, proper punctuation would make the comma following *"Sud"* a period — surely the statements are about different rivers. But Burpee translates, "and the third [flows] to the south at the mouth of the river Blanche. There is a rapid . . ." Since the mouth of the Blanche shows on the map 300 miles from the source of this southern outlet, the rendering is not only contradictory but without meaning. Surely it should read, ". . . to the south. At the mouth of the River Blanche there is a rapid . . ."

In what follows Burpee translates *"haut"* and *"audessus"* as "above," which is proper, but interprets them as meaning "north," which makes everything inconceivable, including his identification of the Woods Crees. But quite clearly Vérendrye uses both words to mean "upstream" — in the direction of the source — and this sometimes makes geographical sense.

The Blanche, which flows out of the lake on the divide in an opposite direction from the River of the West and another river, is the crux. It flows into Lake Winnipeg near the southwest corner, only a little way up the elongated southern shore but beyond Lake Manitoba, which is violently displaced. I conclude that this is the Saskatchewan, which Vérendrye had heard about, but that also he has added to it rumored features of the Red, with whose full course he was not familiar. He was capable of holding different conceptions at the same time, as his ideas about the River of the West show, and as people frequently must when they deal with the geography of hearsay. Well upstream from ("above") a nonexistent or at least unidentifiable (unless it is Lake Dauphin) Lac des Glaizes — "a lake in the clay country" would be the way he heard it — the Blanche forks. Its long western branch flows into Nelson River and here again the text has some details which seem to be from the Red River.

Although Vérendrye had met Crees who had been to the Mandan villages, or in their vicinity, those whom he now quotes had not been. They eked out what they knew with invented details but told him some true things about the country. They said that they had made a five days' journey on the River of the West, by which they meant the Missouri for the Mandans lived on it. On the basis of their information Vérendrye makes a new estimate of distances. The Western Sea may now be 750 miles beyond the height of land, which may be a couple of hundred miles from Fort Maurepas, since the Mandans beyond it are either 200 or 400 miles from Fort St. Charles.

These appear to be reasonable conclusions:

1. In 1736 Vérendrye had no dependable knowledge west and northwest of the mouth of Red River, and no knowledge of the Red upstream from the mouth of the Roseau.

2. His distances north and west were extremely foreshortened.

3. He both did and did not distinguish between the River of the Mandans (which was actually the Missouri) and the River of the West.

4. He had no conception whatever of the Rocky Mountains; if he had heard about them, and it seems impossible that he had not, he understood nothing of what he heard. Burpee's identification of the Rockies as Vérendrye's height of land will not hold: both the map and the text make it untenable. (Note that for a long time after Vérendrye, in fact as late as Arrowsmith, an east-west

ch'ain of mountains across Canada periodically reappeared on maps.)

5. Burpee's suggestion that the Indians north and northwest of the height of land may be Blackfeet is untenable.

8. I follow Upham, "The Explorations of Vérendrye and His Sons," MVHA *Proceedings*, I (1908), quoted, Burpee (2), pp. 311–12. A radically different route which includes the loop of the Souris was worked out by Robinson and De Land, "The Vérendrye Explorations."

9. Vérendrye is now calling them Mandans, instead of the various names he had previously heard. F. O. Libby once argued that the tribe he met in 1738 were not the Mandans but the neighboring Minnetarees ("Some Vérendrye Enigmas"). But the Assiniboins, whose designation the word "Mandan" was, never had any doubt whom they were taking him to nor which language the interpreter would have to speak. They may be presumed to have known who the Mandans were, and so may the Mandans.

10. The Heart empties into the Missouri at Mandan, North Dakota, opposite Bismarck. The geographical center of North America is a little west of Devil's Lake, a hundred miles northeast of there.

11. Burpee (2), pp. 335–38 and 344–46. I cannot accept Burpee's decision that what the Mandans say about the white men below the Arikaras is "rather . . . the testimony of eyewitnesses than that acquired at second-hand." We know that the Mandans had never seen the French of the Illinois. Though there is no direct evidence to show whether or not they had met the Spanish of New Mexico, it is inconceivable. Vérendrye presently gets better information, but this first batch seems to a jumble of distorted and misunderstood hearsay. To make an extreme point, when he says that these white men had iron bucklers, sabers, and lances, we must remember that he was learning about them by means of sign talk. What would be an unmistakably specific sign for "bucklers of iron (pare fleche de fer)," or for "sabres" as certainly to be differentiated from "war clubs"? Could not the bucklers be the parfleche shields of the Plains tribes? The report that you could not kill these white men because they wore iron armor but could capture them by killing their horses appears to have a fine ring of truth, which might have reached them from the Pawnees or Kiowas, perhaps the Crows, or possibly the Comanches. But so far as I have been able to learn the only armor worn by New Mexicans at this time was an occasional shirt of mail or leather cuirass. It was certainly possible to kill a man who was so clad with either a bow, especially the Plains horn bow, or the wretched trade musket.

Indians had been killing armored Spaniards ever since De Soto and we have seen that the Pawnees easily disposed of Villasur's force without, or with very few, guns.

12. Historical treatments of the Vérendryes — Brebner's is much the best — contradict one another at important points. Some of the trouble, I believe, has resulted from occasional failure to keep in mind what the facts actually were. It is a mistake to assume that everything Vérendrye heard from the Indians had at least some factual reality behind it, albeit distorted or misunderstood. Thus, New Mexico, which was unquestionably the abode of the white men thus reported, had no seacoast and there were no Spanish settlements on the Pacific coast north of Mexico. Nowhere in New Mexico were there bearded black men, as the Indians said there were. Nowhere were there such palatial dwellings and furnishings. Some of these details are certainly invented, though others may signalize honest awe of quite commonplace objects. I take it that the magnificent walled towns built of brick (with drawbridges and iron doors!) were simply the pueblos of the Rio Grande valley. The moats must be rejected as an Indian invention or a French misunderstanding. They are not irrigation ditches for there were none at these pueblos and it is unlikely that the Indians were reporting pueblos farther to the southwest where there were.

It may be that the *"Gens du Serpent"* of this passage, who must be avoided by a long detour, are Comanches. By detouring the Comanches, one did get to New Mexico. This added to the confusion which the Chevalier had to grope through later on. The *"Gens du Serpent"* who barred access to the sea when he made his journey were far to the north, and the sea he thought he was traveling toward was off the Pacific coast somewhere in the Northwest. But he assumed that he was hearing about the same tribe and the same part of the world.

Sign language was suggestive, rather than fixed. A sign which meant "snake" could always be used to mean "enemy." The Sioux, for instance, used the same sign for both conceptions. But there were also precise signs. W. P. Clark, *The Indian Sign Language,* gives signs for "snake" and "Comanche" that are barely distinguishable. He gives others for "Shoshone" and "Kiowa" that could easily be confused with "snake" or "Comanche."

13. Thus most texts, but a safer statement would be that they were in the first period of the horse age when Vérendrye first saw them. The Arikaras had horses not far to the south of them and horse Indians were coming to the villages to trade. That the Vérendryes did not see horses at two villages does not necessarily

mean that there were none at the other villages, or even that there were none at those two.

14. There has been more difference of opinion about the western extent of the Chevalier's journey than about anything else in the French exploration of the West. The relation of distance to season, travel time, and travel method is always the index to the possible. It constrains me to follow Doane Robinson (1), in his determination of the route and farthest west. I do not accept all his identifications of Indians.

The Black Hills are the only mountains which the party could have reached in the time they traveled, considering the season, the long halts, and the fact that the Indians were traveling as villages. Even more obviously, they are the only ones from which they could have returned in the time given. At least six other termini, however, have been suggested. Parkman believed that the mountains were the Big Horns, and these have been named oftener than any others. Granville Stuart believed that they were the Wind Rivers, west of the Big Horns; others have taken the party across the Wind Rivers, whose crest is the Continental Divide, and on to various ranges beyond them. Prud'homme took them still farther, to the main range of the Rockies in Alberta. These and other conclusions are conveniently listed in Burpee (2), pp. 18–20. They are all untenable; the mountains named are too far from the Missouri to have been reached in the elapsed time.

Note that whereas Vérendrye himself was at the Mandan villages in 1738 he did not consider the River of the Mandans the Missouri, in 1741 the Chevalier accepts it as what it is.

15. Usually spelled Hendry and sometimes Hendey in the texts but Tyrrell seems to have established that it is Henday in the Hudson's Bay Company books. (There is often no correct spelling for 18th-century names; a man may spell his own name in several ways.)

16. The summary of La France's account is in Dobbs, pp. 26–45, scattering allusions elsewhere. Dobbs was capable of suppressing any mention of the Vérendryes if La France made any, but presumably they would have alerted Beauharnois if they had heard of him.

## CHAPTER VII: TWELVE-YEAR ARMISTICE

DOCUMENTARY As before: Burpee (3); Coues (1); Dobbs; Tyrrell (1) (Thompson). Lawrence J. Burpee, ed., (4) "An Adventurer from Hudson Bay" (Cocking's journal), Royal Soc. of Can. *Proceedings,* 3d Ser., II (1908). Jonathan Carver, *Travels through the Interior Parts of North America,* 3d ed. (London, 1781); the best ed. for students; must be used with 1st ed. Daniel Coxe, *A Description of the English Province of Carolana;* reprinted in French, Part II; also mimeographed reprint as Project No. 665–08–3–236, Reprint Series No. 11, sponsored by California State Library, 1940. Charles M. Gates, ed., *Five Fur Traders of the Northwest* (Minneapolis, 1933); Pond's "Narrative" and John Macdonnell's journal. Daniel W. Harmon, *A Journal of Voyages and Travel.* Alexander Henry, *Travels and Adventures in Canada and the Indian Territories.* Alexander Mackenzie, *Voyages from Montreal.* L. R. Masson, ed., *Les Bourgeois de la Compagnie du Nord-Ouest* (Quebec, 1890). J. B. Tyrrell, ed.: (2) Samuel Hearne's *A Journey from Prince of Wales Fort in Hudson's Bay to the Northern Ocean* (Toronto, 1911); (3) *Journals of Samuel Hearne and Philip Turnor* (Toronto, 1934). W. Stewart Wallace, ed., (1) *Documents Relating to the North West Company* (Toronto, 1934).

SECONDARY Parkman. As before: Blegen; Brebner; Burpee (1); Gipson; Hanna; Innis (1); Morton (1); Nute (2) and (3). G. C. Davidson, *The Northwest Company* (Berkeley, 1918). Harold A. Innis, (2) *Peter Pond* (Toronto, 1930). Louise P. Kellogg, (3) *The British Regime in Wisconsin and the Northwest* (Madison, 1935). Arthur S. Morton, (2) *Under Western Skies* (Toronto, 1937). Grace Lee Nute: (4) *The Voyageur* (N.Y., 1931); (5) "The British Regime in the Northwest," *Minnesota Alumni Weekly,* Vol. 31, No. 16 (1937). Howard W. Peckham, *Pontiac and the Indian Uprising* (Princeton, 1947). Wayne E. Stevens, *The Northwest Fur Trade, 1763–1800* (Urbana, 1926).

1. *The Conspiracy of Pontiac,* though the last in chronological order, was the part of Parkman's great history that he wrote first. It is the only part that later scholarship has to any considerable extent superseded but remains by far the most vivid account. The authoritative treatment is Peckham; there is a copious monographic literature.

2. *A Summary View of the Rights of British America.*

3. The Virginians who crossed the mountains at mid-century expected to see the Pacific but did not cross for the purpose of finding it.

4. The bracketed and queried word "country" is supplied by T. C. Elliott, Oregon Historical Society *Quarterly,* XXII (June 1921). I point out that a route which leads from the "head of the

Mississippi" definitely is not "the same route" as that of Lewis and Clark, as Mr. Kenneth Roberts holds it to be, *Northwest Passage: Appendix* (Vol. II, Ltd. ed., 1937), p. 62.

5. In 1767 all the rivers shown on maps or described in texts as emptying into the Pacific in the area of the United States or Canada were fictitious, imaginary, or hypothetical. So, above Monterey Bay, were all eastward or "northeastwardly" indentations.

6. Note that Rogers's idea of the Minnesota River is accurate. (By 1772 it would have been hard not to be right about it.) Mr. T. C. Elliott, however, who violently objects to both Rogers and Carver on moral grounds that seem irrelevant, protests that a portage from the Minnesota to the Missouri would have to be 150 miles long. That is true, but the Bix Sioux River, an affluent of the Missouri, was navigable by canoe to within fifty miles of Big Stone Lake, the source of the Minnesota.

A long scholarly controversy has centered on Rogers and a more violent one on Carver. I cannot feel that Mr. Kenneth Roberts has resolved either one in the appendix volume to his fine novel, *Northwest Passage.* I am not concerned in this chapter with the controverted questions except so far as the honesty of Carver's account of his travels has been impugned. In my opinion this question was settled by the work of Louise P. Kellogg, supplemented by that of Dr. Nute and others, and particularly by Dr. Kellogg's "The Mission of Jonathan Carver," *Wisconsin Magazine of History,* Vol. 12 (1928). I follow Kellogg in her statements of his route and of the mission set for him by Rogers. I accept responsibility for everything else said about him in my text. I have compared his geography and ethnology with those of all his predecessors and have appraised them according to the knowledge of his time. He was fully as good an observer and trustworthy a reporter as anyone else of English blood who went west before Alexander Mackenzie, and in my judgment he reported as an honest man.

7. Pond is especially disdainful because Carver fell for one of the tall tales with which the West has always regaled greenhorns and too many students have adopted his attitude. The story described a pet rattlesnake which an Indian carried about with him and which he set free in October with instructions to return the following May at a specified rendezvous; the snake faithfully did so "and of his own accord crawled into the box which was placed ready for him." It was a mild traveler's tale, Carver told it only as "hearsase," not on his own authority, and his credulity is hardly worth laughing at in the 20th century. Seventy-odd years later the great Audubon could draw a tree-climbing, birdsnest-invading rattlesnake (*Birds of America,* Pl. xxi), and I cannot forbear to note

that some historians who have paused to deride Carver have re-counted as fact the fictitious journey of Moncacht-Apé.

8. In the long and almost academic dispute about the actual source of the Mississippi, the U.S. Geological Survey finally settled on Little Elk Lake, which is about five miles beyond Lake Itasca, whose name Schoolcraft derived from *veritas caput*.

CHAPTER VIII: PRIME MERIDIAN

DOCUMENTARY As before: Coues (1); Dobbs; Gates; Harmon; Henry; Mackenzie; Masson; Tyrrell (1) and (2); Wallace (1). *Illinois Historical Collections*, Vols. II and V. Herbert E. Bolton, ed., (3) *Pageant in the Wilderness* (Salt Lake City, 1950); Escalante's journal. Gilbert Chinard, ed., *Le Voyage de La Pérouse Sur Les Côtes de l'Alaska et de la Californie* (Baltimore, 1937). James Cook, *A Voyage to the Pacific Ocean* (London, 1784); 3 vols., the third by James King. Elliott Coues, ed., (2) *On the Trail of a Spanish Pioneer* (N.Y., 1900); Garcés' journal. James Alton James, ed., (1) *George Rogers Clark Papers* (Springfield, 1912–26). John Ledyard, *A Journal of Captain Cook's Last Voyage* (Hartford, 1783).

SECONDARY As before: Bancroft, *California* and *Northwest Coast;* Brebner; Burpee (1); Davidson; Innis (1) and (2); Kellogg (3); Morton (1) and (2); Stevens. Thomas P. Abernethy, *Western Lands and the American Revolution* (N.Y., 1937). Clarence W. Alvord, (1) *The Mississippi Valley in British Politics* (Cleveland, 1917). Helen Augur, *Passage to Glory* (N.Y., 1946). John D. Barnhart, *Henry Hamilton and George Rogers Clark in the American Revolution* (Crawfordsville, 1951); with documents. Samuel F. Bemis: (1) *Pinckney's Treaty* (Baltimore, 1926); (2) *The Diplomacy of the American Revolution* (N.Y., 1935). Robert G. Cleland, *From Wilderness to Empire* (N.Y., 1944). James Alton James, (2) *George Rogers Clark* (Chicago, 1928); (3) "Spanish Influence in the West during the American Revolution," MVHR, IV (1917). Leonard W. Labaree, *Royal Government in America* (New Haven, 1930). Andrew C. McLaughlin, "The Western Posts and the British Debts," American Historical Association (hereafter cited as AHA; the *Review* as AHR) *Annual Report*, 1894. John C. Miller, *Triumph of Freedom* (Boston, 1948). Grace Lee Nute, (6) "A Peter Pond Map," *Minnesota History*, XIV (1933). Paul C. Phillips, *The West in the Diplomacy of the American Revolution* (Urbana, 1913). Albert T. Voliver, *George Croghan and the Westward Movement* (Cleveland, 1926). W. Stewart Wallace, (2) "The Pedlars from Quebec," *Canadian Hist. Rev.*, XIII (1932). Justin Winsor, (4) *The Westward Movement* (Boston, 1897).

1. "Thay are navagated By thirtey Six men who row as maney oarse. Thay Bring in a Boate Sixtey Hogseats of Wine on one ... [marks of ellipsis by the editor, meaning something undecipherable in the ms] Besides Ham, Chese &c." (Gates, ed., "The Narrative of Peter Pond," *Five Fur Traders*, p. 45.) The narrative, which was written in Pond's old age, is also printed in *Wisconsin Historical Collections*, XVIII.

2. Hamilton was a first-rate soldier and as expert a leader of Indians as the British service ever produced. He had a larger share than anyone else in the decision to use Indians as auxiliaries, and he used them with dispassionate understanding of their capabilities, limitations, and cruelty. As an expert leader he took part in the scalp dances and other rituals of Indian warfare and his policy, which was sound and in fact inevitable, was to reward those who brought in American scalps. He vehemently denied that he offered bounties for scalps, the charge made against him throughout the West and officially repeated by Jefferson. He appears to have been telling the truth. Technically there is a difference between offering a bounty for scalps and paying a reward for those that are brought in.

3. The best account of this campaign is James B. Musick, *St. Louis as a Fortified Town* (St. Louis, 1941).

4. Tyrrell (1), *Narrative*, p. 176.

5. Quoted in Carl Van Doren, *Benjamin Franklin*, pp. 617–18.

6. The authorities are divided between Drake's Bay and Bodega Bay. They also disagree about his landfall, though agreed that it cannot have been above 43°. My statement may bring him as much as three degrees too far north. Drake's chaplain claimed 48°.

7. From evidence which could not be misinterpreted Cook learned, however, that the polar lands now called Antarctica must exist.

8. A London map of 1771 called "A Map of the Discoveries made by the Russians on the Northwest Coast of America. Published by the Royal Academy of Sciences at Petersburg" has the River of the West flowing west from Lake Winnipeg to an "Entrance discovered by Martin d'Aigular in 1603" at exactly 45°. In 1768 Jeffreys of London again reprinted a map on which Dobbs had relied, "Carte Générale de Découvertes de l'Amiral de Fonte," which shows the Strait of Juan de Fuca at 47° running to *"le Mer Rinquello,"* north and east of Hudson Bay. Janvier's "L'Amérique Septentrionale" of 1762 has both the Strait of Juan de Fuca and the Entrance of Martin Aguilar leading to a Western Sea, and this idea seems to have flourished widely at mid-century.

9. John Ledyard wrote, "I had no sooner beheld these Americans than I set them down for the same kind of people that inhabit the opposite side of the continent." In her *Passage to Glory* Helen

Augur comments with admiration on his recognition. By 1778 what other conclusion was possible? Indeed at any time what else could anyone who had known Indians think? It never entered Cook's mind that they could be anything else; from his first mention of them they are American Indians.

Ledyard is a fascinating man but his fantastically adventurous career and tragic end make it easy to sentimentalize and overestimate him. Clearly he had a sympathy for primitives unusual in Anglo-Saxons, though common among the French, and his understanding of their thinking and folkways anticipated some theses of modern anthropology. (So did those of many before him.) But both the amount and the importance of this have been exaggerated. Cook's detailed, exact notes on the Northwest Indians are much more valuable than Ledyard's jottings. His book has a few passages of descriptive charm and a few others of narrative vigor, and though his style is rhetorically turgid as compared with Cook's workmanlike prose, he infuses parts of it with his personal charm. But it shakes down to nothing much: it has much less to say about Indians than a good many books already mentioned here, Carver's, say, or Alexander Henry's. It is a curious rather than an important book and has remained unreprinted for cause.

10. He intended to "make another and final attempt to find a northern [Arctic] passage but I must confess I have little hopes of succeeding." Letter to the Admiralty from Unalaska.

11. The epidemic is Robert Cleland's suggestion.

12. Translation by Greenhow, quoted in *Voyage of the Sonora in the Second Bucareli Expedition,* Thomas C. Russell, ed. (San Francisco, 1920), pp. 86–87. The book is a reprint of Daines Barrington's *Miscellanies* (London, 1781). (Bucareli was the viceroy of New Spain.)

13. Garcés' journal is translated in Coues (2). Coues's editing of Garcés is his least satisfactory work. The most charming account of Garcés' journey I have seen is Edwin Corle's in *Listen, Bright Angel.*

The best translation of Escalante's journal is in Bolton (3). This too is a disappointing work by a leading historian. Bolton's editing is perfunctory and although he knows the country by heart many of his geographical identifications are useless to anyone who does know it as well as he, for they cannot be located on any existing maps. He is also perfunctory about the Indians. His map of Escalante's route, however, is very valuable and he also reproduces Miera's map. See also: W. R. Harris, *The Catholic Church in Utah* (Salt Lake City, 1909); J. Cecil Alter, "Father Escalante and the Utah Indians," *Utah Historical Quarterly,* I–II (July 1928–April 1929), and "Father Escalante's Map," UHQ, IX (Jan.–April

1941); Herbert S. Auerbach, "Father Escalante's Route," UHQ, IX, and (an invaluable study) "Father Escalante's Journal with Related Documents and Maps," UHQ, XI (1943), 4 nos. in 1. The documents published in the last item include very illuminating letters by Escalante before and after his journey, which Bolton does not reprint.

14. I do not understand why Bolton (3) twice, on p. 71 and p. 72, explicitly denies that there is a connection between Utah Lake and Great Salt Lake. There certainly is, the river which the Mormons named Jordan as soon as they saw it, and which shows on Bolton's map. If Utah Lake lacked an outlet, it too would be salt.

15. Certain theses in Miss Augur's life of Ledyard, *Passage to Glory*, require comment. They may be conveniently examined on pp. 125–27. Miss Augur undertakes to establish that Ledyard profoundly influenced Jefferson's desire to explore (and eventually to acquire) Louisiana, in so much that by the end of her book she has Ledyard precipitating most of American history both ways from the Louisiana Purchase. She says, p. 125, that Jefferson's "thinking about the West was saturated with Ledyard's own thoughts, and there is no sorting them out." If that is true, no evidence for it has been found. When the two met, Jefferson (frontier-born, son and pupil and friend of Western explorers and pamphleteers, author of *Notes on Virginia*) knew vastly more about the West than Ledyard did and had been thinking about it all his life. As will appear in Chapter X, the powerful and complex motivations that produced the Lewis and Clark expedition had been developing since Jefferson's boyhood. Miss Augur herself notes, p. 174, that Jefferson had already proposed the expedition to George Rogers Clark; how old an idea it was to him then Chapter X shows. She also says, p. 126, that "the excitement caused by Ledyard's book" in 1784 stimulated Peter Pond and Alexander Henry to propose transcontinental explorations. She thus makes Ledyard responsible for Alexander Mackenzie's crossing and all that followed from it. This in spite of the fact that, as she records on the same page, Henry, on the basis of Pond's experiences and ideas, had formally proposed such an expedition in 1781, more than a year before Ledyard returned to America. (He made the proposal to Banks, who was to be Ledyard's patron.) The sole evidence for her belief is an allusion in Pond's memorial of April 1785 to the report that American ships are fitting out for the Northwest coast under men who had been with Cook. (Quoted, Innis (2), pp. 127–28. Note the important fact, disregarded by Miss Augur, that Pond immediately ties this up with British retention of the Western posts and American possession of Grand Portage and the route west.) This is an allusion to Ledyard but

Pond's conviction that a connection with the Pacific was necessary for the Canadian fur trade of the inland Northwest was formed years earlier — even before Cook and Ledyard got to Nootka, perhaps. As soon as the Montreal traders worked north of the Saskatchewan River, if indeed not before then, the increasing cost of transportation west made the Pacific exploration desirable and ultimately inevitable. The weakness of Miss Augur's book is that it ignores the entire development of the continental fur trade, especially the period 1774–85.

As for Pond's knowledge of and speculation about Cook's River, there is no evidence that he read about it in Ledyard. During his 1785 stay in Montreal he could have seen not only Ledyard's book but two others by members of Cook's party that were published before it was. Innis (2) conjectures, p. 126, that his ideas came from the much more specific details of Ellis's *An Authentic Narrative,* which had been published in 1782. As a matter of fact, Pond could have seen the official account itself, which was published in 1784. He must have read it fairly soon for he was familiar with Cook's maps.

In a final passage, pp. 230–31, Miss Augur puts into Pond's mouth a speech to Mackenzie which alludes to Ledyard. It nicely summarizes her thesis but it is entirely imaginary and is justified by nothing that the record shows.

On p. 125, speaking of Ledyard's early proposals for his transcontinental journey, Miss Augur says that he intended to start farther north than Nootka Sound and remarks, "In other words, he planned to save miles by using the high latitudes, just as we are finally [!What about mariners of the 16th century?] using them in our air lines." Great circle courses are of the air and the ocean. They are not possible on land and Ledyard could not have followed one. By going north in order to travel east he would have added "miles," not saved them. Nootka is 126° W., Prince Rupert 130° — and Unalaska, which Ledyard knew, 167°. And the distance to New York, his proposed terminus, increased not only with the longitude but with the latitude as well.

16. So far as I can find no one has bothered to scrutinize the idea. Miss Augur and Miss Jeannette Mirsky (*The Westward Crossings*) regard it as inspired, feasible, and certain to succeed. On the contrary, it was in the highest degree impractical and was probably impossible. No one can say whether Ledyard would or would not have succeeded, but I am unable to imagine his trip ending in any way but starvation or his murder, and that soon.

Miss Augur counts on Ledyard's ability to make friends with the Indians. He was "to merge himself with the country like a very Indian, to make his journey *by means of* the Indians, not in spite of them." (Her italics. Nine-tenths of all North American exploration was by means of Indians, not in spite of them.) He had an "Indian pipe," which she takes to have been a calumet, and "two great dogs," which, she says, "were an inspiration," since they "could kill his meat and keep off prowling beasts by night." Jack-rabbits and ground squirrels? In all primitive America dogs were never used for hunting larger meat, and in fact no dog was fast enough to hunt jackrabbits. Big game was either too fast or too dangerous for dogs. They were not trained even for trailing, though they sometimes did trail wounded animals and occasionally brought down a small deer in circumstances when it could not get away, or a buffalo calf. Ledyard, whether or not he may be presumed to have known so, would have been in no danger from "prowling beasts." There are perhaps a few recorded cases of a cougar attacking a man but attacks on men by other animals are unknown except when the animals are rabid. Though the European wolf in packs was sometimes dangerous, the American wolf never was.

That Ledyard could have crossed Alaska is inconceivable. Cf. Samuel Hearne's two failures on much shorter trips in easier country, with the best preparations, an adequate outfit, and hired Indians. From somewhere between Alaska and Nootka, then, and the farther south the better. If his pipe was indeed a calumet, he would have met no Indians who understood its symbolism till he got across the Rockies. He knew nothing about Western geography. He would first have met the Northwest Indians, from the Tlingits on the north to the Nootkas on the south. Routes across the coastal ranges were known to them and conceivably he could have traveled from tribe to tribe with trading parties, perhaps with war parties, and gone on to cross the Rockies in the same way. I understand this to be what Miss Augur had in mind. The catch is that he would have had no goods for presents: the only basis for friendship with the Indians and the fee always paid for guide service. The continual, fully justified fear of being killed which the expert and experienced Alexander Mackenzie and his even more experienced crew felt is ample evidence of the hazard — and they had goods to pay their way. Without guides Ledyard could probably not have got across the first ranges, for he knew nothing about mountain travel or indeed about wilderness life. From the moment he left the beach he would have been a greenhorn in conditions which men of long experience frequently failed to survive when alone. And though it is possible that he could have made friends of the coastal Indians without presents, it is infinitely more likely that they would have killed him for his gun, knives, and clothes. And the even more difficult Rockies remained.

If he had crossed the Rockies, escaping or charming such tribes as the Blackfeet and Gros Ventres on the way, and had reached Canadian traders, he would have had no difficulty from then on. (Nothing suggests that he thought of them.) He could have traveled with any returning party to Grand Portage, Mackinac, and thereafter at his choice. If, however, he were to cross Louisiana, which was the idea, there would have been more difficulty. The Canadians would presumably have taken him to Portage la Prairie or the Red River posts. Could he have sufficiently mastered geography and travel to get to St. Louis by himself? No one ever had. He would still have been without presents to command the services of several very skeptical tribes.

17. "The tract of transport [from Montreal] occupies an extent of from three to four thousand miles, through upwards of sixty large lakes and numerous rivers . . . those waters are intercepted by more than two hundred rapids, along which the articles of merchandise are chiefly carried on men's backs, and over an hundred and thirty carrying places, from twenty-five paces to thirteen miles in length, where the canoes and cargoes proceed by the same toilsome and perilous operations." Alexander Mackenzie, the last footnote in *Voyages.*

18. The researches of Burpee, Davidson, and Innis led to somewhat confused results but they are all essential to an understanding of Pond. Brebner and Morton have pretty well harmonized them but perhaps the key discovery was (as so often) Dr. Nute's. See Nute (6).

19. In Chapter XI of *Across the Wide Missouri* I have described at length a similar epidemic on the upper Missouri a half-century later.

20. My thanks for this information and citations in the literature to Dr. Hugh Raup, the Director of the Harvard Forest, who has provided me with much other information, botanical, zoological, and topographical, relating to western and northern Canada.

21. Printed in Burpee (1) as an appendix.

22. "Extract of a letter from —— of Quebec to a Friend in London," quoted in Nute (6).

## CHAPTER IX: EQUINOCTIAL TIDE

DOCUMENTARY  As before: Coues (1), for the younger Henry and Thompson; Gates; Harmon; J. A. James (1); Mackenzie; Masson; Tyrrell (1), for Thompson; Wallace (1). Annie Heloise Abel,

ed., (1) *Tabeau's Narrative of Loisel's Expedition* (Norman, 1939).
Frederick L. Billon, *Annals of St. Louis* (St. Louis, 1886). Paul
Leicester Ford, ed., *The Works of Thomas Jefferson*, 12 vol. (N.Y.,
1904–5); cited as Ford. Louis Houck, ed., (1) *The Spanish Regime
in Missouri* (Chicago, 1909); cited as Houck. Frederic W. Howay,
ed., *Voyages of the "Columbia" to the Northwest Coast* (Boston,
1941). Edwin James, (1) *A Narrative of the Captivity and Adventures of John Tanner* (N.Y., 1830). Laurence Kinnaird, ed., *Spain
in the Mississippi Valley, AHA Annual Report*, 1945, III and IV;
cited as Kinnaird. John Meares, *Voyages Made in the Years 1788
and 1789* (London, 1790). Arthur S. Morton, ed., (3) *The Journal
of Duncan M'Gillivray* (Toronto, 1929). C. F. Newcombe, ed., *Menzies' Journal of Vancouver's Voyage* (Victoria, 1923). James Alexander Robertson, ed., *Louisiana under the Rule of Spain, France, and
the United States* (Cleveland, 1911). Doane Robinson, ed., (3) "Trudeau's [Truteau's] Journal," *South Dakota Historical Collections,*
VII (1914). Reuben Gold Thwaites, ed., André Michaux, *Travels
West of the Alleghanies,* reprint (Cleveland, 1904), in *Early Western
Travels,* III. George Vancouver, *A Voyage of Discovery to the
North Pacific Ocean* (London, 1798). Note that various works
listed as secondary print source documents.

SECONDARY   Bancroft: *California* and *Northwest Coast*. As before: Alvord (1); Brebner; Davidson; Gayarré (2); Innis (1); McLaughlin; Morton (1); Stevens; Winsor (4). Annie Louise Abel: (2) "Trudeau's Description of the Upper Missouri," MVHR, VIII (1921–22);
(3) "Mackay's Table of Distances," MVHR, X (1923–24). Samuel
F. Bemis, (3) *Jay's Treaty* (N.Y., 1924). John Walton Caughey, ed.,
*McGillivray of the Creeks* (Norman, 1938). Raphael N. Hamilton,
"Early Cartography of the Missouri Valley," AHR, XXXIX (1924).
Louis Houck, (2) *History of Missouri* (Chicago, 1908). Ida Johnson,
*The Michigan Fur Trade* (Lansing, 1919). E. Wilson Lyon, *Louisiana in French Diplomacy, 1759–1804* (Norman, 1934). William
R. Manning, "The Nootka Controversy," AHA *Annual Report,*
1904. A. P. Nasatir: (1) "Anglo-Spanish Rivalry on the Upper
Missouri," MVHR, XVI (1929); (2) "Jacques D'Eglise on the Upper
Missouri, 1791–1795," MVHR, XIV (1927); (3) "The Anglo-Spanish
Frontier in the Illinois Country during the American Revolution,"
Illinois State Historical Society *Journal*, XXI (1928); (4) "An Account of Spanish Louisiana," *Missouri Hist. Rev.*, XXIV (1930); (5)
"The Formation of the Missouri Company," *ibid.*, XXV (1930–31);
(6) "John Evans, Explorer and Surveyor," *ibid.* William R. Shepherd,
"Wilkinson and the Beginning of the Spanish Conspiracy," AHR,
IX (1904). Frederick Jackson Turner, *The Significance of Sections*

*in American History* (N.Y., 1932). Jacob Van Der Zee, "Fur Trade Operations in the Eastern Iowa Country under the Spanish Regime," *Iowa Jour. of Hist. and Politics,* XII (1914). Arthur P. Whitaker: (1) "Spanish Intrigue in the Old Southwest," MVHR, XII (1925); (2) *The Mississippi Question, 1795–1803* (N.Y., 1934).

1. The first section of this chapter does not hold to strict chronology. The chapter deals with events of the 1790's; most of the material in this section is drawn from the last few years of the decade and the first few of the next one. It is, however, entirely true to post life everywhere in the country west, southwest, and northwest of Lake Winnipeg throughout the decade; no item is anachronistic.

2. Quoted (from *The Records of the Federal Convention*) in Charles Warren, *The Making of the Constitution,* 717.

3. Morison (2), p. 51.

4. It is possible that a New York ship, the *Eleanora,* may have preceded her. Howay, p. *x.*

5. Cecil Jane, ed., *A Spanish Voyage to Vancouver and the Northwest Coast of America* (London, 1930), p. 54.

6. Many documents in Houck and Kinnaird. Many additional ones are in a large collection, mostly from the archives of the Department of the Indies, edited by A. P. Nasatir, which remains in preparation for the press when this is written but which I have consulted by the kindness of Dr. Nasatir.

7. His specific declaration to just this effect must not be ignored. See the draft of a letter to Carmichael and Short, Ford, VII, 267–68. For the guarantee, *ibid.,* p. 129, and elsewhere.

8. Of a large scholarly literature on Jay's Treaty Bemis (3) is standard. For the point here emphasized see especially his "Jay's Treaty and the Northwest Boundary Gap," AHR, Vol. 22 (1922).

9. Ford, VII, 208–12, dated January on inference from their content but the records of the American Philosophical Society imply April. I applied to Mr. Lyman Butterfield, who was then helping to edit the definitive edition of Jefferson's works. He informed me that the evidence in the possession of the editors is so conclusive that the instructions will be dated April 30 when the volume containing them is published.

10. McKay became a North West Company partner in 1799. Later he changed services, became a partner in Astor's Pacific Fur Company, and was killed on the *Tonquin.* He was the father of Tom McKay who appears in *Across the Wide Missouri* as a Hudson's Bay Company brigade leader. Mackenzie first offered John McDonald of Garth the chance to make the exploration with him but he declined.

11. Nasatir (4). Letter also printed in Kinnaird.

12. Travel up the Missouri from St. Louis is now to be by pirogue as well as dugout. The term covers a wide variety of craft but they may be generalized as being made of lumber and having neither keel nor cabin. They were equipped with oars, poles, cordelle, and a small sail, sometimes two sails. All were necessarily of slight draft and probably flatbottomed. They could be fairly large, up to fifty feet long and carrying ten tons or more of lading.

13. Relevant documents accompany Nasatir (2); some supplementary ones scattered in Houck, II, and Kinnaird, IV, indexed as III. Most of my citations for Truteau and Mackay, below, have some bearing here. Nasatir (2) gives 1790 as the date of D'Eglise's first Missouri voyage but makes it 1791 in the title. Abel (1), a most important and very careful research as well as the source for Tabeau and Loisel, accepts 1790. D'Eglise's license was issued in August 1790; it is unlikely that so experienced a trader would have started for the Mandans so late in the year. Indeed there is no indication that he went in 1791; he appears to have made the strenuous, but presently not unusual, round trip between St. Louis and the Mandans in a single season, that of 1792. Nothing in the record suggests that he spent a winter upriver and in his 1794 application for a monopoly he says that he made his first trip in 1792.

14. As late as 1847 the best recorded time from Santa Fe to Independence (250 miles west of St. Louis) was twenty-five days. This was by forced riding, which Vial would have had neither reason nor equipment and preparation to do. Two months from Independence to Santa Fe was good traveling time for a wagon train after the trail was well established and all the watering places known. Three men on horseback would normally take more time, not less.

15. For Fidler, J. B. Tyrrell, "Peter Fidler, Trader and Surveyor, 1769–1822," *Royal Soc. of Can. Proceedings,* 3d Ser., VII (1913); also Tyrrell (3) and Burpee (1). Fidler's most important work was done in the far North, outside the interest of this narrative. There is no foundation whatever for the statement sometimes made by Western antiquarians that he traveled Montana and Wyoming east of the Rockies and even west of them and that Lewis and Clark had information about or from him apart from the Arrowsmith maps.

16. Trudeau, the lieutenant-governor at St. Louis, reporting to Carondelet, the governor at New Orleans, June 8, 1794, says both Sioux and Arikaras. Later statements appear to rest on this single phrase. It is unlikely. As subsequent events make clear, either tribe would probably have stopped D'Eglise, but the Sioux were currently harrying the Arikaras and the war would hardly have been interrupted for an alliance against him.

17. Not to be confused with the lieutenant-governor, Zenon Trudeau. They appear to have been relatives.

18. During Zenon Trudeau's examination of these men, cited later, one of them made a startling statement that no one seems to have commented on. He said that a bond-servant of Garreau's had deserted and made his way north to Fort Esperance, a North West Company post on the Qu'Appelle River. If the statement is true this man, whose name is given in the testimony as Lauson, was the first white man to reach Manitoba by way of the Missouri River.

19. The Arikaras were agriculturists as well as hunters and builders of earth lodges and came by it naturally, being related to the populous Pawnee group. But the earth lodges of the Pawnees and other tribes of the central Missouri region were smaller, simpler, and less defensible than those of the Mandans and Minnetarees. The ones which the Arikaras had built near the Grand Detour may have been the most elaborate that any tribe ever had.

20. Robinson (2); Abel (2); Nasatir (1) and (2). For the instructions given Truteau by the Missouri Company, Houck, II, 164–72.

21. Jonquard's statement is ambiguous but appears to mean 1794. John Macdonnell's journal has him leaving Fort Esperance for the Missouri in December 1793, Masson, I, 286.

22. Macdonnell's diary, Gates, p. 112 and note 105. See citations for Evans, below. David Thompson, who made the trip in 1797, says that Jessaume had lived eight years in the Mandan villages, which would have made his first year 1789, Tyrrell (1), *Narrative,* p. 209. Almost all his contemporaries and almost all modern writers say that Jessaume was illiterate. Yet occasional journal entries seem to mean that a letter from him has been received and the Meriwether Lewis papers at the Missouri Historical Society contain a holograph receipt signed, not crossed, by him. Dated May 20 1807, it acknowledges the receipt of 100 piasters from Pierre Dorion in full settlement of a debt.

His name is spelled in at least a dozen different ways in the literature, with greatest virtuosity by William Clark.

23. Zenon Trudeau's examination of the two is published among the documents accompanying Nasatir (2). Fotman speaks of the Arikaras as the "Pawnee Hocas." Some of his remarks include with them both the Poncas and the Minnetarees.

24. All my Missouri River distances are from *Missouri River Basin, River Mileage and Drainage Areas, June 1949,* by the Missouri River Division of the Army Engineers.

The Yellowstone's mountains are the Absaroka, Snowy, Madison, and Gallatin ranges. Conceivably behind all this geographical confusion there may be an allusion to the Missouri River canyon that

Meriwether Lewis was to call the Gate of the Mountain.

25. Clamorgan and Reyhle to, presumably, Zenon Trudeau, July 8 1795, four days after the interrogation of Fotman and Jonquard. Houck, II, 173–77.

26. "When I was in the North West I saw the Priest (or rather Conjuror) of the Piegan nation at the Rocky Mountains Perform one of his Miracles. . . ." Note 4 of ms, "Indian Tribes," described below. The "at" of this quotation must be understood as "of," for Mackay's ideas about the Rockies show that he had not seen them.

27. For the Madoc story, Stephens, *Madoc,* cited in notes to my Chapter II. For Evans, Nasatir (6) and David Williams, "John Evans' Strange Journey," AHR, Vol. 54 (Jan. and April 1949). The latter corrects and extends Stephens in certain important matters relating to Madoc and is the first extended biographical treatment of Evans. Williams is not at home in the fur trade and makes a number of unimportant mistakes about various Indian tribes, but the article is invaluable. "Extracts from Capt. Mackay's Journal," unsatisfactorily edited by Milo M. Quaife, is in Wisconsin Historical Society *Proceedings,* 1915. It contains parts of Evans's journal (reprinted by Nasatir, *op. cit.*) and annotations, extremely ambiguous ones, made on it by a fur trader named John Hay, whom I try to identify when he reappears in my Chapter X. Mackay's journal is a crucial document. Unfortunately neither Quaife nor Nasatir gives its provenance and I cannot learn where it is now located. Also extremely important is a manuscript called "Indian Tribes" in the E. G. Voorhis Memorial Collection at the Missouri Historical Society. I believe that I was the first to identify this as Mackay's, but Dr. Williams did so independently. The first page of this manuscript carries the notation "Evins Notes" in William Clark's unmistakable handwriting. An astonishing number of writers have described it as by Evans; none who did so could have read past the first page for Mackay's authorship is self-evident and he not only describes his hiring of Evans but alludes to his death. See also Abel (1), (2), and (3), and Diller "Maps of the Missouri River before Lewis and Clark," cited in notes to Chapter II, above. Mackay's report to the Missouri Company is in Houck, II, 182 ff.

Nasatir, Williams, and Abel provide abundant bibliographical notes for anyone who wants fuller detail about Evans. (Be warned that most of the material, including all of Evans's letters, is in Welsh, a lugubrious fact for me, since I had paid stiff prices to have much of it translated when Dr. Williams's article, which I read in manuscript, was submitted to the AHR.) Abel, especially in (1), resolves much earlier confusion, most valuably that in Teggart's

"Notes Supplementary to Any Edition of Lewis and Clark." Evans's impact on the British traders can be followed a little farther in Masson and in Gates.

It is obviously of great importance that what survives of Mackay's journal with the accompanying extracts from Evans's journal was first published in the *Medical Repository,* New York, in 1807 (Second Hexade, IV, 27–36). It ends, "Written in French but a few Years before the Surrender of the Country to the United States, in 1804, and communicated to Dr. Mitchill." The text differs in slight details from that edited by Quaife. This publication appears to have been unnoticed by Quaife and Nasatir. It is discussed at greater length in Chapter X.

28. The text in Houck reads "Panis and Layos." Houck's footnote makes the latter "Lobos or Wolves." Obviously the Pawnee Loups, on Loup Fork of the Platte.

29. The earliest I have found and nearly a year earlier than David Thompson's reference, which is frequently said to have been the first. We must assume that so many traders could not have been so long on the Souris without hearing at least vaguely of a big southern affluent of the Missouri and that the French translation comes from them — somehow — but no records yet found say so. The name is commonly said to be derived from the predominant color of the Grand Canyon of the Yellowstone, though the terraces of Yellowstone Valley have a drabber shade of the same color. Had the Arikaras, from whom the name probably comes, met any Indians who had been to the canyon?

30. This is the best I can do with the assumptions in his letters and reports and with Evans's maps. His instructions to Evans are printed as Document V in Nasatir (6). I treat as part of them the supplementary letter to Evans which Nasatir prints as Document VI.

31. When Evans started out in February he went by land. Nothing in his journal or in Mackay's papers indicates how he traveled in June. But I cannot reconcile either his travel time or the accuracy of his map with land travel and assume that he went by pirogue. The size of his party cannot be determined but it was small.

32. Williams, *op. cit.,* p. 522, assumes that there were, and perhaps the Tremblante post could have been reached by October 8, the date of Cuthbert Grant's first letter, Nasatir (6) Part II, p. 458. This, however, would have been a very fast trip. Evans's journal does not say that any British traders were at the villages when he got there. What it does say indicates, I think, that there were none and that those who first appeared were Northwesters, presumably

from the Souris post, and not HBC men as Williams takes them to have been. My reading requires the assumption that Grant misdated his letter. Considering the distance to his post as well as the various contexts, this assumption, though precarious, seems less precarious than Dr. Williams's reading.

Evans says, "I instantly hoisted the Spanish flag which seemed very much to please the Indians." His party was too small to oppose the group, obviously small, which arrived on October 8. If any Englishmen had been present would there not have been opposition to his hoisting the flag, or at least protest, and if there had been would he not have mentioned it? And he speaks of the fort as belonging to "the Canada Traders" and repeats the phrase when the first group arrives. Is not the natural reading of this the customary designation of the Northwesters, which he would have learned from Mackay? Against this, however, must be set the fact that he calls Jessaume, who was a Northwester, an *engagé* of "English traders," the usual designation of the HBC.

## CHAPTER X: A MORE PERFECT UNION

In this chapter it has seemed advisable to document a number of statements at the time they are made. Most of the documentary and secondary sources that would normally be grouped together here are therefore given in individual notes. I apologize to the reader for the annoying frequency of indices.

Two works first cited in this chapter are used repeatedly in the rest of the book. They are the basic documents for Lewis and Clark. One is *History of the Expedition under the Command of Captains Lewis and Clark,* written by Nicholas Biddle and first published in 1814. I use it in the edition magnificently edited by Elliott Coues and published in 1893. To maintain uniform usage I should cite it as Coues (3) but typographical difficulties are insuperable and I make it simply Coues. The other is *Original Journals of the Lewis and Clark Expedition,* edited by Reuben Gold Thwaites and published in 1904. I cite it as *Journals.*

DOCUMENTARY  As before: Abel (1); Billon; Coues (1); Ford; Gates; Houck (1); Edwin James (1); Kinnaird; Mackenzie; Masson; Morton (3); Robinson (2); Robertson; Tyrrell (1); Wallace (1). Elliott Coues, ed., (4) *The Expeditions of Zebulon Montgomery Pike* (N.Y., 1895). Edwin James, (2) *Account of an Expedition from Pittsburgh to the Rocky Mountains,* reprint in Thwaites's *Early Western Travels.* Amos Stoddard, *Sketches Historical and Descrip-*

*tive of Louisiana* (Philadelphia, 1812). M. Catherine White, ed., *David Thompson's Journals Relating to Montana and Adjacent Regions* (Missoula, 1950). (An absurd situation in Montana has prevented the sale of Miss White's exceedingly valuable book, the introduction to which is indispensable to a proper understanding of Thompson. No one can write justly about the subject introduced toward the end of this chapter and developed in the succeeding chapters without taking into account these portions of Thompson's journals and what Miss White says about them.)

SECONDARY  As before: Bemis (3); Gayarré (2); Houck (2); Innis (1); Lyon; Nasatir (2) and (5); Shepherd; Turner; Williams; Whitaker (2). Henry Adams, *History of the United States During the First Administration of Thomas Jefferson,* reprint edition (N.Y., 1903).

1. Robertson, I, 361–76; II, 29–59. Also E. P. Renaut, "La Question de la Louisiane," *Revue de l'Histoire des Colonies Françaises* (Paris, 1918), and citations from *Archives du Ministre de la Marine, Série A,* III and IV, and *Archives des Affaires Etrangères,* XXXVII and XLII; André Laforgue, "Pierre Clement Laussat," *Louisiana Historical Quarterly,* XX (1937). Also Lyon, *op. cit.,* Chap. V. My quotation from Decrés is taken from Renaut, p. 140.

2. *Correspondence de Napoléon Ier,* VIII and IX (Paris, 1861). *Oeuvres de Napoléon à Sainte Hélène.* F. Barbé-Marbois, *Histoire de la Louisiane* (Paris, 1829). A. Thiers, *Histoire du Consulat,* IV and V (Paris, 1874). Adams, I and II. Renaut, *op. cit.* Lyon, *op. cit.* Laforgue, *op. cit.* Baron Marc de Villiers, *Les Dernières Années de la Louisiane Française* (Paris, 1904). Fletcher Pratt, *The Empire and the Glory* (N.Y., 1949). Ralph Korngold, *Citizen Toussaint* (Boston, 1944).

3. Barbé-Marbois, p. 282. The preceding quotation, p. 298.

4. Livingston to Madison, April 24 1802 *(American State Papers: Foreign Relations,* II, 515–16; cited henceforth as ASP).

5. Livingston to Madison, April 11 1803 (ASP, II, 552); see also his summary, II, 557. Barbé-Marbois, p. 301.

6. Ford, IX, 364, 365.

7. Thornton to Hawkesbury, May 30 1803 (Robertson, II, 21).

8. See Chap. IX. "Whenever they [the Westerners who use the Mississippi for commerce] shall say, 'We cannot, we will not be longer shut up,' the United States will be reduced to the following dilemma: 1. To force them to acquiescence. 2. To separate from them, rather than take part in a war against Spain. 3. Or to preserve them in our Union by joining them in the war. . . . The third

is the alternative we must adopt." (Ford, VI, 125–26.) As for the British possibility, he had written the year before, "Weigh the evil of this new accumulation of debt Against the loss of market and eternal danger and expence of such a neighbor. But no need to take a part as yet. We may choose our own time for that. Delay gives us many chances to avoid it. . . ." (*Ibid.,* 91.)

9. Adams, II, 55: "President Jefferson had chiefly reckoned on this possibility [that Napoleon would not provoke the United States to a hostile alliance] as his hope of getting Louisiana; and slight as the chance seemed, he was right." Coming at the end of a hundred pages designed to show that in the questions which the Louisiana Purchase involved Jefferson's policy was naïve, unrealistic, blindly foolish, and intellectually dishonest, this is one of the most ironical sentences ever written by a historian. Note that so careful a writer as Adams would not have made it "Louisiana" instead of "New Orleans" inadvertently.

10. Jefferson quotation, to John Bacon, April 30 1803 (Ford, IX, 464). Laussat to Decres April 18 1803 (Robertson, II, 33–36).

11. To Monroe, January 13 1803 (Ford, IX, 419).

12. To Monroe, January 10 1803, three days before the above (Ford, IX, 416).

13. This letter to Claiborne is not printed in Ford. It is cited in Vol. II of Randall's *Life of Thomas Jefferson,* p. 62, and quoted in W. P. Cresson, *James Monroe,* p. 187. The letter to Livingston is the one cited in note 6 above and the sentence is on p. 365. The letter to Priestley is January 29 1804 (Ford, X, 69–72).

14. See for example Livingston to King, March 10 1802 (ASP, II, 515), and King to Madison April 2 1803 (p. 551). Cf. Thornton above.

15. Madison to Livingston and Monroe, July 29 1803 (ASP, II, 566).

16. To Monroe, November 24 1801 (Ford, IX, 317).

17. ASP, II: 530, 531, 533, 535, 551, 552, 554, 557, 566, etc.

18. Ford, X, 3.

19. Barbé-Marbois, p. 312.

20. *The Medical Repository,* Second Hexade, I (1804), 406.

21. Edward M. Douglas, *Boundaries, Areas, Geographic Centers and Altitudes of the United States and the Several States. Geological Survey Bulletin,* No. 817 (Washington, 1939).

22. Barbé-Marbois, p. 335.

23. *Ibid.,* p. 300.

24. Ford, X, 6. But note, four sentences later, "When we shall be full on this bank we may [that is, with the method just described we will be free to] lay off a range of States on the Western bank

from the head to the mouth, & so, range after range, advancing compactly as we multiply." The second quotation in my text is from the previously cited letter to Priestley (*ibid.*, 71).

25. As my text turns to a study of the Lewis and Clark expedition, the reader must be notified that impressive authority dissents in part from my reading of its imperial purposes. In deference to that authority I have phrased portions of this chapter and the next one more tentatively than, in my judgment, the visible facts would warrant. Evidence is supplied in the text. The fullest and best statement of the opposed view is Ralph B. Guinness, "The Purpose of the Lewis and Clark Expedition," MVHR, XX (June 1933).

26. This is the conclusion of John Bakeless and in my judgment it cannot be questioned. See his *Lewis and Clark*, pp. 3–5 and *passim*. For Lewis's desire to join Michaux, Jefferson's letter, printed as "Memoir of Meriwether Lewis," Coues, I, xix.

27. *Journals*, I, 285.

28. J. Stoddard Johnston, ed., *First Explorations of Kentucky* (Louisville, 1898); Thomas P. Abernethy, *op. cit.;* Robert L. Kincaid, *The Wilderness Road* (Indianapolis, 1947).

29. Maury's letter is printed in Ann Maury, *Memoirs of a Huguenot Family* (N.Y., 1872). The date is there given erroneously as January. The correct date was supplied from the manuscript by Mr. Henry Reck, who at my request made a search of the Maury papers in the Alderman Library, University of Virginia. He was unable to find the earlier letter to which Maury alludes in this one. I am much indebted to him and to Mr. Francis L. Berkeley, Jr., of the Alderman Library. The significance of Maury's letter was first pointed out to me by Mr. John Dos Passos, who had encountered it in manuscript without realizing that it had been published. For Peter Jefferson: Dumas Malone, *Jefferson the Virginian*, Chap. II; Marie Kimball, *Jefferson: the Road to Glory*, Chap. II. For the Evans map, Lawrence H. Gipson, *Lewis Evans*.

30. To George Rogers Clark, Annapolis, December 4 1783 (*Journals*, VII, 193).

31. The relevant portions of Jefferson's autobiography and of his letters are reprinted in *Journals*, VII. The letter to Stiles in itself shows that Helen Augur's correction, *Passage to Glory*, Chap. XIII, is right.

32. Colton Storm, "Lt. Armstrong's Expedition to the Missouri River, 1790," with documents, *Mid-America*, Vol. 25, N.S. (1943), p. 14. Thwaites, who printed a few documents, *Journals*, VII, App. III, erroneously believed that Armstrong never got started.

33. See Coues's note, Coues, I, xx.

34. Quoted in Adams, II, 61.

35. The first title is Coues's, who dates it "within a year after September 1806," the second Thwaites's, who dates it 1809. Close reading of the text shows that Coues is right.

36. See Jefferson's "The Northern Boundary of *Louisiana,* Coterminous with the possessions of *England," Documents Relating to the Purchase and Exploration of Louisiana* (Boston, 1904), p. 42.

37. See White, pp. lxxxiv ff. Also Morton (1) and (3) "The Northwest Columbia Enterprise and David Thompson," *Canadian Hist. Rev.,* XVII (Sept. 1936); "Did Duncan M'Gillivray and David Thompson Cross the Rockies in 1801?" *ibid.,* XVIII (June 1937); J. B. Tyrrell, "David Thompson and the Columbia River," *ibid.* (Mar. 1937).

38. *Voyages from Montreal on the River St. Laurence through the Continent of North America to the Frozen and Pacific Oceans in the Years 1789 and 1793 with a Preliminary Account of the Rise, Progress, and Present State of the Fur Trade in that Country.*

39. Mackenzie outlines his plan in the concluding section of his book. In the two-volume New York reprint of 1902 which I have at hand, with an introduction by Robert Waite, the passage begins on p. 353 of Vol. II. The remarks about Lake of the Woods in the first chapter are also in point. Finally, see his letter of January 7 1802 to Lord Hobart, the colonial secretary, and the accompanying "Preliminaries," *Report on Canadian Archives, 1892,* pp. 147–51. Here he explicitly demands "A supreme civil and military establishment on Nootka Island" with a "subordinate" one on the Columbia and enters into detail about the necessary changes in the charters of the Hudson's Bay Company and the East India Company. Note that Jefferson cites Mackenzie in "The Northern Boundary of Louisiana." I believe that a copy of the *Voyages* was one of the books which Lewis and Clark took with them up the Missouri. The *Journals* do not mention it by name, however, and various of Mackenzie's names which they use, most strikingly the Columbia River as the "Tacoutche Tesse" could have come from the Arrowsmith map. But the *Journals* repeatedly show that, whether or not they had the book with them, they were letter-perfect in its ideas.

40. Practically all students of Lewis and Clark since Teggart have identified Truteau as "the agent of the trading company of St. Louis up the Missouri" whose journal Jefferson first summarized for Lewis and then sent to him entire. So far as I can see the identification rests on Teggart's unsupported statement in "Notes Supplementary to Any Edition of Lewis and Clark," 1908. The identification is open to some doubt, in spite of the allusion to "Printeau's journal" in Jefferson's letter of November 16. Truteau's journal is long and bulky, Mackay's is short and includes an ac-

count of and notes by Evans, whose map Jefferson forwards to Lewis two months later. By 1807 at the latest Mackay's had been translated and it was known at least a year earlier. The translator was Samuel L. Mitchill, who published it in the Second Hexade of the *Medical Repository*, IV (1807). Yet when Mitchill published a short notice of Truteau's journal the year before, in Vol. III, he had spoken of Mackay's journal as already so well known as to render Truteau's superfluous. Truteau's had obviously come to him later than Mackay's.

(Both journals were in French. Presumably Jefferson sent Lewis a translation. If he did, this might be a further indication that the journal was Mackay's much shorter one. Yet this is predicated on the universally accepted statement that Lewis could not speak or presumably read French. Perhaps he did not, though French was frequently spoken in Jefferson's household and he must have picked up a little in two years. Among the Lewis papers at the Missouri Historical Society is a document written in French and in Lewis's handwriting, addressed in his capacity as Governor of Louisiana to one Delauny, perhaps the one mentioned in a letter from Harrison to Wilkinson, *Territorial Papers of the United States*, VIII, 134. Of course this may be Lewis's fair copy of a document prepared for him by a bilingual clerk, though there are persuasive misspellings.)

If the journal which the captains took with them was Truteau's, one might perhaps expect more reflection of its greater detail in the *Journals*. And Evans, whose map Jefferson sent to them, was after all Mackay's *engagé*. How Jefferson got it and how Lewis and Clark got Mackay's map (covering the Missouri from the mouth, whereas Evans's begins at Fort Charles) which they certainly had do not appear. Nor is there any suggestion how or when Clark got Mackay's "Indian Tribes," which has his handwriting on the first page. Any or all of these could have been secured at St. Louis during the winter of 1803–4, perhaps from Mackay himself though the *Journals* do not say they met him but only that he lived at St. Charles, and the last might have come into their possession after their return in 1806. But it is also possible that Jefferson had seen any or all of them by April 1803 when he drafted his instructions.

A conjecture is in order. I am willing to guess that before 1803 Mitchill was in correspondence with Mackay, or that acquaintances or correspondents of his were, and that Mackay's papers reached Jefferson through Mitchill. Possibly Mitchill and Mackay had met: Mackay was in New York in 1794. (Note 4 of "Indian Tribes.")

The John Hay who came to the camp on Wood River April 2 1804 (*Journals*, I, 4) en route "to his winter station on the Assiniboine River" is presumably the same John Hay who annotated Mackay's journal as published by Quaife. Conceivably he annotated it at this time, Lewis and Clark having shown him a translation in their possession. His home was Cahokia and a few days before he came to camp Lewis sent him a draft for $159.81½, but whether for services rendered or goods sold the expedition does not appear (VI, 270).

There is a tantalizing suggestion that they had more of Evans's notes or manuscript than has survived. See *Journals*, I, 198 and 200, entries for October 18 and 20. Nothing in Mackay's quotations from Evans or identifiable on the Evans map corresponds to "those remarkable places mentioned by evins."

41. One of the party, mentioned but not named by Whitehouse, had traded up the Kansas River. Cruzatte, the head boatman, had spent two years in the Omaha trade; his understudy Labiche was also an old hand in the same or related trade. Drewyer had wintered on the Platte and had gone pretty far up it; clearly too he had had experience elsewhere along the Missouri. We do not know what maps of the Missouri the expedition had except for Mackay's, Evans's, the sketch Hay drew, and some anonymous sketches. But the *Journals* below the mouth of the Platte contain many more recognitions and identifications than can be accounted for by maps or by notes gathered in St. Louis. I think it evident that others besides the four men named knew the lower river at first hand — members of the party itself and so at hand in the boats in which the captains traveled.

The greater activity of the trade is evident in the *Journals* and will be alluded to in Chapter XI. But see, for instance, *Territorial Papers*, VIII, *passim;* Nasatir, various articles already cited and "Jacques Clamorgan," NMHR, XVII (1942); Abel (1) and (2); articles cited in note 47 below.

42. Everything that follows is from the *Medical Repository* for 1803 and 1804.

43. Here I rest entirely on the work of others. This is Thwaites's statement, *Journals*, VII, 229. I have found nothing in my reading of contemporary material or of modern studies of the Louisiana Purchase that contradicts it. No ship bearing news of the negotiators' success is known to have reached the United States by June 19. Is this to be taken as an additional suggestion that the administration had always regarded the purchase of all Louisiana as possible?

44. Masson, I, 311.

45. Though 45 is the number everyone gives, no one can quite prove it. I say this in full and even weary knowledge of the evidence

marshaled by Coues and of the statements of Biddle and Patrick Gass. When the expedition starts west from Fort Mandan next spring we can count it, but not when it starts up the Missouri in 1804. The *Journals* can be read to justify a count of 44, 45, 46, or 47. To be certain one would have to answer at least these questions: What is meant by "4 sergeants, 3 Intptrs." and "9 or 10 French" in "our party"? (I, 16.) What is the meaning of "Francis Rivet &c," in Sergeant Floyd's mess? (I, 16.) Was "La Liberty" also "Jo. Barter," was he a member of the "patroon's" crew, and where is he enumerated? (See Chap. XI.) Why do Floyd and Whitehouse both make the number 44 with York or 43 without him?

46. Delassus to Casa Calvo, Houck (1), II, 357.

47. *Loc. cit.* and *passim;* Wilkinson to Dearborn, August 10 1805 *(Territorial Papers,* VIII, 183). See also: H. E. Bolton, "New Light on Manuel Lisa and the Spanish Fur Trade," *Southwestern Historical Quarterly,* XVII (1913); Isaac J. Cox, "Opening the Santa Fe Trail," *Missouri Hist. Rev.,* XXV (1930); Lansing B. Bloom, "The Death of Jacques D'Eglise," NMHR, II (1927).

Some statements which Cox makes about these first ventures are demonstrably wrong and so I am loath to follow him when he interprets the Delassus letter which I cite in the preceding note to mean that Gervais, who was to guide one of the 1804 parties, had reached New Mexico in 1803. Nevertheless his article is brilliantly illuminating, indispensable on Pike, and fascinating in the vista it opens when it links Auguste Chouteau to Wilkinson and the whole Burr conspiracy.

48. Robertson, II, 342. On the authorship of this document, James Ripley Jacobs, *Tarnished Warrior,* pp. 205–6. Spanish official correspondence after the Purchase, Houck (1), II.

## CHAPTER XI: WESTWARD THE COURSE OF EMPIRE

The matrix in which this chapter and the next one are set has been sufficiently indicated in the notes to the two chapters immediately preceding. Before beginning to write I tried to cover the entire literature relating to Lewis and Clark. I may have missed some vitally important source or some scholarly study that would have fundamentally changed my results, I may have missed a number or many, but I do not believe I have. I accept responsibility for everything said in my text about Lewis and Clark except what is covered by the next paragraph here, I will direct any serious inquirer to the evidence if I have failed to cite it, and anything I say that may be erroneous is open to correction but was not written without cause.

In biographical statements about Lewis and Clark I follow John Bakeless, *Lewis and Clark*. I follow Abel in all statements about Loisel, Tabeau, and their business history before 1804; I disagree with some of the inferences she makes from Tabeau's text. I accept Aubrey Diller's identification of Evans's map as Plates 5–11 of the Atlas volume of the *Journals*, which Thwaites ascribes to Clark. Mr. Diller's statement about the copy of Mackay's map which Lewis and Clark used is, however, incomplete.

I reject the supposed journal of Charles Le Raye as spurious. Mr. George E. Hyde preceded me in this judgment, saying that Le Raye's account of the Teton Sioux is manifestly untrue and that none of the white men mentioned in it are known elsewhere in the literature. I add that little of the country it describes can be recognized — with certainty, perhaps, only that in the vicinity of the James River — and that the Sioux-Osage wars with which the journal opens and in the course of which the narrator is captured are unknown to history.

It is worth pointing out that the editing of both Coues and Thwaites has suffered erosion. The annotation of the *Journals* by Thwaites was frequently careless to begin with and Coues has stood up much better than he, but a good deal that neither could know when he wrote has since been learned.

The journals of Floyd and Whitehouse are printed in *Journals*, VII. Gass's journal, the first narrative of the expedition that was published, is of course separate and in various editions. Lewis's journal of his descent of the Ohio River and Ordway's journal, edited by Milo M. Quaife, are in *Wisconsin Historical Collections*, XXII (Madison, 1916). I several times cite Olin D. Wheeler, *The Trail of Lewis and Clark* (N.Y., 1904). However, exact determination of the route is important in only three places in my text and in these places I rely on my own study, supported by repeated examination of the terrain and assisted by a good many local antiquarians and geographers and by the U.S. Forest Service and the U.S. Geological Survey.

1. The only description of the boat is Biddle's: "a keel-boat 55 feet long, drawing three feet of water, carrying one large square-sail and 22 oars. A deck of ten feet in the bow and stern formed a forecastle and cabin, while the middle was covered by lockers, which might be raised so as to form a breastwork in case of attack." Many kinds of craft that plied the western waters were called keelboats; perhaps the only feature common to them all was a keel. But the expedition's boat differed from the type that was becoming standard on the Missouri. She has exceptional draft and there is neither an amidships cabin (or cargo box) nor *passe-avants*, the narrow catwalks along which the crew walked while poling the

boat. Indeed, there is nowhere any suggestion that the boat was ever poled. *Journals*, I, 32, provides that the "sergeant at the bow" will have a "setting-pole" but since his duty is to assist the all-important bowsmen, who never poled, it must have been meant for fending and similar uses. In the various journals the boat is rowed, sailed, and cordelled — or at least towed, for there is no suggestion, either, that the towrope is attached to the mast or that it is equipped with a bridle, as a cordelle had to be. Nor is there any echo of the characteristic language of poling. At various times the mast is replaced after accidents and details are sent to cut timber and make new oars, but there is no mention of replacing poles or the sockets and shoulder-knobs with which they were equipped. Finally, I do not believe that a boat of the design which Biddle briefly but clearly describes could be poled as a means of locomotion.

On the way to the Mandan villages there is one allusion to poling a pirogue. The next year there is frequent mention of poling the pirogues or dugouts and the term "setting-pole" is used several times. But poling a pirogue from necessarily fixed positions is a different matter from poling a keelboat — it was not decked and there was no way of walking while poling. It would seem that each pirogue was equipped with two or three poles and that they were used while the boat was also being towed.

We may confidently infer that the keelboat used on the Missouri is the one which Lewis had had built at Pittsburgh and in which he descended the Ohio. He bought one pirogue at Pittsburgh and another one at Wheeling but there is no way of knowing whether or not these are the two used on the Missouri.

2. The status of Frazier, Robertson, and "Boley" (*Journals*, I: 11, 12, 13, 14) is not clear but I add them with Corporal Warfington to "the detachment from Captain Stoddard's Company" and treat them as enlisted men. (See I, 12.) The others were John Dame, Ebenezer Tuttle, and Isaac White. Thus there appear to have been seven soldiers besides the permanent detachment. I, 283, gives "six soldiers" as the crew of the boat when it left for St. Louis in the spring. That would be Warfington and all those just named except Frazier, who had been enlisted for the "corps of Discovery." This same entry assigns to the keelboat "two [blank space in manuscript] Frenchmen." I take it that the blank space was not intended to number the Frenchmen, who are otherwise accounted for. Perhaps the "two" should be followed by "discharged soldiers," Newman and Reed.

The bourgeois of the boatmen was Baptiste Deschamps. Six others are named. But I, 111, speaks of "La Liberty" (who is not

named as a boatman) as "one of the deserters" and previous entries
have treated him as a member of the party though he is nowhere
listed as one. Biddle, who calls the other deserter, Reed, "a soldier," twice speaks of "our man Liberté" (Coues, I, 64, 69) and says
(Coues, I, 62) that "one of our own party" was sent with the Missouri
Indian to invite the Otos to a council. In I, 93, this is "Sent a french
man la *Liberty* with the Indian," and just possibly the previous
entry, "[blank] miles further where there was a french man" could
be read to mean "la Liberty." But I do not think so; the reference,
instead, is almost certainly to the "Mr. Fairfong" of I, 97.

Ordway, p. 102, makes him "Jo Barter," not Liberté.

The list of Floyd's mess, I, 30, includes "Francis Rivet &" and
Biddle, presumably on Shannon's identification, has added the explanation, *"French."* I cannot account for him. But see Thwaites's
note, I, 282, which identifies the "Reeved" of V, 350, as Francis
Rivet and says he was a member of the expedition. But what Clark
actually says on V, 350, is that "Reeved & Greinea [Grenier] wintered with us at the Mandans in 1804." This certainly means that
"Reeved" was one of the Frenchmen who camped near or inside
Fort Mandan; the trouble is that some were members of the boat
crew staying over for the winter and some were independent hunters
seeking protection. Abel (1), p. 168, changing Clark's spelling to
"Reevey," identifies him as one of the hunters mentioned (but not
named) by Ordway, as having been robbed by the Mandans. They
are also mentioned but not named in the *Journals*. Whoever the
Rivet of I, 30, was, Thwaites's identification of him appears to be
wrong. He rests it on Eva E. Dye, many of whose statements I have
found to be overconfident. Thwaites quotes her once as making
it "François" and again as "Francis" Rivet. The other of two names
she gives as Frenchmen who claimed to have belonged to the expedition, Philippe Degie, is not mentioned in the *Journals*. The monument to these two claimants is in a cemetery at St. Paul, Oregon, not
Champoeg as Thwaites intimates.

On June 12, the day on which the expedition meets Dorion and
his party with two "Chaussies," Whitehouse writes in his journal,
"Sent One of Our Men belonging to the white pierouge back that
Belong^d to Cap^tn Stodders Company of Artilery." No other journal
makes any mention of this and I cannot understand it. It would
appear certain that at least Clark, who is writing the official log,
would record so important an event if in fact it occurred. The
white pirogue was the one commanded by Corporal Warfington and
manned by the soldiers detached from Captain Stoddard's company.

3. *Tabeau's Narrative*, superbly edited by Abel, is a prime source.

4. Lewis and Clark promptly found out that the Sioux were supplied by British competitors of the North West Company, who had recently moved the annual trading fair from the Des Moines River to the James. Abel does not mention the importance of this competition as an intertribal irritant.

5. Clark's courses make it a little more than twenty-five miles above Council Bluffs, Iowa, and it was on the Nebraska (western) side. The site was long ago obliterated by the river's surgery.

6. The grave and a monument are at Sioux City, Iowa. The bluff on whose summit Floyd was originally buried was long since undercut and washed away.

7. Spirit Mound, not far from Vermillion, South Dakota, and one of the few landmarks that can be identified in the supposed journal of Charles Le Raye. The legend still faintly exists. In the fall of 1948 I asked a rancher who lived near the hill whether he had ever seen any dwarfs there. He said that he had — often. He explained them as an effect of plains mirage.

8. Meanwhile one geographical crux had got itself written down correctly at last. (Truteau must have known it but hadn't stated it.) Dorion told them that the headwaters of the James were close to those of the Minnesota and the Red. The vague and contradictory rumors of Carver, perhaps even those of Vérendrye, were thus screened to the basic fact.

Presumably this was the place where, if there had been any feasibility in the plans of Rogers and Carver, their search for the Northwest Passage would have reached the Missouri from the Great Lakes.

9. "Said to be in this neighborhood by Mr. J. McKey of St. Charles." Since the *Journals* do not mention meeting Mackay, this may mean that the volcano was indicated on the copy of his map Clark had been using. It is not on the one based on Mackay in Perrin du Lac.

10. It is impossible to be exact about the sequence of events September 25–28. Clark's account has to be supplemented by all the other journals and the most clarifying information is in Ordway's.

11. It is not within my province to determine whether or how far Lewis and Clark shared the evening pleasures of the men which the *Journals* freely record. The folklore may be mentioned, however, and there is a lot of it. The West is well stocked with people who claim to have descended from Lewis and women of most tribes the expedition met, and they are not unknown in the East. I have corresponded with gentlemen in Virginia and Florida who have elaborate genealogical charts which convince them, and have talked

with a number of professed descendants in the West. Curiously, I have met no one who claims Clark as an ancestor. But literature long ago filled that gap, bestowing on him a love affair with Sacajawea which, on no evidence, has steadily gained ground even in histories. The DAB says that more statues have been erected to Sacajawea than to any other American woman. Mrs. Dye and Miss Hebard began a trend that may give her more husbands too.

12. I summarize the most important misapprehensions presently. Note a detail here. One of Tabeau's men, Vallé, who was in great part right about the Black Hills, had certainly not wintered 300 leagues (750 miles) up the Cheyenne River and its forks were not 100 leagues (250 miles) upstream, as he said. The forks are 115 miles from the mouth. From the mouth to the head of the south fork is just under 300 miles, to the head of the Belle Fourche (the north fork) 431. The Belle Fourche does not flow through the Black Hills, as he said it did, or through any mountainous country.

Tabeau says repeatedly that both the upper Missouri and its affluents are so poor in beaver that they can never provide a trade worth having. Yet for more than a month Clark has been commenting on the abundance of beaver, and from the Platte on Drewyer, who as a hunter is one of the busiest men of the party, has been trapping them in the main stem of the Missouri, practically every night. It may be advisable to point out that beaver do not build dams in such streams as the Missouri. Their houses are on the edge of the bank.

13. *Journals,* I, 190, *passim.* See also "Lewis's Map," actually drawn by Clark and sent to Jefferson next spring (reproduced in Coues, IV) where some of these tribes reappear. Of the ten tribes named as trading with the Rees, some are certainly subdivisions or mere bands; some others cannot be identified. The "Paducars" are not Comanches, as Abel says, but Pawnees. Abel (1), p. 190, identifies "Alitanes" as probably the Shoshones or possibly the Sisseton Sioux. They were the Ietans, that is the Utes, and the Rees could not have known them at first hand.

14. Clark's courses for October 26 and his later interlineation show three miles downstream from the lower town (*Journals,* I, 207). Wheeler calculates the distance by land from the village to the site of the later Fort Clark, almost directly opposite Fort Mandan, at three-quarters of a mile. This must be too short; the valley is so narrow here that the river cannot have straightened so much as the figure would require. Fort Mandan was on the left bank — the north bank since the river is here flowing approximately east. The monument that now marks the site is on the high bluff of the valley wall but the fort was in the bottom.

15. The two tribes are confused innumerable times in French, Canadian, and American accounts before and after the period we deal with here (even by David Thompson) and on maps (Arrowsmith, for example) and by modern writers. That the Crees called the Atsina "willow people" and the Minnetarees sometimes referred to themselves by a word which must be translated with the same phrase didn't help any. Canadian traders called the Atsina Rapid Indians and Fall or Falls Indians and Tongue Indians — all from their earliest known country, between the forks of the Saskatchewan — and all these names got attached to the Minnetarees. The Mandan name for the Hidatsa, Minnetarees (which ethnologists like to spell Minitarí), was inevitably applied to the Atsina in turn. Confusion was increased by the fact that a common name for the Minnetarees' relatives and neighbors the Amahami, "Shoe Indians," could be mistaken for a reference to the Blackfeet, who had once worn (or were said to have worn) moccasins colored black, and so to the Blackfeet's neighbors, the Atsina. But "Souliers," shoemen, could easily be corrupted to "Saulteurs" and so both the Hidatsa and the Atsina were frequently confused with the Chippewas.

16. The tropical emotion that has created a legendary Sacajawea awaits study by some connoisseur of American sentiments. I have referred to the statement in the DAB (the entry is by W. J. Ghent) that more statues have been erected to her than to any other American woman. Few others have had so much sentimental fantasy expended on them. A good many men who have written about her, including a couple with some standing as historians, have obviously fallen in love with her. Almost every woman who has written about her has become Sacajawea in her inner reverie. And she has received what in the United States counts as canonization if not deification: she has become an object of state pride and interstate rivalry. Grace Hebard's *Sacajawea* is in some degree a product of this mawkishness. Most of the book deals with Sacajawea's life after the return of Lewis and Clark and so I need not express an opinion about the odyssean and frequently herculean story it tells. But the greater part, indeed the vital part, of its identification of Sacajawea as the woman who is its heroine, consists of two statements that must be examined here.

One is the statement that Charbonneau had three wives, not the two attributed to him by Clark and all the other diarists who enumerate his wives. The other is the statement that a marriage ceremony, "doubtless required by either Lewis or Clark," joined Sacajawea and Charbonneau in legal marriage on February 8 1805 (Hebard, *Sacajawea*, pp. 49–52). Miss Hebard's evidence for the first statement consists of a quotation from Whitehouse's journal for Christmas Day 1804, which she makes "three squaws, the interpreter's wives." Whitehouse is saying that the Christmas festivities have been held "without the comp^y of the female Seck," and Thwaites prints what follows as "except three Squaws the Intreptirs wives." *(Journals,* VII, 72.) Thwaites was right to insert no apostrophe in "Intreptirs"; there is none in the manuscript, which is at the Newberry Library, Chicago, and Miss Hebard provided the one she printed. She disregarded Gass's entry, which might seem to confirm her, "except three squaws, wives to our interpreter." This was judicious disregard, for not only did the diarists frequently copy each other but Gass's manuscript has disappeared and the printed version has been so heavily rewritten by David McKeehan, the schoolmaster who originally edited and published it, that there is simply no way of knowing what Gass wrote or what he meant. The proper place for this crucial apostrophe depends on whether the wife of the other interpreter, Jessaume, was present at Fort Mandan on Christmas. If she was, then the three wives of the interpreters would be just what they appear to be throughout all the journals, the two wives of Charbonneau whom Clark and the others count and Jessaume's wife. There is no reason to suppose that she was not present. There is no reason, either, to suppose that Jessaume was not there. His goings and comings are freely mentioned in the journals all winter, he was certainly there on December 4 for he is mentioned on that day, and nothing suggests that he or his wife left before Christmas. I think we may safely make Whitehouse read, "three Squaws the Intreptirs' wives."

Miss Hebard's evidence for the alleged marriage ceremony is an item she found in *Dictionnaire Historique des Canadiens et des Métis Français,* by the apologist for the Métis, Father A. G. Morice. I find a Quebec and a Kamloops, B.C., edition of this book, both of 1908 and a Quebec edition of 1912. I have exhausted the ingenuity of American and Canadian research librarians without finding any other. It is a history of the French in Canada arranged in the form of a dictionary. The author says that he corresponded with thirty or more persons over a long period and examined over 150 books in order to make his account as accurate as possible. His note on Charbonneau appears on pages 64 and 65. He cites no source and gives no evidence for what he says. He says that Lewis and Clark lent Charbonneau (as an interpreter) to François Larocque, which is true. He goes on: *"Avant de partir avec son nouveau maître, il dut aller avec le capitaine Clarke [sic], accompagné de 25 hommes et un parti de Mandanes, essayer de punir des Sioux, qui avaient tué un Mandane. Pendant qu'il servait sous Laroque il s'unit, le 8 février 1805, à une femme de la tribu des Gens-des-*

*Serpents qui avait été faite prisonnière par d'autres Indiens."* He also says, *"Les deux explorateurs américains disent qu'il était bigame."*

This woman of the Gens-des-Serpents, who is otherwise unknown, being mentioned by no other person, becomes for Miss Hebard a third wife for Charbonneau and so permits her to substitute another wife for the one whose early and well-attested death has usually been taken to be Sacajawea's, and so frees her heroine to all the adventures chronicled in her book. Since this is the only known allusion to her, I suppose we need not go beyond the date given by Morice and accepted by Miss Hebard, February 8. But on that date Charbonneau was still out with Clark — on a hunting party, not a punitive expedition as Morice says, having confused this trip with one presently described in my text. He did not return till February 10 (*Journals*, I, 257). Larocque had left Fort Mandan February 2 (I, 252) and the village February 9 (Masson, I, 311). There is no suggestion in any of the journals, Larocque's, Clark's, or those of the sergeants and Whitehouse that a wedding ceremony was ever performed on anyone. There was indeed no one who was empowered to perform either a secular or an ecclesiastical marriage. Anyone, of course, could have performed a ceremony that would be binding on the frontier, but Charbonneau was already married to Sacajawea by Indian custom and there is no suggestion anywhere that either Lewis or Clark found any flaw in or made any objection to it. As for either of them having called Charbonneau a bigamist, they were entirely familiar with Indian and squawman customs and there is nowhere in the *Journals* any hint that they considered them immoral. I cannot believe that in their eyes he was a bigamist; he was simply a man who had two squaws. Miss Hebard attributes this sudden and unique prudery to a desire to legitimatize the child presently to be born. (She also says that the birth of the child "is not mentioned in the Biddle narrative." Biddle writes under date of February 11, "About five o'clock one of the wives of Charbonneau was delivered of a boy," and proceeds to summarize Lewis's entry of the same date. See Coues I, 232, and *Journals*, I, 257.)

There is no evidence that Charbonneau took a third wife this winter and there is no evidence that any kind of ceremony married him this winter to Sacajawea.

17. *Across the Wide Missouri*, pp. 22–23, 27–28, *passim; The Year of Decision*, Chap. II; any account of Manuel Lisa, Andrew Henry, the Astorians, and William Ashley.

18. John McDonald of Garth, Masson, II, 31–34; Charles MacKenzie, *ibid.*, p. 329.

19. *Journals*, VI, 51. Larocque's statement, Masson, I, 311. His "Qu'Appelle" may be a slip for "Catepoi," the Souris.

20. *Journals*, VI, 52–53. There was no imperialistic purpose behind the Lewis and Clark expedition?

21. This is not to disparage Thompson's great work. He laid the basis of geographical knowledge of western and northern Canada; parts of his surveys have not yet been superseded, apparently some parts have not even been retraced. But they are just that, surveys, and except for the courses of the day's travel, Lewis and Clark did nothing that resembles a survey. Thompson's notebooks, in great part still unpublished, are very little more than a surveyor's working notes. Apart from his own empirical knowledge immediately at hand he is in the main incurious and frequently misinformed. The notes show little geographical speculation or general geography. His *Narrative* is one of the most valuable of all books about the Western wilderness. But though it was written much later, after a great bulk of additional knowledge was available, it has much fewer geographical details and generalizations than the *Journals* and in what it has is sometimes less reliable.

A single example. Thompson estimates the length of the Missouri at 3560 miles. I do not know whether this estimate was made in 1797 when he first saw the Missouri or when he wrote the *Narrative* at some time after 1826 which Tyrrell does not date. At any rate, he saw no reason to alter it. At the Mandan villages in the winter of 1804–5, when he had traveled the river no higher, Clark estimated its length at 2435 miles. *(Journals,* VI: to Fort Mandan, p. 59, 1600 miles; to the Three Forks, p. 60, 835 miles.) The length today is generalized at 2464 miles; the Corps of Engineers makes it 2519 as of 1890; perhaps it was a hundred miles longer still in 1804, or 2600 miles plus.

Thompson's *Narrative* contains many valuable pictures of Indian life and some sound ethnological analysis. Yet with all the experience of the North West Company to call on he did not seek information so widely as Lewis and Clark, nor analyze evidence so soundly, nor put related fragments together so surely, nor construct so comprehensive a descriptive picture.

22. Scholarship badly needs a comprehensive study of what Lewis and Clark learned, what they accomplished as a "literary" expedition. Oddly, the most valuable study is incidental, appearing in E. H. Criswell's *Lewis and Clark: Linguistic Pioneers (University of Missouri Studies,* XV, 1940). (Much more than a linguistic study, this is the most useful book yet written about the expedition.) Botany has been more adequately treated than any other part of their work, probably because Coues laid the foundations, but it

has not been fully treated. There are several articles on medicine and medical practice but this interesting if rather unimportant subject has not been exhausted. Most urgently needed is a comprehensive ethnological study. Ethnologists who deal separately with Indian tribes which Lewis and Clark met, appraise their work (sometimes, I think, not too well) but there is no general treatment.

What my text now proceeds to say is winnowed from the daily entries in *Journals,* I, II, III, and V; the "Statistical View" and the related matter which Thwaites has headed "Ethnology" in VI; Lewis's "Summary View of Rivers and Creeks, Etc.," Clark's "Summary Statement . . ." tables, and miscellaneous items, all in VI; various letters in VII; a few items in the other journals; and "Lewis's Map" which, as I have said, was drawn by Clark at the villages.

23. Inevitably they classified as tribes a number of what were certainly tribal divisions or even mere bands. Some of the tribes they named must be adjudged nonexistent, a failure of the St. Louis traders to understand what the Indians were telling them. Though both Abel and Thwaites provisionally identify such designations as Cataka, Nemousin, Dotame, and Castahana, I believe there is no way of telling what they mean. A few names are simply reproduced from Arrowsmith's map. Note that Lewis believed that "the once powerful" Padoucas had been split up into tribal fragments and had disappeared. This was because by now everyone was using the name for different Indians and he could not locate them as an entity anywhere. Originally applied to the Comanches before the French had seen any of them, the term had by now been attached to various Pawnee divisions, Arikaras, Kiowas, and tribes or bands these happened to be in contact with.

24. The crucial rectangle, the one first mentioned, lies below 39°, which is actually too far south for both the Arkansas and the South Platte. It is west of 110°, four degrees too far west for them.

The failure to understand the nature of the mountains is entirely conclusive, as well for Lewis and Clark as for their predecessors. It will not do to say that the sources of the South Platte and the Arkansas do in fact lie very close together, or that the headwaters of the Grand (now officially the Colorado) are not very far from both, or that the Green (always the Colorado till government decided otherwise) does come close to the Yellowstone. Two facts must be kept in mind: that the separations are by mountain range, and that everyone who reports these headwaters is thinking of navigable streams. In the case of the Grand and the Green the separating range is the highest and one of the widest in the United States, in the case of the Rio Grande the only a little smaller Sangre de

Cristo. There is no portaging such ranges, by canyon or otherwise, and no navigating the upper reaches of mountain rivers; few can be navigated at all. We must accept the fact that Lewis and Clark, like all their predecessors, did not know and *were not told* what mountain rivers are. Precisely the same accumulation of misconception — it was of course entirely unavoidable — is to be seen in Alexander Mackenzie's second exploration.

25. After the expedition gets started in the spring, whenever the Indians whose information is under discussion are named, they are the Minnetarees, never once the Mandans. I am convinced that we must scale down the Mandans' claims about their fighting in and acquaintance with the farther parts of the West. How far beyond the Black Hills did they raid or trade?

26. This is what I make of *Journals,* VI, 24–26, 28, 72–74; Clark's map, which is revealing here; their dissatisfaction at the Salmon River and the questions they asked the Snakes; III, 50–61; various conversations with the Nez Percés; V, 175–78; and the earlier portions of Lewis's journey to Marias River, V, 183–89. Note that on Clark's map, which was drawn at the villages, the indicated "war path of the Big Bellies Nation," does not lead west from the Three Forks but crosses the mountains well to the north. Bring to bear on this III, 34 and 60.

27. Again it will not do to suppose that the Indians were telling them how close the sources of the Yellowstone lie to those of the Green. (Though with a high range between.) The Minnetarees had never been to the Green and there is no evidence that they knew any Indians who had. And no Indians knew that the Green of the Green River Valley was the Colorado below the Grand Canyon.

28. During the winter Lewis and Clark learned that both the Flatheads and the Snakes had horses. On Clark's map the Flatheads are shown on the branch of "the south branch of the Columbia" that actually issues from the pass which crosses the Divide, at the end of the "war path of the Big Bellies Nation." This is Lemhi Pass in the understanding of the captains, but is actually a blending of Cadotte's or Lewis and Clark's Pass across the Divide and Lolo Pass across the Bitterroot Mountains. The Snakes are shown in the vicinity of the Three Forks. This is additional evidence that the captains were told about both routes but did not differentiate them. Note that on Arrowsmith's map, which was so important to their thinking, there is a fragmentary river fork west of the Rocky Mountain range. The river breaks off and a dotted line, like that of the "war path," goes on to the Columbia. It is captioned "The Indians say they sleep 8 Nights before they get to the sea." This is without provenance but perhaps it represents

something in Peter Fidler's data, which were given to Arrowsmith. If so, a further guess would be that the Blackfeet, whom Fidler knew east of the Rockies, had told him that the Kutenais or some other transmontane tribe so reported the route to the Columbia (or to the Fraser or perhaps some more northerly river) and to the sea. Alternatively, such a Blackfoot account might have reached Arrowsmith from David Thompson's notes, which were North West Company property. Whether this notation on Arrowsmith's map affected the thinking of Lewis and Clark is anyone's guess.

29. For the horses of the Snakes, see works cited in note 6, Chap. V. For the Blackfoot wars, Thompson, Henry, Duncan M'Gillivray. The rest of this paragraph rests on incidental facts and inference in a large number of ethnological papers; ethnologists seem not to be interested in this period of Shoshone history.

30. The people called the Wind River Shoshones are tolerably spectral, hard to identify and to locate. They may have moved east of the Divide later than 1804.

31. I can make nothing of what Clark's map does to the Jefferson. He brings its source to about the site of Winnemucca, Nevada, which resembles what he does with that of the Yellowstone but corresponds to nothing in his notes and contradicts his table of distances.

I give the understanding of Missouri distances the expedition had in April 1806 and the actual distances:

|  |  |  |  |  | CLARK | ACTUAL |
|---|---|---|---|---|---|---|
| Fort Mandan | to | the | Little Missouri |  | 100 mi. | 56 mi. |
| " | " | " | " | Muddy River (Clark's White Earth) | 220 | 201 |
| " | " | " | " | River that Scolds | 370 | 747 (Marias) |
| " | " | " | " | Musselshell | 490 | 554 |
| " | " | " | " | Great Falls | 610 | 890 |
| " | " | " | " | Medicine (Sun) River | 625 | 812 |
| " | " | " | " | Three Forks | 835 | 1024 |
| " | " | " | " | northward-flowing river | 900 | 1150 *(circa)* |

The distances to the White Earth and the River that Scolds account for much of the confusion about the Marias when the expedition reached it.

32. See pages 418–20 and 429–30, above.

33. The Welsh Indian story was first published in the *Kentucky Palladium,* December 12 1804. The author is identified in Stephens, *Madoc,* p. 59. Copious extracts from the story in Benjamin F. Bowen, *America Discovered by the Welsh,* pp. 96 ff. Bowen's book, published in 1876, repeats and accepts (without identifying by

name) Stoddard's argument about a different branch of the Missouri, which occurs pp. 484–85 of *Sketches of Louisiana*. The 1951 believer is Zella Armstrong, *Who Discovered America? The Amazing Story of Madoc*. The unsigned story about the bearded Indians, *The Medical Repository*, 1805, pp. 113–14. Bruff's letter to Wilkinson, *Territorial Papers*, VIII, 56–61.

34. There is no way of reconciling what the various diarists say about the number of men who went on the boat, but if Gravelines be counted as one of Tabeau's men at least Lewis's "two, perhaps four" harmonizes with Clark's "4 hands and himself" and Clark's total crew of "15 strong" is attained. Coues, I believe, is wrong (I, 258) to count Reed and Newman among the "six soldiers." This gives one too many. See note 2 above.

So far as I know no student has ever learned the name of the Arikara chief; I have searched all likely places but have not found it. His adventures, however, can be followed in *Territorial Papers*, VIII, *passim*. When the boat reached St. Louis Wilkinson was arranging to send a sizable delegation of lower Missouri and upper Mississippi Indians to Washington in charge of Amos Stoddard. The Arikara was to travel with it but fell sick. So did an Oto who had gone to St. Louis in accord with Lewis's arrangements. (In whose charge does not appear and in fact the *Journals* do not mention him at all. Possibly Gravelines picked him up on the way down, possibly "Mr. Fairfong" accompanied him.) They blamed their sickness on their folly in leaving home and demanded, "loudly & incessantly" Wilkinson says, to be taken home. Wilkinson sent them back with Gravelines and a tiny military escort. (He was right to do so but Dearborn appears to have suspected that he was serving some private purpose as well, probably because he gave the command to his son and probably correctly too, and censured him for acting without orders.) The Kansas Indians, playing Sioux, stopped the party and it returned to St. Louis. (This may be why Lewis added the Kansas to his list of river pirates who must be coerced.) Eventually, with Gravelines, the Arikara did go on to Washington, where he was put through the stately ceremonies which were being worked up as an instrument of Indian policy — and which Dearborn feelingly remarked had henceforth better be staged when Congress was not in session. (*Territorial Papers*, VIII, 491, prints the President's set speech "to the chiefs & people of the Mandane nation" and says that the same one was made to the Arikaras. The date is April 19 1806 and this is inexplicable for so far as any record shows the first Mandan who made that journey was Big White, whom the captains took with them in August 1806.) Early in 1806 Stoddard's embassy was sent on a kind of grand tour of Eastern cities. The Arikara, whom Dearborn thought "amiable" and "an interesting character" was to join it but he died in Washington April 7. Gravelines was dispatched to take back his effects and "two to three hundred Dollars" in goods to the tribe, to cover his body. Lewis and Clark met him, only a few days out from St. Louis, September 12. Whether the chief's death played a part in the subsequent hostility of the Arikaras I do not know but by 1807 they had begun to make trouble.

35. Jefferson to Congress, January 18 1803: "The river Missouri and the Indians inhabiting it are not as well known as is rendered desirable by their connection with the Mississippi, and consequently with us. It is, however, understood that the country on that river is inhabited by numerous tribes, who furnish great supplies of furs and peltry to the trade of another nation, carried on in a high latitude through an infinite number of portages and lakes [Mackenzie, *Voyages,* Chap. I] shut up by ice through a long season. The Missouri, traversing a moderate climate, offering, according to the best accounts, a continued navigation from its source, and possibly with a single portage from the Western Ocean, and finding to the Atlantic a choice of channels through the Illinois or Wabash, the Lakes and Hudson, through the Ohio and Susquehanna, or Potomac or James rivers, and through the Tennessee and Savannah rivers. [Sentence has no main verb.] . . . The interests of commerce place the principal objects [of the expedition] within the constitutional powers and care of Congress, and that it should incidentally advance the geographical knowledge of our own continent can not but be an additional gratification." Jefferson to the American ministers to France and Spain, January 15 1804, Louisiana having been meanwhile acquired: "For England holding nothing in that quarter Southward of 49° the line proposed in the Vth. article [of a convention for the survey of the boundary], from the North Western point of the lake of the Woods Southwardly to the nearest source of the Misipi, is through a country, not belonging to her, but now to the US. Consequently the consent of no other nation can now be necessary to authorize it. It may be run, or not, and in any direction which suits ourselves. It has become a merely municipal object respecting the line of division which we may chuse to establish between two of our territories. It follows then that the Vth. Article of the Convention of London of May 12. 1803. should be expunged, as nugatory; and that instead of it, should be substituted one declaring that the dividing line between Louisiana & the British possessions adjacent to it, shall be from the North Western point of the Lake of the Woods, along the water edge Westwardly to it's intersection with the parallel of 49°. North from

the Equator, then along that parallel (as established by the treaty of Utrecht between Gr. Britain & France) until it shall meet the limits of the Spanish province next adajcent. And it would be desirable to agree further that, if that parallel shall, in any part, *intersect any waters of the Missouri, then the dividing line shall pass round all those waters to the North until it shall again fall into the same parallel* [my italics], or meet the limits of the Spanish province next adjacent. Or, unapprized that Spain has any right as far North as that, & Westward of Louisiana [Livingston, Mitchill], it may be as well to leave the extent of the boundary of 49°. indefinite, as was done on the former occasion."

36. Coues, I, 345, points out that the Milk River watershed comes in between the north fork (Marias River) and the Saskatchewan. That is true but he is only technically correct in saying further that the headwaters of the Marias are in the Rockies. Only one of its constituent streams flows from deeper in them than the eastern slope. The deduction of Lewis and Clark that it does not "penetrate" the Rockies is a more correct statement.

## CHAPTER XII: THE PASSAGE TO INDIA

David Thompson's *Narrative* is Tyrrell (1). His journals are the basis of hundreds of footnotes in Coues (1). White, *Thompson's Journals,* prints some two hundred pages of the journals. Shorter portions are in T. C. Elliott, "David Thompson's Journeys in Idaho," *Washington Historical Quarterly,* XI (1920); "David Thompson's Journeys in the Pend Oreille Country," *ibid.,* XXIII (1932); "David Thompson's Journeys in the Spokane Country," *ibid.,* VIII (1917); "Journal of David Thompson," Oregon Hist. Soc. *Quarterly,* XV (1914). See also Morton (1), (2), and (3); Wallace (1). Also citations in note 39, Chap. X.

1. The streams that I call the Hellgate, the Missoula, and Clark's Fork are all a single river. Those are the names used locally today and they have the sanction of historical usage. Officially, however, they are Clarksfork River, so spelled, and this name applies to the Silverbow till the Deer Lodge joins it and to the Deer Lodge from there on to the junction with the Blackfoot. This Clarksfork River is also the one I call Clark's Fork from here on till it joins the Columbia, including the stretch below Pend Oreille Lake that on many maps and in some localities is called the Pend Oreille River. The stream which Lewis and Clark first called Flathead River and then Clark's River is the Bitterroot, which joins Clark's Fork (or the Missoula) just west of Missoula.

Montana. The body that passes official judgment on nomenclature is the Board of Geographic Names, U.S. Department of the Interior.

2. I tell the story as it was printed by Bradbury. The site of this amazing run has been disputed, some students locating it as far away as the Beaverhead but I see no reason to amend Bradbury. He had it straight from Colter and we may assume that the first mountain man knew where he was. He said the Jefferson and the lay of the land makes it, therefore, the Jefferson east of the mountains and northeast of the Madison.

3. They said "Pahkees," meaning the Fall Indians, the Atsina. But the Atsina were leagued with the Blackfeet and were frequently called Blackfeet, and all the allied tribes had been raiding the Snakes for two generations.

4. The lava plains, *Journals,* II, 381, make this certain. And after the expedition reaches the Columbia every succeeding allusion to westward- or southward-dwelling Shoshones that is an allusion to actual Shoshones, not a mistaken identification, means the Bannocks or their local affiliates.

5. It must be made entirely clear that the guide's distances were of Idaho only, not of New Mexico or California. His assertion that the Spaniards could be reached from the Yellowstone River in ten days enormously buttressed the misconception that gave their Southwest so small a scale. Almost certainly this old man and probably no Indian he had ever met had been so far up the Yellowstone as Yellowstone Lake but, taking it from as far south as that, ten days' travel would bring a war party to somewhere in the upper valley of Green River, if they rode the mountains hard; it would bring a migrating village perhaps twenty-five miles beyond the southern boundary of Yellowstone Park. It is a tolerably safe assumption that neither the old man nor any other member of this band of Snakes had any empirical knowledge of the Yellowstone River.

6. I believe that the ubiquitous foldboat sometimes runs the Salmon canyons too, but not upstream.

7. The last quotation is from Clark, September 4 (*Journals,* III, 53). Clark's journal, the official log, does not mention the momentary suspicion that the party might have encountered the Welsh Indians but Ordway and Whitehouse are explicit. Ordway says, p. 282, "We think perhaps that they are the welch Indians, &.C." Whitehouse, VII, 150: "We take these Savages to be the Welsh Indians if there be any Such from the Language. So Capᵗ. Lewis took down the names of everry thing in their Language, in order that it may be found out whether they are or whether they Sprang or origenated first from the welch or not."

8. Clark says "large pine" (*Journals,* III, 88), which should be the Ponderosa or Western Yellow Pine. Note that the Bitterroot Mountains, which were empty of game in 1805, are so heavily stocked with it now that in recent years the Forest Service has regarded them as overpopulated. Such changes as these are the result in the change of cover, and so of browse and forage, caused by great forest fires.

9. This statement, which appears to rest more on their geographical analysis than on direct Indian testimony, is true but the same qualification must be noted that I have pointed out in regard to the Platte and the Colorado or the Green: a maze of peaks and ridges makes communication between them possible only on paper. The Jefferson rises in Red Rock Lake, in Montana, west of Yellowstone Park. This is not far from the Idaho source of Henry's Fork of the Snake. (Clearly Henry's Fork is the basis of some of the data they got from the Indians but they nowhere differentiate it from the main stem of the Snake itself, Lewis's River.) The Snake rises in Yellowstone Park only a short distance — in miles — from the source of that once great trout stream, the Madison.

10. *Journals,* IV, 68–72, and sketch, p. 318. These pages are fully illuminated by maps in Vol. VIII. In order to understand the distances worked out, see "Clark's Summary Statement of Rivers, Creeks, and Most Remarkable Places." VI, 63–64: from the mouth of Dearborn River to where they left the canoes, 420 miles; to Traveller's Rest 220 miles more, making 640 miles. VI, 72: from the mouth of Medicine River to Traveller's Rest, 840 miles. VI, 63–65: from the mouth of Medicine River to the Clearwater, 901 miles. This is a difference of 561 miles, corresponding to Lewis's "further by 500 miles" of IV, 68, and Clark's "further by abt 600 miles," IV, 70.

11. Jefferson's instructions allowed for the possibility that it might prove wise for the whole party to return by sea. When he started west Lewis had thought that they might go from the mouth of the Columbia to Nootka Sound and take ship there, but he abandoned the idea with no further mention.

12. Lewis thought that this was a band of the affiliated Fall Indians, the Atsina. They appear, however, to have belonged to one of the three Blackfoot tribes, the Piegans. See Wheeler, II, 311.

13. *Journals,* VI, 74, but the related passages, *passim* and in Vols. IV and V, must be added. Excellent maps and pretty good verbal descriptions of the various trails, the mountains, passes, prairies, and rivers can be found in Vol. XII (Stevens's Survey) of

the Pacific Railroad Report, 36th Congress, 1st Session, H. R. Doc. No. 56, or Senate Doc. not numbered (Washington, 1860).

Note that at one place, *Journals,* IV, 331, Clark suggests going by way of Smith River instead of the Dearborn.

14. Lewis to Jefferson, September 23 1806 (*ibid.,* VII, 334 ff.).

15. They must have been Crows and yet even this could be questioned. Larocque's journal of the trip is one of the most irritatingly vague documents in fur trade literature. Little of his route and very few places can be positively identified by the text. It has had only perfunctory editing. Burpee published it in the original French in *Publications of the Archives of Canada* (1911); Ruth Hazlitt, in translation in *Frontier,* XIV (1933–34), and XV (1934–35).

16. See p. 420.

17. And may presently have been the author of two threatening letters which warned Thompson off American soil and which were delivered to him by Indians. See T. C. Elliott, "The Strange Case of David Thompson and Jeremy Pinch," Oregon Hist. Soc. *Quarterly,* XL (1939), and W. J. Ghent, "Jeremy Pinch Again," *ibid.* The letters were signed by imaginary officers of the American army. (Elliott's speculation that Wilkinson, as well as Lisa, could have written them is plausible, but his further speculation that Wilkinson may have had a fur trade venture in the upper Missouri country is not. Though that notion is renewed at intervals, it has been completely exploded.) This incident has but one importance: it is additional evidence that, from the beginning, both the Americans and the Canadians were fully aware of what their rivals were doing in the West.

18. Miss White, *Thompson's Journal,* vigorously defending Thompson's against Morton's strictures, which came close to attributing to him full responsibility for the British loss of the Columbia and Oregon, has taken most of the sting out of them — but not all. It is quite true that Thompson's first business was to develop the trade, not to descend the Columbia, and it is also true that Mackenzie's and M'Gillivray's plan could not be fully worked out while the monopolies of the Hudson's Bay Company and the East India Company remained intact. But (to concentrate in three items an examination that could be made in detail), the North West Company was no respecter of charters and could have confronted the East India Company with a *fait accompli* and then made at least as satisfactory terms with it as the maritime traders did, and most probably much better terms; it is possible, and seems plausible, that the premature death of M'Gillivray, as fiery an advocate of the great plan as Mackenzie and a man who used force as he saw

fit, temporarily prevented the company from any more forcible action than its efforts to get Parliament to stop Astor or seize the mouth of the Columbia before he could; and, whatever held Thompson back, it is hard if not impossible to imagine any other Northwester or any American explorer who, being in Thompson's position, would not have gone down the Columbia as soon as he learned what it was. Thompson himself seems to have felt that he should have done so; in the *Narrative* he implies, and apparently without justification, that John McDonald of Garth either sabotaged the Columbia enterprise or remained so indifferent to it that it could not get proper backing.

19. See his journal entries at Hell Gate and the mouth of the Bitterroot, White, pp. 204 and 206. T. C. Elliott, "The Fur Trade in the Columbia River Basin Prior to 1811," Oregon Hist. Soc. *Quarterly*, XV (1914), says flatly that Thompson had a copy of Gass with him but there appears to be no way of establishing this. But see Miss White's suggestion, p. 206, n. 94.

20. The letter to Lewis, Ford, XI, 37. The second quotation is from a letter to Astor, April 1808, not printed in Ford but quoted in Grace H. Flandreau, *Astor and the Oregon Country,* p. 7. For the Pacific Fur Company to Irving and Chittenden add Kenneth W. Porter, *John Jacob Astor* (Cambridge, Mass., 1931), Chaps. III, VI, and VII, and Morton (1).

Parts of Fraser's journal are in Masson and *Report of the Public Archives for the Year 1929* (Ottawa, 1930). Morton (1) is the best treatment of Fraser.

21. Wheeler, II, 190–93.

# Bibliography, Annotation, and Minutiae

I HAVE NOT compiled a bibliography for this book. Excellent bibliographies of all the subjects it deals with have been made by people whose business it is to facilitate scholarship, and every new book adds to them in its notes. I have listed two kinds of material before the notes for each chapter. In no case do these lists constitute a catalogue of the sources and secondary works I have used. For all chapters such a catalogue would have to include scores of additional items; for some chapters, hundreds.

In "Documentary" I list source material covering the basic subjects treated in the chapter. Since all the original source documents for my first eight chapters are in print, my references are usually to collections and usually under the names of editors. Spot and *ad hoc* citations are in the notes. I do not give full bibliographical detail about books that have had many editions.

In "Secondary" I list works on which I rely for statements not based on my own work in the sources and those to which I feel a considerable debt for clarification, enlightenment, and insight. In a book which treats so many subjects, events, and persons as this one does, a writer must repeatedly substitute second-hand for first-hand knowledge. For every topic my purpose includes I have gone to the original sources, subjected my results to the criticism of the leading authorities, and brought in findings according to my lights. But the connective tissue, the background of historical judgment, and the peripheral history are so extensive that no one could master them all at the sources. No one can pretend, either, that more than a small part of his own judgment is his private achievement.

I owe a great deal to an appallingly large number of historians but I am glad to name those from whom I have taken most or on whom I have principally relied: foremost and always Parkman; Herbert E. Bolton and the well-known group of his disciples who have followed the trails he blazed; Louise Phelps Kellogg, whose determination of a pattern in a key area and period has been, I think, insufficiently praised; and Grace Lee Nute, whose massive researches have splendidly illuminated areas that were dim or dark. I would add a school of historians, among whom I have most often cited Justin Winsor, who in recent years have been too much ignored because parts of their work have been superseded but who both understood the past and were able to write prose about it in ways we would be wise to restore. For the rest, I have listed principal authorities and those from whose findings I have found reason to dissent. If I have failed to name someone from whom I have taken something important, it is by inadvertence; I have no wish to pass off the work of others as my own or to seem to know more than I do. But I have not tried to account for my use of peripheral facts or to acknowledge trivial borrowings.

I want to express particular indebtedness to Canadian historians of the fur trade, especially Harold Innis. They have treated the subject much more authoritatively and exhaustively than our historians have done. This is natural enough, since the fur trade is a much larger part of Canadian than it is of American history. But also as a group they have versatility and breadth of interest beyond what we are used to here and their prose has not been discolored by the Ph.D. style.

I wish that books could be written without footnotes. Since they cannot be I have written most of my notes to amplify for students matters which are carried no farther in the text than the interest of the general reader requires. Perhaps a dozen notes, and these the longest ones, are obligatory discussions of matters important to specialists. Some notes are intended to remind students of what other discussions have left out of account. Some are the protectiveness common in historical works, notifying the reader that if a statement is in error someone else is to blame.

In my text the spelling of proper names is usually that of the commonest usage in the United States. I have had the reader's comfort in mind. He will not be confused by "Jolliet," though in the American one of the *l*'s is deleted, but he might be confused by "du Lhut" and annoyed by "de la Salle." Geographical names are in their modern form except when something important in the text depends on an earlier form. Here too I have usually followed the commonest usage, though it may sometimes have required me to sacrifice consistency or to ignore the United States government. A few Canadian forms differ and I have let the context decide.

The maps are primarily to help a reader visualize geographical relationshins. That visualization is not easy, nor is the orientation of the text he reads with his mental map of the United States, and I have tried hard to assist both. I have identified routes and sites by means of modern cities. I have used round numbers when close approximation was not a necessity of the text. Where travel time or distance traveled is important, there are usually two distances, the actual one and the traveler's estimate. When I give the former, I distinguish between the map-distance and that of possible or feasible routes, and I hope I have made clear that commonly the traveler's estimate is wrong. Where I make estimates and they differ from those of other students, it is for cause and only after careful study. Determinations of distances are among the most onerous jobs of one who writes about primitive America. The nature of the terrain, the mode of travel, the season and the weather, the times of other travelers when they are known, the company, known or inferred reasons for haste or delay — these are only some of the factors that must be considered. As for routes traveled, it is usually enough to know that the party passed somewhere in this vicinity; we need not know that it passed exactly here. That is just as well, for though devoted men have earnestly and ingeniously labored to determine our historic routes precisely, part of their effort was always hopeless. The firstcomers were usually too occupied with more pressing matters to make notes which would enable historians to follow them, and in the notes they did make they were often deceived. Only in mountainous country, furthermore, can we count on the landmarks being what they were, and not there unless the builders of dams and highways have withheld their hands. Elsewhere settlement may have obliterated landmarks and nature may have so altered the topography that the place where we fought Indians is now ten miles on the other side of the river and the lake we guided by has dried up. I have done my full share of field work, perhaps more than was called for, and I have spent a great deal of time following my characters across the land. But it must be candidly admitted that in most statements of routes some portions are by grace of historical convention.

# Index